Theory and Practice
of Infrared Technology
for Nondestructive
Testing

Theory and Practice of Infrared Technology for Nondestructive Testing

XAVIER P. V. MALDAGUE

Electrical and Computing Engineering Department
Université Laval

A WILEY-INTERSCIENCE PUBLICATION

JOHN WILEY & SONS, INC.

NEW YORK / CHICHESTER / WEINHEIM / BRISBANE / SINGAPORE / TORONTO

Library of Congress Cataloging-in-Publication Data:

Maldague, Xavier P. V.
    Theory and practice of infrared technology for nondestructive testing / Xavier P. V. Maldague.
        p.   cm.
    ISBN 0-471-18190-0 (cloth : alk. paper)
    1. Nondestructive testing. 2. Infrared technology. I. Title.
TA417.5 .M35 2001
621.1'127—dc21   00-063344

10 9 8 7 6 5 4 3 2 1

Ainsi des sophistes célèbres
Dissipant les fausses clartés,
Tu tires du sein des ténèbres
D'éblouissantes vérités.

—Alphonse de Lamartine
Méditations **22**, *Le génie*

To my family,
Marthe, Lucas, Marti, and Flavie

Contents

Preface

At the beginning of the new millenium, with the generalized trend to free trade and the growth of exchange stimulated by the emergence of the Internet, a global competitive Earth-wide market is now a reality. The question of quality control is thus an essential issue for the 2000s. The time when the promise was to replace a product that has failed seems gone; what is essential now is not so much a reduction in what is going wrong as an increase in what is going right the first time. This new trend, sometimes referred to as *total quality*, is tracked by a new credo, ISO certification (9000).

Among the many advantages of this zero-defect manufacturing policy are superior marketability of wholly dependable products, an enormous gain in productivity, elimination of wasteful cost in replacing poor-quality work, and retrofitting products rejected from the field. Total quality is now an established concept for mass products such as cars, consumer electronics, and personal computers. In many fields, primarily aerospace and military, it has been the rule for years, for security reasons.

A major effort to reach this quality concept is to implement inspection tasks along the production line through machine vision. Whereas manual inspection is a variable process prone to fatigue and lack of motivation, machine vision can be used in hostile environments and allows for uniform and repeatable judgment. The increase in quality resulting from careful inspection leads to a reduction in rejection of defective parts passed to subsequent operations. The quality is hence incorporated directly into the product. Since fewer unreliable and defective products are delivered, the customer's satisfaction grows and the market share stabilizes and increases. In this respect, in the field of thermographic nondestructive testing (NDT or TNDT), due to an integrated approach combining an IR image sensor, a computer processor to interpret data, and a machine control interface to provide feedback, a new approach has emerged, *infrared machine vision* (Foucher, 1999; Maldague, 2000).

This book centers on the use of infrared machine vision for industrial processes with automatic inspection being distanced increasingly from traditional methods, based on personal expertise acquired after many years of practice. More specifically, the scope of this book is infrared thermography inspection, theory, and practice, which is an important technique in NDT. Nondestructive testing (NDT), nondestructive evaluation (NDE), and nondestructive inspection (NDI) are terms describing methods for testing without damage. In fact, these terms have much the same meaning and attempts to differentiate have never been successful (Scott, 1990): NDT (discovery of defects, practical aspects of the techniques), NDI (quantification of defects), NDE (assessment of the importance of defects). In this book the term NDT is used.

Unlike visible spectra images, which are produced by reflection and reflectivity differences, infrared images are produced by a self-emission phenomenon and by variations in emissivity. Consequently, inspection processes are different than for traditional processes suitable to inspection based on interpretation and analysis of visible spectrum images. The military were among the first interested in the use of infrared (see the historical notes in Chapter 1). The main idea is to have a vision system capable of revealing the presence of potential targets in the darkness, or during conditions of bad visibility (e.g., fog). Temperature sensing itself in NDT has been around "forever." For example, in his foreword to Rastogi et al. (2000), L. Pflug mentions that in the 1940s, when he took a train to attend school every day, he was surprised that the train engineers touched the wheels of every car axle. Inquiring about this practice, he was informed that it was necessary to check for bearing overheating before a train left on a long-distance trip (his school was located at the last stop before the long-distance trip). Although simple, temperature sensing with the bare hand is convenient and readily available since human skin has a resolution of less than 1°C from ambient temperature. I had a similar experience myself when after the purchase of a small sailboat trailer, I noticed with my hand, several hundred kilometers from home, that one of the two wheels was getting hot. It turned out that the person who assembled the trailer simply forgot to fill the bearings with grease, thus causing severe overheating (fortunately, it could be fixed with a $20 bearing kit and grease).

The potential of TNDT and civil applications of infrared have grown since the availability, in the 1960s, of commercial infrared cameras whose video signals were compatible with the black-and-white television standard, so that now TNDT is used routinely in many different industries using either the NDT technique alone or in combination with other NDT methods. The growing interest in TNDT is particularly evident when attending conferences in the field, such as SPIE-Thermosense* or QIRT,[†] where growing audiences are learning about improving and expanding research efforts involving TNDT.

In this book we present readers with wide coverage of infrared thermog-

---

* www.thermosense.org.
† www.gel.ulaval.ca/qirt.

raphy technology applied in the context of industrial inspection. Particular emphasis is given to fundamental concepts of active and passive thermography as well as industrial case studies so that a comprehensive panorama of the field is available. Three fundamental aspects of TNDT are covered: thermal methods (passive and active), signal (image) processing, and quantitative characterization. The TNDT procedure can be thought of as a two-step procedure. First, thermal methods are deployed to perform thermal inspection, and second, detection and characterization of defect (or abnomaly) take place. Interestingly, for a majority of industrial applications (80% of cases), the detection step is sufficient. The primary classical methods needed to detect defects and flaws in infrared images are thus covered, including several dedicated image-processing techniques needed either to detect defects, enhance their visibility, or reduce infrared image noise. For other applications the aspect of quantitative characterization is of relevance and is discussed as well.

The intended audience is threefold. The book is intended for students who want to learn in depth about TNDT. In this respect this publication can complement a curriculum in advanced material engineering. Examples within chapters and end-of-chapter problems are included with complete solutions, so that the book can serve, in essence, as a textbook on TNDT. Due to the complete coverage of practical aspects, laboratory sessions can be scheduled as well. Industrial engineers will find valuable information regarding possible use of TNDT as an alternative inspection tool, for actual deployment on the plant floor, or as a reference tool on TNDT. Finally, researchers will appreciate extended coverage of the subject in one volume with a large bibliographical section. Some readers may find surprising the many occurrences where the discussion jumps from fundamental aspects to very practical concerns. This was done intentionally since applications of scientific principles require experimental capabilities supported by appropriate theory and practical capabilities. Particular effort was also paid to providing an easily accessible content, since a large portion of the intended audience is not necessarily familiar with complex mathematics.

One interesting aspect of this book is that it brings together diverse topics essential to TNDT. The book starts in Chapter 1 with a brief introduction to TNDT as well as a presentation of pros and cons and a current list of applications. Historical notes complete the first chapter. Next follow the three parts of the book: fundamental concepts, active thermography, and case studies. In Part I the essential concepts of TNDT are presented: thermal emission in Chapter 2, heat transfer in Chapter 3, infrared sensors and optics fundamentals (including infrared optics) in Chapter 4, images in Chapter 5, automated image analysis in Chapter 6, materials in Chapter 7, and experimental concepts in Chapter 8.

In Part II, active thermography which requires thermal stimulation of the surface being inspected is covered, including pulsed, step, lock-in, and vibro-thermography. Following this presentation, data analysis, which is essential for the interpretation of measurements, constitutes the topic of Chapter 10. Finally, in Part III, case studies involving both passive and active thermography are discussed, including a section on evaluation of material thermal diffusivity. Following this, a large bibliographical section is presented, combining references for Chapters 1 to 11 and numerous additional citations. It is planned soon to offer this list for on-line search on the Web.*

Several appendixes follow, including a simple thermal model suitable for initial work on both direct and inverse problems, and a list of material properties (metal and nonmetal). Also provided is a list of Matlab M-scripts for many of the computational tasks discussed in the book. This should help readers to quickly test the concepts discussed. Comments are included with the sources so that readers can adapt these routines to their own image format requirements. The mathematical package Matlab was selected because it is widely available and is particularly easy to handle. When developing the M-scripts, particular attention was paid to avoiding the use of special toolboxes.

In a recent conference, "Trends in Optical NDT and Inspection," the "stairs to heaven" were discussed (see the accompanying figure). I sincerely hope that this book will help those working in this field to climb these steps and thus help to forge a better world in the new millenium.

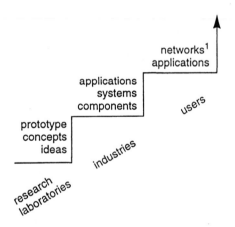

[1]Example: This applies to networks of industries with distributed NDT sensors. In these cases measurements at various places permit rapid decisions/actions following detection of trends, etc.

"Stairs to heaven" (From D. Inaudi in Rastogi and Inaudi, 2000).

*For an update check at: www.gel.ulaval.ca/~maldagx/IRNDT.html.

Acknowledgments

The book takes its origin from the book *Nondestructive Evaluation of Materials by Infrared Thermography* (Maldague, 1993), which was published in 1993 by Springer-Verlag and which is now out of print. From that base, expanded and updated, this book was prepared. The idea for this new book was proposed in 1996 by Kai Chang, Texas A&M University, Series Editor of the Wiley Series in Microwave and Optical Engineering. Through e-mail and phone calls, George J. Telecki of the Wiley staff helped to remotivate me periodically to complete this long, long adventure.

The book would not have been possible without interaction with many people. I am indebted to Paolo Cielo of the Industrial Materials Institute (IMI) in Montréal, a division of the National Research Council of Canada, who introduced me to the field of TNDT inspection in 1984. After several years at IMI, I took a tenured position at Université Laval in Québec City in 1989, where I started research activities in the field of TNDT.

Special thanks are also due Nathalie Beaulieu and Gaston Guay of the Computer and Electrical Engineering Department and to Alain Laverdière of the Sciences and Engineering Faculty. They helped with word processing and prepared several of the drawings. I also wish to thank the students of the Laboratoire de Vision et Systèmes Numériques (LVSN)* who helped prepare some of the material.

Support by the Computer and Electrical Engineering Department is acknowledged, as well as the financial support of the National Sciences and Engineering Research Council of Canada and the Fonds pour la Formation des Chercheurs et l'Aide à la Recherche of Québec province. This essential support is greatly appreciated. Finally, the invaluable understanding of my family, especially of my wife, Marthe, was essential to the success of this work.

XAVIER P. V. MALDAGUE

Electrical and Computing Engineering Department
Université Laval
Sainte-Foy, Québec, Canada
June 2000

* www.gel.ulaval.ca/~vision.

CHAPTER ONE

Getting Started with Thermography for Nondestructive Testing

The purpose of this first short chapter is to introduce *thermography for NDT* (TNDT) concepts, thus enabling readers to better understand subsequent chapters. The chapter concludes with historical notes. *Thermography* is one of many techniques used to "see the unseen." As the name implies, it uses the distribution (suffix *-graphy*) of surface temperatures (prefix *thermo-*) to assess the structure or behavior of what is under the surface. Strictly speaking, *thermography* is a contact technique to record a distribution of surface temperature; and *infrared thermography* is a contactless technique with distinct advantages (see Chapter 2). At present, thermography is generally utilized contact-free, as discussed in this book.

1.1 PASSIVE THERMOGRAPHY TESTING PROCEDURE

The *first law of thermodynamics* concerns the principle of energy conservation and states that an important quantity of heat is released by any (industrial) process consuming energy because of the law of entropy. Temperature is thus an essential parameter to measure in order to assess proper operation (Figure 1.1). In *passive thermography*, abnormal temperature profiles indicate a potential problem, and a key term is *temperature difference* with respect to a reference, often referred to as the *delta-T* (ΔT) value or the *hot spot*. A ΔT value of a few degrees ($>5°C$) is generally suspicious, while greater values indicate strong evidences of abnormal behavior. Generally, passive thermography is rather qualitative since the goal is simply to pinpoint anomalies. However, some investigations provide quantitative measurements if thermal modeling is

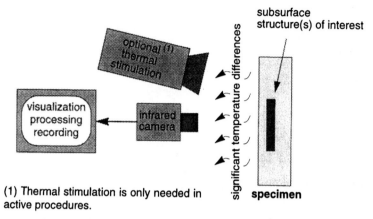

FIGURE 1.1 Schematic setup of infrared thermography for nondestructive testing (TNDT). Drawing shows the reflective scheme.

available, so that measured surface temperature (isotherms) can be related to specific behaviors or subsurface flaws (see Chapter 3). For instance, such dedicated modeling helps us to understand needle heating during high-speed sewing in the automobile industry, and this makes it possible to optimize sewing operations through needle redesign and needle cooling, with significant economic and quality benefits, due to the millions of products sewed daily (seat cushions, airbags, etc.; Q. Li et al., 1999). The fundamental concepts described in Part I are useful for passive thermography. In addition to the many applications discussed through the book, in Chapter 11 we present various applications of passive thermography in greater detail.

1.2 ACTIVE THERMOGRAPHY TESTING PROCEDURES

In *active thermography* it is necessary to bring some energy to the specimen inspected in order to obtain significant temperature differences witnessing the presence of subsurface anomalies. Various testing procedures are deployed, as reviewed briefly below. The fundamental concepts described in Part I are useful for active thermography. In Chapter 9 we discuss procedures and in Chapter 10, data analysis in active thermography. Data analysis enables quantitative information extraction about the specimen inspected. In addition to the many applications discussed through the book, in Chapter 11 we present various applications of active thermography in greater detail.

1.2.1 Pulsed Thermography

Pulsed thermography (PT) is one of the most common thermal stimulation methods used in TNDT. One reason for this is the quickness of the inspection, in which a short thermal stimulation pulse lasting from a few milliseconds for high-conductivity material (such as metal) to a few seconds for low-conductivity specimens (such as plastics, graphite-epoxy laminates) is used. Basically, PT consists of heating the specimen briefly and then recording the temperature decay curve (Figure 1.1). Qualitatively, the phenomenon is as follows. The temperature of the material changes rapidly after the initial thermal pulse because the thermal front propagates by diffusion under the surface and also because of radiation and convection losses. The presence of a subsurface defect modifies the diffusion rate so that when observing the surface temperature, a different temperature with respect to the surrounding sound area appears over a subsurface defect once the thermal front has reached it. As for the detection depth, it is limited since TNDT is a "border technique," but often, anomalies (such as cracks, etc.) start close to the surface.

1.2.2   Step Heating (Long Pulse)

In this case, the increase in surface temperature is monitored during application of a *step heating* (SH) pulse (Figure 1.1). Step heating finds many applications, such as for coating thickness evaluation (including multilayered coatings), inspection of coating–substrate bonds or evaluation of composite structures.

1.2.3   Lock-in Thermography

Lock-in thermography (LT) is based on thermal waves generated inside a specimen and detected remotely. Wave generation, for example, is performed by periodic deposition of heat on a specimen's surface (e.g., through sine-modulated lamp heating) while the resulting oscillating temperature field in the stationary regime is recorded remotely through thermal infrared emission (Figure 1.1). *Lock-in* refers to the necessity to monitor the exact time dependence between the output signal and the reference input signal (i.e., the modulated heating). This is done with a locking amplifier in point-by-point laser heating or by computer in full-field (lamp) deployment so that both phase and magnitude images become available. Phase images are related to the propagation time, and since they are relatively insensitive to local optical surface features (such as nonuniform heating), they are interesting for NDT purposes. The depth range of images is inversely proportional to the modulation frequency, so that higher modulation frequencies restrict the analysis in a near-surface region.

1.2.4   Vibrothermography

Vibrothermography (VT) is an active TNDT technique in which under the effect of mechanical vibrations induced external to the structure (due to direct conversion from mechanical to thermal energy), heat is released by friction precisely at locations where defects such as cracks and delaminations are located.

1.3 TNDT LIMITATIONS AND APPLICATIONS

Limitations and capabilities of TNDT are listed in Table 1.1, and some common applications are given in Table 1.2. In the following chapters we discuss TNDT concepts in greater detail.

1.4 HISTORICAL NOTES

Before we address the technical aspects of the subject, it is worthwhile to turn back a little to some historical detail. For interested readers, R. D. Hudson (1969) presents an exhaustive historical survey of infrared technology.

Over all human history, heat and temperature have been concerns. From the first people living in caverns, discovery of a reliable mean to generate fires brought light, heat, and a way to keep away dangerous beasts such as wolves at night. Later good-quality metal and glass work depended on the ability of skilled artisans to relate, say, the glowing (brightness, color) of a gob of glass to the moment it is ready to be shaped. Craftspeople still use this long intuitively known relationship between target temperature and emitted radiation to accomplish their work (e.g., the Murano glassworkers in Venice). We are still concerned by weather forecasts and temperature scales: Too cold or too warm temperatures makes day-to-day life uncomfortable, and thermal considerations are of interest for a whole spectrum of industries and applications.

From a more quantitative perspective, the first step toward accurate measurement of temperature was achieved by Galileo in 1593, when he designed the first glass thermometer, a liquid-filled glass bulb connected to a partially filled capillary tube. These contact devices are still of use for temperature measurement from -180 to $650°C$. Differential expansion of the liquid with respect to the glass bulb due to a temperature increase causes the liquid to rise in the capillary, thus indicating the temperature on an engraved scale. Various liquids, such as mercury, alcohol, and toluene, are commonly used. However, in Galileo's time, the underlying physics behind temperature was not fully understood.

A few centuries later, in 1800, Herschel* discovered the existence of infrared (IR) rays through a famous experiment. As royal astronomer to King George III of England, Herschel accidentally discovered Uranus on March 13, 1793. This "accident" led to his discovery of infrared rays. Herschel wanted to protect his eyes when observing the sun and used a prism to separate the various colors from blue to red. Using a mercury thermometer, he noted that temperature was still elevated beyond the red band where no radiation was visible (Hershel, 1800a,b). In fact, this experiment had been done before (e.g., by Sir Isaac Newton, 1642–1727), but Herschel was the first to notice that the distance where the heating is greatest has a specific location (i.e., depending on the

---

*The William Herschel Museum is located at 19 New King Street, Bath, BA1 2BL, United Kingdom (tel: 01225 311342).

TABLE 1.1 Limitations and Capabilities of TNDT Procedures

Procedure[a]	Capabilities	Limitations
All procedures (generally)	• Fast, surface inspection • Ease of deployment • One-side-only deployment • Safety (no harmful radiations) • Ease of numerical thermal modeling • Ease of interpretations of thermograms • Great versatility of applications • Sometimes unique tool (corrosion around rivets)	• Variable emissivity • Cooling losses (convection/radiation causing perturbing contrasts) • Absorption of infrared (IR) signals by the atmosphere (especially for distances greater than a few meters) • Difficult to get uniform heating (active procedures) • Transitory nature of thermal contrasts requiring fast-recording IR cameras • Need of straight viewing corridor between IR camera and target (although it could be folded through first surface mirrors) • Limited contrasts and limited signal/noise ratio, causing false alarms (measurement of a few degrees above background at around 300 K) • Observable defects generally shallow • Works only if thermal contrasts naturally present
P	• No interaction with specimen • No physical contact	
A-PT	• No physical contact • Quick (pulsed thermal stimulation: cooling or heating) • Phase and modulation images available with frequency processing (such as in pulsed phase thermography)	• Requires apparatus to induce the pulsed thermal perturbation • Computation of thermal contrasts requires knowledge of an a priori defect-free zone within the field of view • Inspection surface limited (ca. 0.25 m² maximum)
A-SH	• No physical contact	• Requires apparatus to induce the thermal perturbation • Risk of overheating the specimen

(Continued)

TABLE 1.1 (*Continued*)

Procedure[a]	Capabilities	Limitations
A-LT	• No physical contact • Large surface inspected (a few square meters at a time) • Phase and modulation images available • Modulated ultrasonic heating (for some applications, might require physical contact, bath immersion)	• Requires modulated thermal perturbation • Requires observation for at least one modulation cycle (longer observation with respect to APT) • Thickness of inspected layer under the surface related to the modulation frequency (unknown defect depth might require multiple experimentations at different frequencies)
A-VT	• Closed cracks revealed	• Difficult to generate mechanical loading • Thermal patterns appear only at specific frequencies • Physical contact to induce thermal stimulation

Source: Adapted from *Encyclopedia of Materials* (New York: Elsevier, 2001), Table 2 in the article "Thermographic techniques for NDT" by X. Maldague.
[a]P, passive thermography; A-PT, active pulsed thermography; A-SH, active step heating; A-LT, active lock-in thermography; A-VT, active vibrothermography.

TABLE 1.2 Common Applications of TNDT Procedures

Applications	Procedure[a]
Buildings	
• Walls assemblies, moisture evaluation, roofs, liquid level in tanks	• P
• Water entrapment, fresco delamination	• A-PT
Components/processes	
• Carton sealing line inspection, automobile brake system efficiency, heat dissipation of electronic modules, recycling process identification, welding process, printed-circuit boards, glass industry (bottles, bulbs), metal (steel) casting	• P
• Aircraft structural component inspection, solder quality of electronic components, spot welding inspection	• A-PT
• Degradation of EPROM (erasable programmable read-only memory) chips, paper structure (cockling)	• A-SH
• Aircraft structural component inspection, loosening bolt detection, plastic pipe inspection, radar-absorbing structure investigation	• A-LT
Defect detection and characterization	
• Metal corrosion, crack detection, disbonding, impact damage in carbon fiber–reinforced plastic (CFRP), turbine blades, subsurface defect characterization (depth, size, properties) in composites, wood, metal, plastics	• A-PT
• Defects in adhesive and spot-welded lap joints	• A-SH
• Crack identification, disbonding, impact damage in CFRP	• A-LT
• Coating wear, fatigue test, closed-crack detection	• A-VT
Maintenance	
• Bearings, fan and compressors, pipelines, steam traps, refractory lining, rotating kilns, turbine blades, electric installations, gas leaks	• P
Medical/veterinary	
• Thermal coronary angiography, allergen reactions, human breast tumors, rheumatology, neuromuscular disorders, soft-tissue injuries	• P
• Blood vessel flow	• A-LT
Properties	
• Glaze thickness on ceramics, crush test investigation	• P
• Thermophysical properties (diffusivity, etc.), underalloyed and overalloyed phases in coatings on steel, moisture, anisotropic material characterization	• A-PT
• Thermal conductivity measurement in CFRP	• A-SH
• Adhesion strength, anisotropic material characterization, coating thickness in ceramics, depth profile of thermal conductivity or diffusivity, moisture	• A-LT
Public services	
• Forest fire detection, people localization in fires or at night, monitoring of road traffic, target detection (military)	• P

Source: Adapted from *Encyclopedia of Materials* (New York: Elsevier, 2001), Table 1 in the article "Thermographic techniques for NDT" by X. Maldague.

[a] For an explanation of abbreviations, see Table 1.1.

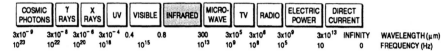

FIGURE 1.2 Electromagnetic spectrum. (From Maldague, 1994b, with permission from Gordon and Breach Publishers.)

wavelength). We now know that this is related to Planck's and Stefan's laws (Chapter 2).

Hershel's findings can be summarized as follows:

1. He was concerned with the similarity between light and heat. He called his discovery "invisible rays" or "rays that occasion heat." As we know now, "light" and "heat" (or infrared) are both forms of electromagnetic radiation of different wavelength and frequency. Figure 1.2 shows the electromagnetic spectrum, including the visible and infrared.

2. He also demonstrated that quantitative measurements are possible in this newly discovered part of the electromagnetic spectrum using the mercury thermometer.

3. He was first to discover that transmission of infrared radiation is different from material to material, which explains, for example, why germanium is widely used in infrared optics rather than conventional optical glass, which transmits poorly in the infrared band (Chapter 4). He discovered, for example, that table salt (NaCl) is a good infrared transmitter. He also noticed that the same laws for reflection and transmission as those that govern visible rays govern infrared radiation.

Following Hershel, milestones continued to be set:

- *1829:* Nobili invented the first thermocouple (based on the thermoelectric effect discovered in 1821 by Seebeck, known as the *Seebeck effect*). A thermocouple is a contact sensor formed of two distinct metal junctions. When one junction is set at a different temperature with respect to the other, a proportional difference in voltage is generated, related to the temperature difference between the two junctions (Section 4.3).

- *1833:* Melloni made the first thermopile by connecting many thermocouples. By focusing the incoming radiation on one side of the junctions, the increased sensitivity achieved allowed detection of the presence of a person at a 10-m distance (the focused radiation heats the junction).

- *1840:* Hershel's son John produced the first infrared image using an evaporograph. In this device, an infrared image is formed by differential evaporation of a thin film of oil.

- *1880:* The bolometer was invented by A. Longley and perfected by Abbot, who used it to sense the heat from a cow some 400 m away. A bolometer is a thermal detector whose electrical conductivity changes when heated by impinging radiation (Section 4.3.1).

- *1900:* Max Planck's (1858–1947; Nobel prize recipient in 1918) theory of radiation (Chapter 2) clarified Herschel's experiment. The wavelength of maximum radiation λ_m (micrometers) is given by the following equation (Stefan's law), with temperature T expressed in kelvin:

$$\lambda_m = \frac{2898}{T} \tag{1.1}$$

For the sun, whose surface temperature is about 6000 K, λ_m is about 0.5 μm, corresponding to the yellow band, just in the middle of the sensitivity bandwidth of our eyes (roughly 0.4 to 0.75 μm). Looking at the sun, Herschel's readings peaked in the yellow band. As another example, the peak wavelength of the human body at 37°C is $2898/(37 + 273) = 9.3$ μm.

After this, progress in infrared and thermal technologies accelerated:

- *1914–1918:* World War I brought the first photoconductive detector, in 1917. Previous thermal detectors were sensitive to a direct increase in their temperature by impinging radiation; in photoconductive detectors, a change of electrical conductivity is obtained directly by interaction with the photons of the incident radiation, thus making them faster and more sensitive than what was then currently available.
- *1940–1945:* During World War II many patents were registered. Applications included the areas of machinery, ships, iceberg detection, communications, and guidance of flying torpedoes. The latter application is particularly interesting, as we know the impressive efficiency of infrared and photonics devices in offensive attacks during the Persian Gulf War (1991), with more than 80% of the weapons equipped with such detection capabilities. World War II torpedos used an active illumination scheme and were sensitive in the near infrared (wavelengths shorter than 1 μm): A filtered (to block visible radiation) tungsten lamp lit the target, and the reflected infrared radiation served to guide the torpedo. The Germans found that performance could be improved by cooling the detectors. This is of major importance; cooling is used in many infrared detection devices (Section 4.6).

The postwar period was very fruitful in research and development, as the war clearly demonstrated the usefulness and great potential of infrared. Although many spin-offs found applications in other areas, it is estimated than 80% of the IR market continues to be for military applications. Nevertheless, several infrared technologies developed during wars passed to civilian applications thereafter. This is, for example, the case with commercially available high-quality interference filters used to select working spectral bands, and also for numerous detectors. For instance, the first quantum detector [lead sulfite (PbS)] is an achievement of that period. Quantum detectors do not depend on heating of their sensitive surfaces to produce an output signal related to the target temperature; they are thus faster than the thermal detectors such as bolometers or thermopiles available previously. Indium antimonide (InSb) detectors followed (in the 1940s and 1950s) and mercury tallium telluride (HgCdTe) detectors in the 1960s.

In the 1960s and 1970s, the first infrared cameras became commercially available. Among the world largest commercial infrared camera providers, Inframetrics was set up in 1975 in the United States, while Agema of Sweden (known originally as AGA) marketed the first commercial infrared camera system in 1965 (the companies later merged and are now known as FLIR Systems). Other important companies also emerged during these years.

Early IR cameras used pyroelectric tube technology. Pyroelectric tubes are thermal detectors similar to vidicon TV cameras but with an IR-transmitting face plate and specific pyroelectric target material such as triglyceride sulfate (Section 4.3.4). Images are produced by an electron beam scanning the target. Other early infrared cameras contained only a single detector, and image generation (either line or bidimensional imaging) was made possible by a rotating electro-optical system (prisms, mirrors). On some models, during the retrace, it was possible for the detector to be in the viewing corridor of an internal blackbody, a calibrated source of infrared radiation (Section 2.2). This feature enables direct calibration of the output signal. For this reason, such cameras are often referred to as scanning radiometers.

A detector known as SPRITE (signal processing in the element) was developed in the United Kingdom by Elliot at the Royal Signals and Radar Establishment, makes it possible to circumvent some of the disadvantages of scanning radiometers. For example, signal is integrated on the focal plane itself, reducing the noise level and timing considerations. In the early 1980s a small revolution occurred in the IR community with the development of the first focal plane array (FPA) IR cameras. FPAs are similar in structure to solid-state CCD (charge-coupled device) video cameras but are sensitive in the IR. Basically, an FPA camera comprises a chip (the focal plane array, which is a matrix of detectors) with associated electronics (amplifiers, analog-to-digital converters). Since no scanning mechanism is needed for image formation, these cameras offer several advantages, such as higher spatial resolution, less noise, and more rugged structure. Moreover, direct digital output suppresses the need for TV-compatible signals, thus enabling acquisition of fast thermal events. Full image acquisition at over 400 Hz and subimage acquisition at over 30 kHz are now available (e.g., products from Santa Barbara Infrared or Raytheon). A drawback is the uncalibrated output signal since without a scanning mechanism, no internal blackbody is seen as in scanning radiometers.

In the late 1970s and early 1980s, with the broader availability of large integrated-circuit (IC) fabrication technology, it was possible to make large uncooled bidimensional arrays based on pyroelectric-effects thermopiles and microbolometers (Sections 4.3.4 and 4.5). As we have seen, IR has a long history with succeeding milestones for always-improved performance, now combined with more complex and powerful signal processing techniques packaged in always-smaller computers and processors.

Back on the NDT side, one of the earliest infrared applications dates back to 1935, when Nichols employed a radiometer to verify the uniformity at which steel slabs are reheated in a steel-rolling mill (Nichols, 1935). As applications proved to pay themselves back, infrared NDT technology became widespread. Other early NDT investigations dealt with analysis of temperature distribution in brake shoes (R. C. Parker and Marshall, 1948), inspection of soldered seams on a tin can (Gorrill, 1949), power transmission line surveys (Leslie and Wait, 1949), and detection of overheated components on circuit boards (*Electronic Design*, 1961). More complete reviews of these pioneers in infrared NDE applications can be found in Wilburn (1961a,b) and McGonnagle et al. (1964). The availability of commercial infrared cameras emerging from unclassified military technology in the mid-1960s saw a florishing of NDT applications, some of which are discussed in detail in this book.

PART I

FUNDAMENTAL CONCEPTS

CHAPTER TWO

Introduction to Thermal Emission

This chapter has been inspired by many sources, among which DeWitt and Nutter (1988) and Gaussorgues (1989) were found particularly helpful, and we invite interested readers to refer to these documents, which include much more detail than is possible in a single chapter.

2.1 BASIC CONCEPTS

Let's start our discussion with a consideration of temperature units. (Table 2.1 lists main physical quantities with their accepted values.) Today, three units are used for temperature T: kelvin, degrees Celsius (formerly called centigrade), and degrees Fahrenheit, and temperature scales are set by international standards. The relationships among these systems are as follows:

$$T(°C) = [T(°F) - 32] \times \tfrac{5}{9} \tag{2.1}$$

$$T(°F) = T(°C) \times \tfrac{9}{5} + 32 \tag{2.2}$$

$$T(K) = T(°C) + 273.15 \tag{2.3}$$

Interestingly, 273.15 K corresponds by definition to the triple point of water (i.e., the temperature at which water, ice, and water vapor coexist). The triple point of water can be observed in a cooled evacuated cell containing pure water; it corresponds to 0.01°C, a temperature of 0°C thus equal to exactly 273.15 K. The international temperature scale (ITS) established in 1927 being was renamed the *international practical temperature scale* (ITPS) in 1960. The freezing points of a group of pure materials (such as gold) are highly reproducible and were substituted for the temperature of the triple point of water,

TABLE 2.1 Values of Physical Constants of Interest

Quantity	Symbol	Value	Units
Speed of light in vacuum	c_0	2.99792458×10^8	m s^{-1}
Permeability of vacuum		$4\pi \times 10^{-7}$	N A^{-2}
		$= 12.566370614 \times 10^{-7}$	
Permittivity of vacuum, $1/\mu_0 c_0$	ε_0	$8.854187817 \times 10^{-12}$	F m^{-1}
Planck constant	h	6.626076×10^{-34}	J s
Boltzmann constant	K	1.380658×10^{-23}	J K^{-1}
Stefan–Boltzmann constant	σ	5.67051×10^{-8}	W m^{-2} K^{-4}
First radiation constant, $c_1 = 2hc_0^2$	c_1	$1.1910439 \times 10^{-16}$	W m^2
Second radiation constant, $c_2 = hc_0/K$	c_2	1.438769×10^4	μm K
Third radiation constant	c_3	2897.7	μm K

Source: Adapted from DeWitt and Nutter (1988).

which was the first basis of temperature scales, due to the high cost and complexity of such measurements.

The ITPS is based on measurements of a few materials at a specific known temperature, such as the freezing point of gold (1064.43°C). In this case, a blackbody (Section 2.2) is immersed in a bath of freezing gold, and once the thermodynamic equilibrium is reached, the radiation exiting the blackbody is measured with a disappearing-filament optical pyrometer, made of a telescope containing a red filter (around 0.653 μm) and a small tungsten filament lamp. The filament is located in the image plane of the objective lens such that it is superimposed on the image of the target. The user then adjusts the filament current in order to have its brightness equal that of the target. In this condition, the filament seems to disappear when the target temperature is reached. Knowing the temperature of the target (here 1064.43°C), it is possible to calibrate the instrument. If manual adjustment is a concern, automatically adjusted disappearing-filament optical pyrometers are also available. Such instruments have been on the market for a long time (initially patented in 1899 by Morse). Measuring the gold point is a primary standard and is made by national standardizing laboratories such as the National Research Council, Division of Physics in Canada and the National Bureau of Standards (NBS) in the United States. Platinum (Pt) resistance thermometers are used to establish the temperature scale from the hydrogen point (13.81 K) to the aluminum point (630.74 K). Pt–10% rhodium thermocouples are then used up to the gold point. For higher temperatures, pyrometers in conjunction with Planck's law (Section 2.5) are the rule.

The energy released by particles in oscillation produces thermal emission; such oscillations are themselves caused by the temperature of the matter. Particles can be electrons, ions, atoms, or molecules. In fact, all matter produces thermal emission. A distinction exists, however, between gases and a liquid or solid. In the case of gases, thermal emission is volumetric; that is, particles within the gas contribute to the emission, whereas for a solid and liquid the phenomenon is a surface phenomenon. In the following discussion, we restrict ourselves to surface thermal emission, which is pertinent to most NDT applications.

An important aspect is that thermal emission does not need matter for transportation. In fact, there are two different ways to explain thermal emission transportation. One is to consider *photon emission*. Photons are energy particles with zero mass at rest and with a discrete quantity of energy called *quantum*. The other approach is to consider an electromagnetic wave with a specific frequency and energy. In fact, these two concepts are related since for a given wavelength λ, the liberated photonic energy W due to the oscillatory nature of particles inside matter is given by

$$W = \frac{hc}{\lambda} \quad \text{joules} \tag{2.4}$$

where h is the Planck constant and c is the speed of light (see Table 2.1). Another useful relationship links the wavelength with the frequency f in hertz) of the electromagnetic wave:

$$\lambda = \frac{c}{f} \tag{2.5}$$

Wavelengths are generally expressed in micrometers (μm; $1 \ \mu$m $= 10^{-6}$ m), nanometers (nm; 1 nm $= 10^{-9}$ m), or angstroms (Å; 1 Å $= 10^{-10}$ m). In spectroscopy it is common to work with the wave number ($= 1/\lambda$) expressed in cm^{-1}: for example, $10 \ \mu$m $= 1000$ cm^{-1}.

Figure 1.2 shows the electromagnetic spectrum. As shown, with wavelength increasing, the spectrum is divided into specific bands (mainly on the basis of their utility and their source of emission): gamma rays, x-rays [also known as Roentgen rays; see Berger (1995) for an excellent article about this great discovery], ultraviolet rays, visible rays, infrared, radio waves, and microwaves. The IR spectrum can be split further: near IR (for wavelengths from 0.78 to 1.5 μm), middle IR (1.5 to 20 μm), and far IR (20 to 1000 μm). These three IR bands are divided with respect to IR detector capabilities. For instance, quantum, photoemissive, and special photoemulsive films work in the close IR, thermal and quantum detectors in the middle IR, and thermal detectors in the far IR (more details on detectors in Chapter 4). From an NDT perspective, middle IR is of most interest.

Energy emitted from a surface has various features: Many wavelengths are emitted with an unequal distribution of energy among them; moreover, emission is not uniform in all directions (Figure 2.1). To explain these emission phenomena more fully, it is useful to introduce the concept of a blackbody or a perfect radiator.

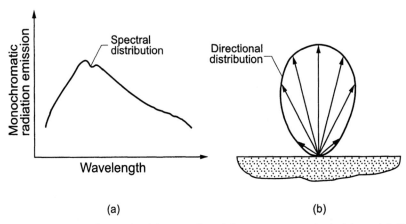

FIGURE 2.1 Radiation emitted by a surface (planar representation): (a) spatial distribution; (b) directional distribution. (Data from Dewitt and Nutter, 1988.)

2.2 BLACKBODY

A blackbody is an instrument that absorbs irradiated energy totally from any direction and wavelength. Since it also has the property (following Kirchhoff's laws, Section 2.9.5) of reemitting this energy until thermodynamic equilibrium is reached with the surrounding environment, it is referred to as a *perfect radiator*. The radiation emitted from the blackbody is a function of the blackbody temperature, and its magnitude depends on the wavelength. Since it is also independent of the direction, it is called *isotropically diffuse* (or *Lambertian*). Another feature is that no other surface can emit more energy than a blackbody for a given temperature and wavelength.

Although a useful mathematical concept, blackbodies do exist. For instance, they can be made from a cavity that has a small aperture (Figure 2.2). A flat surface with a perfectly absorbing coating can also be considered a blackbody. Practically, real blackbodies are different from perfect blackbodies. For example, for a cavity-type blackbody, the emission is not isotropically diffuse on the entire hemisphere but rather is limited to a vertical cone of about 20° (Figure 2.1b).

2.3 SOLID ANGLE, COORDINATE SYSTEM, EMITTED AND INCIDENT RADIATION, AND UNITS

To introduce the concepts of emitted and incident radiation and the corresponding laws, it is necessary to introduce some concepts and notation [which more or less follow the notation used by DeWitt and Nutter (1988) or Spiegel (1974)]. A spherical system of coordinates is convenient for a more formal de-

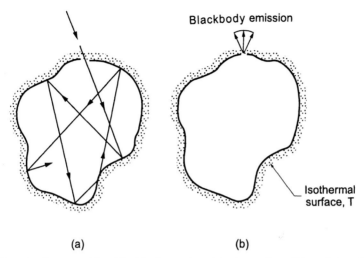

FIGURE 2.2 Features of a blackbody cavity: (a) absorption; (b) emission from an aperture. (Adapted from Dewitt and Nutter, 1988.)

scription of radiation interaction with a surface. Such a system of coordinates (Figure 2.3) requires three variables to locate a point on a sphere surface: the radius r, which is the distance from the sphere center located at $(0,0,0)$; the azimuthal angle θ, and the zenith angle ϕ. Point P is thus located in (r,θ,ϕ).

Another important definition is the solid angle (Figure 2.4). Let's first define the plane angle (Figure 2.4a). Suppose that we have two lines connected in a point O; the plane angle $d\alpha$ is thus given by the ratio of the length of the arc

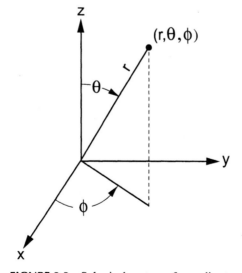

FIGURE 2.3 Spherical system of coordinates.

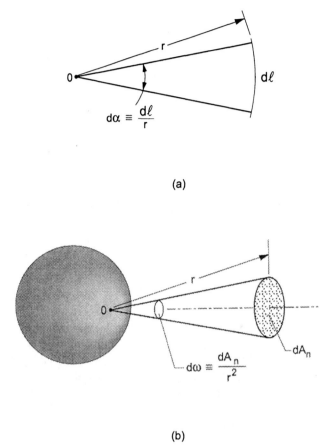

(a)

(b)

FIGURE 2.4   Definition of (a) plane; (b) solid angle.

obtained from the intersection of a circle of radius r centered in O with the two lines to this radius: $d\alpha = l/r$. The units for the plane angle are radians (rad). The plane angle thus corresponds to the fraction of the infinite plane defined by these two lines, which meet in O. When the length of the arc circle is equal to its radius, the sustended angle is equal to 1 rad.

The solid angle is the extension of the plane angle definition to the space, since it is the fraction of the space contained within an infinite cone centered in O (Figure 2.4b). Next, imagine the intersection of a sphere centered in O with radius r and a cone of infinite extent. In this case, the ratio of the intercepted surface dA_n on the sphere to the square of radius r is the measure of the solid angle: $d\omega = dA_n/r^2$ (the subscript n indicates that the differential surface element is *normal* to the cone axis); the units for a solid angle are steradians (sr). The differential notation is used here since we have the element of a solid angle.

Example 2.1   Find the solid angle subtended by a sphere.

SOLUTION

$$\omega = \frac{A}{R^2} = \frac{\text{sphere surface}}{R^2} = \frac{4\pi R^2}{R^2} = 4\pi \quad \text{steradians}$$

Example 2.2 Compute the subtended solid angle of the small surface A_1 when viewed from A_0 in the following configuration:

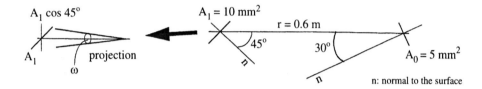

SOLUTION   In this case the projected surface of A_1 in a direction perpendicular to A_0 is equal to $A_1 \cos 45°$ (see Problem 2.3). Since the surfaces are small with respect to involved distances, the subtended solid angle ω_{10} is

$$\omega_{10} = \frac{A}{r^2} = \frac{\text{projected surface}}{r^2} = \frac{(10 \times 10^{-6}) \cos 45°}{(0.6)^2} = 1.9 \times 10^{-5} \text{ sr}$$

2.4 RADIANCE, EXITANCE, AND IRRADIANCE

Energy can be either emitted from a surface or incident on a surface. To distinguish between these two situations, we will affix variables related to the emitted radiation with a prime. If we are interested to the energy leaving the differential surface patch dA located in $(0, 0, 0)$ and passing through point P, the prime notation will thus be used (r, θ', ϕ').

The *spectral radiance* $L'(\lambda, \theta', \phi')$ is the rate at which energy or emitted flux Φ' (emitted at a given wavelength λ) is emitted from surface patch dA in the direction (θ', ϕ') and passes through dA_n' (Figure 2.5). Based on previous definitions, the spectral radiance L_λ' is [DeWitt and Nutter, 1988, eq. (1-5)]

$$L'(\lambda, \theta', \phi') = \frac{d^3\Phi'}{dA \cos \theta' \, d\omega' \, d\lambda} \quad \text{W m}^2 \text{ sr}^{-1} \text{ μm}^{-1} \quad (2.6)$$

This equation expresses the spectral radiance as the ratio of the emitted flux Φ' to the surface element subtended by the solid angle $d\omega$ and immediately surrounding the flux direction (θ', ϕ') per unit of wavelength interval $d\lambda$. This surface element is, in fact, the projected surface element perpendicular to the emitted radiation $(dA \cos \theta')$.

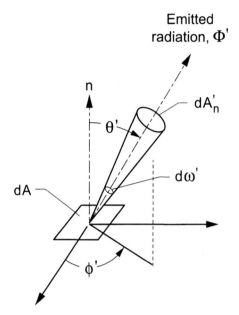

FIGURE 2.5   Emitted radiation.

The radiant power per unit area for an emitting surface can be computed if the spectral and directional distributions of the spectral radiance are known, by integrating $L'(\lambda, \theta', \phi')$ over any finite angle and finite wavelength interval. For example, if an isotropically diffuse emitter (a Lambertian emitter) is considered, that is, a surface for which the emitted radiation is independent of the direction $[L'(\lambda, \theta', \phi') = L'(\lambda)]$, it is possible to compute the spectral radiant power per unit area associated with emission in the hemisphere above a dA patch located on this surface. This quantity is known as the spectral *exitance* $M(\lambda)$ with units $\text{W m}^{-2} \, \mu\text{m}^{-1}$. For a Lambertian surface

$$M'(\lambda) = \pi L'(\lambda) \tag{2.7}$$

Interestingly, a Lambertian emitter approximates reasonably well many real surfaces. Finally, the *total exitance* is the radiant power per unit area emitted over all wavelengths and all directions, symbol M with units of W m^{-2} and

$$M' = \pi L' \tag{2.8}$$

In these two last equations, π is a constant with units of steradians.

The radiation incident on a surface either from reflection or emission at other surfaces also has spectral and directional distributions. Ideas discussed for radiation emitted from a surface can be adapted to incident radiation, and the *incident spectral radiance* $L(\lambda, \theta, \phi)$ is thus introduced with the following definition (notice here that unprimed variables are used to denote incident conditions and the similarity with eq. (2.6):

$$L(\lambda, \theta, \phi) = \frac{d^3\Phi}{dA \cos\theta \, d\omega \, d\lambda} \qquad \text{W m}^2 \text{ sr}^{-1} \, \mu\text{m}^{-1} \qquad (2.9)$$

This equation expresses the incident spectral radiance as the ratio of the incident flux Φ for the wavelength interval $d\lambda$ from the direction (θ, ϕ) to the projected surface element ($dA \cos\theta$, perpendicular to the incident radiation) subtended by the solid angle $d\omega$ and immediately surrounding the flux.

The *spectral irradiance* $E(\lambda)$ is defined as the spectral radiant power at the wavelength λ incident on a surface per unit area of the surface from all directions in the hemispheric space above the surface considered:

$$E(\lambda) = \frac{d\Phi(\lambda)}{dA} \qquad (2.10)$$

and the total irradiance is thus defined as the radiant power at all wavelengths incident on a surface per unit area and from all directions E (W m^{-2}):

$$E = \int_0^\infty E(\lambda) \, d\lambda \qquad (2.11)$$

If the incident radiation is isotropically diffuse, that is, independent of the direction (θ, ϕ), then $E(\lambda) = \pi L$. It is important to note that for the irradiance, the surface considered is the actual surface dA, while for the radiance, this is the projected surface $dA \cos\theta'$.

Example 2.3 Assuming that the small surface A_0 of Example 2.2 is a Lambertian emitter with a total radiance $L' = 1000$ W m^{-2} sr^{-1}, compute **(a)** the radiant flux in watts emitted from A_0 and reaching A_1, and **(b)** the irradiance on A_1 due to the emission from A_0.

SOLUTION   **(a)** Equation (2.6) can be rewritten as

$$d\Phi = L'(\lambda, \theta', \phi') \, dA \cos\theta' \, d\omega'$$

and since the total radiance here is independent of the direction and with small involved surfaces and solid angles (thus $dA \approx A$, $d\omega \approx \omega$), the radiant flux Φ in this case becomes

$$\Phi = L' A_0 \cos\theta' \omega'_{10}$$

and with ω_{10} as computed in Example 2.1, the radiant flux

$$\Phi = 1000 \text{ W m}^{-2} \text{ sr}^{-1} \times 5 \times 10^{-6} \text{ m}^2 \times \cos 30° \times 1.9 \times 10^{-5} \text{ sr} = 82.3 \text{ nW}$$

(b) The irradiance is computed with the same assumptions as for the radiant flux:

$$E = \frac{\Phi}{A_1} = \frac{82.3 \text{ nW}}{10 \times 10^{-6} \text{ m}^2} = 8.23 \text{ mW m}^{-2}$$

To conclude this section, it is interesting to note that the concepts introduced here were originally developed for the visible spectrum (since the human eye was historically the only sensor available) with a specific nomenclature that still prevails in the machine vision and television fields, for example. Units such as candelas per square meter for *luminance* and lux (lumens per square meter) for *illumination* are thus related to the radiance and irradiance, respectively:

visible spectrum (subscript v) $\leftrightarrow$ electromagnetic spectrum as a function of λ

Emitted radiation:

$$\text{luminance } L_v \quad 1 \, \frac{\text{cd}}{\text{m}^2} = 1 \text{ lm m}^{-2} \text{ sr}^{-1} \leftrightarrow \text{radiance } L(\lambda, \theta, \phi) \tag{2.12}$$
$$1 \text{ W m}^{-2} \text{ sr}^{-1} \text{ }\mu\text{m}^{-1}$$

Incident radiation:

$$\text{illuminance } E_v \quad 1 \text{ cd m}^{-2} = 1 \text{ lm m}^{-2} \text{ sr}^{-1} \leftrightarrow \text{irradiance } E \quad 1 \text{ W m}^{-2} \tag{2.13}$$

As stated above, the irradiance is concerned with incident radiation coming from all directions on a surface, thus explaining the lack of directional variables (θ, ϕ) in eq. (2.10).

2.5 PLANCK'S LAW

Planck's law is one of the most important laws governing thermal emission. It describes the distribution of emitted energy as a function of the wavelength for a given temperature. The spectral radiance, the power irradiated by a blackbody (subscript b) per unit surface and per unit of solid angle,

$$L_{\lambda,b}(\lambda, T) = \frac{2hc_0^2}{\lambda^5[\exp(hc/\lambda KT) - 1]} \quad \text{W m}^2 \text{ }\mu\text{m}^{-1} \text{ sr}^{-1} \tag{2.14}$$

It is common practice to represent Planck's law with a curve family (Figure 2.6). From such a graphic, some conclusions can be drawn. For a given temperature, the magnitude of the emitted radiation varies with wavelength.

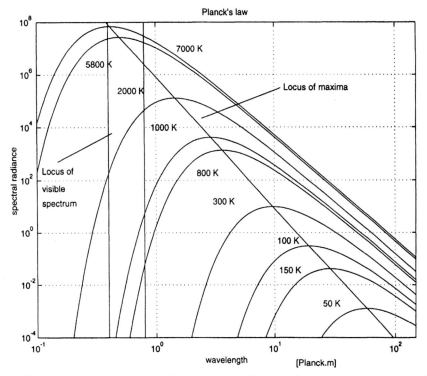

FIGURE 2.6 Spectral radiance of a blackbody (Planck's law; could be plotted with Matlab script: Planck.m).

More conveniently, eq. (2.14) can be rewritten using the first and second radiation constants (Table 2.1):

$$L_{\lambda,b}(\lambda, T) = \frac{c_1}{\lambda^5 [\exp(c_2/\lambda T) - 1]} \qquad \text{W m}^2 \text{ μm}^{-1} \text{ sr}^{-1} \qquad (2.15)$$

Before Planck, two relationships were developed for specific wavelength bands. The first is the *Wien law*, which is valid for short wavelengths $\lambda T \ll c_2$, in which case the "−1" can be dropped in eq. (2.14), yielding

$$L_{\lambda,b}(\lambda, T) = \frac{c_1}{\lambda^5} \exp\left(-\frac{c_2}{\lambda T}\right) \qquad \text{W m}^2 \text{ μm}^{-1} \text{ sr}^{-1} \qquad (2.16)$$

For long wavelengths, when $\lambda T \gg c_2$, development in series of the exponential term of eq. (2.14) provides the *Rayleigh–Jeans law*:

$$\exp(a) - 1 \approx (1 + a + \cdots) - 1 \approx a \qquad \text{for small values of } a$$

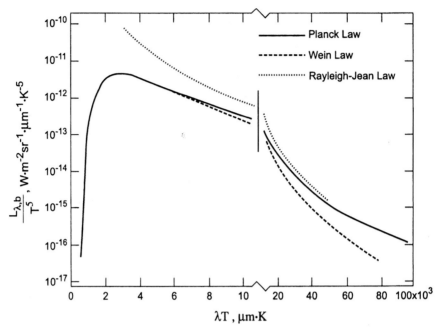

FIGURE 2.7   Comparison of Planck spectral distribution with approximation given by Wien law and Rayleigh–Jeans law. (Data from Dewitt and Nutter, 1988.)

and thus

$$L_{\lambda,b}(\lambda, T) = \frac{c_1}{\lambda^5(c_2/\lambda T)} = \frac{c_1}{c_2}\frac{T}{\lambda^4} = 2c_0 k\frac{T}{\lambda^4} \qquad \text{W m}^2 \, \mu\text{m}^{-1} \, \text{sr}^{-1} \qquad (2.17)$$

This equation shows that at long wavelengths, a linear dependence of the spectral radiance with temperature is observed. Figure 2.7 illustrates the validity of these approximations with respect to Planck's law.

2.6   WIEN DISPLACEMENT LAW

As seen from Figure 2.6, the locus of the maximum spectral radiance for a given temperature is a line given by the following expression, which is obtained by derivation of Planck's law:

$$\lambda_{\max} = \frac{2897.7}{T} = \frac{c_3}{T} \qquad (2.18)$$

where c_3 is the third radiation constant, with units of μm K (Table 2.1). A few examples can illustrate the usefulness of eq. (2.18). For instance, assuming that

FIGURE 2.8 Baths of liquid aluminum (alloy 65054) not emitting much at this temperature of about 659°C. Notice, for instance, the small extracted metal sample on the bottom right, which has already solidified and exhibits similar reflective behavior. This is not the case of steel, as shown in Figure 4.26.

the sun surface temperature is at about 6000 K, from eq. (2.18) it is seen immediately that the radiation emitted from the sun peaks at 0.5 μm, right in the middle of the visible spectrum. If we use a range heating plate as an example, when the unit is activated, IR radiation is first sensed above the plate with one's hand although the plate is still dark. As the plate temperature increases, it starts to glow, first dark red (at a temperature of about 400°C; see Problem 2.5), then increasing up to bright orange. Interestingly, even if a significant amount of radiation appears in the visible spectrum, the peak emission is still in the IR at a maximum temperature of about 600°C (μm). Interestingly, liquid aluminum with a melting point of 658°C is still seen as shiny under a plant-floor light (although it glows red in the dark; Figure 2.8). As a final example, liquid nitrogen at a temperature of 77 K ($-196°C$) is commonly used for IR detector cooling (Chapter 4) and has a peak emission in the far-IR band (at 38 μm).

2.7 STEFAN–BOLTZMANN LAW

This law is obtained by integration of Planck's law for all wavelengths ($0 \leq \lambda \leq \infty$). It provides a total exitance M_b for a blackbody at temperature T:

$$M_b = \sigma T^4 \tag{2.19}$$

where σ depends on c_1 and c_2 ($\sigma = \pi^5 c_1/15c_2^4$) and is known as the *Stefan–Boltzmann* constant with units of W m^2 K^{-4} (Table 2.1).

Example 2.4 Determine the power dissipated by the human body.

SOLUTION   The average skin surface of male 1.80 m (6 ft) tall is 1.86 m^2 at a normal temperature of 37°C. The total exitance M computed at this temperature is $\sigma T^4 = 5.67 \times 10^{-8} \times (37 + 273)^4 = 524$ W m^{-2}. The power lost is thus 524 W m$^{-2} \times 1.86$ m$^2 = 974$ W or about 1 kW. This power loss is compensated by radiation absorption from the surroundings, clothes in particular. If we repeat these computations for a person with a fever who has a body temperature of 42°C, the power lost becomes about 1.04 kW, an increase of 64 W, which would cause the person to feel chilly unless additional clothes or blankets were added to compensate for the increased dissipation.

Example 2.5   Compute the power lost by a 1000-W polished stainless steel kettle 30 cm in diameter filled with boiling water.

SOLUTION   The surface of a sphere equals $4\pi r^2$, and as the kettle is about a half-sphere, its surface is 0.14 m^2. The total exitance M for an hemisphere filled with boiling water (100°C = 373 K) is 1100 W m^{-2}, and thus the power lost is 155 W. However, a polished stainless steel surface cannot pretend to be a blackbody, and as eq. (2.19) is for a blackbody, a correction factor must be applied (Section 2.9 and Table 8.1).

2.8   RADIATION EMITTED IN A SPECTRAL BAND

The fraction $F(\lambda_1, \lambda_2)$ of the total emission in a spectral band (λ_1 to λ_2) is obtained by integration of Planck's law between these two limits (Figure 2.9):

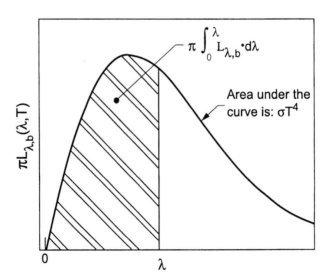

FIGURE 2.9   Fraction of the blackbody emission in the wavelength band (0 to λ). (Data from Dewitt and Nutter, 1988.)

TABLE 2.2 Blackbody Radiation Functions

λT (μm · K)	$F(0 \to \lambda)$	$L_{\lambda,b}(\lambda, T)/\sigma T^5$ [(μm · K · sr)$^{-1}$]	λT (μm · K)	$F(0 \to \lambda)$	$L_{\lambda,b}(\lambda, T)/\sigma T^5$ [(μm · K · sr)$^{-1}$]
200	0.000000	0.375195×10^{-27}	6,200	0.754187	0.249710×10^{-4}
400	0.000000	0.490424×10^{-13}	6,400	0.769282	0.230973×10^{-4}
600	0.000000	0.104056×10^{-8}	6,600	0.783248	0.213775×10^{-4}
800	0.000016	0.991183×10^{-7}	6,800	0.796180	0.197997×10^{-4}
1,000	0.000321	0.118508×10^{-5}	7,000	0.808160	0.183524×10^{-4}
1,200	0.002134	0.523935×10^{-5}	7,200	0.819720	0.170247×10^{-4}
1,400	0.007791	0.134411×10^{-4}	7,400	0.829580	0.158065×10^{-4}
1,600	0.019720	0.249128×10^{-4}	7,600	0.839157	0.146883×10^{-4}
1,800	0.039345	0.375563×10^{-4}	7,800	0.848060	0.136614×10^{-4}
2,000	0.066735	0.493422×10^{-4}	8,000	0.856344	0.127177×10^{-4}
2,200	0.100897	0.589636×10^{-4}	8,500	0.874666	0.106766×10^{-4}
2,400	0.140268	0.658848×10^{-4}	9,000	0.890090	0.901414×10^{-5}
2,600	0.183135	0.701271×10^{-4}	9,500	0.903147	0.765296×10^{-5}
2,800	0.227908	0.720216×10^{-4}	10,000	0.914263	0.653243×10^{-5}
2,898	0.250126	0.722294×10^{-4}	10,500	0.923755	0.560490×10^{-5}
3,000	0.273252	0.720229×10^{-4}	11,000	0.931956	0.483294×10^{-5}
3,200	0.318124	0.705948×10^{-4}	11,500	0.939027	0.418701×10^{-5}
3,400	0.361760	0.681517×10^{-4}	12,000	0.945167	0.364373×10^{-5}
3,600	0.403633	0.650369×10^{-4}	13,000	0.955210	0.279441×10^{-5}
3,800	0.443411	0.615199×10^{-4}	14,000	0.962970	0.217628×10^{-5}
4,000	0.480907	0.578040×10^{-4}	15,000	0.969056	0.171855×10^{-5}
4,200	0.516046	0.540370×10^{-4}	16,000	0.973890	0.137421×10^{-5}
4,400	0.548830	0.503231×10^{-4}	18,000	0.980939	0.908187×10^{-6}
4,600	0.579316	0.467321×10^{-4}	20,000	0.985683	0.623273×10^{-6}
4,800	0.607597	0.433089×10^{-4}	25,000	0.992299	0.276458×10^{-6}
5,000	0.633786	0.400794×10^{-4}	30,000	0.995427	0.140461×10^{-6}
5,200	0.658011	0.370562×10^{-4}	40,000	0.998057	0.473862×10^{-7}
5,400	0.680402	0.342428×10^{-4}	50,000	0.999045	0.201592×10^{-7}
5,600	0.701090	0.316361×10^{-4}	75,000	0.999807	0.418572×10^{-8}
5,800	0.720203	0.292287×10^{-4}	100,000	1.000000	0.135744×10^{-8}
6,000	0.737864	0.270108×10^{-4}			

Source: Adapted from DeWitt and Nutter (1988).

$$F(\lambda_1, \lambda_2) = \frac{\int_{\lambda_1}^{\lambda_2} L_{\lambda,b}(\lambda)\, d\lambda}{\int_0^{\infty} L_{\lambda,b}(\lambda)\, d\lambda} = \frac{\int_{\lambda_1}^{\lambda_2} L_{\lambda,b}(\lambda)\, d\lambda}{\sigma T^4} = F(0, \lambda_2) - F(0, \lambda_1) \qquad (2.20)$$

For convenience, instead of computing these integrals, look-up tables are available with universal functions with the product λT as entries. In Table 2.2, values of the spectral radiance are also provided.

Example 2.6 What fraction of blackbody radiation is emitted by an over-heating transformer with a casing surface temperature of 80°C in the ranges 0.4

to 0.8, 3 to 5, and 8 to 12 μm? For the sake of simplicity, assume that the transformer casing behaves as a perfect emitter (as a blackbody).

SOLUTION $80°C = 353$ K. With respect to Table 2.2 and eq. (2.20), the following values for $F(0 \rightarrow \lambda)$ are obtained through linear interpolation with adjacent values in the table:

λ_2, λ_1 (μm)	T (K)	$\lambda_2 T$ (μm · K)	$F(0 \rightarrow \lambda_2)$	$\lambda_1 T$ (μm · K)	$F(0 \rightarrow \lambda_1)$	$100 \times F(\lambda_1 \rightarrow \lambda_2)$ (%)
0.8, 0.4	353	282	0.000	141	0.000	0
5, 3	353	1765	0.0359	1059	0.000856	3.5
12, 8	353	4236	0.522	2824	0.233	29

Comment: It is not a surprise that the fraction of blackbody radiation emitted is null in the visible spectrum (0.4 to 0.8 μm), and this is precisely the reason that infrared surveys of electric utilities are routine. The bands from 3 to 5 and 8 to 5 μm are known, respectively, in the IR community as short wave (SW) and long wave (LW). In this example, although it seems to be more advantageous to operate in the LW, other factors might be relevant, such as the detector sensitivity, which is generally higher in the SW (more decision criteria are discussed in Section 4.7).

Example 2.7 Find the wavelength and the corresponding magnitude for the peak of the spectral radiance for the transformer of Example 2.6.

SOLUTION Wien's displacement law, eq. (2.18), provides the spectral location for the peak of the spectral radiance: $\lambda_{max} = 2898/T$; thus $\lambda_{max} = 2898/353 = 8.2$ μm. Table 2.2 provides $L_{\lambda,b}(\lambda, T)/\sigma T^5$ [0.722294×10^{-4} $(\mu m \cdot K \cdot sr)^{-1}$] directly, and thus the magnitude of the maximum spectral radiance is 22.5 W m^{-2} sr^{-1} μm^{-1} ($= 0.722294 \times 10^{-4} \times 353^5 \times 5.67 \times 10^{-8}$ W m^{-2} sr^{-1} μm^{-1}). The total exitance is given by the *Stefan–Boltzmann law*, eq. (2.19), for which we find that $M_b = \sigma T^4 = 880$ W m^{-2} ($= 5.67 \times 10^{-8} \times 353^4$ W m^{-2}).

2.9 REFLECTION, ABSORPTION, AND TRANSMISSION

Real surfaces do not behave as blackbodies since, for instance, they absorb only a part of the incident flux Φ_i, with another part reflected and another transmitted. Moreover, in the general case, these absorbed, reflected, and transmitted fractions of the incident flux depend on the wavelength (λ), orientation (ϕ, θ), and temperature (T) but also on the surface quality, such as smoothness and presence of contaminants (e.g., grease, oil patches). Reflected flux does not affect the object, and absorbed flux increases the internal thermal energy and

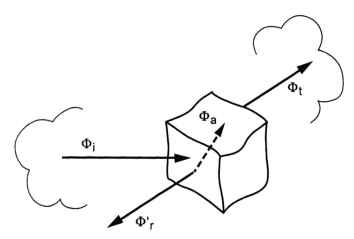

FIGURE 2.10   The flux incident Φ_i is equal to the flux reflected Φ'_r, absorbed Φ_a, and transmitted Φ_t.

thus the temperature of the medium, and at thermal equilibrium, following the energy conservation law, all energy exchange is compensated mutually: The flux incident Φ_i is equal to the flux reflected Φ'_r, absorbed Φ_a, and transmitted Φ_t (Figure 2.10):

$$\Phi_i = \Phi'_r + \Phi_a + \Phi_t \tag{2.21}$$

Some media have particular values. For instance, for an opaque medium $\Phi_t = 0$ and reflection and absorption are surface phenomena [these processes occur within a limited (less than 1 μm) depth under the surface]. For perfect mirrors, Φ_a and Φ_t equal zero. For a blackbody, all the incident flux is absorbed: $\Phi_i = \Phi_a$. In general cases, terms on the right side of eq. (2.21) are specifically weighted following particular radiative properties related to reflection, absorption, and transmission. The first factor that we study is the emissivity.

2.9.1   Emissivity

In the case of a real surface, the blackbody emission we discussed earlier must be corrected. Such a correction factor is the emissivity, ε. In fact, emissivity is a surface property that states the ability to emit energy; it is expressed as the ratio of the radiation emitted by a surface to the radiation emitted by the blackbody (in the same conditions of temperature, direction, and spectral band of interest). Emissivity is a unitless quantity and spans from 0 to 1.

It is important to point out that generally, emissivity is not a constant, as it depends on several parameters. The emission of a real surface is thus different from the emission of the blackbody (Figure 2.11). Due to these dependencies, the concept of spectral-directional emissivity $\varepsilon(\lambda, T, \theta', \phi')$ is introduced. The

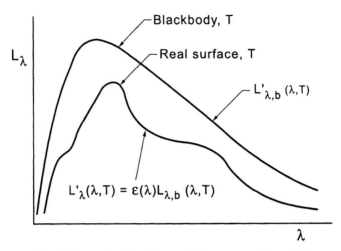

FIGURE 2.11   Blackbody and real surface emission; comparison of the spectral distribution of the radiance. (Data from Dewitt and Nutter, 1988.)

spectral-directional emissivity of a surface at temperature T emitting at a wavelength λ in the direction (θ', ϕ') is defined as

$$\varepsilon(\lambda, T, \theta', \phi') = \frac{L'_\lambda(\lambda, T, \theta', \phi')}{L_{\lambda,b}(\lambda, T)} \tag{2.22}$$

For metals, emissivity increases with temperature and is rather constant up to an important value of θ, while for nonmetallic materials, ε increases with θ. Figure 8.31 shows polar diagrams of ε for metallic and nonmetallic materials.

In practical situations, it is not always necessary to carry all dependencies of the emissivity, and some specific cases occur. For instance, objects whose emissivity is independent of wavelength are called *gray bodies*: $\varepsilon(T, \theta', \phi')$, while the expression of *colored bodies* refers to the full dependence emissivity $\varepsilon(\lambda, T, \theta', \phi')$. For optically homogeneous materials with a smooth surface, emissivity is isotropic in the azimuth $\varepsilon(\lambda, T, \theta')$. Also, if only small solid angles are involved and if the emissivity is constant in the spectral and temperature band of interest, reference to emissivity is made as being a simple calibrated constant ε. Finally, *spectral-hemispheric emissivity* $\varepsilon(\lambda, 2\pi)$ refers to the average emissivity over all directions $(\theta : [0 \to \pi/2], \phi : [0 \to 2\pi])$ within the hemispherical space above a surface, and *total hemisperic emissivity* $\varepsilon(t, 2\pi)$ is for the average emissivity over all possible wavelengths (spectral range $\lambda : [0 \to \infty]$) and directions:

$$\varepsilon(t, 2\pi) = \frac{M(T)}{M_b(T)} = \frac{\int_0^\infty \varepsilon(\lambda, 2\pi) M_b(\lambda, T)\, d\lambda}{M_b(T)} \tag{2.23}$$

Uncertainty about emissivity is a major concern in infrared radiometry, which is based on the measurement of radiated energy since emissivity and temperature are linked together $L'(\lambda, T) = \varepsilon L_b(\lambda, T)$. Measurement of T requires knowledge of ε, and vice versa. Tabulated values of emissivity are provided for normal incidence over the total spectral range and for a given temperature $\varepsilon(t, 0)$, with 0 designating normal incidence. Table 8.1 lists $\varepsilon(t, 0)$ for various surfaces and temperatures.

Example 2.8 Consider an emitter at a temperature of 1000°C. The spectral-hemisperic emissivity has the following distribution: 0.1 from 0 to 5 μm, 0.4 from 5 to 50 μm, and 0 thereafter. Draw a plot of the spectral radiance $L'_\lambda(\lambda, T = 1000°\text{C})$ for this emitter.

SOLUTION   Following the definition of the emissivity given by eq. (2.22), it follows that $L'_\lambda(\lambda, T) = \varepsilon(\lambda, 2\pi) L_{\lambda,b}(\lambda, T)$. It suffices to multiply the spectral radiance of the blackbody for that temperature by the spectral distribution of the spectral-hemisperic emissivity at that temperature (1000°C = 1273 K). This is done in Figure 2.12 with Matlab script PlanEmis.m. Interestingly here, due to

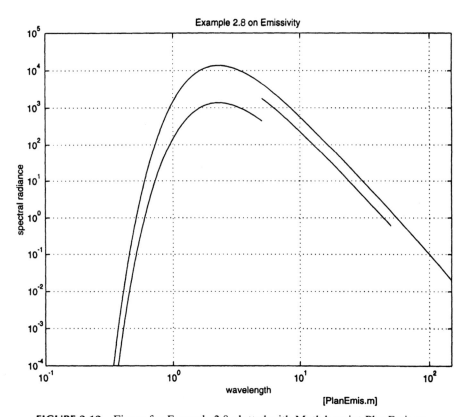

FIGURE 2.12   Figure for Example 2.8 plotted with Matlab script PlanEmis.m.

emissivity variations, the maximum spectral radiance for the emitter does not correspond to that of the corresponding blackbody curve, which occurs at $2898/1273 \sim 2.2$ μm, as stated by the Wien displacement law.

2.9.2 Absorption

The ratio of the incident flux to the absorbed flux is called *absorbance*:

$$\alpha = \frac{d\Phi_a}{d\Phi_i} \qquad (2.24)$$

As for emissivity, absorbance has dependences on wavelength, orientation, and temperature: $\alpha(\lambda, \theta, \phi, T)$, and *spectral-hemisperic* $\alpha(\lambda, 2\pi, T)$ and *total-hemisperic* $\alpha(t, 2\pi, T)$ *absorbance* are also introduced. Spectral-hemisperic absorbance $\alpha(\lambda, 2\pi, T)$ represents the average absorbance over the hemisperic space above the surface. The total hemisperic absorbance $\alpha(t, 2\pi, T)$ is for the spectral and directional average.

Example 2.9 An opaque surface has a spectral-hemispheric absorbance $\alpha(\lambda, 2\pi) = (\frac{1}{2}\lambda + 0.4)$ μm in the range $[0.4, 0.8]$ μm. An isotropic spectral irradiance $\acute{E}(\lambda) = 10^4$ W m^{-2} μm^{-1} in the range $[0.4, 0.8]$ μm is directed toward the surface. **(a)** Determine the irradiance E[W m^{-2}] impinging on the surface. **(b)** Compute the incident irradiance E_a absorbed by this surface. **(c)** Find the total hemisperic absorbance for this surface.

SOLUTION   **(a)** The irradiance is calculated from the spectral irradiance using eq. (2.11):

$$E = \int_0^\infty E(\lambda)\, d\lambda = \int_{0.4}^{0.8} 10^4 \, d\lambda = 10^4 (0.8 - 0.4) = 4000 \text{ W m}^{-2}$$

(b) Let's first obtain the total irradiance E_a absorbed by the surface from the definition of the spectral absorbance:

$$E_a = \int_0^\infty \alpha(\lambda, 2\pi) E(\lambda)\, d\lambda = \int_{0.4}^{0.8} \left(\frac{1}{2}\lambda + 0.4\right) \times 10^4 \, d\lambda$$

$$= 10^4 \left(\frac{\lambda^2}{4} + 0.4\lambda\right)\Bigg|_{0.4}^{0.8} = 2800 \text{ W m}^{-2} \text{ μm}^{-1}$$

(c) The total hemisperic absorbance for this surface is evaluated as the ratio of the absorbed irradiance to the irradiance with the values computed before:

$$\alpha(t, 2\pi) = \frac{E_a}{E} = \frac{2800}{4000} = 0.7$$

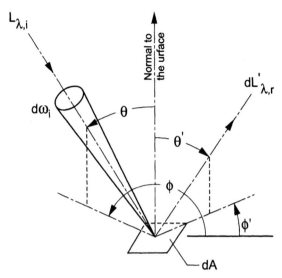

FIGURE 2.13 Spectral-bidirectional reflectance distribution function (BRDF) defined in terms of the spectral radiance of the incident and reflected radiation. (Data from Dewitt and Nutter, 1988.)

It is important to realize that the total hemisperic absorbance is not an intrinsic surface property, as it is related to the spectral and directional distribution of irradiance.

2.9.3 Reflection

This property is characterized by directional conditions of both incident (θ, ϕ) and reflected (θ', ϕ') flux. The spectral-bidirectional reflectance distribution function (or BRDF) is expressed as the ratio of differential radiance to irradiance thus has sr^{-1} units (Figure 2.13):

$$f_r(\lambda, \theta, \phi, \theta', \phi') = \frac{dL'_{\lambda,r}}{dE_\lambda} \qquad (2.25)$$

Similarly, in terms of incident and reflected flux, the spectral-bidirectional reflectance is introduced:

$$\rho(\lambda, \theta, \phi, \theta', \phi') = \frac{d\Phi'_{\lambda,r}}{d\Phi_{\lambda,i}} \qquad (2.26)$$

Obviously, use of eq. (2.26) becomes complex for the most general case, since for a given incidence, all possible reflected directions have to be taken into account. In some situations, such as for isotropically diffuse surfaces (i.e.,

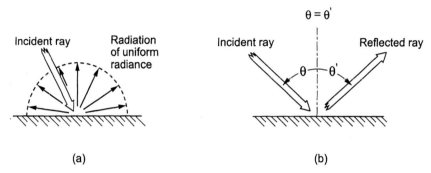

FIGURE 2.14 Reflection: (a) isotropically diffuse; (b) perfectly specular. (Data from Dewitt and Nutter, 1988.)

Lambertian reflectors), the reflected radiance is uniform within the hemisperical space above the surface, being independent of the reflected direction. For ideal (lossless) diffuse reflector, $\rho = 1$, since the incident is totally reflected.

In fact, for most practical cases, surfaces are considered either as perfectly specular (as a polished metal surface or a mirror) or isotropically diffuse (as for a rough or scattering surface). Figure 2.14 presents these two alternatives. Moreover, the isotropically diffuse assumption is reasonable in many instances, especially since it allows us to keep computations simpler.

2.9.4 Transmission

Transmission through a semitransparent body is not an easy question, due to its bidirectional nature and also to possible interreflections which imply a need to consider material thickness and properties. In fact, this aspect of energy transmission is the object of Chapter 3. Nevertheless, from a macroscopic point of view, it is possible to define spectral–directional–hemispheric transmittance as the ratio of the incident flux to the transmitted flux in the hemisperic space at the exit surface (Figure 2.15):

$$\tau(\lambda, \theta, \phi, 2\pi) = \frac{d\Phi'_{\lambda, t}}{d\Phi_{\lambda, i}} \tag{2.27}$$

As seen before for other properties, it is possible to average the spectral–directional–hemispheric transmittance for all the wavelength; this is called the total directional hemispheric transmittance $\tau(t, \theta, \phi, 2\pi)$.

2.9.5 Flux Exchanges and Kirchhoff's Law

Properties described earlier are linked together considering the flux exchanges on a semitransparent object in its environment for which (Figure 2.10)

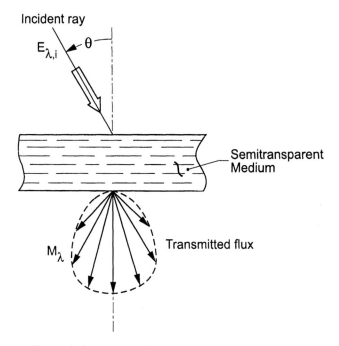

FIGURE 2.15 Transmission process illustrated as directional irradiance and exitance. (Data from Dewitt and Nutter, 1988.)

$$\rho + \alpha + \tau = 1 \qquad (2.28)$$

Of course, this relation is valid for similar spectral [or total spectral, $\lambda(0 \rightarrow \infty)$] and directional conditions [incidence conditions (θ, ϕ) and hemisperic conditions for transmission and reflection]. For an opaque object without transmission, the preceding relation simplifies to

$$\rho + \alpha = 1 \qquad (2.29)$$

The general form of Kirchhoff's law (named after Gustav Kirchhoff, German physicist, 1824–1887) provides a link between the absorption and emission processes and thus between emissivity and absorbance since

$$\varepsilon(\lambda, \theta', \phi') = \alpha(\lambda, \theta, \phi) \qquad (2.30)$$

This is demonstrated by considering a small object within a closed opaque cavity which can be considered a blackbody, as discussed previously. At thermal equilibrium, all involved fluxes are compensated since temperatures are equal. In particular, the absorbed flux is equal to the emitted flux, and thus conditions of eq. (2.30) prevail.

Although there is no particular restriction for eq. (2.30) to be valid (other than the same considered direction and wavelengths), it is important to notice that it cannot be said that emissivity is always equal to absorbance. For instance, total hemisperic emissivity $\varepsilon(t, 2\pi, T)$ is not always equal to total hemisperic absorbance $\alpha(t, 2\pi, T)$. Consider, for example, a surface that absorbs strongly in a particular spectral band and just a little in another spectral region. Since the total hemisperic absorbance depends on the spectral distribution of the incident radiation, and since $\varepsilon(t, 2\pi, T)$ is independent on the irradiation, the equality will not occur.

If for a given surface, the directional-spectral emissivity and absorbance do not depend on the direction and wavelength conditions, this surface is then called *Lambertian, gray* or *isotropically diffuse, gray*: Lambertian due to direction independence and gray due to spectral independence.

PROBLEMS

2.1 **[electromagnetic spectrum]** Give the wavenumber for a wavelength of 5 μm. (*Ans.:* 2000 cm^{-1}.)

2.2 **[radiance, exitance, irradiance]** How many degrees equal an angle of 2π radians? (*Ans.:* Exactly 360°.)

2.3 **[radiance, exitance, irradiance]** Demonstrate that $dA \cos \theta$ corresponds to the surface patch dA projected in direction θ.

2.4 **[radiance, exitance, irradiance]** What is the total self-exitance M for a Lambertian surface? (*Ans.:* $\pi L'$.)

2.5 **[radiance, exitance, irradiance]** The small surface A_8 illustrated below is a Lambertian emitter with a total radiance $L' = 8500$ W m^{-2} sr^{-1}. Compute **(a)** the radiant flux Φ in watts emitted from A_8 and falling upon A_{12}, and **(b)** the irradiance E upon A_{12} due to the emission from A_8. (*Ans.:* **(a)** $\Phi = 6.5$ μW; **(b)** $E = 0.32$ W m^{-2}.)

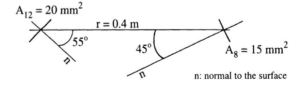

2.6 **[radiance, exitance, irradiance]** Why is the radiance spatially invariant? (*Ans.:* Because the radiance is defined in terms of the projected surface.)

2.7 **[Planck's law]** Show that an oven heating unit at about 400°C will be visible with a dark red color.

Solution 400°C corresponds to about 700 K, and from Figure 2.6 it is seen that a surface at 700 K barely emits significant radiation in the visible spectrum and in the red band. It thus appears in the dark red. It is also possible to compute the emission in the visible spectrum. Refer to Example 2.6.

2.8 **[blackbody radiation]** Compute the fraction of blackbody radiation emitted by the sun (about 5500°C) and a tungsten lamp (about 2600°C) in the visible spectrum. Also find the wavelength for the peak of the spectral radiance, the corresponding magnitude, and the total exitance.

Solution Repeat the computations of Example 2.6 to obtain the following values:

Object	Fraction (%)	Peak Wavelength (μm)	Peak Magnitude ($W\,m^{-2}\,sr^{-1}\,\mu m^{-1}$)	Total Exitance ($W\,m^{-2}$)
Sun	47	0.5	270×10^5	65×10^6
Lamp	11	1	8.4×10^5	4×10^6

It is interesting to see that lamp emission in the visible spectrum is relatively small.

2.9 **[blackbody radiation]** Consider a blackbody at a temperature of 500°C. For which wavelength is the spectral radiance maximum?

Solution The Wien displacement law gives the wavelength for peak emission. For this problem we have $\lambda_{\max} = 2898/(500 + 273)$ K = 3.7 μm. This result is confirmed by looking at the 800 K plot of Figure 2.6.

2.10 **[emissivity]** Consider an emitter at a temperature of 500°C. The spectral-hemisperic emissivity has the following distribution: 0.1 from 0 to 8 μm, 0.6 from 8 to 30 μm, and 0 thereafter. For which wavelength is the spectral radiance maximum?

Solution 1 Modify the PlanEmis.m Mathlab script provided for the corresponding emissivity and spectral values: the plot gives the answer immediately: 8 μm.

Solution 2 Use values of Table 2.2 to compute spectral radiance L_b for the blackbody at 3.7 and 8 μm (respectively, 1300 and 390 W m^2 μm^{-1} sr^{-1}) and then multiply by the corresponding emissivity values of 0.1 and 0.6 to obtain L' (respectively, 130 and 230 W m^2 μm^{-1} sr^{-1}) to realize that the peak emission is at 8 μm. [*Hint:* Use of the F.m (temperature, lambda) Matlab function provided greatly facilitates these computations.]

2.11 **[irradiance]** An opaque surface has a spectral-hemispheric absorbance $\alpha(\lambda, 2\pi) = (\frac{1}{4}\lambda + 0.15)$ μm in the range $[1, 2]$ μm. An isotropic spectral irradiance $E(\lambda) = 10^3$ W m^{-2} μm^{-1} in the range $[1, 2]$ μm. **(a)** Determine the irradiance E (W m^{-2}) impinging on the surface. **(b)** Compute the incident irradiance absorbed E_a by this surface. **(c)** What is the total hemisperic absorbance for this surface?

Solution Repeat the computations described in Example 2.9 to find the following answers: **(a)** $E = 1000$ W m^{-2}; **(b)** $E_a = 525$ W m^{-2}; **(c)** $\lambda(t, 2\pi) = 0.53$.

CHAPTER THREE

Introduction to Heat Transfer

Heat transfer allows us to predict the energy transfer taking place between two bodies due only to a temperature difference. This science is important for all energy-related applications, such as in power plants, industrial processes, refrigeration, and electronics. Heat transfer problems are thus very important in TNDT, helping to explain observed phenomena such as abnormal temperature patterns. For instance, in studying building envelopes, it helps to compute heat losses and insulation factors. In power plants, heat transfer is used to evaluate the size and position of heat sinks on power transformers, for example. Due to the many applications of this field of knowledge in TNDT, an entire chapter is necessary to bring underlying concepts. However, extensive study of heat transfer is an enormous task and we refer readers to the many textbooks that are available on this topic, especially since heat transfer is part of many curricula of various discipline [see, e.g., Holman (1981) and L. C. Thomas (1980), from which part of the material presented here is adapted with same nomenclature used to facilitate cross-references].

An often asked question is the difference between *thermodynamics* and *heat transfer*. While thermodynamics refers to the study of systems in temperature equilibrium, heat transfer adds the time variable and is thus more complete. More specifically, heat transfer is concerned with calculations of temperature distribution and heat transfer exchanges in a given system, knowing the operating conditions and also the opposite: finding the operating conditions from known temperature distribution and heat transfer exchanges.

In this chapter we restrict ourselves to what is essential for TNDT; in particular, we introduce the three exchange mechanisms: conduction (within solids), convection (between a solid surface and a moving fluid), and radiation (between two solids) with a special focus on conduction, due to its wide use in TNDT (unidimensional cases and finite-difference modeling are included). The study starts with the fundamental laws.

3.1 BASIC CONCEPTS

3.1.1 First Laws and Definitions

The *first law of thermodynamics* is the *law of energy conservation*:

$$\text{energy creation} = 0 \tag{3.1}$$

and more specifically,

$$\sum E_o - \sum E_i + \Delta E_s = 0 \tag{3.2}$$

where the term $\sum E_o$ refers to the total energy output from the system, $\sum E_i$ the total energy input by the system, and ΔE_s the energy stored in the system (including internal, kinetic, and potential energy, all storable forms of energy):

$$\Delta E_s = \Delta U + \Delta \text{KE} + \Delta \text{PE} \tag{3.3}$$

The temperature is a property corresponding to the kinetic energy of particles composing a substance (atoms, molecules, etc.). Particles with greater agitation, that is, greater kinetic energy, cause a greater temperature to occur. The conduction mechanism in heat transfer refers to energy transfer caused by physical interaction between neighboring particles in a substance having different kinetic energy. The *Fourier law of conduction* (1822), based on experimental evidence, formalizes this heat transfer mechanism. Let's starts with the unidimensional form. The rate q_x of heat transferred by conduction in the x-direction through a finite surface A_x is given by

$$q_x = -kA_x\frac{dT}{dx} \qquad \text{watts} \tag{3.4}$$

where A_x is normal to the heat transfer direction x, k the thermal conductivity (W m^{-1} °C^{-1}), and T the temperature. The negative sign in eq. (3.4) is related to the second law of thermodynamics, which says that energy is transferred in the direction of decreasing temperature. For example, in the case of a plate with surface temperatures T_1 and T_2, assuming no temperature variation in y and z ($q_y = q_z = 0$), since T_1 is greater than T_2, gradient q_x will be negative (Figure 3.1). If temperature depends upon time t, eq. (3.4) is written as a partial differential equation):

$$q_x = -kA_x\frac{\partial T}{\partial x} \tag{3.5}$$

If heat also flows in y and z directions, equations similar to eqs. (3.4) and (3.5) are written for these variables.

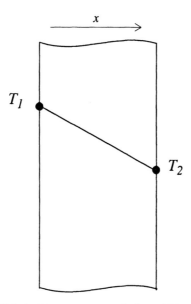

FIGURE 3.1   Temperature variation within a plate.

3.1.2 Thermal Conductivity

The thermal conductivity (symbol k, units W m $°C^{-1}$) is an important thermal property that affects heat transfer, as seen in eqs. (3.4) and (3.5). It represents the conduction rate of a conductive medium per surface unit and for a temperature gradient of 1°C. Table 3.1 lists value of k for several common substances (see also Appendix E). Exhaustive tables exist (see, e.g., Touloukian and DeWitt, 1970). As seen from this table, good thermal insulators are characterized by thermal conductivity lower than 1 W m $°C^{-1}$, while substances with k greater than 100 are considered as good thermal conductors (interestingly, these substances are metal and are also characterized by high electrical conductivity). Depending on their k value, material will have specific uses such as metal pans to cook food (high value of k) and glass wool to thermally insulate walls (low value of k).

For a given substance, the thermal conductivity depends on the temperature. Consider, for example, nitrogen (gas), which has a k value of 0.1 W m $°C^{-1}$ at 1000 K and 0.01 W m $°C^{-1}$ at 100 K. Aluminum also exhibits huge thermal conductivity variations: $k = 4000$ W m $°C^{-1}$ at 1 K, 20,000 W m $°C^{-1}$ at 10 K, flattening to 300 W m $°C^{-1}$ in the temperature range 100 to 1000 K. Material with such a huge thermal conductivity at low temperatures are referred to as *superthermal conductors*. In homogeneous material, the thermal conductivity is isotropic (i.e., identical in all directions); this is not the case of a substance such as wood or carbon fiber–reinforced plastic (CFRP), for which k is higher along the fibers with respect to the perpendicular direction.

TABLE 3.1 Thermal Conductivity Value k of Common Materials at Room Temperature

Material	k (W m^{-1} °C^{-1})	Material	k (W m^{-1} °C^{-1})
Metals		Asphalt	0.75
Silver (pure)	410	Asbestos–cement boards	0.7
Copper (pure)	385	Brick	0.3
Gold	320	Maple or oak	0.17
Aluminum (pure)	202	Asbestos, loosely packed	0.15
Silicon	150	Plaster	0.13
Nickel (pure)	93	Sawdust	0.059
Chromium	90	Cork	0.040
Iron (pure)	73	Glass wool	0.038
Germanium	60	Liquids	
Carbon steel, 1C	43	Mercury	8.7
Lead (pure)	35	Water	0.65
Chrome–nickel steel	16.3	Ammonia	0.540
(18% Cr, 8% Ni)		Lubricating oil, SAE 50	0.147
Nonmetal solids		Freon F-12, CCl$_2$F$_2$	0.073
Diamond, type 2A	2300	Gases	
Diamond, type 1	900	Hydrogen	0.18
Sapphire (Al$_2$O$_3$)	46	Helium	0.141
Quartz, parallel to axis	41.6	Air	0.026
Magnesite	4.15	Nitrogen	0.026
Marble	2.08–2.94	Water vapor	0.0206
Sandstone	1.83	(saturated)	
Limestone	1.5	Steam	0.018
Glass (Pyrex 7740)	1.0	Carbon dioxide	0.0146
Glass, window	0.78	Freon F-12	0.0097

Source: Adapted from L. C. Thomas (1980) and Holman (1981).

In computations such as conduction heat transfer (see the next section), thermal conductivity plays a paramount role, but to ease calculations when involved temperature ranges are small, the conductivity value for a given substance of interest is assumed constant. In the case of larger temperature ranges, this is not possible, and variations of k with T must be taken into account. As a final example, silicon is very important in integrated-circuit fabrication and has k ($T = -30$°C) $= 200$ W m^{-1} °C^{-1}, reducing down to $k = 100$ W m^{-1} °C^{-1} at $T = +130$°C. In such cases, k is often expressed as a linear relationship:

$$k = k_0(1 + \beta_T T) \tag{3.6}$$

with β_T being the temperature coefficient of thermal conductivity in units of °C^{-1}.

TABLE 3.2 Common Shape Factors (Unidimensional Conduction)

Particular Situation	Shape Factor (m)
One-dimensional conduction: flat plate	$S = \dfrac{A}{L}$
One-dimensional radial conduction: hollow cylinder	$S = \dfrac{2\pi r_1 r_2}{\ln(r_1/r_2)}$
One-dimensional radial conduction: hollow sphere	$S = \dfrac{4\pi r_1 r_2}{r_1 - r_2}$

3.1.3 Conduction Heat Transfer

In a permanent regime (no variation with time; see below when t is to be taken into account), conduction heat transfer is expressed with the following relationship:

$$q = kS(T_1 - T_2) \tag{3.7}$$

where q in watts is the rate of heat transferred from one surface at temperature T_1 to another surface at temperature T_2, and S is the conduction shape factor in meters, which depends on the particular geometry used. Table 3.2 lists common shape factors.

Example 3.1 Compute the rate of heat transfer of a concrete wall 20 cm thick with wall temperatures of $+20°C$ and $-20°C$. Assume that $k = 0.76$ W m^{-1} °C^{-1}.

SOLUTION   Recalling eq. (3.7) and Table 3.2, we get

$$q = \frac{kA}{L}(T_1 - T_2) = 0.76 \text{ W m}^{-1}\,°C^{-1}\,\frac{A}{0.2 \text{ m}}[20°C - (-20°C)]$$

$$\frac{q}{A} = 152 \text{ W m}^{-2}$$

As known, glass wool is a better insulating material ($k = 0.038$ W m^{-1} °C^{-1}). Repeating the calculation with such a k value gives a reduced heat loss rate of 8 W m^{-2}, a 19-fold lower value!

3.1.4 Radiation Heat Transfer

This heat transfer mechanism makes use of electromagnetic energy transfer. Energy is emitted from a material due to the rotational movements of its constituent particles (atoms, molecules, etc.). Since these particles are always moving (for a material at temperature higher than 0 K), energy (thermal radiation) is always emitted. Moreover, as particle agitation increases, temperature builds up (due to friction) and thermal emission augments (Planck's law, Section 2.5). Such electromagnetic propagates in any medium with transmissive properties (with respect to incident thermal radiation), such as void, transparent gases, liquid, and even solids. Specific factors define fractions of incident radiation absorbed, reflected, and transmitted; these are, respectively, α (absorptivity or absorbance), ρ (reflectivity or reflectance), and τ (transmitivity or transmittance), with the relationship that [from Section 2.9.5, eq. (2.28)]

$$\alpha + \rho + \tau = 1 \tag{3.8}$$

Consider three different situations:

1. A soot-covered surface has $\alpha \sim 0.97$, $\rho \sim 0.003$, and $\tau \sim 0$.
2. Polished aluminum foil has $\alpha \sim 0.1$, $\rho \sim 0.9$, and $\tau \sim 0$.
3. A nonparticipative gas has $\alpha \sim 0$, $\rho \sim 0$, and $\tau \sim 1$ (e.g., air on small distances).

A gas or liquid absorbing a nonnegligible fraction of the incident energy is called a *participating fluid*.

A body emits energy depending on both its surface temperature and conditions. As discussed in Chapter 1, a blackbody for which $\alpha = 1$ absorbs all radiative energy falling on its surface. A blackbody also emits energy (total emissive power) following the Stefan–Boltzmann law (Section 2.7):

$$E_b = \sigma T^4 \tag{3.9}$$

with $\sigma = 5.67 \times 10^{-8}$ W m^{-2} K^{-4}. Nonblackbody surfaces emit a fraction of the blackbody energy depending on their emissivity ε:

$$E = \varepsilon E_b \tag{3.10}$$

The rate of radiative heat transfer q_r between two surfaces is equal to the net thermal radiation exchange rate between these. For two parallel blackbody plates separated by a nonparticipative gas, q_r is thus defined by, for energy from surface R to surface S,

$$q_R = \sigma A_R F_{R-S}(T_R^4 - T_S^4) \tag{3.11}$$

and from surface S to surface R,

$$q_R = \sigma A_S F_{S-R}(T_S^4 - T_R^4) \tag{3.12}$$

Due to the reciprocity principle, $A_R F_{R-S} = A_S F_{S-R}$ and both eqs. (3.11) and (3.12) are identical to a sign [see Chapter 4 in Thomas (1980)]. F_{S-R} and F_{R-S} are the thermal radiation shape factors and specify the part of the energy leaving one surface and reaching the other. To have a similar relationship with respect to the conduction mechanism, q_r is sometimes written as follows:

$$q_R = \overline{h_R} A_S (T_S - T_R) \tag{3.13}$$

with

$$\overline{h_R} = \sigma F_{S-R} \frac{T_S^4 - T_R^4}{T_S - T_R} \tag{3.14}$$

This last form is often used in practical computations.

Example 3.2 Calculate the rate of radiation heat transfer of a pump cabinet located in a room with walls at 10°C. The cabinet is at temperature of 100°C and of size 50 cm × 30 cm × 50 cm.

SOLUTION   The total cabinet surface is $0.5 \times 0.3 \times 4 + 0.5 \times 0.5 \times 2$ m^2 = 1.1 m^2.

The rate of radiation heat transfer assuming a radiation shape factor of F_{S-R} of 1 is given by eq. (3.12):

$$
\begin{aligned}
q_R &= \sigma A_S F_{S-R} (T_S^4 - T_R^4) \\
&= 5.67 \times 10^{-8} \text{ W m}^{-2} \text{ K}^{-4} \, (1.1 \text{ m}^2)(1)(373^4 \text{ K}^4 - 283^4 \text{ K}^4) \\
&= 807 \text{ W}
\end{aligned}
$$

If these operating conditions are in a permanent regime, this means that the power consumption of the cabinet is 807 W. This calculation is valid only if the in-between medium is nonparticipating; otherwise, convective effects are to be considered.

3.1.5 Convection Heat Transfer

In the convective heat transfer mechanism, heat is transferred from a surface by a moving fluid. In that case, the convection is the main heat transfer mode and radiation is neglected. Exhaustive analysis of convection does require the study of mass, momentum, and energy laws in addition to laws of viscous shear in order to predict velocity and temperature distribution in the fluid. Such a study is out of our scope; we restrict ourselves to the essential practical results.

A useful relationship to compute the convection heat transfer rate q_c is (Figure 3.2)

$$q_c = \overline{h} A_S (T_S - T_F) \tag{3.15}$$

where q_c is the rate of heat transferred from a surface at temperature uniform

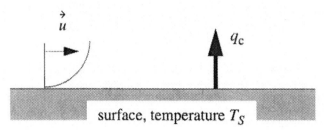

moving fluid (velocity $\vec{u}$, temperature T_F)

FIGURE 3.2   Explicative geometry for convection mechanism.

T_s to a fluid at temperature T_F, A_s is the surface, and $\bar{h}$ is the average convective heat transfer coefficient in W m^{-2} °C^{-1}. Equation (3.15) is often referred to as the *Newton law of cooling*. Table 3.3 lists common values of $\bar{h}$ for natural (due to temperature gradient in the fluid, generating density differences and thus turbulences) and forced convection (provoked by pumps, fans, etc.).

TABLE 3.3   Common Values of Convective Heat Transfer Coefficient $\bar{h}$ for Natural and Forced Convection

Mode	W m^{-2} °C^{-1} [a]
Natural convection	
Air	5–30
Vertical plate 30 cm high in air, $T = 30$°C	4.5
Horizontal cylinder, 5 cm in diameter, in air, $T = 30$°C	6.5
Water	200–600
Horizontal cylinder, 2 cm in diameter, in water, $T = 30$°C	890
Forced convection	
Air	10–500
Airflow at 2 m s^{-1} over 0.2-m^2 plate	12
Airflow at 35 m s^{-1} over 0.75-m^2 plate	75
Air at 2 atm flowing in 2.5-cm-diameter tube at 10 m s^{-1}	65
Oil	60–2,000
Water	100–20,000
Water at 0.5 kg s^{-1} flowing in 2.5-cm-diameter tube	3,500
Boiling water (at 1 atm)	2,000–50,000
In a pool or container	2,500–35,000
Flowing in a tube	5,000–100,000
Condensation of water vapor, 1 atm	
Vertical surfaces	4,000–11,300
Outside horizontal tubes	9,500–25,000

Source: Adapted from L. C. Thomas (1980) and Holman (1981).

[a] 1 W m^{-2} °C^{-1} = 0.1761 Btu/(h · ft^2 · °F), 1 Btu (British thermal unit) = 1055 J; 1 kcal = 4182 J.

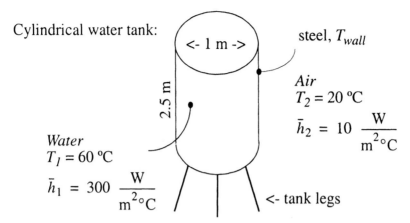

Cylindrical water tank:

<- 1 m ->

steel, T_{wall}

2.5 m

Air
$T_2 = 20\ ^\circ\text{C}$

$\bar{h}_2 = 10\ \dfrac{\text{W}}{\text{m}^2\,^\circ\text{C}}$

Water
$T_1 = 60\ ^\circ\text{C}$

$\bar{h}_1 = 300\ \dfrac{\text{W}}{\text{m}^2\,^\circ\text{C}}$

<- tank legs

FIGURE 3.3   Explicative geometry for Example 3.3.

Computation of $\bar{h}$ values involves knowledge of hydrodynamic conditions and thermodynamical and thermophysical properties which will not be covered here [see, e.g., Chapters 5 to 10 in Thomas (1980) for more details].

Example 3.3   Consider a steel cylindrically shaped noninsulated water tank of 2.5 m height and 1 m diameter (Figure 3.3). The air temperature is $T_2 = 20°\text{C}$ and the water temperature is $T_1 = 60°\text{C}$. With $\bar{h}_1 = 300$ W m^{-2} $°\text{C}^{-1}$ for the water inside the tank and $\bar{h}_2 = 10$ W m^{-2} $°\text{C}^{-1}$ for the surrounding air, compute the convection heat transfer in the permanent regime.

SOLUTION   Hypothesis here are that radiation transfers are negligible and that the temperature is uniform across the steel wall $(=T_{wall})$. In the next section we consider combined heat transfer mechanisms.
  In the permanent regime, following the first law of thermodynamic, we have that

$$q_{c1} = q_{c2} = q$$
$$q = \bar{h}_1 A_S (T_1 - T_{wall}) = \bar{h}_2 A_S (T_{wall} - T_2)$$

Solving for T_{wall} leads to

$$T_{wall} = \frac{h_1 T_1 + h_2 T_2}{h_1 + h_2} = 58.7°\text{C}$$

and then with $A_S \doteq 9.42$ m^2 (total tank surface),

$$q = \bar{h}_1 A_S (T_1 - T_{wall}) = 3676 \text{ W}$$

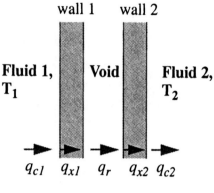

wall 1   wall 2

Fluid 1, | Void | Fluid 2,
T_1 | | T_2

q_{c1} q_{x1} q_r q_{x2} q_{c2}

Note: here $T_1 > T_2$

FIGURE 3.4 Combined heat transfer mode for a Thermos wall.

Such heat loss indicates that it is certainly worthwhile to insulate the tank for power saving.

3.1.6 Combined Heat Transfer Mechanisms

In practical situations, combinations of conduction, convection and radiation do occur. For instance, let's considering what is happening within a Thermos wall (Figure 3.4). With T_1 greater than T_2, the heat flows as indicated, for example, to keep coffee warm. As seen in the figure, all three heat transfer mechanisms here are involved with a combined apparent heat conductivity on the order of 0.0003 W m^{-1} °C^{-1}. In this section we show how computations take place in these situations.

Example 3.4 Compute the total heat transfer for the pump cabinet of Example 3.2, assuming that the pump cabinet is painted with a high-emissivity paint, $\bar{h} = 10$ W m^{-2} °C^{-1}, the ambient air temperature is 10°C, and the cabinet wall temperature is maintained at 100°C.

SOLUTION   In that case, in the permanent regime, we have $q = q_c + q_r$. Blackbody behavior is assumed, due to the high-emissivity paint. For radiation (Example 3.3),

$$q_R = \sigma A_S F_{S-R}(T_S^4 - T_R^4) = 807 \text{ W}$$

and for convection,

$$q_c = \bar{h} A_S (T_S - T_F) = (10)(1.1)(100 - 10) \text{ W} = 990 \text{ W}$$

and thus the total heat transfer for the pump cabinet is $q = q_c + q_r = 1797$ W.

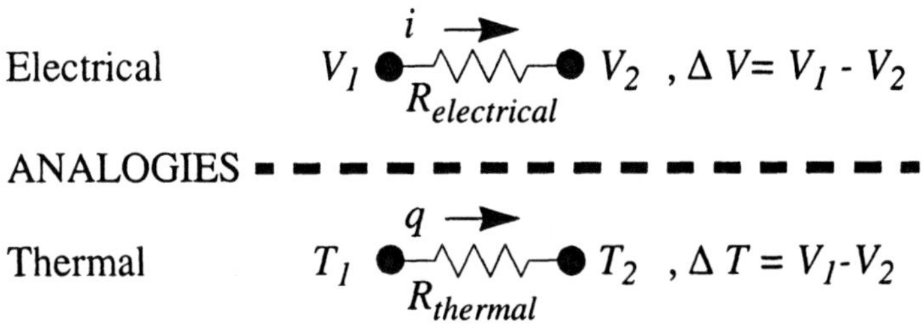

Electrical

$$V_1 \bullet \!\!\! \underset{R_{electrical}}{-\!\!\!\wedge\!\!\wedge\!\!\wedge\!\!-} \!\!\! \bullet V_2 \ , \Delta V = V_1 - V_2$$

$$\overset{i \ \longrightarrow}{}$$

ANALOGIES ▬ ▬ ▬ ▬ ▬ ▬ ▬ ▬ ▬ ▬ ▬ ▬ ▬ ▬ ▬ ▬ ▬

Thermal

$$T_1 \bullet \!\!\! \underset{R_{thermal}}{-\!\!\!\wedge\!\!\wedge\!\!\wedge\!\!-} \!\!\! \bullet T_2 \ , \Delta T = V_1 - V_2$$

$$\overset{q \ \longrightarrow}{}$$

FIGURE 3.5   Electrical analogy.

We notice in this situation, involving natural convection, that radiation represents a significant part of the total exchange. In the case of forced convection or if surface emissivity is low (case of shiny metal surfaces), radiation is small or insignificant.

3.1.7 Electrical Analogy

Conveniently, the fundamental laws of conduction, convection, and radiation can be associated with fundamental laws of electricity, so that heat transfer problems can be associated with electric circuits for simpler understanding and solving. These associations, which work in both unidimensional and multidimensional permanent and transient regimes, are based on the following similarity (Figure 3.5):

$$\text{heat transfer side} \leftrightarrow \text{electrical side}$$

$$q = \frac{\Delta T}{R_{th}} \leftrightarrow i = \frac{\Delta V}{R} \tag{3.16}$$

where i, ΔV, and R stand, respectively, for the current in amperes (A), the voltage difference in volts (V), and the electrical resistance in ohms (Ω). The electrical side of eq. (3.16) is known as *Ohm's law* (1827; after Georg Ohm, German physicist, 1789–1854). It follows from eq. (3.16) that units of R_{th} are °C W^{-1}. Thermal problems can be converted to their electrical analog, for which practical methods are deployed for solving. We present some examples that follow such an approach.

Example 3.5   Imagine the thermal system of Figure 3.6, with $T_1 = 100°C$, $T_2 = 80°C$, $R_1 = 100°C$ W^{-1}, and $R_2 = 60°C$ W^{-1}. Find the heat transfer.

SOLUTION   The total thermal resistance is $R = R_1 + R_2 = 100 + 60 = 160°C$ W^{-1} (resistors in series configuration are summed together). From eq. (3.16), we have

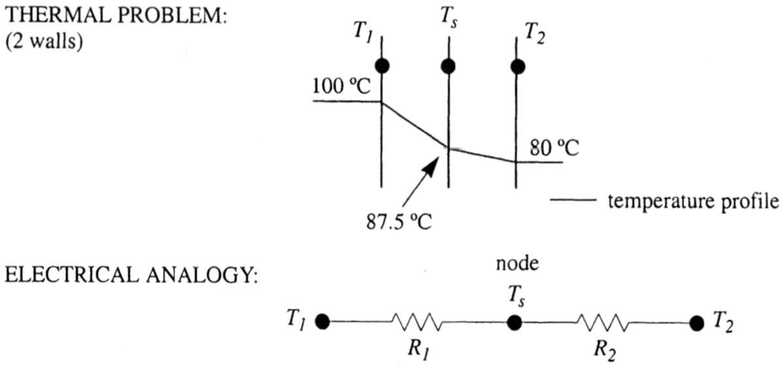

THERMAL PROBLEM:
(2 walls)

ELECTRICAL ANALOGY:

FIGURE 3.6   Figure for Example 3.6.

$$q = \frac{\Delta T}{R_{th}} = \frac{T_1 - T_2}{R_{th}} = \frac{(100 - 80)^\circ C}{160^\circ C/W} = 0.125 \text{ W}$$

Following the first law of thermodynamic, we also have that energy creation in node T_s is null, so that

$$q = \frac{T_1 - T_s}{R_1} \rightarrow T_s = -qR_1 + T_1 = (100 - 0.125 \times 100)^\circ C = 87.5^\circ C$$

The electric analogy here is Kirchhoff's current law, which says that the algebraic summation of all currents entering and exiting a node is null. If only two currents are involved, as here, this means that the current entering the node is equal to the current exiting, or in terms of heat transfer, $q_1 = q_2$.

Instead of a serial combinations of resistors, as in Example 3.5, resistors can also be configured in parallel or serial/parallel (in that case, serial branches are considered first). Table 3.4 lists these configurations with corresponding association relationships.

Following our analysis of heat transfer mechanisms, we can derive expressions for R_{th}, Table 3.5 lists them.

3.2   ONE-DIMENSIONAL HEAT TRANSFER

Since basic concepts were introduced in Section 3.1, we can now turn to more formal one-dimensional heat transfer, which is applicable to one-dimensional problems, including composite and multimechanism problems. Such one-dimensional analysis is also helpful in understanding multidimensional problems that we introduce in the next section, that is, for problems with more than one variable changing, including time t.

TABLE 3.4   Resistor Configurations

Configuration	Schematic	Association Relationships for Equivalent Total Resistor R
Serial	R_1 R_2	$R = R_1 + R_2$ With n resistors in series: $R = R_1 + R_2 + \cdots + R_{n-1} + R_n$
Parallel	R_1 R_2	$\dfrac{1}{R} = \dfrac{1}{R_1} + \dfrac{1}{R_2} \rightarrow R = \dfrac{R_1 R_2}{R_1 + R_2}$ With n resistors in parallel: $\dfrac{1}{R} = \dfrac{1}{R_1} + \dfrac{1}{R_2} + \cdots + \dfrac{1}{R_{n-1}} + \dfrac{1}{R_n}$
Example of a serial + parallel configuration	R_a R_b R_c	$R = \dfrac{(R_a + R_b) R_c}{(R_a + R_b) + R_c}$

TABLE 3.5   Equivalent Thermal Resistance $(R = \Delta T / q)$ for the Three Heat Transfer Modes

Heat Transfer Mode	Thermal Resistance Expression
Conduction	$R_{th} = \dfrac{L}{kS}$ $T_1 \multimap\!\!\bigvee\!\!\!\multimap T_2$
Radiation, with $\overline{h_R}$:	$R_{th} = \dfrac{1}{\overline{h_r} A_s}$
$\overline{h_R} = \sigma F_{S-R} \dfrac{T_S^4 - T_R^4}{T_S - T_R}$	$T_1 \multimap\!\!\bigvee\!\!\!\multimap T_2$
Convection	$R_{th} = \dfrac{1}{\overline{h} S}$ $T_1 \multimap\!\!\bigvee\!\!\!\multimap T_2$

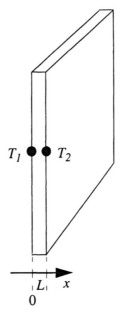

FIGURE 3.7   Conduction heat transfer in a slab with fixed conditions at borders.

3.2.1   Conduction Heat Transfer in a Semi-infinite Slab with Fixed Conditions at Borders

Lets first consider the classical example of a semi-infinite slab of thickness L with fixed conditions at border, $T(x = 0) = T_1$ and $T(x = L) = T_2$ (Figure 3.7). In this case, a semi-infinite slab is a slab that is very large with respect to its thickness, so that edges effects that would add another dimension in this heat transfer problem can be neglected. As evidence suggests, due to uniform slab property (mainly, thermal conductivity), temperature distribution is linear. It is obtained by differentiation of the first law of thermodynamics, eq. (3.4):

$$\frac{d}{dx}\left(-kA_x\frac{dT}{dx}\right) = 0 \qquad (3.17)$$

$$\frac{d^2T}{dx} = 0 \qquad \text{with initial conditions } T(0) = T_1 \text{ and } T(L) = T_2 \qquad (3.18)$$

This differential equation can be solved using simple calculus (see, e.g., the appendix at the end of this chapter): $T = c_1x + c_2$ [differentiating this last expression twice yields zero, as stipulated by eq. (3.18)]. The constants c_1 and c_2 are evaluated with initial conditions at the border to obtain (check the T value for $x = 0$ and L)

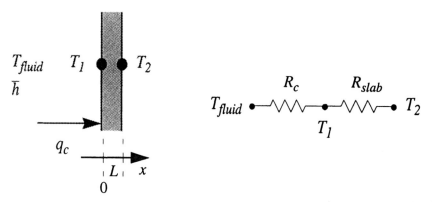

FIGURE 3.8 Figure for Example 3.6 (multi-heat-transfer mechanisms): conduction/convection; solution with electrical analogy.

$$T = (T_2 - T_1)\frac{x}{L} + T_1 \tag{3.19}$$

Equation (3.19) provides a complete temperature expression for all points within the slab provided that both T_1 and T_2 are known. If this is not the case, T_1 and T_2 must first be computed. This could be the case if other heat transfer mechanisms (convection, radiation) are involved. In such a case, either a differential analysis such as the one presented here briefly is to be used, or the method of the electrical analogy of Section 3.1.7 could be used to ease computations. Finally, it is interesting to replace the expression found for T in eq. (3.19) in eq. (3.17), which yields the conductivity expression given earlier in eq. (3.7):

$$q_x = -kA_x\frac{dT}{dx} = \frac{kA_x}{L}(T_1 - T_2) = \frac{kA_x}{L}\Delta T \quad \text{and thus} \quad R = \frac{\Delta T}{q_x} = \frac{L}{kA}$$

Example 3.6 Here we are dealing with convection at the border and solution by electrical analogy (Figure 3.8). A semi-infinite slab (or a wall) is submitted to convection by a fluid at known temperature T_{fluid} and convection heat transfer $\bar{h}$. The other side of the slab, at $x = L$, is at known temperature T_{slab}. Find the expression for the temperature T_1 of the external slab surface at $x = 0$.

SOLUTION   From Tables 3.2 and 3.5 we have that

$$R_{slab} = \frac{L}{kA} \quad \text{and} \quad R_c = \frac{1}{\bar{h}A}$$

and also, $q_c = q_x$, due to the first law of thermodynamics. Using the resistor

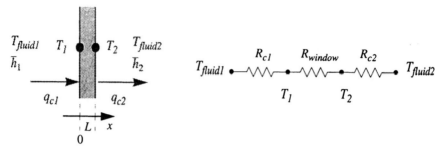

FIGURE 3.9   Figure for Example 3.7 with electrical analogy.

serial configuration of Table 3.4 and with $R = \Delta T/q$, we thus have

$$\frac{T_{\text{fluid}} - T_1}{R_c} = \frac{T_1 - T_2}{R_{\text{slab}}}$$

which allows us to compute T_1.

Example 3.7   Consider a house with a single-pane window of 2 mm thickness and an air temperature on both sides of $+25°C$ (with $\bar{h} = 15$ W m^{-2} °C^{-1}) and $-25°C$ (with $\bar{h} = 30$ W m^{-2} °C^{-1}), an extreme northern winter weather situation. Compute the temperature of both faces of the glass pane of thermal conductivity $k = 0.78$ W m^{-1} °C^{-1} and the heat transfer rate (neglecting radiation).

SOLUTION   An electrical analogy can be used to ease the problem (Figure 3.9). Assuming a surface of 1 m^2, we can compute the thermal resistances (Tables 3.2 and 3.5):

$$R_{c1} = \frac{1}{\bar{h}A} = \frac{1}{15} = 0.0667°C \text{ W}^{-1}$$

$$R_{c2} = \frac{1}{\bar{h}A} = \frac{1}{30} = 0.0333°C \text{ W}^{-1}$$

$$R_{\text{window}} = \frac{L}{kA} = \frac{0.002}{0.78 \times 1} = 0.0256°C \text{ W}^{-1}$$

Using the resistor serial configuration of Table 3.4, we have

$$R_{\text{total}} = R_{c1} + R_{c2} + R_{\text{window}} = 0.00667 + 0.0333 + 0.0256 = 0.102256°C \text{ W}^{-1}$$

and then

$$q = \frac{T_{\text{fluid1}} - T_{\text{fluid2}}}{R_{\text{total}}} = \frac{25 + 25}{0.10256} = 488 \text{ W m}^{-2}$$

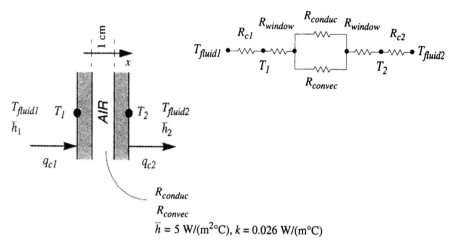

$\bar{h} = 5$ W/(m²°C), $k = 0.026$ W/(m°C)

FIGURE 3.10   Figure for Example 3.9 with electrical analogy.

which indicates a large energy loss; in fact, calculating temperature on both faces of the glass pane, we get

$$q = \frac{T_{\text{fluid1}} - T_1}{R_{c1}} \rightarrow T_1 = T_{\text{fluid1}} - qR_{c1} = 25 - 488 \times 0.0667 = -7.5°C$$

$$q = \frac{T_2 - T_{\text{fluid2}}}{R_{c2}} \rightarrow T_2 = qR_{c2} + T_{\text{fluid2}} = 488 \times 0.0333 + (-25) = -8.7°C$$

indicating that it is freezing inside the house, with ice accumulation due to heavy condensation on the window. Although houses used to have such windows in the past, it is now more common to build windows having two- or even three-pane windows, for which the heat losses are much less (see Example 3.8).

Example 3.8   We repeat Example 3.7, but this time the window has a double-pane window (panes 1 cm apart; see Figure 3.10). With such a window, both convection and conduction have to be considered between the window panes ($k_{\text{air}} = 0.26$ W m⁻¹ °C⁻¹, $\bar{h} = 5$ W m⁻² °C⁻¹).

SOLUTION   Here again an electrical analogy eases the computations. Let's first calculate the added thermal resistors R_{conduc} and R_{convec} in between the glass panes:

$$R_{\text{convec}} = \frac{1}{\bar{h}A} = \frac{1}{5 \times 1} = 0.2°C \text{ W}^{-1}$$

$$R_{\text{conduc}} = \frac{L}{kA} = \frac{0.01}{0.026 \times 1} = 0.385°C \text{ W}^{-1}$$

Using the resistor configuration formulas of Table 3.4, we compute R_{total} as follows:

$$R_{total} = 0.067 + 0.0256 + \frac{0.2 \times 0.385}{0.2 + 0.385} + 0.0256 + 0.0333 = 0.283°C \ W^{-1}$$

and

$$q = \frac{T_{fluid1} - T_{fluid2}}{R_{total}} = \frac{25 + 25}{0.283} = 177 \ W$$

Finally,

$$T_1 = T_{fluid1} - qR_{c1} = 25 - 177 \times 0.0667 = +13°C$$

$$T_2 = qR_{c2} + T_{fluid2} = 177 \times 0.0333 + (-25) = -19°C$$

This time no freezing occurs inside the house and heating losses are reduced almost threefold.

3.2.2 Composite Systems

In many situations, various materials are bonded together in various ways and it is of interest to compute the overall heat transfer across all these bonded materials. Figure 3.11 shows two typical configurations: with perfect thermal contact in part (a) and imperfect contact in part (b), due, for example, to surface rugosity, contact pressure, or temperature at the bonding interface. Imperfect thermal contacts may occur, for example, in bonded laminates, used widely in

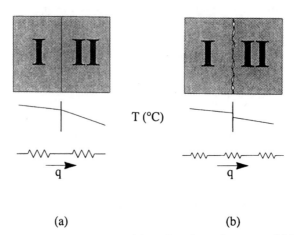

(a)   (b)

FIGURE 3.11 Composite wall system: (a) perfect thermal contact; (b) imperfect thermal contact. Also shown are the electrical analogy and temperature profiles.

the aircraft industry. If the interface thermal resistance is significant, TNDT can be aimed at the detection of such a problem by detecting differential surface temperature related to the heat transfer rate between sound and defective areas. However, in many instances, contact is present between materials without an actual bonding force and with a marginal interface thermal resistance. These situations are very difficult to detect by NDT, although some successful attempts used a laser beam to lift one side through local heating. Measurement of this lifting using a second laser mounted as an Michelson interferometer allows detection of defective bondings [see Cielo et al. (1985a) for more details]. If needed, interface thermal resistance can be reduced by inserting special substances between materials, such as silicon grease (with a typical R value of $0.5°C$ W^{-1}) or a thin foil. Silicon grease is often used in mounting heat sink in electronic circuits to maximize heat transfer (see Example 3.10).

Serial Configuration. Figure 3.11 is an example of serial configuration with the heat flow passing through all materials one after the other. Examples below illustrate such common configuration.

Example 3.9 Hot water (90°C) circulates in one side of a 2-cm marble wall of $k = 2$ W m^{-1} °C^{-1} in an old Roman spa. To minimize heat losses, it is decided to cover this wall with glass wool of $k = 0.038$ W m^{-1} °C^{-1}. Temperature on the dry side of the marble wall is maintained at 10°C. Calculate the thickness of the glass wool layer in order for the heat losses not to exceed 50 W m^{-2}. Figure 3.12 shows the configuration. Let's assume a negligible interface thermal resistance thanks to the intended proper bonding of the glass wool.

SOLUTION   Let's first calculate R_{mar}:

$$R_{mar} = \frac{L}{kA} = \frac{0.02}{2 \times 1} = 0.01°C\ W^{-1}$$

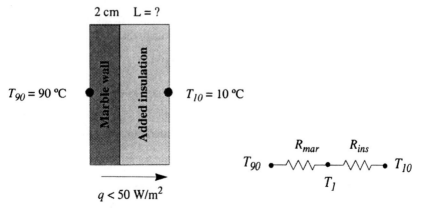

FIGURE 3.12 Figure for Example 3.9 with electrical analogy.

The total heat transfer is expressed as follows:

$$q = \frac{T_{90} - T_{10}}{R_{\text{total}}} = \frac{T_{90} - T_{10}}{R_{\text{mar}} + R_{\text{ins}}}$$

and thus

$$R_{\text{ins}} = \frac{T_{90} - T_{10}}{q} - R_{\text{mar}} = \frac{90 - 10}{50} - 0.01 = 1.59°C \ W^{-1}$$

and finally, the insulation thickness is computed:

$$L = R_{\text{ins}}kA = 1.59 \times 0.038 \times 1 = 6 \ cm$$

Example 3.10   A power rectifier is mounted on a black anodized aluminum plate of surface 1000 mm^2 acting as a heat sink. Power rectifier specifications are $T_{\text{max}} = 70°C$ (343 K) and $R_{\text{casing}} = 50°C \ W^{-1}$ (this is the thermal resistance of the power rectifier to ambient). The interface thermal resistance between the power rectifier and the aluminum plate is $R_{\text{interface}} = 0.5°C \ W^{-1}$. [It is achieved by covering the base of the electronic device with thermal compound, such as Gardtec Inc. (available, e.g., from Future Electronics); a thin mica gasket is thus added for electrical insulation with the same grease on the heat-sink side]. The maximum ambient air temperature is $T_{\text{air}} = 20°C$ with a convection heat transfer $\bar{h} = 15 \ W \ m^{-2} \ °C^{-1}$. Compute the maximum power that can be dissipated in these conditions by the power rectifier.

SOLUTION   Heat is dissipated by both casing and heat sink so that heat directly transferred by the power rectifier itself is

$$q_{\text{casing}} = \frac{T_{\text{max}} - T_{\text{air}}}{R_{\text{casing}}} = \frac{70 - 20}{50} = 1 \ W$$

and q_{heatsink} is computed using Figure 3.13, so that we first compute individual thermal resistors, $R_{\text{interface}}$, R_{radia} (related to radiated energy by the heat sink), and R_{convec} (related to convected energy by the heat sink):

$$R_{\text{interface}} = 0.5°C \ W^{-1}$$

$$R_{\text{convec}} = \frac{1}{\bar{h}A} = \frac{1}{15 \times 0.001} = 67°C \ W^{-1}$$

Computation of R_{radia} is a little more complicated since it requires knowledge of the unknown temperature T_1:

$$R_{\text{radia}} = \frac{1}{h_r A_s} = \frac{1}{\left(\sigma F_{S-R} \dfrac{T_1^4 - T_{\text{air}}^4}{T_1 - T_{\text{air}}}\right) A_s}$$

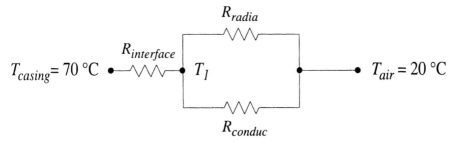

FIGURE 3.13 Electrical analogy for Example 3.10.

This can be solved by iteration, starting with, say, $T_1 = T_{\text{casing}} = 70°C = 343$ K:

$$R_{\text{radia}} = \frac{1}{\left(\sigma F_{S-R} \dfrac{T_1^4 - T_{\text{air}}^4}{T_1 - T_{\text{air}}}\right) A_s} = \frac{1}{\left(5.67 \times 10^{-8} \times 1 \times \dfrac{343^4 - 293^4}{343 - 293}\right)}$$

$$R_{\text{radia}} = 136°C \ W^{-1}$$

and R_{total} is computed next (Table 3.4):

$$R_{\text{total}} = R_{\text{interface}} + R_{\text{radia}} \| R_{\text{convec}}$$

$$R_{\text{total}} = R_{\text{interface}} + \frac{R_{\text{radia}} R_{\text{convec}}}{R_{\text{radia}} + R_{\text{convec}}} = 0.5 + \frac{136 \times 67}{136 + 67} = 45.3°C \ W^{-1}$$

giving q_{heatsink}:

$$q_{\text{heatsink}} = \frac{T_{\text{casing}} - T_{\text{air}}}{R_{\text{total}}} = \frac{70 - 20}{45.3} = 1.10 \ W$$

Interface temperature is then computed as

$$q_{\text{heatsink}} = \frac{T_{\text{casing}} - T_1}{R_{\text{interface}}} \rightarrow T_1 = T_{\text{casing}} - q_{\text{heatsink}} R_{\text{interface}}$$

$$T_1 = 70 - 1.10 \times 0.5 = 69.4°C$$

Our assumption of $T_1 = T_{\text{casing}}$ was close so that we do not need to repeat the previous calculations. The total dissipated power is finally obtained as follows:

$$q = q_{\text{casing}} + q_{\text{heatsink}} = 1.10 + 1.10 \ W$$

Interestingly here, both contributions are equal; changing of ambient temperature changes this, as shown on Table 3.6. One should thus be aware of

TABLE 3.6 Effect of Ambient Temperature on Heat Sink Dissipation

$T_{ambient}$ (°C)	$q_{heatsink}$ (W)	q_{total} (W)
−10	1.7	2.8
0	1.5	2.6
10	1.3	2.4
20	**1.1**	**2.2**
30	0.9	2.0
40	0.7	1.8
50	0.5	1.6
60	0.2	1.3
70	0	1.1

these dramatic variation efficiencies. As ambient temperature increases, over-engineering is a good practice in these situations. A common way to increase a heat sink surface is to form it with fins (see below).

Parallel Configuration. In some cases, two (thermally) different materials are bonded together parallel to the heat propagation. This is, for example, the case of a roof supporting a beam made of a steel base topped with laminated wood. Such a situation is depicted on Figure 3.14 for materials A and B. Following our previous discussions, we can write for this configuration:

$$q_x = \frac{T_1 - T_2}{R_{parallel}} \quad \text{with} \quad R_{parallel} = \frac{R_1 R_2}{R_1 + R_2} \quad \text{(from Table 3.4)} \qquad (3.20)$$

Moreover, with q_A and q_B being the heat transfer rate through materials A and B, respectively, it is seen that

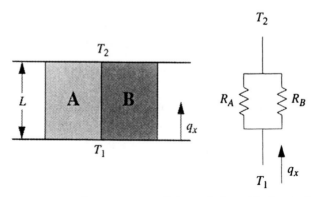

FIGURE 3.14 Composite systems: parallel association with electrical analogy.

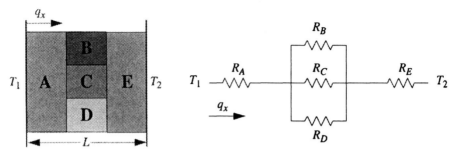

FIGURE 3.15 Composite systems: parallel–serial association with electrical analogy.

$$q_x = q_A + q_B \tag{3.21}$$

which is another way to look at such a configuration.

Parallel–Serial Configuration. Parallel–serial configurations as shown on Figure 3.15 are more complex to analyze since they are no longer one-dimensional heat transfer problem due to lateral heating through the material assembly. A complete analysis thus requires more complex two- or even three-dimensional analysis. However, under some hypothesis and in particular circumstances, it is still possible to reduce the problem to a one-dimensional one. For instance, in the case of Figure 3.15, if thermal conductivities of materials A, B, and C are very different, two-dimensional effects must be taken into account since a more complex heat flow path is taking place within the assembly.

Example 3.11 Consider a composite slab made of a steel supporting base topped with concrete. This composite slab has one side in a cold room maintained at $0°C$ while the other side ends up in a $25°C$ solar-heated environment. Figure 3.16 shows the configuration. Compute the heat transfer rate through this slab.

SOLUTION   Let's first compute individual resistances:

$$R_{x,\,\text{steel}} = \frac{L}{k_{\text{steel}} A_{\text{steel}}} = \frac{0.3}{70 \times 0.1} = 0.04°C\ W^{-1}$$

$$R_{x,\,\text{concrete}} = \frac{L}{k_{\text{concrete}} A_{\text{concrete}}} = \frac{0.3}{0.76 \times 0.5} = 0.79°C\ W^{-1}$$

$$R_{c,\,\text{steel}} = \frac{1}{\bar{h} A_{\text{steel}}} = \frac{1}{10 \times 0.1} = 1°C\ W^{-1}$$

$$R_{c,\,\text{concrete}} = \frac{1}{\bar{h} A_{\text{concrete}}} = \frac{1}{10 \times 0.5} = 0.2°C\ W^{-1}$$

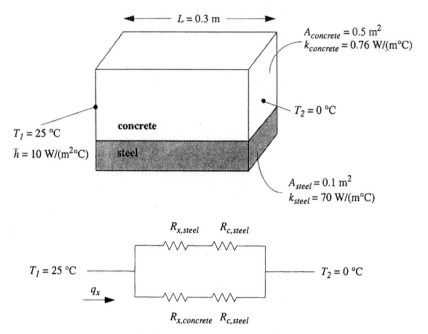

FIGURE 3.16   Figure for Example 3.11 with electrical analogy.

Then the equivalent total resistance R_{total} is (Table 3.4)

$$R_{\text{total}} = (R_{x,\text{steel}} + R_{c,\text{steel}}) \| (R_{x,\text{concrete}} + R_{c,\text{concrete}})$$

$$= (0.04 + 1) \| (0.79 + 0.2) = 1.04 \| 0.99 = \frac{1.04 \times 0.99}{1.04 + 0.99} = 0.51°C \ W^{-1}$$

and finally, the total heat transfer rate is obtained:

$$q = \frac{T_1 - T_2}{R_{\text{total}}} = \frac{25 - 0}{0.51} = 49 \ W$$

However, it should be noted such computation is done using a one-dimensional approximation. We will see later some criteria that can be used to judge whether or not this can be justified.

3.2.3   Radial Systems

Other common systems for TNDT are pipes and cylinder shapes. These are studied in this section on radial systems using the natural symmetry of these objects along their main rotation axis. The Fourier law of conduction writes itself in cylinder coordinates (Figure 3.17):

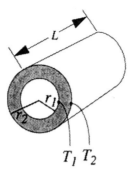

FIGURE 3.17   Radial system: hollow cylinder.

$$q_r = -kA_r \frac{\partial T}{\partial r} \tag{3.22}$$

which in the case of a hollow cylinder, can be solved as (L is the cylinder length)

$$-k \int_{T_1}^{T_2} dT = q_r \int_{r_1}^{r_2} \frac{1}{A_r} dr \rightarrow q_r = \frac{2\pi L k}{\ln(r_1/r_2)}(T_1 - T_2) \tag{3.23}$$

Recalling the conduction heat transfer equation for a flat plate equation (3.7) and the thermal resistance definition (Table 3.4), we have immediately that the thermal conduction resistance associated to an hollow cylinder is given by

$$R_{x_{\text{rad}},\text{cyl}} = \frac{\Delta T}{q_r} = \frac{\ln(r_1/r_2)}{2\pi L k} \tag{3.24}$$

Finally, the temperature T within the hollow cylinder for any radius r ($r_1 < r < r_2$) is obtained from eq. (3.23) by forming the ratio with ($T - T_1$) since

$$\frac{T - T_1}{T_2 - T_1} = \frac{\ln(r_1/r)}{\ln(r_1/r_2)} = \frac{\ln(r/r_1)}{\ln(r_2/r_1)} \tag{3.25}$$

A similar analysis performed in the case of an hollow sphere gives the following expressions:

$$q_r = \frac{4\pi r_1 r_2 k}{r_2 - r_1}(T_1 - T_2) \tag{3.26}$$

$$R_{x_{\text{rad}},\text{sph}} = \frac{\Delta T}{q_r} = \frac{r_2 - r_1}{4\pi r_1 r_2 k} \tag{3.27}$$

$$\frac{T - T_1}{T_2 - T_1} = \frac{r - r_1}{r_2 - r_1} \frac{r_2}{r} \tag{3.28}$$

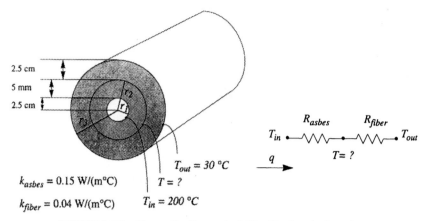

2.5 cm

5 mm

2.5 cm

R_{asbes} R_{fiber}

T_{in} ——/\/\/——•——/\/\/——• T_{out}

$T = ?$

q ⟶

$T_{out} = 30\ °C$

$k_{asbes} = 0.15\ W/(m°C)$ $T = ?$

$k_{fiber} = 0.04\ W/(m°C)$ $T_{in} = 200\ °C$

FIGURE 3.18   Figure for Example 3.12 with electrical analogy.

Example 3.12   Consider a copper pipe covered with an asbestos layer for insulation purposes and finally covered with fiberglass. Figure 3.18 shows the configuration. Calculate the temperature at the asbestos–fiberglass interface.

SOLUTION   Let's first calculate radial conduction resistances using eq. (3.24) for $L = 1$ m and with (from Figure 3.18) $r_1 = 0.025$ m, $r_2 = 0.03$ m, and $r_3 = 0.055$ m:

$$R_{x_{\text{rad}}, \text{asbes}} = \frac{\ln(r_2/r_1)}{2\pi L k_{\text{asbes}}} = \frac{\ln(0.03/0.025)}{2\pi(0.15 \times 1)} = 0.193°C\ W^{-1}$$

$$R_{x_{\text{rad}}, \text{fiber}} = \frac{\ln(r_3/r_2)}{2\pi L k_{\text{fiber}}} = \frac{\ln(0.055/0.03)}{2\pi(0.04 \times 1)} = 2.412°C\ W^{-1}$$

The total resistance (Figure 3.18 and Table 3.4)

$$R_{\text{total}} = R_{x_{\text{rad}}, \text{asbes}} + R_{x_{\text{rad}}, \text{fiber}} = 0.193 + 2.412 = 2.6°C\ W^{-1}$$

Now we can compute the heat transfer rate,

$$q_r = \frac{\Delta T}{R_{\text{total}}} = \frac{200 - 30}{2.6} = 65.4\ W$$

and the interface temperature T:

$$q_r = \frac{T_{in} - T}{R_{x_{\text{rad}}, \text{asbes}}} \rightarrow T = -q_r R_{x_{\text{rad}}, \text{asbes}} + T_{in} = 200 - (65.4 \times 0.193) = 187°C$$

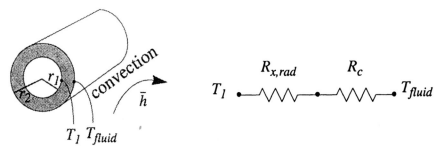

FIGURE 3.19 Critical radius with electrical analogy.

3.2.4 Critical Radius

It is interesting to study what happens when the outside radius r_2 of a hollow cylinder of internal radius r_1 increases (Figure 3.19). In fact, in this case, the conduction resistance $R_{x,rad}$ increases, but on the other hand, the convection resistance R_c decreases, especially in the presence of forced convection. With known convective heat transfer coefficient $\bar{h}$, we have

$$q = \frac{T_{fluid} - T_1}{R_{x,rad} + R_c} = \frac{T_{fluid} - T_1}{\ln(r_2/r_1)/2\pi Lk + 1/\bar{h}2\pi Lr_2} \tag{3.29}$$

Depending on the particular values in eq. (3.29), heat transfer q will increase or decrease. For the limit case, when q is maximum, playing with eq. (3.29) gives the critical radius:

$$r_c = \frac{k}{\bar{h}} \tag{3.30}$$

This results is found by finding the value of r_2 for which $dq/dr = 0$, the classical method to determine the maximum or minimum of a function. In the case of an insulation layer, the *critical radius* is also called a *critical thickness of insulation*. To summarize this concept, if r_2 (outer radius) is greater than r_c, heat transfer q will decrease with increased r_2 (the case of predominant conduction). However, if r_2 is smaller than r_c, heat transfer q will increased with increased r_2; this is often the case with natural convection with a small $\bar{h}$ value (the case of predominant convection). In the case of a hollow sphere, the critical radius is found as

$$r_c = \frac{2k}{\bar{h}} \tag{3.31}$$

3.2.5   Internal Energy Sources

Many heat transfer situations occur that involve an internal energy source. This is, for example, the case for chemical or nuclear reactions and for electric and electronic circuits. In the case of TNDT, an important application is in the inspection of electrical motors and transformers (Chapter 11). In such situations the strength of an internal source is expressed as the rate of energy per unit volume; the symbol is $\dot{q}$. In the case of electric or electronic systems we have

$$\dot{q} = \frac{\text{power}}{\text{volume}} = \frac{I_e^2 R_e}{V} \rightarrow q = \frac{I_e^2 \rho_e}{A^2} \qquad \text{W m}^{-3} \tag{3.32}$$

where I_e is a uniform current and the electrical resistivity is given as

$$\rho_e = \frac{R_e A}{L} \qquad \Omega \cdot \text{m} \tag{3.33}$$

Often, these situations can be solved using superposition. For instance, in the case of a flat plate, the total heat transfer is obtained as the summation both for no internal energy source but known face temperatures T_1 and T_2 and energy generation with known face temperatures T_1 and T_2.

Example 3.14   A 4 A current is circulating in a copper wire with $k = 386$ W m^{-1} °C^{-1}, electrical resistivity $\rho = 30$ µ$\Omega \cdot$ cm, and diameter 2 mm. Blackbody radiation cooling does occur at $T = 20$°C. Compute the surface temperature of the wire. Figure 3.20 shows the configuration.

SOLUTION   Let's first calculate the power generated within the wire:

$$\dot{W} = I_e^2 R_e = \frac{I_e^2 \rho_e L}{A^2} = \frac{4^2 \times 30 \times 10^{-6} \times 0.01 \times L}{\pi \times (1 \times 10^{-3})^2} \rightarrow \frac{\dot{W}}{L} = 1.53 \text{ W m}^{-1}$$

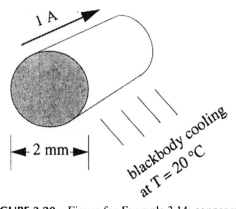

FIGURE 3.20   Figure for Example 3.14: copper wire.

Since all energy is transferred by radiation, we can write, with eq. (3.11),

$$\dot{W} = q_R = \sigma A_R F_{S-R}(T_S^4 - T_R^4) = 1.53L \text{ W m}^{-1}$$

in which we can solve for T_s:

$$T_S^4 = \frac{1}{\sigma A_R F_{S-R}} q_R + T_R^4 = \frac{1.53L}{5.67 \times 10^{-8} \times 1 \times 10^{-3}L \times 2\pi \times 1} + (20 + 273)^4$$

$$T_S = 55.6°C$$

3.2.6 Fins and Biot Number

Fins are used in various electronic and electrical applications in cooling applications, such as on the large power transformers present in all electric plants and various industrial plants. It is thus important to understand their behavior for proper TNDT deployment in these contexts. In Example 3.10 we studied heat exchanges for a power electronic device and concluded that the efficiency of the used heat sink was dramatically depended on ambient temperature. One way of contouring this problem is to maximize convected heat transfer by increasing useful exchange surfaces. Fins are deployed for such purpose. This is easily seen recalling the Newton law of cooling:

$$q_c = \bar{h}A_s(T_S - T_F) \tag{3.34}$$

An increase in the surface A_s augments q_c. In the case of fins, the initial surface is extended (Figure 3.21) so that energy is transferred by conduction to the fin and then by convection to the environment. Of course, to be efficient, the convection resistance R_c must be smaller than the conduction resistance R_x (practically, $R_c \ll R_x$). A important parameter in case of fins is the characteristic length l, which is defined as the ratio of its volume on its surface:

$$l = \frac{V_{\text{fin}}}{A_{\text{fin}}} \tag{3.35}$$

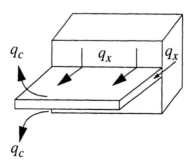

FIGURE 3.21 Surface extension: fin principle.

This short analysis of fins brings us to the Biot number Bi (named after Jean-Baptiste Biot, French physicist, 1774–1862). This is a dimensionless number that has many uses in TNDT, as we will see later. For instance, one use establishes the fin efficiency. As a rule of thumb, it is defined as

$$\text{Bi} \sim \frac{R_x}{R_c} \tag{3.36}$$

A more formal definition is

$$\text{Bi} = \frac{\bar{h}}{k} l \tag{3.37}$$

Efficient fins have a small Bi value, typically smaller than 0.1. Moreover although the heat transfer analysis of a fin is truly a two-dimensional problem (Figure 3.21), it can be demonstrated that if Bi is smaller than 0.1, the error is performing a one-dimensional analysis for fin heat transfer computation is typically less than 5%. In fact, fin efficiency η_{fin} is defined as

$$\eta_{\text{fin}} = \frac{q_{\text{fin}}}{q_{\text{max}}} \tag{3.38}$$

where q_{fin} is the actual heat transfer from the fin and q_{max} would be the heat transfer in the best conditions [i.e., when $\text{Bi} = 0$, when $R_x = 0$ in eq. (3.36)]. For thin fins,

$$\eta_{\text{fin}} = \frac{A}{\rho L \sqrt{\text{Bi}}} \tag{3.39}$$

It is noted that efficiency decreases as the length of the fin L increases; this is explained physically by the fact that fin temperature tends to reach T_{fluid} at its extremity.

From our previous analysis of thermal resistances, the thermal resistance of a fin can be expressed as (Figure 3.22)

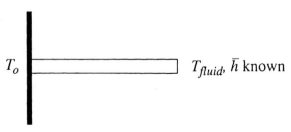

FIGURE 3.22   Thermal resistance of a narrow fin.

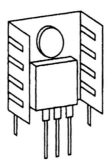

FIGURE 3.23 Heat sink model 6021B de AAVID, Thermalloy, Inc., used to mount power transistor type TO-220.

$$q_{\text{fin}} = \frac{T_o - T_{\text{fluid}}}{R_{\text{fin}}} \qquad (3.40)$$

and for a narrow circular fin the following expression has been found for R_{fin}:

$$R_{\text{fin}} = \frac{1}{\sqrt{\bar{h}\rho k A}} \qquad (3.41)$$

Heat sinks have a complex shape with the thermal resistance actually measured by manufacturers using eq. (3.40), which takes into account all radiative, capacitive, and conductive effects. A complete catalog does exist, with heatsinks available for all sort of electronic device packaging; see, for example, the one shown on Figure 3.23, which has thermal resistance $R = 12.5°\text{C W}^{-1}$.

Example 3.14 Consider a circular nickel fin of diameter $D = 1$ cm, 10 cm in length, with $k = 91$ W m^{-2} °C^{-1}, transferring the heat from a 100°C surface of a power transformer to a 10°C cooled oil pan with $\bar{h} = 100$ W m^{-2} °C^{-1}. Compute the heat transfer assuming negligible radiative effects (low emissivity of bare nickel).

SOLUTION We first compute the Biot number to validate our one-dimensional analysis. Since Bi < 0, we can go ahead:

$$\text{Bi} = \frac{\bar{h}}{k}l = \frac{\bar{h}V}{kA} = \frac{\bar{h}}{k}\frac{L\pi(D^2/4)}{kL\pi D} = \frac{\bar{h}}{k}\frac{D}{4} < 0$$

From the previous analysis, we have

$$R_{\text{fin}} = \frac{1}{\sqrt{\bar{h}\rho k A}} = \frac{1}{\sqrt{100 \times 91 \times \pi \times (0.01/2)^2}} = 1.18°\text{C W}^{-1}$$

and

$$q_{\text{fin}} = \frac{T_o - T_{\text{fluid}}}{R_{\text{fin}}} = \frac{100 - 10}{1.18} = 76 \text{ W}$$

3.3 UNSTEADY HEAT TRANSFER

Up to now, we have neglected the time variation of the heat transfer. Now we concentrate our analysis in a particular class of such situations. Obviously, these situations are truly multidimensional since we are considering time t and at least one dimension; however, in the case of situations with small Biot numbers (Bi < 1), it is possible to study only the time variable t and the problem becomes one-dimensional, as in our previous discussions. An example of this situation is the case of a hot part (say, a screwdriver) quenched by dropping it in a large bath of liquid such as water. These situations have known values of T_{fluid}, $\bar{h}$, and T_i, the initial part temperature, which is uniform, and thus the relationship

$$\frac{T - T_{\text{fluid}}}{T_i - T_{\text{fluid}}} = \exp\left(\frac{-\bar{h}A_s t}{\rho V c}\right) \tag{3.42}$$

where c is the specific heat of the part and the other symbols are as defined previously. Demonstration of eq. (3.42) is demonstrated in Example 3.16. In eq. (3.42) it is assumed that T_{fluid} does not change in time, which is, in Example 3.15, more or less the case if the part is quenched in a large bath of liquid. The total heat convected from the surface at any time t is thus

$$q_c = \bar{h}A_s(T - T_{\text{fluid}}) = \bar{h}A_s(T_i - T_{\text{fluid}}) \exp\left(\frac{-\bar{h}A_s t}{\rho V c}\right) \tag{3.43}$$

Performing time integration over a length of time τ on equation (3.43) provides the energy Q transferred during that period [L. C. Thomas, 1980, p. 99, eq. (2-137)]:

$$Q = \rho V C_v (T_i - T_f)\left[1 - \exp\left(\frac{-hA_s \tau}{\rho V c}\right)\right] \tag{3.44}$$

and finally, the total energy convected to the fluid is obtained by letting $\tau \to \infty$, which gives

$$q_{\max} = \rho V c (T_i - T_{\text{fluid}}) \tag{3.45}$$

which, in fact, corresponds to the total energy owned by the part at time $t = 0$.

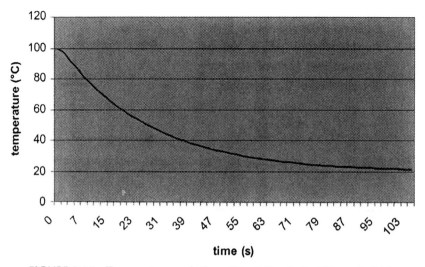

FIGURE 3.24 Temperature evolution of the falling ball of Example 3.16.

Example 3.15 A brass ball of diameter 1 cm, thermal conductivity $k = 130$ W m^{-1} °C^{-1}, mass density $\rho = 8400$ kg m^{-3}, and specific heat $c = 380$ J kg^{-1} °C^{-1} is falling in a oil bath of fixed temperature $T_f = 20$°C. The initial ball temperature T_i is 100°C. It is asked to compute the temperature history of the ball starting at time $t = 0$ s, when the ball makes contact with the oil. The convective heat transfer coefficient of the oil is $\bar{h} = 200$ W m^{-2} °C^{-1}. Compute the temperature history of the ball.

SOLUTION   The Biot number is first computed using eq. (3.37), and it is found as Bi = 0.00256. Since Bi $\ll$ 0.1 and assuming negligible radiation effect, eq. (3.42) can be used to compute the temperature evolution of the ball:

$$T(t) = (100 - 20)\exp(-0.3759t) + 20 \qquad °C$$

$T(t)$ is plotted in Figure 3.24.

3.3.1 Electrical Analogy

The one-dimensional unsteady situations presented in this section offer an analogy with some simple electric circuit of the first order, that is, electrical circuits formed of resistors and only one capacitor (or inductor coil). Considering a capacitor, it can be said that there is an analogy between the electrical capacity of the capacitor within the electrical circuit and the thermal capacity of a mass in the heat transfer problem. We recall that for a capacitor, the following relationships describe the E_e (voltage) $- I_e$ (current) behavior:

$$I_e = +C_e\frac{d}{dt}E_e \qquad\qquad (3.46)$$

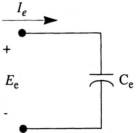

Important: the convention is that the relationship I_e-E_e is positive if the current I_e enters the positive terminal of the capacitor:

$$I_e = +C_e \frac{d}{dt} E_e$$

FIGURE 3.25 I_e–E_e relationship and sign convention used in capacitor circuits.

This is, in fact, the equivalent, well-known Ohm's law ($E_e = RI_e$) of resistors applied to capacitors. Notice the + sign associated to a current entering from the + terminal of the capacitor C_e (Figure 3.25).

Figure 3.26 shows a first-order electrical circuit with capacitor initially charged at potential $V_{initial}$ corresponding to $T_{initial}$. At time $t = 0$, the switch is flipped from position 1 to position 2 so that the capacitor sees all its energy dissipating through resistor R_e to reach ground potential $V_{final} = 0$, corresponding to T_{final}. The analogies are:

$$I_e \leftrightarrow q \tag{3.47}$$

$$E_e \leftrightarrow T \tag{3.48}$$

$$C_e = \rho V c \tag{3.49}$$

Such analogies are useful to heat transfer problems, as seen in the next example.

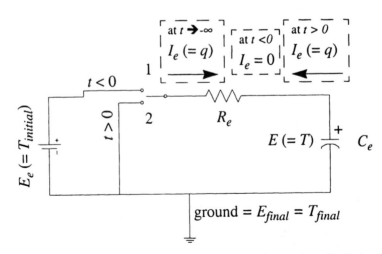

FIGURE 3.26 First-order electrical circuit with charged capacitor C_e discharging at time $t = 0$.

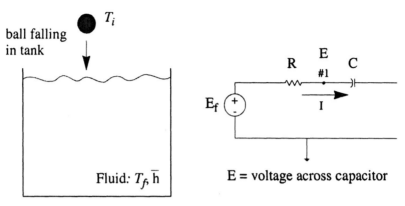

FIGURE 3.27   Figure for Example 3.16 with electrical analogy.

Example 3.16   A ball at initial temperature T_i falls in a large tank containing a fluid at temperature T_f. The convection heat transfer coefficient in the fluid is known at $\bar{h}$. Compute the temperature evolution T of the ball once within the tank. Since the tank is large and the ball is small, T_f is considered constant for the duration of the experiment.

SOLUTION   Figure 3.27 depicts the situation together with the electrical analogy. The symbol analogy is $E \leftrightarrow T$, $E_i \leftrightarrow T_i$, and $E_f \leftrightarrow T_f$. Using Kirchhoff's law of current applied to node 1 (we recall that this law stipulates that the summation of all current entering/leaving a given point of junction or node is always zero), we can write with eq. (3.46) and Ohm's law applied to the resistor R that (by convention a current is considered to be positive when entering a node)

$$\frac{E_f - E}{R} - \left(C\frac{dV}{dt} \right) = 0$$

leading to the following differential equation to solve:

$$C\frac{dE}{dt} + \frac{1}{R}E = \frac{E_f}{R} \tag{3.50}$$

We have a wide choice of methods to solve this equation (see, e.g., Robertson, 1996; Williamson, 1997). A popular technique consists of dividing the task into two simpler problems, solving the homogeneous equation (i.e., the equation with the second term on the right set to zero) first and then solve the particular response with the second term on the right as is. The complete solution is the summation of both responses. Let's first solve the homogeneous equation:

$$C\frac{dE}{dt} + \frac{1}{R}E = 0 \tag{3.51}$$

To solve eq. (3.51) a solution is supposed. Here we need to find an expression that does not change after differentiation. The exponential function is such a function; we thus set $E_h = A \exp(st)$, which is introduced in eq. (3.51) with $dE_h/dt = sA \exp(st)$. Solving for s yields $s = -1/RC$, and we thus have $E_h = A \exp(-t/RC)$.

The particular solution is also found by supposing that the final response of the circuit will be of the same nature as the forced response which appears in the second term of eq. (3.50) and which is a constant. We thus set $E_p = K$ (a constant), which is introduced in eq. (3.50) with $dE_p/dt = 0$. Solving for K yields $K = V_f$.

The complete solution is given by

$$E(t) = E_h + E_p = A \exp\left(-\frac{t}{RC}\right) + E_f \tag{3.52}$$

At this point the initial condition $V(0) = V_i$ is used to find the value of A in eq. (3.52):

$$E(0) = E_i = A \exp\left(-\frac{t}{RC}\right) + E_f$$

Solving for A in this equation gives $A = V_i - V_f$. The final expression for $V(t)$ is then obtained:

$$E(t) = (E_i - E_f) \exp\left(-\frac{t}{RC}\right) + E_f \tag{3.53}$$

which can also be written as

$$\frac{E - E_f}{E_i - E_f} = \exp\left(-\frac{t}{RC}\right) \tag{3.54}$$

and going back to the original variables, the final solution is found with $RC = (1/\bar{h}A_s)\rho Vc$ [eq. (3.49) for C and Table 3.5 for R]:

$$\frac{T - T_f}{T_i - T_f} = \exp\left(\frac{-\bar{h}A_s t}{\rho Vc}\right) \tag{3.55}$$

Equation (3.55) is the same as eq. (3.42). As seen before, opportunity to make use of an electrical analogy provides an efficient approach in heat transfer problem solving.

3.4 MULTIDIMENSIONAL CONDUCTION HEAT TRANSFER

In active thermography (Section 1.2), in order to establish a link between abnormal isotherms recorded on the inspected workpiece surface following the application of an external thermal stimulation and the presence of a subsurface defect, it is necessary to study thermal front propagation inside the material. This is a very complex study if all the variables have to be taken into consideration. The complexity of the analytical approach takes tremendous proportions as soon as we do not consider the ideal case of two flat surfaces of a uniform and isotropic material (i.e., a material having the same physical properties in all directions) and neglecting the heat transfers inside both the defect itself and outer sample surfaces (Siegel and Howell, 1972).

In many practical systems, one-dimensional approximations cannot be assumed valid, and multidimensional methods are thus needed. For instance, in the case of a specimen containing subsurface defects and having an awkward shape, computations of the time–temperature evolution following sudden heating of the specimen surface are truly multidimensional. Finite-element or finite-difference approaches are the two most common methods used to solve these problems since analytical methods become rapidly too complex. Finite elements are very popular, but since it is based on more advanced variation computations, it will not be addressed here. Instead, a finite difference method based on the generalized Fourier equation will be presented after a brief review of the analytical approach.

3.4.1 Analytical Approach

In a solid body, these transfers are governed by the Fourier diffusion equation (3.4), here generalized to x, y, and z dimensions (Arpaci, 1966):

$$\frac{\partial T}{\partial t} = \frac{k}{\rho c}\left(\frac{\partial^2 T}{\partial x^2} + \frac{\partial^2 T}{\partial y^2} + \frac{\partial^2 T}{\partial z^2}\right) \tag{3.56}$$

where T is the temperature, t the time, ρ the density, k the thermal conductivity, and c the specific heat. Table 3.7 (and Appendix E) lists these thermophysical properties for some commonly used materials. To illustrate the analytical approach, we briefly study a simple example [adapted from the work of Williams et al. (1980) and Sayers (1984)]. Equation (3.56) can be simplified and solved analytically if we consider the case of a semi-infinite isotropic plate. This plate spans infinitely in the space along x and y axes, but has finite thickness along the z axis ($0 \leq z \leq l$). This corresponds to a one-dimensional case (S. K. Lau et al., 1990). Both plate surfaces are considered perfectly thermally insulated (no thermal loss), but the side $z = 0$ is heated with an instantaneous and uniform thermal pulse. For the sake of simplicity, we consider that at time $t = 0$, the plate temperature is zero (no temperature differential with respect to

TABLE 3.7 Thermal Properties of Some Materials

Material	Specific Heat $(\mathrm{J\,kg^{-1}\,^{\circ}C^{-1}})$	Density $(\mathrm{kg\,m^{-3}})$	Heat Capacity[d] $(\mathrm{J\,cm^{-3}\,^{\circ}C^{-1}})$	Thermal Conductivity $(\mathrm{W\,m^{-1}\,^{\circ}C^{-1}})$	Thermal Diffusivity[a] $\alpha \times 10^{-6}$ $(\mathrm{m^2\,s^{-1}})$
Air (as defect)	700	1.2	0.8×10^{-3}	0.02	33
Aluminum	880	2,700	2.4	230	95
Brass	380	8,400	3.2	130	32
(65% Cu, 35% Zn)					
CFRP[b]	1,200	1,600	1.9	0.8	0.42
$\perp$ fibers					
$\parallel$ fibers	1,200	1,600	1.9	7	3.7
Concrete	800	2,400	1.9	1	0.53
Copper	380	8,900	3.4	380	110
Epoxy resin	1,700	1,300	2.2	0.2	0.09
Glass	670	2,600	1.7	0.7	0.41
GRP[c]	1,200	1,900	2.3	0.3	0.13
$\perp$ fibers					
$\parallel$ fibers	1,200	1,900	2.3	0.38	0.17
Lead	130	11,300	1.5	35	23
Nickel	440	8,900	3.9	91	23
Plexiglas	667	1,200	0.8	0.2	0.25
Porcelain	1,100	2,300	2.5	1.1	0.43
Steel	440	7,900	3.5	46	13
Mild					
Stainless	440	7,900	3.5	25	7.1
Teflon	—	—	—	0.42	1.59
Titanium	470	4,500	2.1	16	7.6
Uranium	120	18,700	2.2	27	12
Water	4,180	1,000	4.2	0.6	0.14
Zircaloy 2	280	6,600	1.8	13	11

Source: Adapted from Touloukian and DeWitt (1970), Vavilov (1980, p. 182), W. N. Reynolds and Wells (1984, p. 143), and Tretout (1987, p. 49).

[a] Defined as $\alpha = k/\rho c$, where k is thermal conductivity, ρ is mass density, and c is specific heat.
[b] Carbon fiber–reinforced plastic.
[c] Glass-reinforced plastic.
[d] Defined as $C_p = \rho c$, where ρ is mass density, and c is specific heat.

the environment), but in a narrow region ($0 \leq z \leq z_e$, $z_e \ll l$), the temperature is T_0. For $t > 0$, the heat from this narrow region diffuses in the bulk of the plate. Of course, for very long time ($t \to \infty$), the temperature distribution becomes uniform with a final value of $T_f = T_0 z_e / l$.

Taking this geometry into consideration, eq. (3.56) can be reduced to its unidimensional equivalent, with $T(z, t)$ corresponding to temperature distribution at time t and depth z:

$$\frac{\partial^2}{\partial z^2} T(z,t) = \frac{\rho c}{k} \frac{\partial T(z,t)}{\partial t} \tag{3.57}$$

The border conditions are

$$T(z,0) = \begin{cases} 0 & z_e \leqslant z \leqslant l \\ T_0 & 0 \leqslant z \leqslant z_e \end{cases}$$
$$\frac{\partial T(0,t)}{\partial z} = \frac{\partial T(l,t)}{\partial z} = 0 \tag{3.58}$$

This problem has an analytical solution (Carslaw and Jaeger, 1959):

$$T(z,t) = T_f \left[1 + \frac{21}{\pi z_e} \sum_{n=1}^{\infty} \frac{1}{n} \sin \frac{n\pi z_e}{l} \cos \frac{n\pi z}{l} \exp\left(\frac{-t\alpha^2 \pi^2 n^2}{l^2} \right) \right] \tag{3.59}$$

Analytical solutions [such as that of eq. (3.59)] are very useful to check the results obtained by the numerical computations, as we will see in Section 3.4.2.

This example showed that even in the case of simple geometries, the analytical solution is relatively complex. With the introduction of both anisotropic properties and subsurface defects, the analytical approach degenerates rapidly into unworkable situations, especially if unidimensional approximation cannot be retained.

3.4.2 Finite-Difference Modeling

For more complex shape studies, finite differences modeling is a precious tool, especially since it can provide limits to the effectiveness of the TNDT technique and also the possibility of considering different defect geometries and determining their detectability without the expense of making and testing the corresponding specimens (James et al., 1989). In this section we present the basis of a simple "workable" model whose programming translation is given in C language at the Web site as Appendix A. Although less powerful than some commercially available heat transfer packages, the one we introduce in this section has the advantage of being easily understood and adaptable.

In a first step the workpiece to study is represented by a mesh of elementary volumes of material. A node is associated with each volume and is representative of the corresponding thermal behavior of the volume with which it is associated. More specifically, the temperature of a node is considered as being equal to the average temperature of the surrounding material within the corresponding elementary volume.

A practical approach to laying out the mesh is to adopt a cylindrical system of coordinates (Figure 3.28). Using this scheme, it is only necessary to compute results on one pie slice (Figure 3.29), and because of the symmetry of this configuration, the same values can be applied all around ($0 < \theta < 2\pi$). Computa-

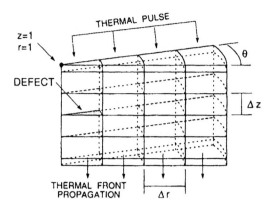

FIGURE 3.28 Cylindrical reference system used in heat transfer modeling. (From Maldague, 1993.)

tions will proceed much more rapidly than for a *true* Cartesian geometry (with coordinates x, y, and z).

Each elementary volume (around a node) can be thought as a box having radial widths Δr and axial depth Δz. There are nz nodes in depth and nr nodes along the radial length. The basic modeling idea is to compute the temperature for all the nodes, taking into account the thermal properties of the material and the thermal exchanges between the nodes themselves and the external world. As noted with eq. (3.56), time is an important variable since we have a dynamic system. In active thermography, transient study after application of the thermal

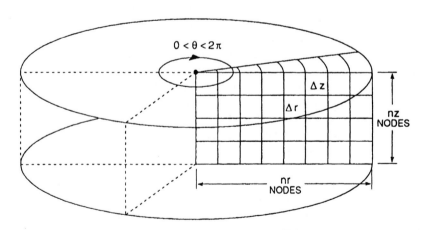

NOTE : The grid represents the computed
area for this cylindrical geometry

FIGURE 3.29 Cylindrical reference system used in heat transfer modeling. Thanks to the symmetry effect, it is only necessary to compute temperature distribution for the nodes shown on the right (matrix nr × nz). (From Maldague, 1993.)

perturbation (at $t = 0$) is fundamental in order to provoke a response from the part inspected. In fact, for active thermography (Section 1.2), the permanent regime does not bring useful information about subsurface defects. The time scale is also divided into small increments Δt, and for each time increment (i.e., for every iteration of the program), the temperature of all the nodes is computed again. This way to proceed allows sufficiently accurate modeling to make possible good insight into the thermal phenomena, although it does not consider all aspects. Notice, for example, that the thermal capacitance between nodes is ignored in this study, as the resistive behavior is prominent (Krapez et al., 1991).

Temperature computations (see, e.g., Thomas, 1980) at all the nodes are done following the *first law of thermodynamic* (energy conservation), the *Fourier law of conduction*, and the external thermal exchanges (losses and energy deposit). For a given instant t,

$$\text{rate of energy creation} = 0$$

$$-\left(\sum \frac{\Delta E_{\text{out}}}{\Delta t} - \sum \frac{\Delta E_{\text{in}}}{\Delta t} \right) = \frac{\Delta E_{\text{s}}}{\Delta t} \tag{3.60}$$

where E_{out} and E_{in} represent energy transferred in and out a particular node, and ΔE_s corresponds to the energy stored in the elementary volume surrounding the node. The minus sign means that the flux exchanged is positive for a negative temperature gradient. This is consistent with the *second law of thermodynamics* (energy transferred in the direction of decreasing temperature), as discussed in Section 3.1.

For our geometry (Figures 3.28 and 3.29) we have both radial (Δq_r flux along radial direction r) and axial (Δq_z flux along the depth) transfers. Equation (3.60) can be rewritten, for the node located at (r, z), as

$$-(\Delta q_r + \Delta q_z) = \frac{\Delta E_{\text{s}}}{\Delta t} = \frac{\rho \Delta V c \Delta T}{\Delta t} \tag{3.61}$$

and thus

$$T_{\text{future}} = T_{\text{present}} - \frac{(\Delta q_r + \Delta q_z) \Delta t}{\rho \Delta V c} \tag{3.62}$$

where ρ is the density, c the specific heat at node (r, z), and ΔV corresponds to the elementary volume through which thermal exchanges occur, for the node in (r, z):

$$\Delta V = \Delta z (S_{\text{crossed}, z}) \tag{3.63}$$

where $S_{\text{crossed}, z}$ is the surface crossed by the thermal flux from z to $(z + \Delta z)$

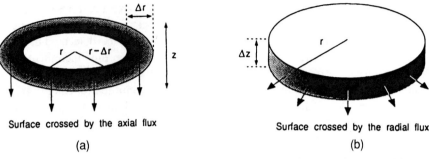

Surface crossed by the axial flux   Surface crossed by the radial flux

(a)   (b)

FIGURE 3.30   Surfaces crossed by thermal flux: (a) axial; (b) radial. (From Maldague, 1993.)

(Figure 3.30a):

$$S_{\text{crossed},z} = \pi[r^2 - (r - \Delta r)^2] \qquad (3.64)$$

Flux values Δq_r and Δq_z are obtained from the *Fourier law of conduction* for the node in (r, z), $\Delta q_r = q_{r,\text{out}} - q_{r,\text{in}}$:

$$\Delta q_r = \frac{k_r S_{\text{crossed},r}}{\Delta r}(T_{\text{present in } r} - T_{\text{present in } r+\Delta r}) \qquad (3.65)$$

where $T_{\text{present in } r}$ and $T_{\text{present in } r+\Delta r}$ are the temperatures at the present time for the adjacent nodes r and $r + \Delta r$ (radial direction), k_r the radial thermal conductivity, and $S_{\text{crossed},r}$ the surface crossed by the thermal front from r to $r + \Delta r$, from Figure 3.30b:

$$S_{\text{crossed},r} = 2\pi r\,\Delta z \qquad (3.66)$$

Similarly, for Δq_r, for the node located in (r, z),

$$\Delta q_z = \frac{k_z S_{\text{crossed},z}}{\Delta z}(T_{\text{present in } z} - T_{\text{present in } z+\Delta z}) \qquad (3.67)$$

where $T_{\text{present in } z}$ and $T_{\text{present in } z+\Delta z}$ are the temperature at the present time for the two adjacent nodes z and $z + \Delta z$ (axial direction), k_z the axial thermal conductivity, and $S_{\text{crossed},z}$ the surface crossed by the thermal front from z to $z + \Delta z$ [eq. (3.64)].

All these computations, eqs. (3.62), (3.65), and (3.67) must be performed along all thermal exchange directions for the nodes. For one *slice* of the defined cylindrical system of coordinates, there are four directions for a given node located in (r, z) : $(r - \Delta r, z)$, $(r, z - \Delta z)$, $(r + \Delta r, z)$, and $(r, z + \Delta z)$. The other two directions of the space are taken into account by the cylindrical geometry:

the temperature distribution is the same for all the slices. Theses effects are considered in computation of the surfaces crossed by the thermal fluxes [eqs. (3.64) and (3.66)].

Let's now consider the initial thermal stimulation. A quantity of energy P (in watts/cm^2) is deposited during a certain number of time steps $(0 < t <$ duration of the initial thermal stimulation) on the front surface (nodes located in $z = 0$). From the surface, the thermal front propagates, by diffusion, inside the whole mesh. The increase of temperature ΔT_p on the front surface $(z = 0)$ corresponding to this stimulation is given by

$$\Delta T_p = \frac{P \Delta t}{\rho c S_{\text{crossed}, z} \Delta z} \tag{3.68}$$

with the parameters defined as previously for eq. (3.62).

Loss effects are very important to consider (Tossel, 1989). In this model, losses (by radiation and convection) are considered only on the external face of the sample (nodes with $z = 0$). The opposite surface $(z = nz)$ is supposed to be well insulated thermally. This hypothesis is close the case of an *infinite* thickness sample. This is also the case of a sample of finite thickness but when the time of observation (following the initial thermal perturbation) is small with respect to the heat front propagation.

Radiative losses are computed after the *Stefan–Boltzmann law*, for the node located in $(x, y, z = 0)$:

$$\Delta q_{\text{rad}} = F_{\text{rad}}(T_s^4 - T_a^4) \tag{3.69}$$

where T_s is the temperature expressed in kelvin for the node on the front surface $(z = 0)$, T_a the ambient temperature, and F_{rad} the empirical factor, which depends on the geometry (see Table 3.8). Convective losses are computed after *Newton's law of cooling*:

TABLE 3.8   Modeling Parameters

Δt	Variable logarithmic progression, 1290 iterations (30 s)
F_{rad}	5.67×10^{-12} W cm^{-2} (Cielo, 1983)
F_{con}	1×10^{-3} W cm^{-2} (Cielo, 1983)
T_a	$0°$C[a]
Δr	2 mm
Δz	200 µm
nr	20
nz	30
Heating duration	2 s
P	1 W cm^{-2}

[a]Equivalent to computing an increase of temperature.

$$\Delta q_{con} = F_{con}(T_s - T_a) \tag{3.70}$$

with T_s and T_a defined as before and where F_{con} is an empirical factor that depends on the geometry (Table 3.8). These losses contribute to reducing the front temperature for the node located in $(r, z = 0)$:

$$T_{\text{present with losses}} = T_{\text{present no loss}} - \frac{(\Delta q_{rad} + \Delta q_{con})\,\Delta t}{\rho c\,\Delta V} \tag{3.71}$$

The parameters are defined as in eq. (3.62). This set of equations is used to simulate various kinds of experimental configurations, as we will see later.

In Figure 3.31 we see the temperature distribution computed using this model for different times (geometries of Figures 3.28 and 3.29). The temperature is given for the first 20 rows and 20 columns of the mesh in the case of a graphite epoxy sample, with the following model parameters: radial conductivity $k_r = 0.05$ W cm^{-1} °C^{-1}, axial conductivity $k_z = 0.01$ W cm^{-1} °C^{-1}, specific heat $c = 1000$ J kg^{-1} °C^{-1}, and mass density $\rho = 2000$ kg m^{-3}. Table 3.8 shows the values used for computation of Figure 3.31.

We remember that for our case of cylindrical geometry, the results printed in Figure 3.31 correspond to one *slice* (Figure 3.28). These results put into evidence some important facts:

- Lateral diffusion of the thermal front around the defect is important (especially at $t = 10.78$ s). This *flow* effect reduces the visibility of the defect.
- Temperature differentials are smaller on the surface than at the interface of the defect itself. This also reduces the visibility of the defect.
- The surface temperature above the defect is not constant and reaches a maximum at a specific time [which depends on the defect depth; eq. (9.7) and Figure 9.2]. Consequently, there is an optimum time of observation. In the case of the geometry of Figure 3.31, the maximum temperature differential is observed at $t = 10.78$ s.

As we will see later, this kind of analysis is very useful for predicting and understanding experimental results, especially in active thermography (Chapter 9). The accuracy that can be obtained with such modeling is strongly related to the size of the elementary volumes considered $(\Delta r, \Delta z, \Delta t)$. With very small values, errors introduced by the finite difference approximations are reduced drastically at the expense of a tremendous increase in computation time and risk of nonconvergence. Of course, a well-balanced compromise must be made between the accuracy necessary for the elementary volume size and the time step size (depending on the infrared camera resolution if comparisons are to be made with experimental results, for instance). To specify values for $(\Delta r, \Delta z, \Delta t)$, we can start with large values and then solve the problem using smaller and smaller values of $(\Delta r, \Delta z, \Delta t)$. Observing the convergence of the distribution of temperature allows us to stop computations when an acceptable accuracy is obtained. A useful approach for added accuracy with reasonable computation time is obtained if a nonregular mesh is used in such a way that a finer mesh is drawn at a subsurface defect tip (Figure 3.32). Of course, such an approach requires a more complex computer program.

In performing these computations, it is also important to specify values with respect to a stability criterion required for the convergence of the computation (Cielo et al., 1986a):

$$(\Delta z)^2 = 6\alpha\,\Delta t$$

$$(\Delta r)^2 = 6\alpha\,\Delta t \qquad \text{with } \alpha = \text{thermal diffusivity of the material} \qquad (3.72)$$

to prevent the temporal instabilities which are otherwise likely to occur in these iterating computation procedures. Other convergence rules are also possible; for example (Incropera and DeWitt, 1990; Rossignol, 2000),

$$\frac{1}{2} \geq \text{Fo}_z + \text{Fo}_r \quad \text{with} \quad \text{Fo}_z = \frac{\alpha_z \Delta z}{(\Delta z)^2} \quad \text{and} \quad \text{Fo}_r = \frac{\alpha_r \Delta r}{(\Delta r)^2} \qquad (3.73)$$

Fo_z and Fo_r being equivalent to Fourier numbers [eq. (10.11)].

Finally, to validate such modeling, two directions can be adopted. The model can be applied to a very simple problem (i.e., an infinite plate) for which an exact analytical solution exists [e.g., eq. (3.59)]. This does not mean, however, that the model will act satisfactorily in complex situations. Probably a better approach would be to compare the model with data acquired from a corresponding experimental setup. Care must be taken when performing such comparisons since experimental values can be corrupted with adverse phenomena not taken into account in the model, such as nonuniform surface thermal stimulation, uneven surface emissivity, complex loss phenomena (e.g., convection on vertical surfaces), variable material properties, and so on. Even if comparisons with experimental data are not straightforward, such modeling brings useful information about typical thermal behavior for specific families of samples. It also helps to optimize TNDT parameters, such as length and strength of thermal stimulation, and the optimum time window of observation expected for best defect visibility.

To help, the source code in C language is provided at the Web site (Appendixes A and F for the Matlab version; Table 3.9). Commercial packages are also available with additional features and improved accuracy. (For example, packages from Innovation Inc.* especially dedicated to TNDT problems are well known within the community).

---

* vavilov@introscopy.tpu.ru.

Time is: 0.50 sec   Delta T max (for row 1, defect - sound area):   0.0 °C

4.224	4.224	4.224	4.224	4.224	4.224	4.224	4.224	4.224	4.224	4.224	4.224
3.286	3.286	3.286	3.286	3.286	3.286	3.286	3.286	3.286	3.286	3.286	3.286
2.120	2.120	2.120	2.120	2.120	2.120	2.120	2.120	2.120	2.120	2.120	2.120
1.296	1.296	1.296	1.296	1.296	1.296	1.296	1.296	1.296	1.296	1.296	1.296
.748	.748	.748	.748	.748	.748	.748	.748	.748	.748	.748	.748
.406	.406	.406	.406	.406	.406	.406	.406	.406	.406	.406	.406
.206	.206	.206	.206	.206	.206	.206	.206	.206	.206	.206	.206
.097	.097	.097	.097	.097	.097	.097	.097	.097	.097	.097	.097
.043	.043	.043	.043	.043	.043	.043	.043	.043	.043	.043	.043
.017	.017	.017	.017	.017	.017	.017	.017	.017	.017	.017	.017
.007	.007	.007	.007	.007	.006	.006	.006	.006	.006	.006	.006
.003	.003	.003	.003	.003	.002	.002	.002	.002	.002	.002	.002
.001	.001	.001	.001	.000	.001	.001	.001	.000	.000	.000	.000
.000	.000	.000	.000	.000	.000	.000	.000	.000	.000	.000	.000
.000	.000	.000	.000	.000	.000	.000	.000	.000	.000	.000	.000
.000	.000	.000	.000	.000	.000	.000	.000	.000	.000	.000	.000
.000	.000	.000	.000	.000	.000	.000	.000	.000	.000	.000	.000
.000	.000	.000	.000	.000	.000	.000	.000	.000	.000	.000	.000
.000	.000	.000	.000	.000	.000	.000	.000	.000	.000	.000	.000
.000	.000	.000	.000	.000	.000	.000	.000	.000	.000	.000	.000

Time is: 2.00 sec   Delta T max (for row 1, defect - sound area):   0.004705 °C

9.669	9.669	9.667	9.665	9.664	9.664	9.664	9.664	9.664	9.664	9.664	9.664
8.542	8.542	8.541	8.540	8.537	8.536	8.536	8.536	8.536	8.536	8.536	8.536
7.006	7.006	7.005	7.003	6.998	6.997	6.997	6.997	6.997	6.997	6.997	6.997
5.678	5.678	5.677	5.674	5.666	5.664	5.663	5.663	5.663	5.663	5.663	5.663
4.548	4.548	4.547	4.542	4.529	4.526	4.525	4.525	4.525	4.525	4.525	4.525
3.604	3.604	3.602	3.594	3.574	3.569	3.568	3.568	3.568	3.568	3.568	3.568
2.831	2.831	2.828	2.816	2.783	2.775	2.774	2.774	2.774	2.774	2.774	2.774
2.213	2.212	2.209	2.192	2.138	2.128	2.127	2.126	2.126	2.126	2.126	2.126
1.737	1.736	1.731	1.707	1.621	1.608	1.606	1.606	1.606	1.606	1.606	1.606
1.387	1.386	1.381	1.349	1.213	1.197	1.195	1.195	1.195	1.195	1.195	1.195
1.155	1.154	1.153	1.147	1.107	.877	.876	.876	.876	.876	.876	.876
1.029	1.029	1.028	1.021	.975	.893	.642	.632	.632	.632	.632	.632
.051	.051	.052	.059	.105	.642	.438	.448	.448	.448	.448	.448
.034	.035	.036	.042	.082	.296	.312	.313	.313	.313	.313	.313
.023	.023	.024	.029	.061	.198	.213	.215	.215	.215	.215	.215
.015	.015	.016	.020	.044	.130	.143	.145	.145	.145	.145	.145
.010	.010	.010	.014	.031	.085	.095	.096	.096	.096	.096	.096
.006	.006	.007	.009	.021	.054	.062	.063	.063	.063	.063	.063
.004	.004	.004	.006	.014	.034	.039	.040	.040	.040	.040	.040
.002	.002	.003	.004	.009	.021	.025	.025	.025	.025	.025	.025

```
Time is: 10.78 sec   Delta T'max (for row 1, defect - sound area):    0.859981 °C

3.280 3.235 3.144 2.838 2.670 2.554 2.487 2.452 2.434 2.426 2.421 2.421 2.420 2.420 2.420 2.420 2.420 2.420 2.420
3.284 3.239 3.147 3.007 2.837 2.667 2.550 2.483 2.448 2.431 2.423 2.420 2.418 2.417 2.417 2.417 2.417 2.417 2.417
3.282 3.237 3.144 3.009 2.837 2.652 2.535 2.468 2.434 2.417 2.410 2.406 2.405 2.404 2.404 2.404 2.404 2.404 2.404
3.275 3.229 3.136 3.005 2.829 2.627 2.508 2.443 2.410 2.394 2.386 2.383 2.382 2.381 2.381 2.381 2.381 2.381 2.381
3.262 3.216 3.122 2.994 2.812 2.590 2.470 2.407 2.375 2.360 2.353 2.350 2.349 2.348 2.348 2.348 2.348 2.348 2.348
3.243 3.197 3.103 2.977 2.788 2.541 2.421 2.361 2.332 2.317 2.311 2.308 2.307 2.306 2.306 2.306 2.306 2.306 2.306
3.219 3.174 3.078 2.955 2.756 2.481 2.362 2.306 2.279 2.266 2.260 2.258 2.256 2.256 2.256 2.256 2.256 2.256 2.256
3.190 3.145 3.050 2.928 2.719 2.409 2.293 2.242 2.218 2.207 2.202 2.199 2.198 2.198 2.198 2.198 2.198 2.198 2.198
3.156 3.112 3.017 2.896 2.677 2.324 2.215 2.171 2.150 2.141 2.136 2.134 2.133 2.133 2.133 2.133 2.133 2.133 2.133
3.118 3.074 2.980 2.862 2.631 2.224 2.128 2.092 2.076 2.068 2.064 2.063 2.062 2.062 2.062 2.062 2.062 2.062 2.062
3.074 3.032 2.940 2.825 2.584 2.109 2.034 2.008 1.996 1.990 1.988 1.986 1.985 1.985 1.985 1.985 1.985 1.985 1.985
3.027 2.985 2.891 2.786 2.538 1.973 1.934 1.920 1.912 1.908 1.906 1.905 1.905 1.905 1.905 1.905 1.905 1.905 1.905
                  2.747 2.499  ——— (flaw line)
 .869  .901  .971 1.316 1.813 1.828 1.830 1.825 1.823 1.822 1.821 1.821 1.821 1.821 1.821 1.821 1.821 1.821 1.821
 .818  .852  .925 1.272 1.673 1.734 1.725 1.736 1.736 1.736 1.736 1.736 1.736 1.736 1.736 1.736 1.736 1.736 1.736
 .769  .803  .878 1.218 1.548 1.620 1.640 1.646 1.648 1.648 1.649 1.649 1.649 1.649 1.649 1.649 1.649 1.649 1.649
 .722  .756  .831 1.158 1.434 1.519 1.546 1.556 1.559 1.561 1.561 1.562 1.562 1.562 1.562 1.562 1.562 1.562 1.562
 .676  .711  .785 1.096 1.331 1.420 1.454 1.467 1.472 1.474 1.475 1.476 1.476 1.476 1.476 1.476 1.476 1.476 1.476
 .633  .667  .740 1.032 1.235 1.326 1.364 1.380 1.387 1.389 1.391 1.391 1.391 1.391 1.391 1.391 1.391 1.391 1.391
 .591  .625  .696  .970 1.147 1.237 1.278 1.296 1.304 1.308 1.309 1.310 1.310 1.310 1.310 1.310 1.310 1.310 1.310
 .552  .585  .654  .909 1.066 1.154 1.197 1.216 1.225 1.229 1.231 1.232 1.232 1.232 1.232 1.232 1.232 1.232 1.232
```

```
Time is: 20.0 sec   Delta T max (for row 1, defect - sound area):    0.559669 °C

2.334 2.302 2.241 2.155 2.056 1.960 1.889 1.844 1.815 1.798 1.788 1.782 1.778 1.776 1.775 1.774 1.774 1.774 1.774
2.338 2.306 2.245 2.159 2.058 1.961 1.891 1.845 1.817 1.800 1.790 1.784 1.781 1.779 1.778 1.777 1.777 1.777 1.777
2.339 2.307 2.246 2.159 2.058 1.959 1.888 1.844 1.816 1.799 1.790 1.784 1.780 1.778 1.777 1.776 1.776 1.776 1.776
2.338 2.306 2.245 2.157 2.054 1.952 1.882 1.838 1.812 1.796 1.786 1.781 1.777 1.775 1.774 1.773 1.773 1.773 1.773
2.334 2.302 2.241 2.153 2.047 1.941 1.872 1.830 1.804 1.789 1.780 1.775 1.772 1.770 1.769 1.768 1.768 1.768 1.768
2.327 2.296 2.234 2.146 2.037 1.926 1.858 1.818 1.794 1.780 1.771 1.766 1.763 1.762 1.761 1.760 1.760 1.760 1.760
2.318 2.286 2.225 2.136 2.025 1.906 1.840 1.803 1.781 1.768 1.760 1.756 1.753 1.751 1.751 1.750 1.750 1.750 1.750
2.306 2.275 2.214 2.125 2.010 1.881 1.818 1.785 1.765 1.754 1.747 1.743 1.740 1.739 1.738 1.737 1.737 1.737 1.737
2.291 2.261 2.201 2.112 1.994 1.851 1.793 1.764 1.747 1.737 1.731 1.727 1.725 1.724 1.723 1.723 1.723 1.723 1.723
2.274 2.245 2.186 2.098 1.977 1.816 1.765 1.741 1.727 1.719 1.714 1.710 1.708 1.707 1.707 1.706 1.706 1.706 1.706
2.255 2.226 2.169 2.082 1.960 1.773 1.734 1.716 1.705 1.698 1.694 1.692 1.690 1.689 1.689 1.688 1.688 1.688 1.688
2.234 2.206 2.150 2.066 1.944 1.723 1.700 1.689 1.682 1.677 1.674 1.672 1.671 1.670 1.670 1.669 1.669 1.669 1.669
                              ——— (flaw line)
1.243 1.263 1.303 1.366 1.464 1.662 1.665 1.661 1.657 1.654 1.652 1.651 1.650 1.649 1.649 1.649 1.649 1.649 1.649
1.220 1.240 1.282 1.347 1.446 1.609 1.629 1.632 1.631 1.630 1.629 1.628 1.628 1.627 1.628 1.627 1.627 1.627 1.627
1.197 1.218 1.261 1.328 1.425 1.563 1.594 1.602 1.605 1.606 1.606 1.606 1.606 1.606 1.606 1.606 1.606 1.606 1.606
1.174 1.196 1.240 1.307 1.402 1.520 1.559 1.573 1.578 1.581 1.582 1.583 1.583 1.583 1.583 1.583 1.583 1.583 1.583
1.152 1.174 1.219 1.286 1.377 1.482 1.525 1.543 1.552 1.556 1.559 1.560 1.561 1.561 1.561 1.561 1.561 1.561 1.561
1.131 1.153 1.198 1.265 1.352 1.446 1.493 1.515 1.526 1.532 1.536 1.537 1.538 1.539 1.539 1.539 1.539 1.539 1.539
1.110 1.133 1.178 1.244 1.327 1.414 1.462 1.487 1.501 1.508 1.513 1.515 1.516 1.517 1.518 1.518 1.518 1.518 1.518
1.091 1.113 1.158 1.223 1.303 1.384 1.433 1.461 1.477 1.486 1.491 1.494 1.495 1.496 1.497 1.497 1.497 1.497 1.497
```

FIGURE 3.31   Temperature distribution computed for a graphite-epoxy specimen. The position of a thin air layer (subsurface flaw) is indicated by a line. The results are shown for four different times (see the text for more details). (From Maldague, 1993.)

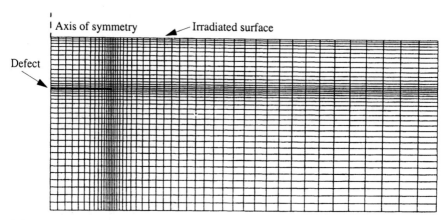

FIGURE 3.32 Finer mesh for added accuracy at subsurface edges. (Adapted from Krapez et al., 1991.)

TABLE 3.9 Matlab m-Script Available in Appendix F

Name of Matlab m-Script	Short Description, Purpose

mod_therm.m
mod_therm;
	Thermal model discussed in Section 3.4.2
	Adjustable modeling parameters (within the script) are:
	Geometry
	  def_nod_pos: position of default nodes
	  rad_nodes: number of nodes along the radial length
	  axe_nodes: number of nodes in depth
	  dr: radial width
	  dz: axial depth
	  frad, fcon: empirical factors that depend on the geometry
	Time and energy pulse
	  dt: small time increments
	  num_it: number of iterations
	  source_power: quantity of energy deposited on the front surface
	  heat_duration: time interval after which source_power = 0
	Thermal parameters
	  rhd: density/1000
	  cp: specific heat/1000
	  ra_conduct: radial thermal conductivity
	  ax_conduct: axial thermal conductivity
	  kint: axial thermal conductivity for def_nod_pos
	  ta: ambient temperature
	  itemp: initial temperature
	Example of use: **mod_therm;**

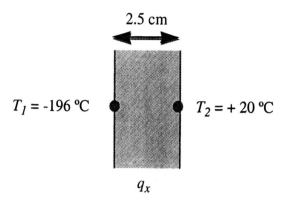

FIGURE 3.33 Figure for Problem 3.1.

PROBLEMS

3.1 **[conduction]** A leg-supported cylindrical tank (1 m high, 0.3 m in diameter) having a combined wall conductivity of $k = 0.0003$ W m^{-1} °C^{-1} is used to keep liquid nitrogen (LN$_2$) at a temperature of -196°C. The energy necessary to evaporate 1 kg of LN$_2$ at this temperature is 200 kJ. The exterior temperature is 20°C (which is also the exterior wall temperature). The wall thickness is 2.5 cm. How much LN$_2$ is vaporized daily? [Problem adapted from Holman (1981, p. 22, Prob. 1.19).]

Solution The situation is depicted in Figure 3.33. The tank surface is given by $\pi DL + 2\pi(D^2/4) = 1.08$ m^2, with D being the diameter and L the height (this is for a leg-supported tank, as shown in Figure 3.3). We next compute the conduction heat transfer as follows:

$$q_x = kS(T_1 - T_2) = (0.0003)(0.65)(20 + 196) \text{ W} = 0.04 \text{ W} = 0.04 \text{ J s}^{-1}$$

with S computed from Table 3.2:

$$S = \frac{2\pi r_1 r_2}{\ln(r_1/r_2)} = \frac{2\pi(0.15 \times 0.125)}{\ln(0.15/0.125)} = 0.65 \text{ m}$$

One day is $24 \times 60 \times 60$ s $= 86,400$ s; thus the daily energy loss is $86,400 \times 0.04$ J $= 3639$ J, corresponding to a loss of $3639/200,000$ kg $= 20$ g lost daily. As we will see in Section 4.6, liquid nitrogen is often used to cool infrared cameras.

3.2 **[convection]** Compute the heat losses for a 1-m-long copper pipe ($\frac{1}{2}$ in. $= 1.25$ cm in diameter). See problem parameters in Figure 3.34.

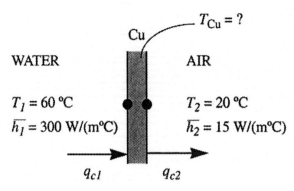

FIGURE 3.34   Figure for Problem 3.2.

Solution   Temperature differences across the pipe and radiation heat transfer can be neglected. Following the first law of thermodynamics, we have that $q_{c1} = q_{c2}$ and thus

$$q_{c1} = \overline{h_1} A_S(T_1 - T_{Cu}) = q_{c2} = \overline{h_2} A_S(T_{Cu} - T_2)$$

Solving for T_{Cu} yields

$$T_{Cu} = \frac{\overline{h_1} T_1 + \overline{h_2} T_2}{\overline{h_1} + \overline{h_2}} = \frac{300 \times 60 + 15 \times 20}{300 + 15} °C = 58.1°C$$

The surface involved is (D = diameter) $A_s = (1) \times \pi D = \pi \times 0.0125 \text{ m}^2 = 0.039 \text{ m}^2$, and then

$$q = q_{c1} = q_{c2} = \overline{h_2} A_S(T_{Cu} - T_2) = 15 \times 0.039 \times (58.1 - 20) = 22.4 \text{ W}$$

This example show clearly the usefulness of thermally insulating hot water pipes with available foam coating.

3.3 **[mechanism combination]** In an industrial process, a spherical tank (diameter = 4 m) contains a liquid maintained at $T = 200°C$. The tank wall thickness is 2.5 cm with $k = 50 \text{ W m}^{-1}\,°C^{-1}$ (see Figure 3.35 for problem parameters). Compute the total heat transfer across the tank wall.

Solution   Following the first law of thermodynamics, we have that the total heat transfer q across the wall is $q = q_x = q_c$. The total surface for the tank is $A = 4\pi R^2$, R = sphere radius, $A = 50.3 \text{ m}^2$. The convective heat transfer q_c is given by eq. (3.15):

$$q_c = \overline{h} A_S(T_{wall} - T_2) = 25 \times 50.3 \times (T_{wall} - 25) \text{ W}$$

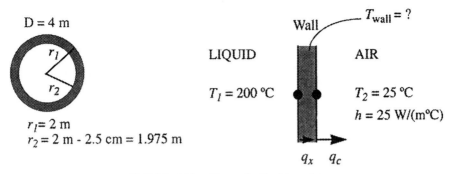

FIGURE 3.35 Figure for Problem 3.3.

and the conduction heat transfer q_x is given by eq. (3.7):

$$q_x = kS(T_1 - T_{wall}) = 50 \times 1985(200 - T_{wall}) \text{ W}$$

with the expression for S given by Table 3.2 for a sphere, giving $S = 1985$ m. Solving for T_{wall} in the last two equations gives $T_{wall} = 197.8°C$ and thus the heat transfer is

$$q = q_c = \bar{h}A_S(T_{wall} - T_2) = 25 \times 50.3 \times (197.8 - 25) \text{ W} = 217 \text{ kW}$$

3.4 **[mechanism combination, electrical analogy]** In an English steel plant, 50 tons of melted steel at 1500°C fills a large pot. It is 16h50, tea time is coming at 17h00, and plant operators have their kettle (of surface emissivity 1, wall thickness of 1 mm) ready on the nearby heater, filled with 100°C boiling water. Assuming a thermal radiation shape factor $F_{S-R} = 1$ between the steel pot and the boiler due to the large pot size, compute the boiler surface temperature (T_1). Figure 3.36 shows the configuration. Assume that only radiation is involved here.

Solution The solution is a little more complicated due to the nonlinear relationship of the radiation equation. We have

$$\frac{kA}{L}(T_1 - T_2) = q_x = q_r = \sigma A F_{S-R}(T_R^4 - T_1^4)$$

$$\frac{k}{L}(T_1 - T_2) = \sigma(T_R^4 - T_1^4)$$

leading to

$$T_1 = \frac{L}{k}\sigma(T_R^4 - T_1^4) + T_2 = \left[\frac{0.001}{55}5.67 \times 10^{-8}(1773^4 - T_1^4) + 20\right]°C$$

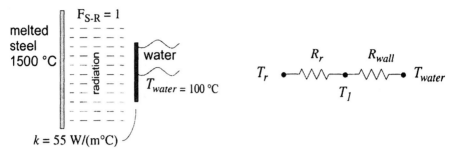

FIGURE 3.36 Figure for Problem 3.4 with electrical analogy.

This expression can be solved by iteration starting with $T_1 = 100°C$ (373 K), the iteration process gives successively $T_1 = 383.19$ K and 383.16 K, which is 110°C. We conclude that these Englishmen could turn off the boiler heater up to 17h00 without any harm for the coming tea since water is not likely to cool down! This situation is not unusual in such a plant, where you can experience a strange feeling—say, in winter, with your body side facing the hot steel literally roasting due to radiation while the other side is freezing.

3.5 **[unsteady heat transfer]** A brass ball of diameter 5 cm, thermal conductivity $k = 130$ W m^{-1} °C^{-1}, mass density $\rho = 8400$ kg m^{-3}, specific heat $c = 380$ J kg^{-1} °C^{-1} is falling in a stirred oil bath of fixed temperature $T_f = 50°C$. The initial ball temperature is 200°C. Compute the temperature history of the ball starting at time $t = 0$ s when the ball makes contact with the oil. The forced convective heat transfer coefficient of the oil is $\bar{h} = 1000$ W m^{-2} °C^{-1} and the boiling temperature of the oil is 300°C.

Solution The Biot number is first computed using eq. (3.37), and it is found as Bi = 0.0641. Since Bi $\ll$ 0.1 and assuming negligible radiation effect, eq. (3.42) can be used to compute the temperature evolution of the ball:

$$T(t) = (200 - 50)\exp(-0.3759t) + 50 \qquad °C$$

$T(t)$ is plotted in Figure 3.37. As expected in the previous expression, it can be checked that $T(t = 0) = 200°C$ and $T(t \to \infty) = 50°C$.

3.6 **[unsteady heat transfer, electrical analogy]** In Example 3.16, instead of a large tank, the ball (which is warmer than the fluid) falls in a small insulated tank. As a consequence, the temperature of the fluid will change over time. Repeat the computation in this situation.

Solution Obviously, the solution will be more complex due to the added complexity of fluid temperature dependence. Thanks to an electrical

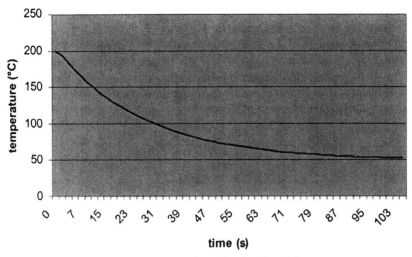

FIGURE 3.37   Figure for Problem 3.5.

analogy [eqs. (3.47) and (3.48)], the problem is solved without much pain. Figure 3.38 illustrates the problem by showing the electrical analogy. Notice, the arbitrary (although logical, since energy is transferred in the direction of decreasing temperature following the second law of thermodynamics) current direction (a current set in the opposite direction leads to the same equations). Since two processes occur here with the ball cooling down and the fluid heating up, two thermal masses are thus implied, which translate as two capacitors in the electrical circuit of the figure, C_1

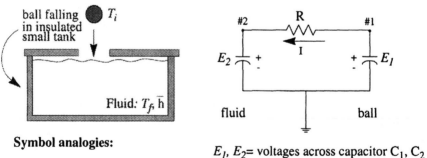

Symbol analogies:

$T_1 \leftrightarrow E_1$, ball

$T_2 \leftrightarrow E_2$, fluid

$I \leftrightarrow q$

$E_1, E_2 =$ voltages across capacitor C_1, C_2

FIGURE 3.38   Figure for Problem 3.6 with electrical analogy.

for the ball and C_2 for the fluid. Convection heat transfer, which is only assumed here, is represented by resistance R (conduction and radiation are neglected here). At both nodes 1 and 2, the Kirchhoff current law [see e.g., D. E. Johnson et al. (1997); $\sum I = 0$] can be written as follows, with the assumption that current circulates as shown on the figure (the sign convention is: a current entering a node is affected with a $+$ sign while a current leaving a node is affected with a $-$ sign): At node 1:

$$-\frac{E_1 - E_2}{R} - C_1 \frac{dE_1}{dt} = 0 \tag{3.74}$$

At node 2:

$$\frac{E_1 - E_2}{R} - C_2 \frac{dE_2}{dt} = 0 \tag{3.75}$$

The minus sign for C_2 in eq. (3.74) is due to the sign convention of Figure 3.24. Solving for E_1 and E_2 gives (notice the use of a prime to designate the derivative d/dt)

$$E_1 = E_2 + RC_2 E_2' \tag{3.76}$$
$$E_2 = E_1 + RC_1 E_1' \tag{3.77}$$

Differentiating eq. (3.77) gives

$$E_2' = E_1' + RC_1 E_1'' \tag{3.78}$$

which can be substituted in eq. (3.76) to provide the second-order differential equation (designated by a double prime) describing the problem:

$$R^2 C_1 C_2 E_1'' + (RC_1 + RC_2)E_1' = 0 \tag{3.79}$$

Let's have $A = R^2 C_1 C_2$ and $B = RC_1 + RC_2$; eq. (3.79) thus becomes $E_1'' + (B/A)E_1' = 0$ and can be written as (by dividing each side with dE_1/dt)

$$\frac{d(dE_1/dt)dt}{dE_1/dt} = -\frac{B}{A}\frac{dE_1/dt}{dE_1/dt} \tag{3.80}$$

thus:

$$\int \frac{d(dE_1/dt)}{(dE_1/dt)} = -\int \frac{B}{A} dt \tag{3.81}$$

Performing the indefinite integration gives [see Spiegel (1974) for a complete integral list; we recall that $\int (du/u) = \ln(u/K_1)$]:

$$\ln \frac{dE_1/dt}{K_1} = -\frac{B}{A}t \qquad (3.82)$$

where K_1 is the integral constant. This expression is transformed knowing that $[(y = \ln u) \leftrightarrow u = \exp(y)]$

$$\frac{dE_1}{dt} = K_1 \exp\left(-\frac{B}{A}t\right) \qquad (3.83)$$

which is solved [see eq. (g) below in the Appendix to this chapter] as

$$E_1(t) = \frac{K_1}{B/A}\left[1 - \exp\left(-\frac{B}{A}t\right)\right] + e_1(0) \qquad (3.84)$$

The constant K_1 is found by differentiating eq. (3.84) with respect to t and isolating $dE_1(t)/dt$ in eq. (3.74) at $t = 0$:

$$\frac{d}{dt}E_1(t) = \frac{K_1}{B/A}\frac{B}{A} = K_1 = \frac{E_{2,t=0} - E_{1,t=0}}{RC_1} \qquad (3.85)$$

which in the temperature domain becomes

$$K_1 = \frac{T_f - T_i}{RC_1} \qquad (3.86)$$

$$T_1(t) = \begin{cases} \dfrac{K_1}{B/A}\left[1 - \exp\left(-\dfrac{B}{A}t\right)\right] + T_i & (3.87) \\[3mm] (T_f - T_i)\dfrac{C_2}{C_1 + C_2}\left[1 - \exp\left(-\dfrac{C_1 + C_2}{RC_1 C_2}t\right)\right] + T_i & (3.88) \end{cases}$$

which with $C_1 = (\rho V c)_{\text{ball}}$, $C_2 = (\rho V c)_{\text{fluid}}$, and $R = 1/\bar{h}A_s$ gives the time evolution of the ball temperature as a function of time. The temperature evolution of the fluid can be found using eq. (3.88) knowing the expression for $T_i(t)$.

3.7 **[unsteady heat transfer, application]** A brass ball of diameter 5 cm, thermal conductivity $k = 130$ W m^{-1} °C^{-1}, mass density $\rho = 8400$ kg m^{-3}, specific heat $c = 380$ J kg^{-1} °C^{-1} is falling in a stirred water–thermal conductivity $k = 0.6$ W m^{-1} °C^{-1}, mass density $\rho = 1000$ kg m^{-3}, specific heat $c = 4180$ J kg^{-1} °C^{-1}—bath of fixed temperature $T_f = 10$°C. The initial ball temperature is 90°C and the bath contains 1 kg of water. The forced

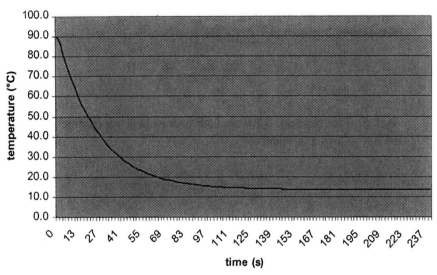

FIGURE 3.39 Temperature evolution of the 90°C ball falling in the 10°C-water-filled insulated small tank of Problem 3.7.

convective heat transfer coefficient of the water is $\bar{h} = 1000$ W m^{-2} °C^{-1}. Compute the temperature history of the ball starting at time $t = 0$ s when the ball makes contact with the water. What is the final temperature?

Solution Equation (3.88) is used with the parameters provided. The temperature history of the ball is plotted in Figure 3.39; it is seen that the final temperature is about 14°C, slightly over the initial bath temperature of 10°C.

APPENDIX: SOLVING DIFFERENTIAL EQUATIONS USING THE LAPLACE TRANSFORM

Solve the following differential equation using the Laplace transform:

$$\frac{d}{dt} E(t) = K \exp(-at) \tag{a}$$

The Laplace transform is an useful tool for such purpose since it allows us to transform a differential equation into an algebraic equation. The principle of using Laplace transforms for differential equation solving involves performing three steps:

1. Transform the differential equation expressed in the time domain (variable t) into an algebraic equation in the Laplace domain (variable s).

TABLE 3.10   Useful Laplace Transforms Used in Differential Equation Solving

Time Domain $f(t) = L^{-1}[F(s)]$	Laplace Domain $F(s) = L[f(t)]$
$f(t)$	$F(s)$
$\dfrac{d}{dt} f(t)$	$sF(s) - f(0)$
$\dfrac{d^2}{dt} f(t)$	$s^2 F(s) - sf(0) - f(0)'$
$\exp(-at)$	$\dfrac{1}{s+a}$
$\delta(t)$, impulse function $\displaystyle\int_{-0}^{+0} \delta(t) = 1$	1
$u(t)$, the step function $\quad u(t) = 1, t > 0$ $\quad u(t) = 0, t < 0$	$\dfrac{1}{s}$
t	$\dfrac{1}{s^2}$

2. Isolate the variable of interest in the Laplace expression obtained in step 1.
3. Transform the s-based expression in step 2 back into the time domain.

See, for example, D. E. Johnson et al. (1997) for additional details on the Laplace transform, especially Chapters 19 and 20. Refer to Table 3.10 for the common direct and inverse Laplace transforms.

Step 1. Express eq. (a) in the Laplace domain (Table 3.10):

$$sE(s) - e(0) = \frac{K}{s+a} \tag{b}$$

Step 2. Isolate the variable of interest, here E:

$$E(s) = \frac{K}{s(s+a)} + \frac{e(0)}{s} \tag{c}$$

Step 3. Back in the time domain. The expression in eq. (c) has to be decomposed in simple terms whose inverses are found in tables such as Table 3.10:

$$E(s) = \frac{A}{s} + \frac{B}{s+a} + \frac{e(0)}{s} \tag{d}$$

A and B are found by simple identification with eq. (c):

$$E(s) = \frac{A}{s} + \frac{B}{s+a} + \frac{e(0)}{s} = \frac{(A+B)s + aA}{s(s+a)} + \frac{e(0)}{s} = \frac{K}{s(s+a)} + \frac{e(0)}{s} \quad \text{(e)}$$

thus $A + B = 0$; $A = K/a$ leading to $B = -A = -K/a$ and finally with $e(t) = L^{-1}[E(s)]$:

$$E(t) = L^{-1}\left[\frac{K/a}{s} + \frac{-K/a}{s+a} + \frac{e(0)}{s}\right] = \frac{K}{a} + \left(-\frac{K}{a}\right)\exp(-at) + e(0) \quad \text{(f)}$$

$$E(t) = \frac{K}{a}[1 - \exp(-at)] + e(0) \quad \text{(g)}$$

The method reviewed here is of interest in solving differential equations in this chapter.

CHAPTER FOUR

Infrared Sensors and Optic Fundamentals*

All thermographical investigations involve some kind of thermal detector. To better predict measurements and understand results, it is important to gain some background on thermal detectors and their ancillary equipment. This chapter is thus dedicated to thermal sensor physical theory and associated parameters such as optics fundamentals, including imaging devices.

4.1 DETECTOR DEFINITIONS AND CHARACTERISTICS

Optical receivers are really the heart of infrared (IR) systems. The worldwide IR-detector market is now estimated to be close to $10 billion (1999) and the rate is increasing by 30% annually. One of the fastest-growing areas is that of focal plane arrays (FPAs), which are covered in Section 4.5, but let's start with detector basics and review the various types available.

4.1.1 Impedance and Responsivity

An infrared detector is a device that converts the radiant infrared energy into a measurable signal, generally electrical in nature. The detector impedance Z is an intrinsic characteristic of detectors measured using Ohm's law ($d\mathbf{V}$, $d\mathbf{I}$ in vector form):

$$Z = \frac{d\mathbf{V}}{d\mathbf{I}} \qquad \text{ohms} \qquad (4.1)$$

*Some of the material in this chapter was originally published by Maldague (1994b); however, such material has been updated, expanded with current technological advances, and especially adapted for this book.

The detector impedance is an important design property since the maximum power we can extract from such a detector is reached when

$$Z = Z_A \tag{4.2}$$

where Z_A is the input impedance of the preamplifier to which the detector is connected. A correct impedance match is important to maximize the output signal since electric signals at the detector output are generally small.

Example 4.1 During design tests, output voltage and associated current from an IR detector are measured as 5 mV and 5 μA, respectively. Calculate the detector impedance Z.

SOLUTION   Remember that eq. (4.1) is in vector form; the trivial answer $Z = 5$ mV$/5$ μA $= 1000 \ \Omega$ is valid only when direct-current (dc) conditions are met or when the phase lag between tension **V** and current **I** is $0°$ ($\cos 0° = 1$). This will be the situation for resistor-type detectors such as bolometers. See also Problem 4.1.

Another important parameter is the responsivity expressed in voltage (R_v) or current (R_i) and which is the transformation ratio of the incident optical flow F (Gaussorgues, 1989, p. 282):

$$R_v = \frac{\partial V}{\partial F} \quad \text{V W}^{-1} \quad \text{or} \quad R_i = \frac{\partial I}{\partial F} \quad \text{A W}^{-1} \tag{4.3}$$

Responsivity is generally not uniform on the detector area; moreover, it also depends on the radiation wavelength λ and the excitation frequency f:

$$R = R(x, y, \lambda, f) \tag{4.4}$$

Figure 4.1 illustrates a typical responsivity curve in which the critical frequency f_c is defined by an observed decrease of 3 dB in the maximum amplitude. Knowledge of this responsivity curve allows us to determine the maximum possible bandwidth of the system, that is, the fastest optical flow variation to which the detector is sensible.

4.1.2   Time Constant

A useful property is also the time constant τ, which specifies the response time of a detector. It is given by

$$\tau = \frac{1}{2\pi f_c} \tag{4.5}$$

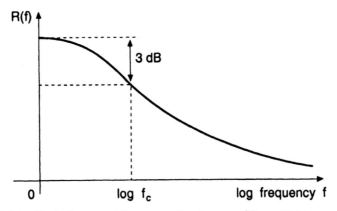

FIGURE 4.1 Typical responsivity curve of a detector. (From Maldague, 1994b, with permission from Gordon and Breach Publishers.)

4.1.3 Noise Equivalent Power

A very important figure of merit for detectors is the noise equivalent power (NEP). This is the amount of power P_s induced by an optical signal whose amplitude is equivalent to the intrinsic noise power P_N present at the detector output without signal, which originates from thermal agitation and optical radiation granularity. This optical flow F_{NEP} is called the *noise equivalent flow*. To be useful the signal $R_V F$ or $R_i F$ should be at least greater than this noise level ($F_{\text{NEP}} R_V$ or $F_{\text{NEP}} R_i$).

4.1.4 Noise Equivalent Temperature Difference

Many detector manufacturers rate their detectors using the noise equivalent temperature difference (NETD) instead of NEP. NETD corresponds to the change in temperature of a large blackbody (i.e., a perfect emitter of infrared radiation, Chapter 2) in the observed scene, causing a change in the signal-to-noise ratio of unity in the output of the detector. For instance, current values for microbolometer arrays (Section 4.5) are about 50 mK; large arrays of InSb (indium–antimonide, Section 4.4) have <20 mK. NETD depends on the F-number of the optics and the pixel size (Kruse, 1995).

Here a little background in optics is of interest. The F-number of a lens is given by its focal distance divided by its effective diameter aperture (Section 4.10). For instance, if the focal distance of a 10-mm aperture lens is 50 mm, the lens is said to be $F/5$.

4.1.5 Detectivity D and D^*

The detectivity D is also a widely used parameter to specify detectors. It is given by

$$D = \frac{1}{F_{\text{NEP}}} \qquad (4.6)$$

D should be high, to obtain a reasonable signal-to-noise ratio. The detectivity depends on many parameters: spectral content and modulation of the incident radiation, receiver system bandwidth, detector temperature, and sensitive surface area A. To be able to compare among various detectors, the specific detectivity D^* is introduced. This is in fact the detectivity D scaled to the unit-sensitive detector area and unit of bandwidth. Units of D^* are W^{-1} cm $\text{Hz}^{1/2}$.

4.1.6 Minimum Resolvable Temperature Difference

The minimum resolvable temperature difference (MRTD) is another parameter used to characterize either a detector or a thermal imaging system. It is defined as the smallest temperature difference observable by an operator when the target is constituted by a repetitive bar pattern over a uniform background (a four-bar pattern, generally). Detector or system gain and offset (thermal range and level settings on infrared cameras) can be adjusted at their best settings to obtain the minimal value of MRTD, which is, in general, on the order of 0.1 to 0.2 K.

4.1.7 Line Spread Function and Slit Response Function

Another important characteristic is the line spread function (LSF), which allows us to evaluate the geometrical quality of a detector (or camera). The LSF measures the ability of the detector to reproduce correctly the spatial frequencies which constitute the observed target. In fact, the LSF specifies the spatial resolution of the detector; this figure of merit determines how a sine pattern of given spatial frequency will be reproduced. Knowing the LSF, the raw image could be restored by dividing (if not zero) its Fourier transform by the detector's LSF expressed in the frequency domain:

$$\text{recorded image} \sim \text{LSF} * \text{original image} \qquad (4.7)$$

where $*$ is the convolution operator. Obviously, besides establishing the LSF (see Ryu, 1991), the main problem in using this relationship comes from the high fluctuations obtained when division occurs with small values.

   For infrared systems, it is important to know the minimum spatial resolution that yields to a signal strong enough to obtain a repetitive measurement. In this respect, the slit response function (SRF) can be measured. Holmsten (1986) propose a simple method to derive the SRF (Figure 4.2). A slit of variable width w is placed at a distance R from the detector. A hotter that ambient background is present behind the slit. The shape of the signal along a line is recorded. A very wide slit will give a profile with a plateau. The amplitude A of the plateau corresponds to a modulation m of 100%. This is the maximum contrast case from all white (hot) to all black (cold) when the infrared image is

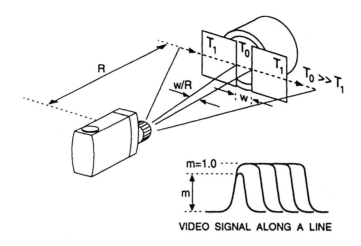

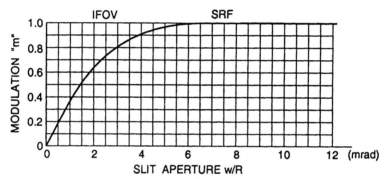

FIGURE 4.2 Simple procedure to obtain the SRF (slit response function curve) experimentally. (From Maldague, 1994b, with permission from Gordon and Breach Publishers. Data from Holmsten, 1986.)

observed on a black-and-white monitor. The slit width w is then reduced and the new peak height normalized with A gives the corresponding m factor. This technique allows us to plot the SRF curve (Figure 4.3). At this point it is interesting to give an example.

Example 4.2 Suppose than an infrared camera that has the SRF curve of Figure 4.3 has a 12° aperture with 105 pixels per line. It is placed 1000 mm from a screen. Establish the spatial resolution for the camera.

SOLUTION   In these conditions, the field of view is 210×210 mm (Figure 4.4). The SRF of Figure 4.3 gives a spatial resolution of 6 mrad at $m = 100\%$ and 2 mrad at $m = 50\%$. If we select $m = 100\%$, that is, we stipulate that a perfect contrast from all white (hot, 255) to all black (cold, 0) is necessary to

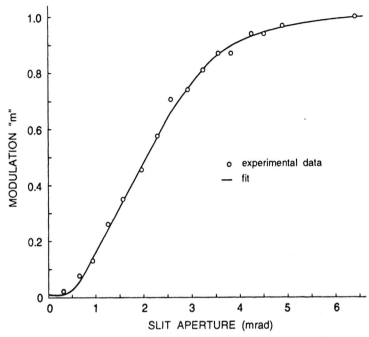

FIGURE 4.3 Experimental SRF curve obtained for an AGEMA 782SW using the procedure of Figure 4.2. (From Maldague, 1994b, with permission from Gordon and Breach Publishers.)

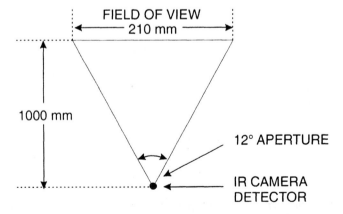

FIGURE 4.4 How to use the SRF curve to evaluate the spatial resolution of an infrared camera. (From Maldague, 1994b, with permission from Gordon and Breach Publishers.)

distinguish between two objects in the image, these objects have to be separated by 1000 tg (6 mrad) = 6 mm. However, if we select a value of $m = 50\%$ (this is generally the case), we consider that a 50% contrast will suffice to distinguish both objects, they will have to be separated by 1000 tg (2 mrad) = 2 mm. This is the *spatial resolution* at m = 50%. Those values can be related to the *apparent resolution*, which for our example is 210 mm/105 pixels = 2 mm/pixel. In this particular example the *apparent resolution* corresponds to the *spatial resolution*; obviously, this is not always the case. Off course, the more pixels available (e.g., in a bidimensional array), the more likely the *spatial resolution* will be of acceptable value. We may also point out that the *apparent resolution* may be different along rows and columns if the number of elements in the array is different along both axes. Finally, it is of interest to note that knowledge of the SRF curve can be used to restore exact values of narrow objects (Beaudoin and Bissieux, 1994, Chap. 2).

4.1.8 Signal Degradation

Before closing this brief overview of detector characteristics, it is interesting to look at signals. As stated previously, the electric signal produced by the detector should reproduce the photonic signal with fidelity. Unfortunately, this is not always the case. In addition to the detector characteristics discussed above (R_V, R_i, F_c, NEP, D, D^*, MRTD, LSF, SRF), there are other phenomena that can degrade the electric signal (mainly drifts, aging, and noise). Drifts cause the signal to fluctuate at a very low frequency around an average value. It depends on external factors such as system temperature and flow variations. Aging of the detector causes fatigue, memory, and training. This phenomenon is particularly evident on pyroelectric surveillance tube cameras, which continuously observe a scene with a nonvarying background. After some thousands hours of operation, the background is essentially frozen. Noise is another important degrading feature; it depends on the granularity essence of photons and atoms and of thermal agitation of particles. Figure 4.5 depicts those degrading processes for square optical incident signals.

4.2 NOISE

At the fundamental limit of all system, there is noise. This subject is so vast that a single book devoted solely to this subject would probably be insufficient to cover the matter completely. A book by Cielo (1988, Chapter 3 in particular) provides an excellent and deep review of this matter. We restrict ourselves to *electronic noise*, which is likely to occur in any thermal imaging system or infrared detector. In addition to the electronic noise, *optical noise* (i.e., random fluctuations of the incident radiation), heating (or illuminating) noise (i.e., noise present in the stimulating devices used in active thermography; Chapter 8), and

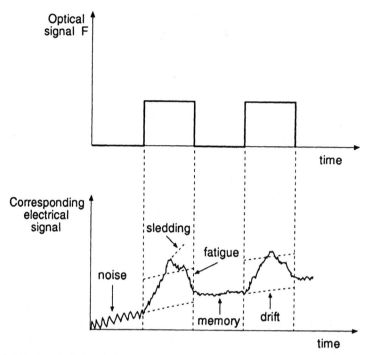

FIGURE 4.5 Typical signal degradation processes. (From Maldague, 1994b, with permission from Gordon and Breach Publishers.)

environmental noise (i.e., electromagnetic interference induced in the detector, wiring, and amplifiers by power lines and by radio and TV broadcast, and by heavy machinery) contribute to degrading the signal. The principal types of noise are reviewed briefly below. Another particularly annoying form of noise in active thermography is *structural noise*, caused by random variability of thermophysical properties of the material under observation (such as thermal conductivity) and which might induce spurious responses from the sample following the thermal stimulation, due to the heat transfer process (Chapter 3). These variations cause flutuations of recorded thermal signals which are not related to the presence of a subsurface defect. For instance, this can be caused by a different epoxy/fiber ratio in a carbon fiber–reinforced plastic (CFRP) specimen, as illustrated in Figure 4.6.

4.2.1 Shot Noise

Shot noise is caused by the arrival of random and discrete photons in the incident radiation. Shot noise is the ultimate limit for radiation detection. When illumination is low or missing entirely, the noise is mainly thermal noise; it is thus possible to determine the prime noise component simply by reducing the incoming radiation that causes shot noise.

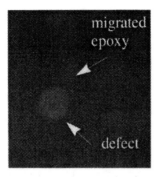

FIGURE 4.6   Structural noise: phase image 0.04 Hz from a graphite-epoxy specimen with an artificial defect located about 2 mm beneath the surface. During the fabrication process it is believed epoxy migrated from defect location (arrow), thus causing a spurious signal. (From Maldague et al., 1997.)

4.2.2   Thermal or Johnson Noise

Thermal or Johnson noise is caused by the random motion of electrons in resistive materials. These fluctuations are given by [Cielo, 1988, p. 122, eq. (3.3)]

$$i_T = \sqrt{\frac{4KTB}{R}} \tag{4.8}$$

where $K = 1.38 \times 10^{-23}$ J K^{-1} is the Boltzmann constant (Table 2.1), T the detector temperature (kelvin), R the load resistance, and B the detector bandwidth. From this equation we see that thermal noise can be reduced by cooling the detector. This is why some detectors must be cooled to achieve acceptable performance (Section 4.6).

4.2.3   Flicker or $1/f$ Noise

Flicker or $1/f$ *noise* is different from shot noise or Johnson noise [which is white noise (frequency independent)] in the sense that it depends on the observation frequency as $1/f^n$, where f is the signal frequency and n is typically between 0.9 and 1.35. Consequently, by reducing the observation frequency to the signal bandwidth, the $1/f$ noise will be reduced. Flicker noise is caused by trapping charge carriers near the sensing surface of semiconductor detectors (Cielo, 1988, p. 122).

4.3   THERMAL DETECTORS

In thermal detectors, the incident radiation heats the surface, and the heating affects a property of the heated material such as the electrical conductivity, which in turn translates into variation of the signal output. One specific characteristic of thermal detectors is that the response is independent of the wavelength. Thus, if a specific wavelength band is desired for a given application, an interference filter rejecting the unwanted radiations will have to be placed in front of the detector to restrict the response.

The limit of sensitivity of thermal detectors is given by the effective conductivity G_R, which is given by (Ballingall, 1990)

$$G_R = 4\sigma T^3 A \tag{4.9}$$

where $\sigma = 5.67 \times 10^{-12}$ W cm^{-2} K^{-4} is the Stefan–Boltzmann constant (Table 2.1), T is the detector temperature, and A is the detector-sensitive area. The detectivity limit is thus

$$D^* = \sqrt{\frac{A}{4kT^2 G}} \tag{4.10}$$

where k (W m^{-1} K^{-1}) is the thermal conductivity and G is the limit of G_R.

Theoretically, a maximum of $D^* = 1.8 \times 10^{10}$ cm Hz$^{1/2}$ W^{-1} can be obtained with a thermal detector; in practice, other factors limit the detector performance. Modern technology currently produces $D^* \sim 10^7$ to 10^9 cm Hz$^{1/2}$ W^{-1}.

4.3.1   Bolometers

For *bolometers*, such as thermistors, the physical property affected is the electrical conductivity. Consequently, a current must cross the element for signal generation. For example, a bridge circuit can be used for signal pickup. It should, however, be noted that this current, having its own fluctuations, adds to the detector noise. One of the main drawbacks of bolometers is their slow response time ($\tau \sim 1$ to 100 ms), which restricts their use to slowly varying processes. The figure of merit of bolometers is the resistive temperature coefficient α_{tc}, defined as (Tissot et al., 1998)

$$\alpha_{tc} = \frac{1}{R_b}\frac{dR_b}{dT} \tag{4.11}$$

The value of this coefficient is either positive or negative, depending on the conduction mechanism taking place within the material of interest. Typical values are 0.1% K^{-1} for metal, 1 to 10% K^{-1} for semiconductors, and higher values for superconductors (Tissot et al., 1998). Recently, arrays of microbolometers have opened a whole new range of infrared imagers (Section 4.5).

As a thermal detector, a blackened platinum stripe is one of the most accurate means to measure temperature. The blackening of the surface increases the emissivity and allows a greater conversion efficiency. Such blackening can be made with soot from a candle flame. Emissivity with values ranging from 0 (for

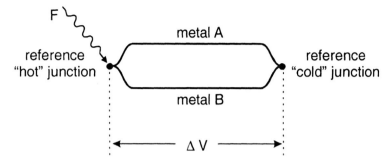

FIGURE 4.7 Principle of operation of a thermocouple. (From Maldague, 1994b, with permission from Gordon and Breach Publishers.)

perfect mirrors) to 1 (for *perfect* emitters, also called *blackbodies*) is a surface property relating to the energy emission/absorption (Section 2.9). A blackbody emits (or absorbs) ideally; a perfect mirror reflects back the incoming radiation (Section 2.2). This is the reason why heat sinks (Section 3.2.6) are black anodized (the high emissivity enabling better heat dissipation by radiation).

4.3.2 Thermopiles

Thermopiles and *thermocouples* generate a voltage difference through the thermoelectrical Thompson effect. Thermocouples are made of two different metals, A and B, connected by two junctions, a reference cold junction and a measuring hot junction (Figure 4.7). The voltage difference ΔV between these junctions is a function of the temperature difference between the two junctions leading to possible temperature measurement.

Thermopiles are constructed from many thermocouples connected serially or in parallel. The serial connection increases both the signal amplitude and the impedance, thus facilitating connection with preamplifiers. Sensitive surface areas are around 50 mm^2 and the time constant τ is a few hundreds of milliseconds. The more commonly used metals for thermocouple fabrication are show in Table 4.1 (among them, types J and K are probably the more common). For example, the company Omega* offers a wide range of thermocouples packaged for different for diverse applications. Due to their low cost, thermocouples are deployed in many NDE contexts. The principal problems are the need of a contact for measurement and the limited number of measurement points. Moreover, messy wire setups are increasing rapidly with the number of involved thermocouples. Powerful computer interfacing is available from various vendors, Omega and Labview[†] among others.

---

* www.omega.com.
[†] www.natinst.com/labview.

TABLE 4.1 Common Thermocouple Configurations

Composition	Type	Operating Temperature (°C)
Fe and Cu–Ni	J	0 to 750
Ni–Cr and Ni–Al	K	−200 to 1250
Ni–Cr and Cu–Ni	E	−200 to 900
Cu and Cu–Ni	T	−200 to 350
Pt–10% Rh and Pt	S	0 to 1450
Pt–13% Rh and Pt	R	0 to 1450
Pt–30% Rh and Pt–6% Rh	B	0 to 1700

4.3.3 Pneumatic Detectors

In *pneumatic detectors* the signal is given by measuring pressure variations of a given volume of gas. These detectors are constituted of two chambers, C_A and C_B, separated by a membrane M. The incident radiation induces heating of element A, thus causing gas expansion in chamber C_A and dilatation of membrane M. The displacement of M is recorded by an optical (e.g., interferometer) or a capacitive scheme. Such devices are used primarily in research applications.

4.3.4 Pyroelectric Detectors

Pyroelectricity is defined as the property of certain crystals to produce a state of electric polarity by a change in temperature (Gränicher, 1984). A broad class of thermal detectors are *pyroelectric* detectors, for which below the Curie temperature (i.e., the temperature below which there is a spontaneous magnetization in the absence of an externally applied magnetic field), electric charges are generated by incident radiation absorption (heating): A change in detector temperature generates a transient change in the surface charges, thus causing a transient current available for pickup by the readout unit. Pyroelectric detectors are sensitive only to temperature variations and thus need a chopper modulating the incoming radiation in order to maintain a signal output, unless the system is panned over the scene. This added complexity is a severe drawback.

The figure of merit of pyroelectric detectors is the pyroelectric coefficient p, defined as the slope of the material polarization P versus temperature T at the operating temperature (Tissot et al., 1998):

$$p = \frac{dP}{dT} \tag{4.12}$$

This effect can be improved with the application of an electrical field bias. This is referred to as the *field-enhanced pyroelectric effect* or *ferroelectric bolometer effect.* In such a case, *p* comes to (Tissot et al., 1998)

$$p = p_0 + \int_0^E \frac{\partial \varepsilon_{dp}}{\partial T} \, dE \qquad (4.13)$$

where p_0 is the pyroelectric coefficient without bias, ε_{dp} the dielectric permittivity, and E the applied electrical field. Pyroelectric elements can be used as point or image detectors (Section 4.5).

4.3.5 Liquid Crystals

Liquid crystals are cholesterol esters which, under a temperature effect, change orientation and reflect colored light from red to violet when illuminated with white light. Depending on their composition, 0.01°C resolution can be obtained. As advantages, they are not expensive, they are sensitive to slight thermal variations, and they make surface measurement possible (e.g., if deployed in paint). The main problems are the contact measurement, the restricted sensitivity span (e.g., 5 to 10°C), the necessity to prepare the surface before application as paint or encapsulated in small balls or film, and the necessity to clean up after the measurement to remove the crystals. Paint coverage is on the order of 1 L per 6 m².

If the required painting step is acceptable, *liquid crystal paints and derivatives* find many uses in TNDT. For example, for refractory vessels, irreversible temperature indicator strips act as watchdogs to detect a rise in surface temperature, where the refractory lining has broken down. In the case of electronic equipment, the same kind of label is fixed to a component and changes color if the rated temperature is exceeded. Another possible application concerns testing of fabrics in the textile industry: A textile strip is attached to the fabric and sent through an oven (a conventional thermocouple is obviously not suitable here). Liquid-crystal paints also find use in aerodynamic studies. For example, a model is painted and installed in a wind tunnel. During airflow experiments, a visible color camera is pointed to the suitably lighted model in order to follow temperature evolution due to a change in reflectance of the applied liquid crystal paint. One advantage of such a practice is that a less expensive color camera can be used instead of an infrared camera; moreover, the temperature ranges are wide (e.g., 160 to 1270°C for Thermax Pirthe). These products come as temperature labels, temperature indicator strips, and thermal paints. For example, the company Thermographic Measurements Inc. of Anaheim Hills, California* has both an irreversible product line, which changes color or state permanently when exposed to temperatures above their rating (available range −17 to 1270°C), and an reversible product line, which changes color continuously and has the temperature range −300 to 150°C.

---

* www.thermax.com/thermax.htm.

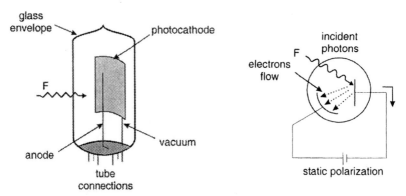

FIGURE 4.8   Schematic diagram of a photoemissive detector. (From Maldague, 1994b, with permission from Gordon and Breach Publishers.)

4.4   PHOTONIC DETECTORS

In those detectors the signal is obtained by direct measurement of the excitation generated by the incident photons. Heating the sensitive surface is unnecessary. Photonics detectors are of two types: photoemissive and quantum (photoelectric, photovoltaic, or photoconducting).

4.4.1   Photoemissive Photonic Detectors

For *photoemissive photonic detectors*, the signal observed is constituted by the measurement of the electron flow (i.e., measurement of a current i) pulled away from the photocathode under the effect of both the incident photons and static polarization (Figure 4.8). The spectral sensitivity depends on the properties of both the material used for the photocathode and the outer envelope infrared transmission (or transmittance, Section 2.9). Typical values span from UV to near infrared: 0.2 to 1 μm. Solid-state photoemissive detectors are also possible (Section 4.5).

In *photomultiplier tubes* (also called *image intensifier tubes*) electrons are accelerated and multiplied by secondary emission using internal plates called *dynodes*. Multiplication factors of 10^5 to 10^7 can be obtained for 10-stage tubes. Those detectors are point detectors, although image converter tubes are made as well, with such uses, for instance, as night-vision image intensifier for military and law enforcement agencies (in some applications, dynodes are now replaced by photodiode or avalanche photodiodes).

The first image intensifier (I^2) tubes were built by RCA in the 1930s (with deployment as night vision equipment in the early 1950s). These tubes, referred to as *generation zero*, incorporated a P-20 phosphor screen still in use today (the human eye is most sensitive in the yellow-green region; Figure 2.6). The first tube generation was replaced in the mid-1960s with the introduction of electrostatically focused cascaded I^2 tubes, which featured insensitivity to power supply variations and automatic brightness control (ABC). Microchannel plates (MCPs) began to appear in the late 1960s; this is a way to multiply the internal current in the tubes. MCP advances correspond to the second generation of I^2 tubes. The third generation incorporated more efficient photocathodes and improvements in the noise, gain, and life characteristics, while being of reduced physical size and weight and thus having more applicability for the military, so that in 1976, a U.S. program was launched to develop high-performance goggles based on the third generation of I^2 tubes. Such a system, the Aviator's Night Vision Imaging System (ANVIS), is still in use (Kraus, 1997). More recently, negative-affinity photocathodes with improved sensitivity and extended spectral range have started to appear (GaAsP, GaAs, GaN) while metal-channel dynodes have made it possible to develop photomulipliers of small size: 50×50 mm, only 10 mm thick (*Photonics Spectra*, 2000a).

4.4.2 Quantum Detectors

Quantum detectors are solid-state detectors in which photon interactions either change conductivity (photoconductor) or generate voltage (photoelectric, also called photovoltaic). Since no heating phenomenon is needed as for a thermal detector, the response time is short while the solid-state structure made these detectors compact, reliable, and robust. Consequently, they are quite popular.

For photoconductive detectors an external current is necessary to measure conductivity change while photoelectric (photovoltaic) detectors acts as a power generator, supplying a signal without need for polarization. Since biasing current is needed for photoconductor detectors, high-charge-capacity alkaline batteries can be used to induce minimum noise and ripple. This is essential to achieve stable results. In this respect, since photovoltaic detectors supply a signal by themselves, they are much more attractive than photoconductor detectors, requiring less complex readout circuits. Common materials used in photoelectric (photodiodes and phototransistors) are Si, InAs, InSb, and HgCdTe [also known as CMT (cadmium mercury telluride)].

Figure 4.9 presents the spectral detectivity curves for the most common detectors. On these curves, a sharp cutoff response is observed at longer wavelengths. This can be explained because of the presence of different energy levels within the atomic structure. At low energy (valence level) the electrons stay close to the nucleus. If enough energy is supplied to electrons, they cross the "forbidden band," which frees them from nucleus attraction and enables them to participate in an electric current. At long wavelengths the photons transmit less energy to electrons [eq. (2.4)]. Consequently, electrons cannot cross the forbidden band and thus stay in the valence band, this off course causes a sharp cutoff in the spectral detectivity curves. At small wavelengths, photons have more and more energy; they penetrate deeper in the substrate, passing through sensitive areas of the semiconductor without interacting. This causes a gradual loss of detectivity, which is also observed in Figure 4.9. Imaging detectors are discussed in the next section.

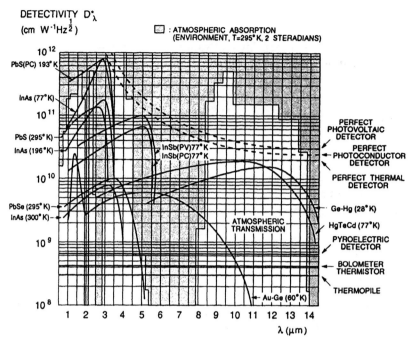

FIGURE 4.9 Spectral detectivity curves for infrared detectors; atmospheric absorption is also indicated. (From Maldague, 1994b, with permission from Gordon and Breach Publishers. Data from Gaussorgues, 1989.)

4.5 INFRARED IMAGING DEVICES

We first cover detectors that lead, by themselves and without mechanical scanning operation, to imaging applications, that is, to the production of an array of points (also called *pixels*, from *picture elements*) in either one dimension (a line array) or in two dimensions (such as for TV image generation). Although infrared films (such as liquid crystals, Section 4.3.5) made from specific dyes sensitive in the range 1 to 3 μm are available, they are not really of practical use in NDT, where real-time imaging is preferred.

In imaging applications the output of the detector device produces a one- or two-dimensional image. Two main image-forming processes exist: direct image formation (with either a detector array, pyroelectric detector, or infrared film) or using a single detector associated with electromechanical scanning of the scene. The latter principle, used in infrared radiometers, is discussed in Section 4.5.3.

4.5.1 Pyroelectric Detectors

Back from the 1960s, the first successful infrared commercial products based on pyroelectric detectors (vidicon tubes and single pyroelectric elements) have been

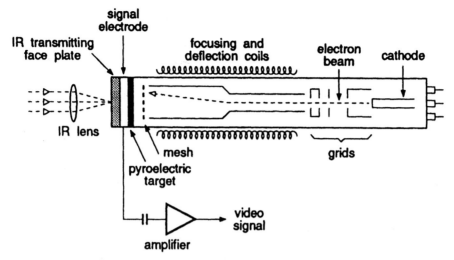

FIGURE 4.10 Schematic diagram of a pyroelectric vidicon tube. (From Maldague, 1994b, with permission from Gordon and Breach Publishers.)

produced (Goss, 1987). The major advantage of this kind of thermal detector (Section 4.3.4) is the ambient or close-to-ambient temperature mode of operation. Pyroelectric tubes are thermal detectors where the signal is proportional to the absorbed energy. For certain ferroelectric crystal materials such as triglycine sulfate (TGS) that have a high pyroelectric coefficient p, heating or cooling of the crystal slice creates an accumulation of charges. This is the pyroelectric effect, which results in a change of polarization.

Pyroelectric tubes are similar to standard vidicon TV camera except for the face plate and target material (Figure 4.10). As the infrared image impinges on the pyroelectric target, a temperature distribution and, in turn, a charge distribution appear on the pyroelectric material. The electron beam scans this material, and two things may happen. If no radiation was absorbed at the scanned spot, there is no charge (no polarization), and the electron beam is redirected toward the mesh. On the contrary, if radiation was absorbed, charges accumulate, allowing the electron beam to reach the signal electrode, thus causing a current to appear. As the electron beam is scanned over the surface, the video signal appears at the tube output.

As charges are released only when the temperature of the pyroelectric material changes, such pyroelectric tubes cannot produce images of a static scene. If only a part of the scene is mobile, a flying aircraft, for example, only the mobile part will be observed. This effect can be used at profit in certain applications, such as in fire detection, where only the fast temperature changes of flames are seen. Another use may be the NDE inspection of parts moving on a conveyor belt. If this effect is undesired, the camera can be panned or the radiation can be modulated using a mechanical chopper. Camera panning has the advantage of simplicity and reliability since, besides the panning mechanism, no moving part is involved. In the chopped (shuttered) mode, a rotating chopper mounted in front of the tube shuts off the scene during one field, causing target cooling; a negative image is then produced. When the target is exposed again, the radiation distribution heats the pyroelectric material and the positive image is reformed. Although a signal is produced continuously, it cannot be displayed on this form since it is composed of alternating fields, causing heavy flickering of the positive (useful) field. Special electronics allow us to overcome this problem by displaying the positive and reverted negative image.

Pyroelectric material absorbs radiation evenly over a broadband wavelength spectrum; consequently, the tube sensitivity is determined mainly by the input window material. Germanium (3 to 5 and 7 to 20 μm) and zinc selenide (ZnSe) (0.6 to 20 μm) are the most commonly used materials for a tube window.

Although the dynamic range is small (30 dB), the linearity is poor (less effective beam readout toward target edge) and the life span is restricted (typically, 10,000 hours: 13 months at 24 h/day). The fact that pyroelectric materials are piezoelectric requires careful mechanical design to avoid microphonic noise, which could result from excessive target vibration; pyroelectric tubes are attractive because they do not need cooling, they do not consume high power, they do not cost much, and they are not restricted to a fixed image format, as in the case of an array of detectors (changing the electron beam scanning rate suffices to obtain another video format).

Recently, new developments in pyroelectric detectors have made it possible to produce pyroelectric solid-state arrays (Tissot et al., 1998). These two-dimensional arrays (32 × 32, 1987; 100 × 100, 1990; 320 × 240, 2000) are made of ferroelectric ceramic thin film (e.g., lead scandium tantalate). They offer high detectivity and produce slow scan images (1 Hz) without the necessity of cooling. Applications include low-cost consumer products for the detection of flames, detection of heat emitted by warm objects such as people (intruder detection), medical thermography, and transport monitoring. Among the advantages of these monolithic chips of pyroelectric ceramic are the ruggedness (with respect to the fragile glass envelope of pyroelectric tubes), ease of operation (no need for high voltage), and little encumbrance (little wafer). For these pyroelectric arrays, element-to-element responsivity variations are typically less than 10%. Absolute temperature measurement thus requires individual calibration of each element, which considerably degrades the image acquisition time.

Ferroelectric materials can be used in two modes: pyroelectric mode or dielectric mode. The more conventional mode is pyroelectric, for which the signal origin is the change of polarization due to the heating of the detector by the incident radiation, as seen above. In the dielectric mode, a change of permittivity with temperature is used: This change of permittivity is sensed as a change in voltage across the detector after a proper bias has been applied. The drawback of the dielectric mode of operation is the necessity to stabilize the detector temperature. Improved spatial resolution can be obtained for pyroelectric imaging detectors by reticulating the detector surface to reduce the sideways heat spread. This can be done by ion-beam machining a pattern of narrow groves at 30 to 40 μm depth all over the pyroelectric target.

In qualitative infrared NDE, pyroelectric-based infrared cameras can be an excellent choice. Quantitative measurement is also possible. In some configurations the rotating chopper, which has a high-emissivity surface, provides a known temperature reference from which scene temperature can be determined. Measurement of the chopper temperature is carried out indirectly by monitoring the signal from a single-element pyroelectric detector, signal caused by a small blackbody target viewed through the chopper. The blackbody temperature is adjusted through a Peltier device so that no temperature differential is observable between the chopper and the blackbody. The blackbody temperature is then measured using a thermistor.

4.5.2 Infrared Focal Plane Arrays

In the 1970s a new kind of imaging device started to appear and revolutionize the infrared community: Large-dimensional infrared arrays simplified infrared camera construction. Using this technology, all that is needed to build an infrared camera is the optics, the focal plane array (FPA), the associated electronics, and for some detector technologies, a cooling unit. Similar to conventional video CCDs, these chips do not require any electromechanical scanning mechanism (no moving parts) for image forming and are of less encumbrance and fragility than pyroelectric tubes. Video signals are obtained directly by on-chip electronics drive. A variety of technologies have emerged. Thermal-based pyroelectric and dielectric arrays were reviewed in the preceding section. In this section we review Schottky barrier, superlattice, intrinsic, and Z-plane technology, and microbolometer arrays.

Although the price of such detectors, especially those based on well-known silicon technology, is falling to a few thousand dollars (in 2000), prices are unlikely to reach those of conventional video CCDs (down to a few hundred dollars in 2000) because of the restricted applications (military, law enforcement, NDE) and the relative exclusion of a mass production consumer market (with applications such as the video camcorder).

Schottky Barrier Detectors. This type of detector was first proposed in 1973 by Shepherd and Yang. Since then, large arrays (512×512) have been introduced commercially by many companies. Infrared cameras are also fabricated as large one-dimensional arrays such as 1024×1 (Marche, 1990). In this design, horizontal scanning is done electronically in the detector plane, while vertical scanning is achieved either by optomechanical means, camera panning, or scene displacement (parts moving on a conveyor belt, air-lifted operation).

The most common type are platinum silicide (PtSi) detectors operating in the photoemissive mode (Section 4.4) and the band 3 to 5 μm (cutoff wavelength: 5.6 μm); GaSi is also available (in the bands at 8 to 14 and 8 to 16 μm). PtSi detectors are often fabricated with an aluminum mirror over the sensitive area while they are illuminated from the back side. The mirror enhances PtSi responsivity but restricts the response to wavelengths greater than 1 μm because of the absorption by the silicon. With respect to intrinsic photon detectors such as InSb (next section), the quantum efficiency of PtSi is rather small (10% vs. 85% for intrinsic photon detectors).

The design is generally similar to video CCDs for both storage and readout circuits. The radiation induces charges that are stored in a capacitorlike insulating layer. Charges are then transferred to the neighbor element under the effect of an electric field. Stored charges are next transferred in this fashion element by element up to the array output where the signal is available. A multiphase (two or three phases) clock performs such a step transfer process. The video signal is generated by sequentially scanning all the rows in the array and by multiplexing each row. In large arrays (512×512), for a compatible TV rate signal, a clock rate greater than 8 MHz is required to read all the detectors during one frame. The fill factor is greatly improved (up to 90%) if the readout circuits are located under the detector array.

Charge injection devices (CIDs) are also used in FPAs. CID devices are similar to CCDs, but generation of the video signal is different. In CID detectors, the video signal is constituted by the substrate current, caused by charge injection (rather than charge transfer as in CCDs). This current is proportional to the photons received. CID detectors are less prone to blooming; moreover, during device reading (after the accumulation interval) charges are less affected by the radiation, and because of the addressing mode, reading of individual cells is possible.

PtSi is the more mature Schottky barrier detector; its main drawback is the necessity of cooling at 77 K (Section 4.6). Other metal silicide arrays have been made to operate at longer wavelength; among them, IrSi, NiSi, and $CoSi_2$ are suitable for imaging applications. Iridium silicide (IrSi) has been reported with a cutoff wavelength of 7.3 μm at 62 K and 9.6 μm at 40 K, 10.7 μm if the detector is biased. IrSi arrays are more difficult to fabricate since impurities have a strong influence on performance. Cooling below 60 K is also a problem for silicon technology, since charge storage is difficult at low temperatures.

Superlattices. In superlattice detector arrays, alternating layers of different semiconductors of different thickness allow us to tune the wavelength of radiation that will be absorbed: Photoconduction occurs in a narrow range of wavelength. One of the most interesting technologies is GaAs/GaAlAs, first proposed by Levine et al. in 1987. Typical cutoff frequencies are between 6 and 11 μm, with detectivities $D^* \sim 10^{10}$ to 10^{11} cm $Hz^{1/2}$ W^{-1} with proper cooling at 50 to 70 K. Other compositions have been reported, such as InAsSb–InSb and InAs–GaInSb. The potential of this technology will evolve notably in the future, especially with the possibility of growing directly on the silicon the detecting superlattice layers that contain readout circuits. InGaAs detectors operate in the band 0.8 to 2.6 μm with important applications in spectroscopy for real-time determination of chemical composition, surveillance, and fiber optic telecommunications (particularly at 1.3 and 1.55 μm). One of the advantages of InGaAs is the room-temperature operation (for intrinsic detectors).

Intrinsic Photon Detectors. Those detectors are, in fact, arrays of photoconductive or photoelectric (photovoltaic) detectors which have been reviewed previously (Section 4.4). Photoelectric (photovoltaic) detectors are more useful since they generate charges spontaneously under illumination by incident radiation. Thermal charge generations due to impurities in the material degrade the signal, mainly because of the Shockley–Reed process and the Auger process (charges generated through impact ionization by carriers whose energy exceeds the forbidden bandgap). Refinement in fabrication processes should improve these aspects.

As mentioned in the case of superlattice detector, hybrid technology where detector layers are manufactured separately and later fused or glued on silicon readout circuits is attractive because the silicon process is a well-established fabricating process. Such hybrid arrays are fabricated with HgCdTe detectors of size 128 × 128 and up for the band at 8 to 12 μm and 256 × 256 and up for the band at 3 to 5 μm. An alternative fabricating technique is to grow HgCdTe cells on GaAs buffers themselves grown on silicon substrate containing bipolar preamplifier transistors and readout circuits. Advantages of this monolithic configuration include improved uniformity, reduced $1/f$ noise, higher operating temperature, and higher thermal insulation, which is essential to obtain high electrooptical performances (Tissot et al., 1998). Monolithic FPAs where HgCdTe diodes are grown directly on Si could also yield to larger arrays. Photovoltaic InSb hybrid FPAs are fabricated as well. For example, the Raytheon[*] radiance high-speed infrared camera is designed around a 256 × 256 InSb array with a high frame rate: 140 per second at 256 × 256 up to 1800 in a 64 × 64 subwindow; other array formats are also available (512 × 512, 640 × 512). Another product, such as the SBF-125 available from Santa Barbara[†] (Lockheed Martin), has a frame rate of 400 Hz (at 320 × 256) up to 1000 Hz (in a subwindow of 128 × 128 pixels).

HgCdTe, InSb, PbSnTe, and InGaAs are the more common types of intrinsic detectors, with newer compositions such as HgMnTe and HgZnTe developing. HgCdTe and its extension, *SPRITE* (signal processing in the element), remain among the most detectors. SPRITE was developed in the United Kingdom at the Royal Signals and Radar Establishment by Ted Elliot. SPRITE is made of a strip of HgCdTe mounted on a substrate such as sapphire and is cooled at cryogenic temperatures (Figure 4.11). The long axis is in the direction of scan and the strip is biased so that the carrier drift velocity matches the speed at which the image is scanned. As a point in the image moves along the strip, the charges carriers it induces move with it and the accumulated charges are then read out at the end of the strip. Consequently, the main

[*] www.amber-infrared.com.
[†] www.sbfp.com.

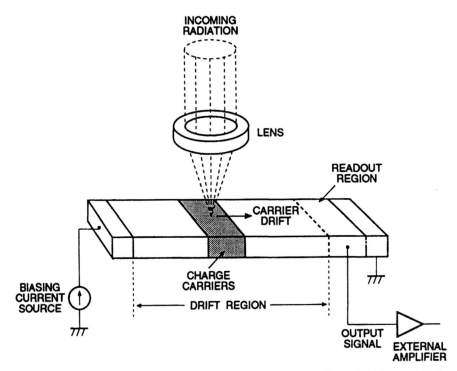

FIGURE 4.11   Principle of operation of a SPRITE detector. (From Maldague, 1994b, with permission from Gordon and Breach Publishers.)

advantage of SPRITE is that the signal is integrated on the focal plane by the detector itself, thus reducing the noise level and the timing considerations. If SPRITE is used and then read out serially, the same scanning rate as that with a single detector element is maintained. However, if several SPRITEs are stacked together in parallel, the scan speed is reduced since many rows can be integrated simultaneously.

Z-Plane Technology. This emerging technology opens new ways to improve infrared camera systems by having the possibility to add processing to the detection function of the array. Silicon circuits are fabricated on a tiny ceramic board (of typically 100 μm thickness) with a detector attached on one edge. These boards are then stacked to form complete arrays. Among possible added processing functions are for instance convolution operations, edge extraction, blobs detection.

Microbolometer Arrays. In the late 1970s and early 1980s, with the broader availability of large integrated-circuit (IC) fabrication technology, it was possible to build large uncooled bidimensional arrays based on pyroelectric effect thermopiles and microbolometers. This research was initially funded by the U.S. Department of Defense (DOD) and pursued by companies such as Honeywell and Texas Instruments. Other developments were also undertaken in other countries: NEC and Mitsubishi in Japan, GEC-Marconi in the United Kingdom, Australia, Defence Science and Technology Office in the National Optic Institute (NOI) and DOD in Canada, and the French DOD. This technology stayed classified up to the late 1990s, when commercial exploitation started (Kruse, 1995). The relatively low cost of these arrays with respect to traditional cooled arrays opened many new fields of activity, such as in law enforcement agencies and fire departments (McCarthy, 1998; Kaplan, 1998c) for night vision in cars (such as the night vision option available in General Motors' Cadillac Deville at a cost of about $2000, which includes a 320 × 240 pixel sensor with a liquid crytal head-up display located on the dashboard), and in the military (to replace image intensification tubes; Section 4.4.1).

In fact, analysts forecast that the microbolometer market will explode in the years to come (*Photonics Spectra*, 1998, 2000b). Bolometers were reviewed in Section 4.3.1. With micromilling technology, it is now possible to constitute arrays with thousands of tiny bolometers called microbolometers, which are, in fact, thermal masses (also called *microbridges*) hung by low-thermal-conductivity arms supported on metal legs anchored in the silicon substrate, thus enabling a direct interface with readout electronics located under the array (Figure 4.12). The IR radiation heats the hanging mass, modifying its electrical conductivity so that a voltage or current variation appears at the device outputs (Section 4.3.1). These arrays are, for example, built using planar MOS or thin coating technologies adapted to micromilling. One common technology is the VO_x, with for example, products from Boeing, Lockheed Martin, and Hughes, among others (these companies possess the Honeywell license discussed above). Nowadays, ferroelectric microbolometer arrays have performances such as time constant τ on the order of a few milliseconds (enough for TV standards), $D^* \sim 10^7$ to 10^9 cm Hz$^{1/2}$ W^{-1}, NETD-50 mK, and a pixel size of 25 μm (for 320 × 240 array size).

In a given design optimized for a response in the band 8 to 14 μm (Tissot et al., 1998), sputtered titanium nitride on the masses made of a thin (0.1-μm) layer of doped amorphous silicon allows for partial infrared radiation absorption. At the final manufacturing stage, etching away of the sacrificial polyimide layer on which masses are built in the first place free them. A reflector (aluminum layer) is located under the hanging masses, reflecting back the transmitted infrared radiation to the microbolometer. Such an arrangement constitutes, in fact, a quarter-wavelength reflective cavity with a maximum spectral response at 10 μm obtained in the case of leg heights of 2.5 μm. In the design discussed, a linear response is obtained in the range ±5 V with a time constant of 4.2 ms, enough to accommodate TV frame rate at 25 Hz (European standard with 256 × 64 pixels).

Uniformity Correction. One problem with large arrays is the nonuniformity of response among detector cells caused by the fabrication process. For qualitative applications this may be of little importance; however, if quantitative data must

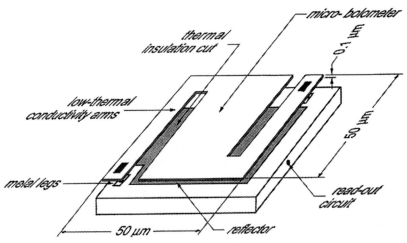

FIGURE 4.12 Picture and drawing of a microbolometer about 50 × 50 μm in size based on the VO_2 technology built by volume micromilling in silicon. (Picture from *NOI Bulletin*, **6**[2]:2, 1995.)

be extracted from the imaging array, some sort of uniformity correction has to be carried out. Although this calibration process can be done on-chip in the case of the Z-plane technology, it is generally executed off-chip. The array is exposed to a scene of uniform temperature and an image is recorded in computer memory. This process is repeated for different scene temperatures. Next, a calibration function that is either linear or of higher order for nonlinear response to the photon flux is computed pixel by pixel (Section 4.8). For linear functions, two coefficients are obtained for every pixel. This is the *two-point method*: gain and offset. Second-order polynomials require three parameters, and more coefficients are needed for multiple-point correction. In single-point compensation, an average of several images recorded over a uniform background with a defocused lens is subtracted (in real time) from live video to remove nonuniforminity. This process is, however, less efficient than other techniques, such as two- and multiple-point methods. During the normal operation of the array, these coefficients serve to correct images. With the present technology, PtSi arrays exhibit the most uniformity. Intrinsic photon detector, although more sensitive than the other types, are less uniform.

The nonuniform response is more easily handled in the case of single-detector-based infrared camera. In this case a more complete correction procedure can be carried out to correct vignetting and radiometric discrepancies (Section 4.8). Because of system instabilities in time, aging, unstable bias voltages, pixel nonlinearities, and $1/f$ noise, periodic recalibration is needed. For static scenes a quantitative measure of image drift can be, for instance, the time evolution of the spatial variance in the image. This could help us in deciding when to recalibrate. When applicable, noise reduction techniques such as running averages contribute to a reduction in the spatial noise level, although periodic recalibration is still required for accurate quantitative measurement.

Conclusions. As conducting remarks on this brief review of infrared detector arrays, we may summarize the various technology characteristics. PtSi is a mature technology; large two-dimensional arrays (512×512) are fabricated and complete infrared camera systems are commercially available. They operate in the band 3 to 5 µm and need cooling at 77 K. This may not necessarily be a problem since a compact Stirling engine cooling unit can be integrated with the detector array. Thermal detectors (such as microbolometer arrays) operate at room temperature, but their performance is not as good as that for quantum detectors. Superlattices and Z-plane technology are more recent and should improve substantially. An operating temperature close to 77 K in the long-wavelength band is possible. Intrinsic photonic detectors such as hybrid Insb and HgCdTe are now common, with operation in both atmospheric bands (3 to 5 and 8 to 12 µm) and with respectable size, although 77 K cooling is required.

4.5.3 Scanning Radiometers

The difference between such an instrument and an ordinary infrared camera is that for a radiometer, the infrared signal is temperature calibrated, thanks to the presence of internal temperature references seen by the detector element during the image formation process. This calibration signal allows us to recover the absolute temperature after proper processing. In such an instrument, the image is electromechanically scanned over the detector surface (single piece or SPRITE) by means of the synchronous rotation of mirrors or prisms (Figure 4.13).

To accommodate the standard video signal format of 30 frames per second (25 in Europe), a very high scanning rate is required. This imposes a wide bandwidth from the associated electronics for the noise level to be kept small. To overcome this problem, some manufacturers use slower scanning rates and have frames made of several fields with one field update at each scan (e.g., 4:1 field interlacing). As a result, the output obtained, comprising several fields, is

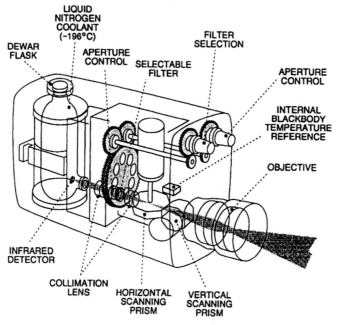

FIGURE 4.13   Internal view of an infrared radiometer. (From Maldague, 1993.)

not updated in real time (i.e., at 30 or 25 Hz). This may causes problems when fast thermal events must be observed since thermal information registered at different times is mixed together. One way to solve this problem is to split the video signal into its basic fields and process them knowing the time interval between them. However, this type of analysis requires careful manipulation of the signal.

4.6   COOLING

As pointed out in Section 4.5 and shown in Figure 4.9, superior detectivity can be achieved, especially for photonics detectors (Section 4.4), if cooling is used. Such cooling is needed to reduce the noise to an acceptable level, as seen in Section 4.5. There are different ways to cool detectors, mainly liquefied gas, cryogenic engine, gas expansion, and thermoelectric effect. Cryogenic cooling employs liquefied gas stored in a vacuum vessel called a *dewar* (after Sir James Dewar, 1842–1923, Scottish physicist; he liquefied hydrogen for the first time in 1892). Dewars are constituted of two-envelope walls with evacuated space maintained between them; moreover, to prevent heat wall losses the surfaces facing the vacuum are heat reflective (Figure 4.14). Metal dewars are unbreakable; however, the relative porosity of metal welds restricts the life span unless vessels are repumped regularly. Generally, manufacturers provide a valve for this purpose, while a cryopump or diffusion pump with vacuum rated

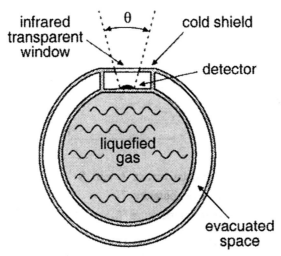

FIGURE 4.14   Schematic diagram of a Dewar. (From Maldague, 1994b, with permission from Gordon and Breach Publishers.)

in the range 5×10^{-6} torr is recommended for the pumping task. Evidence that dewar repumping is required are sweating or condensed water vapor on the outside of the dewar and a rapid boil-off rate, which has the symptom of rapidly escaping nitrogen gas from the fill port (*IRC-160 User Manual*, 1993). Glass dewars do not have this problem; they are, however, very fragile.

In a popular configuration, the detector is mounted directly on the cold surface with a cold shield and a infrared transparent window. Because some detectors (e.g., HgCdTe, Section 4.4) tend to sublimate when exposed to the vacuum, a protective coating such as zinc sulfide can be applied on the sensitive surface to expand the life span. The most commonly used and cheapest liquefied gas is liquid nitrogen at a temperature of −196°C, 77 K (some 70% of Earth's atmosphere is constituted by nitrogen). Liquid hydrogen (−259.2°C, 14 K) and liquid helium (−268.9°C, 4.1 K) are more exotic because of their price and knowing that a typical 1-L dewar will keep liquid gas for 3 to 4 h only, thus requiring regular refilling.

Joule–Thompson gas expansion is another way to cool detectors. In this case the quick expansion of high-pressure gas (such as nitrogen or argon) produces, after a few minutes of operation, droplets of liquid nitrogen or liquid argon (−187°C, 86 K, as a gas argon occurs free in the atmosphere to the extent of 0.935%) at the tip of the expansion nozzle (on which the detector is fastened). This mechanism permits greater autonomy than dewar operation; it is, however, noisy, while the gas tank may be cumbersome.

For applications where refilling is not practical, such as in remote areas or on the production line, cryostat using a closed Stirling cycle engine can be employed. This machine cools through repetitive compression and expansion cycles of gas by a piston: It compresses gas at a low temperature and allows it

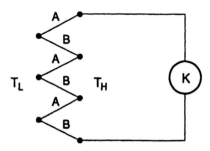

- Dissimilar metals A and B
- Low temperature T_L
- High temperature T_H

→ K = amp.meter (thermocouple, Seebeck effect)
→ K = battery (power generation, Thompson effect)
→ K = generator (cooling, Peltier effect)

FIGURE 4.15 Three processes of direct conversion of heat into electricity, or the reverse. (From Maldague, 1994b, with permission from Gordon and Breach Publishers.)

to expand at a high temperature. Because of the cycling operation, strictly speaking, the cooling is not constant, but temperature variations can be made small (or large) depending on the cycle characteristics. Although either rotary or linear motors can be used, a linear motor positioned at right angles to the detector plane causes less vibration in integral Stirling engines. Split-cycle machines are also available for remote and low-vibration operation. Typical input power involved is around 4 W for 150 mW of cooling for a small engine (0.5 kg), while a large engine (2 kg) will deliver 1 W of cooling for 40 W of input power. Many manufacturers offer IR camera with a Stirling cycle engine, such as among many others, Nikon.*

Another mode of cooling infrared detectors is using thermoelectric elements based on the Peltier effect. This phenomenon was discovered in 1834 by J. C. A. Peltier (French physicist, 1785–1845): At the junction of two dissimilar metals carrying a current, temperature rises or falls depending on the current direction. In Figure 4.15 we show that the Peltier effect is used for cooling, the Seebeck effect for temperature measurement with a thermocouple, and the Thompson effect for power generation. These three phenomena are, in fact, three applications of the same physical phenomenon of direct conversion of heat into electrical energy, or the reverse. Because of the low efficiency of the conversion process, Peltier elements draw high current and are generally stacked upon each other to achieve sufficient heat removal and temperature gradient (Figure 4.16). Peltier elements are, however, unattractive for temperatures below 200 K. For

---

*www.nikon.com.

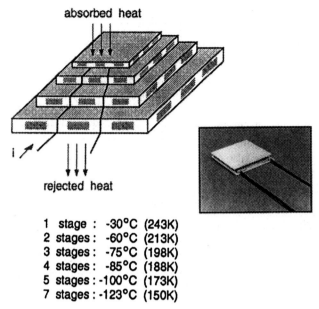

absorbed heat

rejected heat

1 stage :	-30°C (243K)
2 stages :	-60°C (213K)
3 stages :	-75°C (198K)
4 stages :	-85°C (188K)
5 stages :	-100°C (173K)
7 stages :	-123°C (150K)

FIGURE 4.16 Multiple-stage Peltier effect thermoelectric cooling unit. (From Maldague, 1994b, with permission from Gordon and Breach Publishers.) Inset: single-stage thermoelectric cooler from Melcor. (Picture courtesy of Melcor Corp., Trenton, New Jersey.)

example, Komatsu* has a 150-K seven-stages 65-W input power 80-mW head load commercial device. Melcor[†] offers a range of Peltier elements with a temperature differential between faces from 64 to 131°C (for multistage).

Most infrared camera manufacturers offer their products with an optional choice among those various means of cooling. For laboratory operation, nitrogen cooling with a metal dewar is perhaps the best choice because of the low temperature achieved (77 K, −196°C), the quiet mode of operation, and the reliability. For remote operations, thermoelectric cooling will be preferred since it does not need any refilling and its lack of moving parts makes it reliable. For a production-environment Stirling engine, Joule–Thompson gas expansion or thermoelectrical cooling are good choices.

4.7 SELECTION OF AN ATMOSPHERIC BAND

Since the atmosphere has not perfectly flat transmission properties (Figure 4.9), selection of the operating wavelength band will be conditioned by the final application. For the majority of NDT applications, the useful portion of the infrared spectrum lies in the range 0.8 to 20 μm (Figure 1.2); beyond 20 μm,

*www.komatsu.com.
[†] www.melcor.com.

applications are more exotic, such as high-performance Fourier transform spectrometers, which operate around 25 μm. The choice of an operating wavelength band dictates the selection of the detector type, as Figure 4.9 shows. Among the important criteria for band selection are operating distance, indoor–outdoor operation, temperature, and emissivity of the bodies of interest. As Planck's law stipulates (Chapter 2), high-temperature bodies emit more in the short wavelengths; consequently, long wavelengths will be of more interest to observe near room-temperature objects. Emitted radiation from ordinary objects at ambient temperature (300 K) peaks in this long-wavelength range. Long wavelengths are also preferred for outdoor operation where signals are less affected by radiation from the sun. For operating distances restricted to a few meters in the absence of fog or water droplets, atmospheric absorption has little effect (Section 8.3).

Spectral emissivity is also of great importance since it conditions the emitted radiation. Figure 4.17 shows spectral emissivity curves for common materials; also plotted on these curves are the more useful infrared bands of interest. Polished metals with emissivity smaller than 0.2 cannot be observed directly since they reflect more than they emit. A high-emissivity coating (such as "black" painting) or a reflective cavity must be used (Section 8.2).

Although no specific rule can be formulated, generally the most useful bands are 3 to 5 and 8 to 12 μm since they match the atmospheric transmission bands. Most of the infrared commercial products fall in these categories, while the near infrared (0.8 to 1.1 μm) is easily covered by standard ambient-operation-temperature silicon detectors.

Another important point to consider is the detectivity D^* of the detector used. From Figure 4.9, for example, it is seen that a 77 K-cooled InSb detector operating in the range 3 to 5 μm (Section 4.4) has a seven-fold-higher detectivity than a 77 K-cooled HgCdTe detector operating in the range 8 to 12 μm. That means that even if for a specific application, the radiation emitted (temperature of interest, spectral emissivity) is higher in the range 8 to 12 μm. The contrast obtained may be stronger in the range 3 to 5 μm because of the superior D^* of an InSb detector.

As a final notice, we may point that detailed studies (Gaussorgues, 1984, pp. 421–436; Pajani, 1987; Woolaway, 1991) have concluded that for temperature in the −10 to +130°C interval, measurements can be done without much difference in both bands (3 to 5 and 8 to 12 μm). For some special applications (e.g., for the military), bispectral cameras operating simultaneously in both bands have been developed to characterize target thermal signatures more accurately (see Section 4.8.3, two-color pyrometry).

4.8   RADIOMETRIC MEASUREMENTS

Radiometry is concerned with the measurement of radiated electromagnetic energy. Contrary to time measurement of frequency or voltage, where resolution on the order of 1 part to the 10^{10} or even 10^{11} can be obtained, in radio-

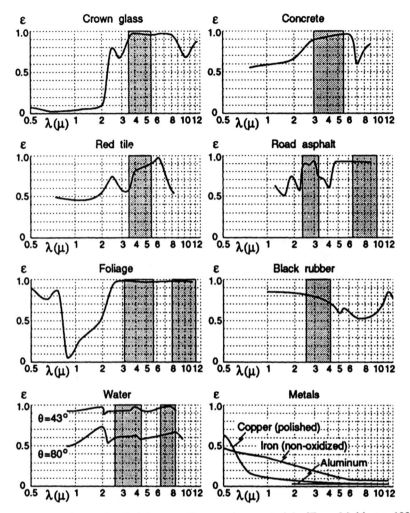

FIGURE 4.17   Spectral emissivity curve from various materials. (From Maldague, 1994b, with permission from Gordon and Breach Publishers. Data from Gaussorgues, 1989.)

metry a resolution of only a few percent is generally the best that can be envisaged and only in very careful experimental conditions. Reaching the fraction of a percent or even the percent itself is not an easy task. The mirage measurement is, however, an exception to this situation, with reported temperature measurement sensitivity of up to 10^{-8} K in rigorous laboratory conditions (temperature measurement of liquid CCl_4; Lepoutre and Roger, 1987; Section 7.5; Lu et al., 1998; Pao et al., 1998).

These poor performances of radiometric measurements are due to numerous factors (Nicodemus, 1967): large time intervals with respect to one period of the radiation frequency; large involved distances with respect to the wavelength; dissipation of the power radiated in all space; great variation in the power radiated, depending on the wavelength, position, direction, and polarization.

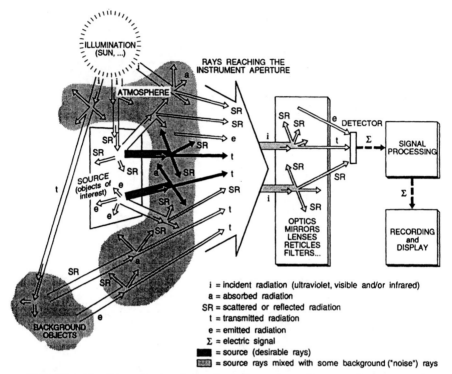

i = incident radiation (ultraviolet, visible and/or infrared)
a = absorbed radiation
SR = scattered or reflected radiation
t = transmitted radiation
e = emitted radiation
Σ = electric signal
■ = source (desirable rays)
▨ = source rays mixed with some background ("noise") rays

FIGURE 4.18   Generalized radiometric configurations. (From Maldague, 1993.)

The liberated photonic energy W, following an increase in temperature (increase in molecular agitation), is expressed by eq. (2.4). This equation pinpoints one potential difficulty of infrared thermography: wavelengths are large (e.g., with respect to visible spectrum wavelengths); consequently, the photonic emitted energy will be small and of the same order of magnitude than radiation emitted by the room-temperature environment. Figure 4.18 illustrates these perturbing undesirable sources of energy.

If some hypotheses are respected and if knowledge of some parameters is available, it thus becomes possible to translate radiometric values registered by the camera into temperature values. Under these circumstances, an infrared camera (at least those deployed for quantitative applications) is not a thermometer but a *radiometer*.

From considerations of thermal emission presented in Chapter 2, in particular the concepts of blackbody (Section 2.2) and Planck's law (Section 2.5), we can introduce the fundamental equation of infrared thermography. The radiance N_{CAM} received by the camera is expressed by

$$N_{\mathrm{CAM}} = \tau_{\mathrm{atm}}\varepsilon N_{\mathrm{obj}} + \tau_{\mathrm{atm}}(1 - \varepsilon)N_{\mathrm{env}} + (1 - \tau_{\mathrm{atm}})N_{\mathrm{atm}} \qquad (4.14)$$

$$\underbrace{\phantom{\tau_{\mathrm{atm}}\varepsilon N_{\mathrm{obj}}}}_{\substack{\text{Object}\\\text{contribution}}} \quad \underbrace{\phantom{\tau_{\mathrm{atm}}(1-\varepsilon)N_{\mathrm{env}}}}_{\substack{\text{Surrounding}\\\text{environment}\\\text{contribution}}} \quad \underbrace{\phantom{(1-\tau_{\mathrm{atm}})N_{\mathrm{atm}}}}_{\substack{\text{Atmosphere}\\\text{contribution}}}$$

where τ_{atm} is the transmission coefficient of the atmosphere in the spectral window of interest, ε the object emissivity (the object is considered opaque), N_{obj} the radiance from the surface of the object, N_{env} the radiance of the surrounding environment considered as a blackbody, and N_{atm} the radiance of the atmosphere, supposed constant. Equation (4.14) can be simplified if we consider the transmission coefficient of the atmosphere as being close to unity. This hypothesis is justified for standard spectral bandwidths of operation (3 to 5 and 8 to 12 µm) over small distances (under 2 m) and if no absorbing gas, dust, or water vapor droplet is present (Section 8.3):

$$N_{CAM} = \varepsilon N_{obj} + (1 - \varepsilon)N_{env} \tag{4.15}$$

Equation (4.15) can be further reduced if the emissivity ε is high and if no high-temperature object is present close to the inspection station which would give rise to parasitic reflections (see Section 8.2 for a discussion concerning the emissivity). Under these conditions,

$$N_{obj} \approx N_{CAM}$$

$$f(N_{obj}) \approx I_{CAM} \tag{4.16}$$

$$S(T_{obj}) \approx I_{CAM}$$

where I_{CAM} corresponds to the radiometric signal obtained on the camera calibration curve (Figure 4.19, $f(\cdot)$ and $S(\cdot)$ are the relationships which allow

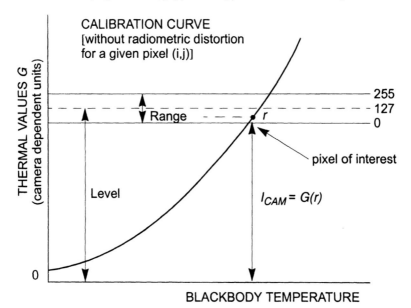

FIGURE 4.19 Infrared camera calibration curve and level and range settings for temperature computations (example of an 8-bit system).

us to convert camera signal to radiance N_{obj} and to temperature T_{obj} values, respectively.

In the case of an ideal instrument, a direct relationship could be derived between the radiometric signal I_{CAM} and the temperature of the object T_{obj}, taking into account the limitations of eq. (4.16). This ideal situation does not exist in general and we must consider the intrinsic limitations of the camera to perform truly quantitative measurements. These limitations are various:

- The vignetting and aberration of the optical system.
- The nonuniform spectral response of the instrument.
- The nonlinear response of the instrument in its dynamic range.
- The mechanical scanning process used for image formation, if present, introduces spatial distortions in the images.
- The infrared detector within the camera is also sensitive to the self-emission of the camera itself. Uncooled optical components such as lenses, mirrors, and prisms are at ambient temperature and emit accordingly.
- The Narcissus effect by which the detector sees itself because its own emission is reflected on the optics.

For instance, if transmittance changes by 5% through the field of view due to vignetting, and if a temperature difference of 20 K exists between the scene and the detector, a shadow effect of about 1 K will be observable (Abel, 1977). As we will see later, following proper calibration these uncertainties can be strongly reduced and then various techniques can be used for temperature measurement. We review two of them: single- and two-color.

Before processing any thermal images, it is necessary to convert the raw image values r into the radiometric signal I_{CAM}. Raw image values r are also called thermal values G(gray level), so that G(gray level) $= I_{CAM}$. In recent IR cameras, pixel coding is performed on 12 or even 14 bits (e.g., Santa Barbara focal plane) yielding, respectively, 4096 (2^{12}) or 16,384 (2^{14}) possible values. Thermograms from such systems do not require further conversion because the available span of values covers directly the entire dynamic range of the IR cameras. Older systems digitizing IR images on, say, 8 bits provide a limited 256 (2^8) possible values do require a particular conversion. In these cases, since the available span of values does not cover the entire dynamic range of the IR camera, manufacturers introduced two settings: *level* and *range*. These settings allow us to cover the entire dynamic span but have to be taken into account to compute the thermal values G. This can be understood with respect to Figure 4.19. The correction is then

$$G = \frac{r - 127}{256} \text{ range} + \text{level} \qquad (4.17)$$

If no corrections are needed, as, for example, in 12- or 14-bit systems, $G = r$.

FIGURE 4.20 Image of a uniform temperature target (50°C) acquired with an FPA camera; the vignetting effect is visible. The three squares correspond to reference points (see the text). (From Marinetti et al., 1997. Copyright © 1997 The American Society for Nondestructive Testing, Inc.; reprinted with permission.)

4.8.1 Single-Color Pyrometry: Correction of the Vignetting Effect

The next step prior to the quantitative analysis of thermograms is to convert the image sequence into temperature images. In the case of FPAs, the image restoration is generally limited to a vignetting effect (if present) since the noise level is generally low (see Chapter 5 on basic image processing, including noise smoothing).

An example of vignetting is shown in Figure 4.20. This phenomenon is explained as follows (see Section 4.10.3). If we form a cone of rays from a point in the object space, limited by the diaphragm of the lens, and intercept this cone with the image plane that is perpendicular to the lens axis, we find that the intercept is a circle if the object lies on the optical axis, and more generally, an ellipse if the object is laterally displaced. Moreover, for many lenses, the front and rear apertures are too small to transmit oblique rays fully, and a part of the light cone may be cut off, causing an amplitude reduction at the edges of the image. Finally, this effect is also caused by antireflection coatings of camera optics, which are optimized for normal incidence, thus explaining reduced distortions for central pixels.

In Figure 4.20, the central bright area corresponds to the portion of sensor fully reached by radiation, and the dark area is due to a loss of radiation caused by the limited lens aperture (the three dots in Figure 4.20 correspond to loca-

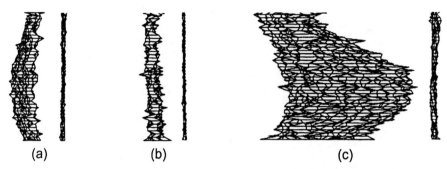

(a)   (b)   (c)

FIGURE 4.21 Side-view wireframe representations of raw and corrected images showing the vignetting effect at (a) 9°C; (b) 22°C; (c) 50°C. The left drawings are before correction for vignetting; the right drawings are after correction. (From Marinetti et al., 1997. Copyright © 1997 The American Society for Nondestructive Testing, Inc.; reprinted with permission.)

tions of the three reference points used below). Vignetting is also more severe if expansion rings are used to restrict the field of view due to the limited effective aperture obtained in this case (Maldague et al., 1991b).

As predicted by the theory (Kingslake, 1965), experiments carried out with the FPA camera show that this effect depends on both the pixel location and temperature difference between the target and ambient temperatures. Figure 4.21 clearly illustrates the vignetting effect on three thermograms recorded at a uniform temperature [respectively 9, 22, and 50°C for parts (a), (b), and (c)]. It is noted that below ambient temperature (22°C), vignetting has the opposite behavior to that above ambient temperature: The curvature direction changes, while at ambient temperature vignetting is not visible (Figure 4.21b). In Figure 4.21, the left drawings are uncorrected plots, while the right drawings are obtained after correction for vignetting (as explained below).

In Figure 4.22, the difference between the signal at the central reference point (corresponding to the center of the brightness area in Figure 4.20) and three points placed at various distances from it is shown (these three points are the three dots shown in the *, +, o plots of Figure 4.22), corresponding, respectively, going away from the image center to

$$d_{i,j}(G^t_{xref,yref}) = G^t_{xref,yref} - G^t_{i,j} \qquad (4.18)$$

where $G^t_{xref,yref}$ is the gray level of the reference at temperature t and $G^t_{i,j}$ is the gray level of the pixel (i,j) at temperature t.

Now considering a sequence of thermograms taken at different temperatures, it is noted that this difference is a linear function of the temperature in gray level, which may then be expressed as

$$d_{i,j}(G^t_{xref,yref}) = a_{i,j}G^t_{xref,yref} + b_{i,j} \qquad (4.19)$$

In the visible spectrum, a possible hardware solution to this vignetting problem

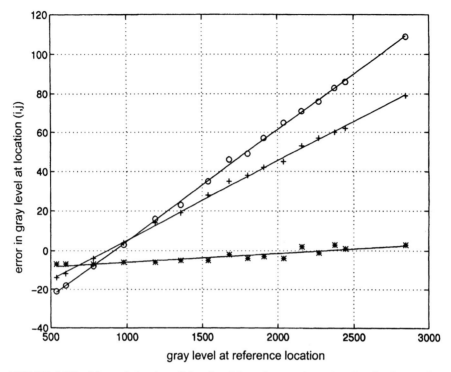

FIGURE 4.22 Linear behavior of the signal loss due to vignetting for the three reference points of Figure 4.20 (see the text). (From Marinetti et al., 1997. Copyright © 1997 The American Society for Nondestructive Testing, Inc.; reprinted with permission.)

is to add an additional lens in front of the objective (Cielo, 1988). If the original optics of the camera introduce a distortion F, then using a lens with a distortion function F^{-1}, the response of the global system is corrected. Of course, such an approach would attenuate the signal. This approach is possible if the function F does not depend on the features of the scene (i.e., temperature due to the self-emission of the optical elements; Hamrelius, 1991). Since in infrared thermography F depends on the temperature, a software approach is to be privileged. The idea is to create a $(M \times N \times 2)$ matrix file (where $M \times N$ is the image format) containing the coefficients $a_{i,j}$ and $b_{i,j}$ [for every location (i,j) in the image]. The correction formula is then

$$G_{i,j}^{corr} = \frac{G_{i,j} - a_{i,j}(G_{env}^{ref} - G_{env,0}^{ref}) + b_{i,j}}{1 - a_{i,j}} \qquad (4.20)$$

where G_{env}^{ref} is the gray level at the reference point location corresponding to the actual ambient temperature (at the time the correction is computed) and $G_{env,0}^{ref}$ is the temperature (in gray level) of the room when the correction matrix was created. As an illustration of the effectiveness of this procedure, it was applied to the left plots of Figure 4.21. Plots on the right of Figure 4.21 are obtained after application of eq. (4.20). The improvement obtained for Figure 4.21b (ambient temperature case) is due to the noise-filtering effect of the fitting process with eq. (4.20).

4.8.2 Single-Color Pyrometry: Temperature Calibration

After assessment of vignetting, the next step consists of converting gray-level values into temperature. The common procedure is as follows. A blackbody is set at a given temperature $(10, 25, \ldots {}^{\circ}C)$ and positioned in front of the camera. For each blackbody temperature, an image is recorded and readings are obtained in a restricted area at the center of the image, where radiometric distortion effects are less important. Values in a central subwindow of, say, 10×10 pixels are averaged together and plotted as a function of blackbody temperatures (Figure 4.23). A polynomial fitted on these values provides the calibration curve. For example, in the case of the FPA camera of Figure 4.23, the following relationship is obtained, where g represents the gray-level values (linear best fits using a third-order polynomial function available in Matlab, for example, with the function *polyfit*):

$$T({}^{\circ}C) = -13.4 + 0.05g - 1.6 \times 10^{-5}g^2 + 2.2 \times 10^{-9}g^3 \qquad (4.21)$$

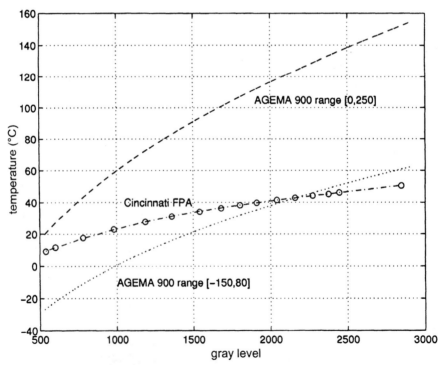

FIGURE 4.23 Calibration functions for three infrared cameras. For the FPA camera, the experimental points are plotted as well. (From Marinetti et al., 1997. Copyright © 1997 The American Society for Nondestructive Testing, Inc.; reprinted with permission.)

Obviously, such a calibration procedure is valid only for a specific experimental setup. If the experimental conditions change, it is necessary to repeat the process.

If instead of the third-order polynomial of eq. (4.21), a line fit is used, the calibration process is referred to as a *two-point calibration process*. To speed up the calibration process, conversion values might be precomputed and saved in a look-up table (LUT) for quick access: input of a gray value G directly provides the corresponding temperature T. If an hardware LUT is available, the processing time is very fast. Digital frame grabbers often include such a functionality.

4.8.3 Two-Color Pyrometry: Temperature Calibration

As recalled from Chapter 2, the spectral surface radiance emitted from an incandescent graybody depends only on its temperature T, its emissivity ε, and the spectral bandpass of the observation λ_i. The term ε is properly named the *spectral emittance*, referring to the property of a particular surface, in contrast *spectral emissivity*, which is the intrinsic property of an uncontaminated, optically smooth surface. However, to avoid the possibility of confusion with the radiant energy flux that also uses emittance, emissivity is generally used.

The main source of uncertainty in radiometric temperature measurement is the unknown emissivity ε. The problem is worse because of variations of the emissivity with wavelength and temperature $\varepsilon(\lambda_i, T)$. It is, however, possible to assume a simple relationship between emissivity and wavelength to calculate the temperature from measurements done at different wavelengths (DeWitt, 1986). Pyrometers based on that principle are called *multiwavelength pyrometers*. The signal measured in channel i of such a pyrometer is given by

$$I_i = K_i \lambda_i^{-5} \varepsilon(\lambda_i, T) \exp\left(-\frac{C_2}{\lambda_i T}\right) \tag{4.22}$$

where K_i is a constant for this channel and C_2 is the second radiation constant (Table 2.1). Equation (4.22) is known as Planck's law under Wien approximation ($\lambda T < 10^5$; Chapter 2).

One simple assumption about the variation of emissivity ε and wavelength λ is that a smooth curve exists between these two variables over the wavelength range of interest (Hunter et al., 1985–1986). This assumption holds for *graybodies* (for which emissivity is constant with wavelength) and to some extent to *colored bodies* (for which emissivity varies with wavelength). For two-color pyrometers, an easy way to compute the temperature is to form the ratio R of the signals, since for a given set of wavelengths λ_1 and λ_2 and temperature T, the spectral emittance ratio I_1/I_2 is unique:

$$R = \frac{I_1}{I_2} \frac{K_1}{K_2} \frac{\lambda_1^{-5}}{\lambda_2^{-5}} \frac{\varepsilon_1(\lambda_1, T)}{\varepsilon_2(\lambda_2, T)} \exp\left[\frac{C_2}{T}\left(\frac{1}{\lambda_1} - \frac{1}{\lambda_2}\right)\right] \tag{4.23}$$

which simplifies to

$$R = K_1 \exp\left[\frac{C_2}{T}\left(\frac{1}{\lambda_1} - \frac{1}{\lambda_2}\right)\right] \tag{4.24}$$

K_1 regroups all the constant terms, and finally,

$$T_1 = P(R) \tag{4.25}$$

A calibration curve $P(R)$ that relates the measured ratio R to the calculated temperature T_1 can be obtained experimentally. Although Planck's law is the underlying physical basis, it does not appear explicitly in eq. (4.25).

The error on the calculated temperature T_1 is given by the following formula (P. M. Reynolds, 1964):

$$C_2\left(\frac{1}{T} - \frac{1}{T_1}\right) = \frac{\lambda_1\lambda_2}{\lambda_2 - \lambda_1} \ln \frac{\varepsilon_1}{\varepsilon_2} \tag{4.26}$$

Zero error is obtained with the condition $\varepsilon_1 = \varepsilon_2$, which corresponds to a ratio of 1 in eq. (4.23). Equation (4.26) assumes no measurement errors in either channel. If any such errors (random errors from detector and photon noise, detector nonlinearity, drift in channel calibration, error in nominal channel wavelengths), are present the equation becomes (Coates, 1981)

$$C_2\frac{\Delta T}{T_2} = \frac{\lambda_1\lambda_2}{\lambda_2 - \lambda_1}\left(\frac{\Delta\varepsilon_1}{\varepsilon_1} - \frac{\Delta\varepsilon_2}{\varepsilon_2}\right) \tag{4.27}$$

For example, a 1% error in one channel leads to an error in the temperature of about 15 K (at 1000°C for the wavelengths selected; see the example below).

For restricted applications, where high-temperature materials emit sufficiently in the visible and near infrared, such as in the steel industry, a dual video camera system can be adapted for temperature measurement since spectral response of such standard CCD sensors extend up to the near-infrared range (common values are operating sensitivity in the range 0.4 to 1.1 μm). Figure 4.24 shows such an apparatus. It is not necessary for the wavelengths of the two channels to be close together. As shown in eq. (4.27), the error is proportional to the slope $\varepsilon_1/\varepsilon_2$ and is independent of the channel separation. In fact, separating the channels improves the sensitivity of the pyrometer, as the ratio of the signals then changes more rapidly with temperature. Recalling Figure 2.6, it is noted that the spectral dependence is different on each side of the maximum because of slope variations. This can be put in evidence by differentiation of Planck's law. Higher sensitivity is obtained with channel wavelengths below that maximum. In this application, for the temperature range of interest (800 to 1500°C), this occurs in the near infrared, where the CCDs are still sensitive and

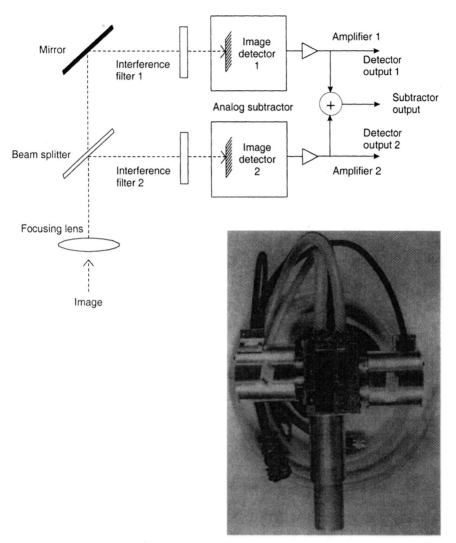

FIGURE 4.24 Schematic diagram and picture of the dual imager used in high-temperature pyrometry. (From Maldague and Dufour, 1989.)

hot bodies emit sufficiently. Although large separation is better, to obtain a greater range of temperatures for a given set of filters and taking into account the limited dynamic range of the two cameras, it is better not to separate the channels too much to ensure overlapping of the effective range of each channel. Moreover, the emissivity slope often increases with increasing wavelength separation and that is why a narrow separation is often used. For this application, filters are selected with the following characteristics for the two channels: 52% transmission for $\lambda_1 = 0.89$ μm (with 12-nm half-peak width) and 58% transmission for $\lambda_2 = 0.94$ μm (with 12-nm half-peak width). The calibration func-

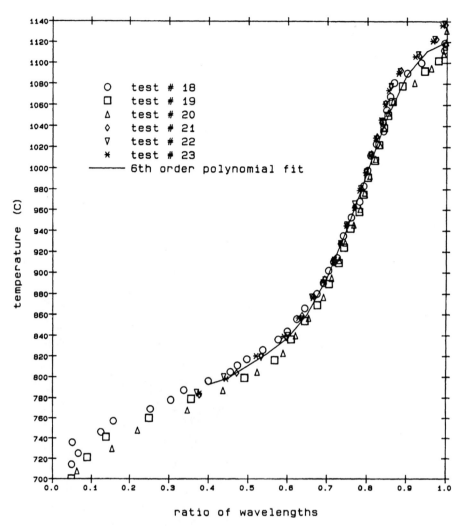

FIGURE 4.25 Calibration curve at $\lambda_1 = 0.89$ mm and $\lambda_2 = 0.94$ mm for the dual-channel pyrometer of Figure 4.24. (From Maldague and Dufour, 1989.)

tion specified by eq. (4.25) is computed by pointing the dual imager toward a high-temperature blackbody source (e.g., Mikron 330). Figure 4.25 shows the calibration curve. The ratio comes dose to 1 at high temperatures because of the light saturation of the two CCDs. Notice that here the calibration is performed on a pixel-per-pixel basis. Laboratory investigations performed over steel pieces indicated a 1% accuracy in the temperature range 800 to 1350°C (steel temperature was both measured with the dual imager and checked with a thermocouple). An example of the use of this pyrometer in the style industry is shown in Figure 4.26: temperature over a continuous casting steel slab was measured

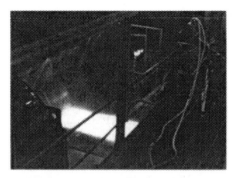 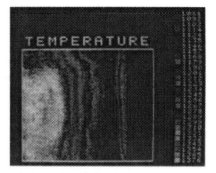

FIGURE 4.26 Dual-channel pyrometer of Figure 4.24 used in the steel industry with temperature image. Coolant is used to avoid overheating of the apparatus head (white pipes on left image). (Right: from Maldague and Dufour, 1989.)

indicating a 150°C difference between the edge and the center (steel is still in the melting stage under the surface at about 1500°C).

4.9 INFRARED OPTICS

In this section we cover the essential material necessary to understand infrared optics and underlying theory, including electromagnetic theory. Since our coverage is limited and textbooks on such topics are numerous, we refer interested readers to the literature for additional details (see, e.g., Ulaby, 1999).

Propagation of electromagnetic energy and infrared rays in a medium is described by many laws of physics. Of prime importance are the Maxwell equations, which show that such energy is composed of both electric and magnetic fields propagating in phase and in orthogonal planes. This means that an electromagnetic wave propagating in a medium makes electrons of this medium oscillate at the same frequency as that of the propagating wave. In the case of a conductive medium for which electrons are quite mobile, this results in strong absorption at all wavelengths. For dielectric, however, electrons are strongly linked to the atomic structure, and less absorption occurs, but at the resonance frequency of the electrons, thus resulting in a very narrow absorption band at a specific wavelength of the propagating wave. Semiconductors exhibit particular behaviors, depending on their own properties.

As we have seen earlier, what differentiates infrared from x-ray or visible radiation is the wavelength. Frequency ω_λ and period T_λ of the radiation are given by the relations

$$\omega_\lambda = \frac{2\pi c}{\lambda} \quad \text{and} \quad T_\lambda = \frac{\lambda}{c} \tag{4.28}$$

where c is the velocity of light in vacuum (Table 2.1). Another quantity of

TABLE 4.2 Optical Properties for Selected Materials

Material	Average Refractive Index n	Knopp Hardness (kg mm^{-2})	Melting Temperature (°C)	Water Solubility (g/100 g of Water)	Transmission Band in the Infrared Spectrum (µm)
Acrylic	1.5	10	127	Insoluble	0.3–30[a]
Air	1	—	—	—	0.3–14[a]
CaF$_2$	1.4	100	1357	0.0017	0.3–12
Fused silica	1.45	480	1477	Insoluble	0.3–5
Ge	4.0	500	940	Insoluble	1.8–23
Glass (common)	1.5	300–600	600	Insoluble	0.3–2.7
KRS-5 (thallium bromoiodide)	2.4	40	415	0.05	
NaCl (table salt)	2.5	20	800	36	0.2–25
Sapphire (Al$_2$O$_3$)	1.7	1500	2027	Insoluble	0.17–6.5
Si (silicon)	3.4	1150	1420	Insoluble	1.2–15[a]
Water	1.33	—	0	—	0.3–14[a]
ZnSe	2.42	137	1460	Insoluble	0.4–20

Source: Adapted from Kogelnik (1984), Cielo (1988), and Gaussorgues (1989).
[a] With specific strong absorption bands.

interest is the complex refractive index N, which describes all optical properties of a medium and is expressed as a complex number:

$$N = n - ie \tag{4.29}$$

where n is the classical optical refractive index, e designates the extinctive index related to the absorbing power of the medium, and i is the imaginary number $i = \sqrt{-1}$. The refractive index n is defined as the ratio of the velocity of light c in vacuum to the phase velocity in the material ($n = c/v$). Table 4.2 lists a few materials with their average refractive index.

Refractive indexes vary with the wavelength λ, this is called the *dispersion* (Figure 4.27). In design of optical elements, dispersion determines the extent of the chromatic aberration and other optical parameters such as the focal length are also affected. For instance, sapphire has good transmissive properties in visible and infrared spectra, thus, visible light can be used to perform some alignments. However, if ones tries to focus infrared light in the range 4 to 5 µm with a sapphire lens, it is important to note that due to a 0.12 variation in the refractive index between the infrared and visible bands, the focal length is changed by 20% (Cielo, 1988).

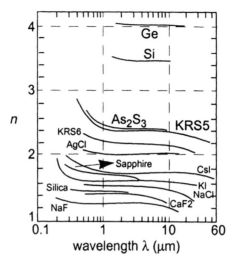

FIGURE 4.27 Effects of the wavelength on the refractive index n for a few common optic materials. (Data from Gaussorgues, 1989.)

4.9.1 Reflection

Reflection is defined as the return of electromagnetic energy by a surface upon which the radiation is incident (Kogelnik, 1984). In the case of plane waves of radiation, the angle of reflection θ_{ref} equals the angle of incidence θ_{inc} (Figure 4.28):

$$\theta_{ref} = \theta_{inc} \qquad (4.30)$$

The reflection is said to be *specular* when the surface is smooth (small surface irregularities with respect to the wavelength) and *diffuse* on rough surfaces. An example of specular reflection are the images of a candle observed when it is brought close to a window: Multiple reflections permit to determine the number of window panes. On the other hand, a diffuse image of the same candle appears on a white plaster ceiling. The reflectance R of a surface is a measure of the

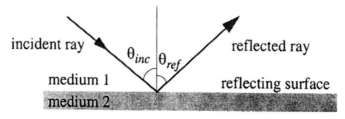

FIGURE 4.28   Reflection on a surface.

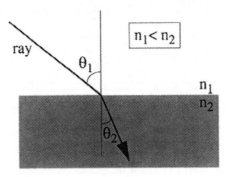

FIGURE 4.29   Refraction at the interface of two media.

amount of reflected radiation and depends on the angle of incidence (it increases at higher incidence), the polarization of the radiation, and the electromagnetic properties of the material forming the boundary surface (Kogelnik, 1984). These properties also change with wavelength. Materials can be divided by their reflectivities. In the case of opaque conducting metals, reflectance is high when highly polished (98% in the case of copper in the band 8 to 12 μm). Dielectric materials, on the other hand, exhibit transparency. At normal incidence ($\theta_{inc} = 0$), the reflectance R is expressed by (subscripts as shown on Figure 4.29)

$$R = \left(\frac{n_2 - n_1}{n_2 + n_1}\right)^2 \qquad (4.31)$$

Example 4.3   Determine the reflectance at an air–glass interface.

SOLUTION   Following eq. (4.31) and with $n_1 = 1$ and $n_2 = 1.5$, we obtain $R = 0.04$, which means that 4% of the radiation is reflected and 96% is transmitted through the glass. This figure makes important the addition of a reflective coating (such as silver) in order to limit thermal reflection losses in windows. Interestingly, materials with $n \approx 1.5$, such as glass, can be used without a coating since the reflection losses are limited (8% per lens, 4% × two faces). Nevertheless, if needed, in the visible spectrum these reflection losses can be reduced further through surface coating. For example, deposition of a single-layer quarter-wave thickness of magnesium fluoride (MgF_2) decreases reflectance by up to 1.5% (per face) at 0.55 μm and normal incidence (this substance is much used, thanks to its durability and limited refractive index variation in visible spectrum).

Example 4.4   What is the reflectance at an air–germanium interface?

SOLUTION   Following eq. (4.31) and with $n_1 = 1$ and $n_2 = 4$, we obtain $R = 0.36$, which means that 36% of the radiation is reflected and 64% is trans-

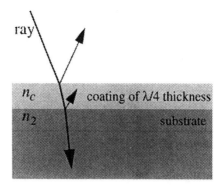

FIGURE 4.30 Coating of $\lambda/4$ makes it possible to decrease reflectivity.

mitted through the piece of germanium. Germanium is much used to make infrared optic elements. However, in such a situation, it is coated to reduce the reflectance.

In some circumstances it is necessary to decrease the reflectance to allow sufficient transmission of energy through an optic element. This is possible by applying on this element a coating of index of refraction n_c of thickness $\lambda/4$. In this condition, eq. (4.31) is modified in such a way that the reflectance at the coating–substrate interface becomes null due to the destructive interference if $n_c = \sqrt{n_2}$ (Figure 4.30); n is the index of refraction (see Example 4.5).

Example 4.5   Find the reflectance coefficient at $\lambda = 10$ μm if the germanium piece of Example 4.4 is covered with a $\lambda/4$ coating of zinc sulfide (ZnS) of $n_c = 2.2$.

SOLUTION   In such a situation the destructive interference takes place since $2.2 \approx \sqrt{4}$, and eq. (4.31) is modified as

$$R = \left(\frac{n_2 - n_c^2}{n_2 + n_c^2} \right)^2 \tag{4.32}$$

which allows us to compute R of the coated germanium, which becomes 0.9%, a better figure that allows to make use of such a piece in infrared optic assemblies. In the range 3 to 5 μm, the coating is made of SiO_2. Table 4.3 lists other materials used for optic coatings.

4.9.2 Refraction

Refraction is defined as the change in direction of propagation of any wave phenomenon that occurs when the wave velocity changes. This apply to all electromagnetic waves, sound waves, and water waves (Kogelnik, 1984). Snell's

TABLE 4.3 Materials Commonly Used for Coatings

Coating	Chemical Composition	Refractive Index n
Cerium oxide	CeO_2	2.2
Cryolite	AlF_3-NaF	1.3
Magnesium fluoride	MgF_2	1.38
Silicon monoxide	SiO	1.6–1.9
Zinc sulfide	ZnS	2.2

Source: Adapted from Gaussorgues (1989).

law governs refraction (Figure 4.29):

$$n_1 \sin \theta_1 = n_2 \sin \theta_2 \qquad (4.33)$$

4.9.3 Properties of Optical Materials

Several properties of materials determine whether or not they are suitable in the manufacture of infrared optical elements, among which water solubility, hardness, melting temperature, thermal conductivity, and thermal expansion are of prime importance.

Water solubility informs us whether or not a given material can be in contact with water or water vapor without degradation. For example, table salt (NaCl) has good transmission in the infrared spectrum (from about 0.3 to 22 μm), but the high solubility makes this material unsuitable for lens making unless covered with coatings (Table 4.2).

Hardness determines how a given material resists scratches and if it can be cleaned by sheet wiping. Moreover, a material too hard is very difficult to polish to make lenses. Hardness is measured by the Knopp indentation method, which consists of hitting the material of interest with a diamond-shaped-tip hammer of a given weight (Chapter 7). The Knopp hardness is then given by the ratio of the hammer weight to the surface of indentation (Table 4.2).

High melting temperature is necessary if the material is going to be exposed to a high-temperature environment such as that involved in furnace monitoring (Table 4.2). High thermal conductivity (k) allows cooling of optical pieces through a forced coolant along the rim.

Thermal expansion is important to know in order to limit mechanical constraints and possible misalignment along rims due to temperature variations. For example, k is 0.5 W cm^{-1} °C^{-1} for Ge, 0.25 for sapphire, but 0.003 for acrylic (Cielo, 1988).

4.9.4 Materials in Infrared Optics

Important materials in infrared optics are glasses, crystals (such as NaCl, CaF_2, KRS 5, KRS 6), semiconductors (Si, Ge), plastics, and metals. Glasses generally have a cutoff transmission for wavelengths greater than 2.7 μm, fused silica

2-5 μm   8-12 μm

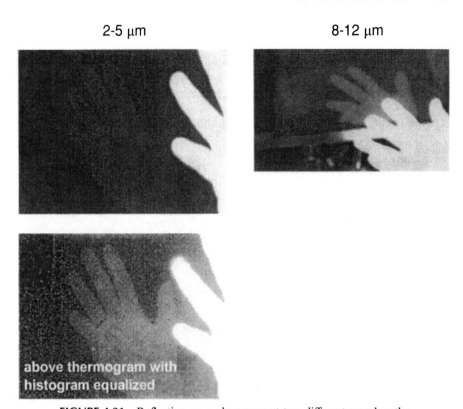

FIGURE 4.31   Reflection on a glass pane at two different wavelengths.

extends the cutoff point up to 5 μm, and special glasses offer transmission at greater wavelengths: for example, CaF_2 up to 11 μm and ZnSe up to 20 μm.

Figure 4.31 shows the reflection of a warmer-than-ambient object on a glass pane at two different wavelengths. The reflection is strong in the band 8 to 12 μm and barely visible in the band 2 to 5 μm, although an histogram equalization reveals the reflected image (Chapter 5 presents some image enhancement techniques; more general techniques are available from packages such as Paint Shop Pro or Adobe Photo Shop). With transmission at short wavelengths, the reflection is low; for longer wavelengths, cutoff transmission increases the reflection (Section 2.9.5). Both cameras were set so that the hand at about 34°C appears white in the thermograms.

Silicon (Si) and germanium (Ge) are materials of great interest in the manufacture of infrared optics elements, due to their good optical properties (Table 4.2). As seen previously, coatings are necessary to increase transmission due to their high refractive index. Above 150°C, transmission in Ge becomes negligible; polycrystal Ge elements reduce cost.

Due to poor thermomechanical properties, plastics are not considered to make optical elements. However, the low cost associated with molding manufacturing make plastic attractive for some applications, such as aspheric optics

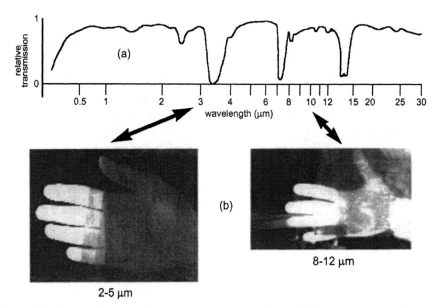

FIGURE 4.32   (a) Transmission properties of polyethylene; (b) see-through experiments in bands at 2 to 5 and 8 to 12 μm performed over a warm background. Gray scale is from black (17°C) to white (hand at 37°C). [(a) Data from Gaussorgues, 1989.]

or windows made of polyethylene or Plexiglas having good transmission in the infrared. For instance, Figure 4.32a shows the relative transmission of poly-ethylene, used for instance to make garbage bags. Transmission is higher in the band 8 to 12 μm than at 2 to 5 μm, due to a strong absorption band. This fact is set in evidence in Figure 4.32b. A white polyethylene garbage bag is folded so that total thickness increases sequentially from left to right as one, two, and three plies (it is opaque in the visible spectrum). The band 8 to 12 μm reveals good transmission; in the band 2 to 5 μm, transmission is reduced as the number of plies increase.

Low-cost plastic containers are widely used in the food industry, production volume is high, and if the manufacturing process is not maintained carefully, holes may appear during molding, especially on the sides. The infrared trans-parency just discussed offers interesting detecting possibilities. A dark brown plastic cooky container is shown in Figure 4.33. Transparency testing from the side over a warm background clearly shows a hole (total container thickness from side to side 10 cm, plastic thickness: 400 μm). The visible band is not useful here since this plastic is opaque. This is an interesting application of passive thermography. The operating wavelength of interest should be selected carefully, depending on the application. For instance, even though in the last transparency examples, the long-wavelength band offered higher transmission, it really depends on the material of interest, as Figure 4.34 shows just the opposite in the case of the 1.5-mm-thick plastic CD holder (higher transmission in the band 2 to 5 μm).

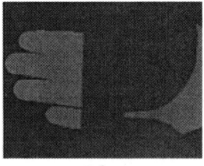

8-12 µm

2-5 µm

visible picture

FIGURE 4.33 See-through experiment set to find an edge-hole defect in this dark polyethylene brown cooky container.

Some metals, such as gold and aluminum, have good reflectance in the infrared, making them attractive for mirrors. For example, first-surface aluminum or gold mirrors are made by vacuum deposition outside the component; common substrates are ceramic or glass. Second-surface mirrors are made by deposition on the internal surface, as in the case of prisms. These mirrors are often protected from abrasion and tarnish with a hard dielectric coating of half-wavelength optical thickness (at 0.55 µm), such as of silicon monoxide coating (SiO). Aluminum has a 90% reflectance in the visible up to the infrared spectrum, silver 98%, and gold 99%. Aluminum oxidizes, but its oxide is tough and

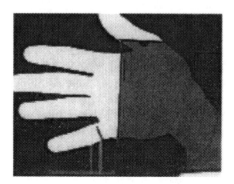

8-12 µm

2-5 µm

FIGURE 4.34 See-through experiment with an acrylic CD holder.

corrosion resistant (the oxide causes light scattering through the spectrum). Silver has a higher reflectance than aluminum (in the visible up to the mid-infrared), but this advantage disappears rapidly after deposition, due to oxidation; silver is thus more often used for internal reflection. In the near infrared, silver is often used instead of aluminum, which exhibits a small dip in reflectance at these wavelengths. Good tarnish resistance and high reflectance in the near, middle, and far infrared makes of gold the most widely used material in these spectral regions. To clean first-surface mirrors, it is recommended loose particles be blown off using pressurized dry gas and then the surface be cleaned gently with a swab using deionized water, mild detergent, or alcohol. Since bare gold is soft, low-pressure blowing is better. Metals are also used as rims and holders for infrared optical elements.

4.10 OPTICS FUNDAMENTALS

One of the most important parts of any infrared system is the camera head, which allows image registration. With respect to standard visible objectives and lenses, infrared objectives made of infrared transparent material (Section 4.9) are quite expensive, due to the limited market compared to the huge consumer-photo market. This makes appropriate selection of the objective to be used in conjunction with an infrared camera even more important. Moreover, as in any camera system, the objective selected will affect the final image output. This is thus an important topic to cover.

In this section we present briefly some important practical concepts of classical optics. Since our coverage is limited and textbooks on such topics are numerous, we refer interested readers to the literature for additional details (see, e.g., Meyer-Arendt, 1972; Jenkins, 1976; brochures of optics manufacturers, such as *Optics Guide 3* from Melles-Griot.* Material more specific to infrared optics was discussed in Section 4.9. What is of interest here are the general concepts, so we review optical elements of interest in infrared imaging systems, primarily mirrors, lenses, and filters.

4.10.1 Mirrors

A mirror is a surface reflecting light and making images of objects. Mirrors find many uses in optics and in imaging systems. For example, some scanning infrared radiometers use rotating mirrors in image formation. Mirrors are also used to fold the optical path, as when inspecting objects with a dewar-equipped infrared camera. In that case, folding the optical path by 90° prevents coolant spilling (Section 8.1). Figure 4.35 shows a curved mirror and the main parameters, such as radius of curvature R and center of curvature C. The object O and image I, with object and image distances o and i, as well as equal angles of

---

* www.mellesgriot.com.

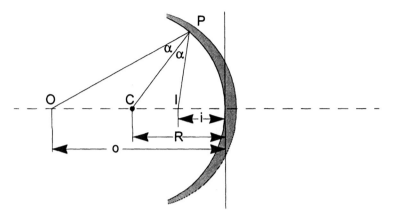

FIGURE 4.35   Concave mirror.

reflection α are also shown. From this drawing, the spherical mirror equation is derived:

$$\frac{1}{o} + \frac{1}{i} = \frac{1}{f} = \frac{2}{R} \tag{4.34}$$

The parameter f is the focal length. Equation (4.34) holds for plane, concave, and convex mirrors.

In the case of a plane mirror having $R = \infty$, eq. (4.34) gives $o = -i$, which means that plane mirrors form virtual images located symmetrically behind the mirror. Figure 4.36 illustrates the situation as it is known; the left and right

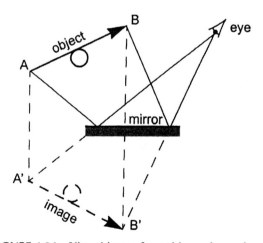

FIGURE 4.36   Virtual image formed by a plane mirror.

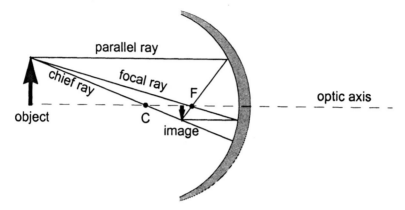

FIGURE 4.37 Ray tracing in concave mirror configuration

image are reversed (while it is right side up). Ray tracing in mirrors is performed with the help of the chief, focal, and parallel rays, as shown in Figure 4.37. The chief ray is the one passing the center of curvature. It hits the mirror surface at normal incidence and is thus reflected back along its original path. The focal ray is the one passing the focal point F, striking the mirror surface so that it becomes parallel to the optic axis. Finally, the parallel ray is the one originating from the object parallel to the optic axis; it hits the mirror surface and passes by focal point F. To locate images, it is sufficient to draw two of these rays.

Four cases are distinguished as to image size.

1. With an object located outside the center of curvature, the image will be real, inverted, and reduced in size (concave mirror).
2. With an object located between the center of curvature and the focal plane, the image will be real, inverted, and magnified (concave mirror).
3. With an object located between inside the focal plane, the image will be virtual, upright, and magnified (concave mirror).
4. For convex mirrors, the image will be virtual, upright, and reduced in size.

Example 4.6 We want an object to appear three times larger using a concave mirror of radius of curvature $R = 30$ cm. Where is the object to be located? (Consider only case 3 above.)

SOLUTION In case 3 the image is virtual, meaning that it is located behind the mirror. The parameters of interest are shown in Figure 4.38. From eq. (4.34) we find that the object distance o is $o = Ri/(2i - R)$. Another equation is found from the similar triangles of Figure 4.38, yielding equal ratios:

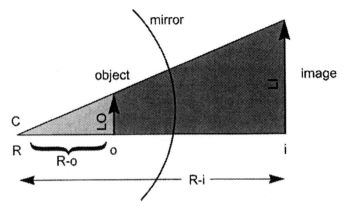

FIGURE 4.38   Figure for Example 4.8 with two similar triangles used in computations (shaded areas).

$$\frac{R-o}{R-i} = \frac{LO}{LI} = \frac{1}{L} = \frac{1}{3}$$

Notice that since the image is virtual, the image distance i is negative. Solving for i in these two equations leads to the following quadratic equation: $2i^2/3 - 600 = 0$, and this gives the values $i_1 = 30$ cm and $i_2 = -30$ cm. The value i_1 is rejected since it does not conform to case 3 (moreover, with $i_1 = 30$ cm, a value of $o = 30$ cm is found). The solution is thus $i_2 = -30$ cm, for which a corresponding value of $o = 10$ cm is found. Problem 4.4 explores the other possible situation (case 2).

4.10.2   Thin Lenses

A *lens* is defined as a piece of a glass or transparent substance curved on one or both sides. The name comes from the lentil seed, with its known characteristic lens shape. In fact, a single word designates both in some other languages, such as in French (*la lentille*) and German (*die Linse*).

A lens is called *thin* if a ray passing through it experiences little spatial translation as it emerges on the other face. Figure 4.39 shows the basic parameters of a thin lens with focal point, nodal planes, and focal distance. The *focal point* is the point where parallel rays coming from the infinite converge, due to refraction on the curved lens surface. The *nodal plane* is the plane where rays seem to curve themselves toward the focal point. In the case of a nontheoretical thin lens, it is possible to have several nodal planes. One important specification in addition to lens diameter and the material the lens is made of is the *focal distance f*, which is related to the radius of curvature R_1 and R_2 of both lens surfaces by the *lens-maker formula*:

$$\frac{1}{f} = \Delta n \left(\frac{1}{R_1} + \frac{1}{R_2} \right) \tag{4.35}$$

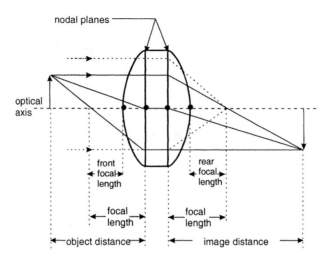

FIGURE 4.39 Basic parameters for a thin lens.

where Δn is the difference between the refractive index of the lens and the surrounding environment. As we saw in Section 4.9, the refractive index depends on the wavelength of interest. The F-number of a lens is given by the focal length divided by the diameter of the lens aperture.

Example 4.7 Calculate the F-number of a 10-mm-diameter lens with $f = 50$ mm.

SOLUTION Following the definition, the F-number is $50/10 = 5$. One thus talks about an F/5 lens.

The F-number is useful to determine the amount of light reaching the image. The mechanical stops of photographic objectives change the illuminance by a factor of about 2. For example, between F/8 (square $= 64$) and F/11 (square $= 121$, thus about 64×2) there is grossly a twofold change. Common F-numbers of photo objectives are: 1.4, 2, 2.8, 4, 5.6, 8, 11, and 16. As seen in Chapter 2, the illuminance depends on the aperture; reducing the aperture will increase the F-number, as noted on reflex cameras.

There are two classes of lenses, based on whether parallel light rays striking them converge or diverge. A *converging* (or *positive*) *lens* is generally thicker at the center; the opposite is true for a *diverging* (or *negative*) *lens*. We say "generally" because this is true only if the refractive index of the material the lens is made of is greater than that of the surrounding medium (e.g., air and germanium). Figure 4.40 shows various types of lenses. Meniscus is, for example, used in glasses.

The *Gaussian form of the thin lens equation* relates the object distance o to the image distance i:

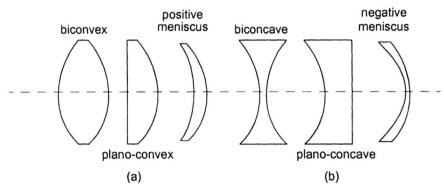

FIGURE 4.40 Types of lenses: (a) converging; (b) diverging.

$$\frac{1}{f} = \frac{1}{i} + \frac{1}{o} \tag{4.36}$$

Although eq. (4.36) assumes a theoretical lens of null thickness, it is sufficient for practical computations. Equation (4.36) also holds for a thick lens if distances are measured from nodal planes, where it is assumed that refraction takes place. The image distance can be either positive or negative. A positive distance indicates a real image that can be projected on a screen; a negative distance is related to a virtual image, which cannot be projected.

The lateral magnification M_{lateral} is given by the ratio

$$M_{\text{lateral}} = \frac{i}{o} \tag{4.37}$$

and the longitudinal magnification is M_{lateral}^2.

Example 4.8 The focal length of a germanium lens is 50 mm. The lens is located 70 mm in front of a sensor of size 10 mm × 10 mm. Determine the size of the field of view of this infrared imaging system.

SOLUTION The system is depicted in Figure 4.41. Equation (4.36) gives an object distance of 175 mm:

$$\frac{1}{50} - \frac{1}{70} = \frac{1}{\text{distance object}} = \frac{1}{175}$$

The magnification is thus $M_{\text{lateral}} = 70/175 = 0.4$, giving a field of view of 10 mm/0.4 = 25 mm. With an image distance of 30 mm, the object distance becomes −75 mm; that is, 45 mm behind the sensor, this virtual image is useless.

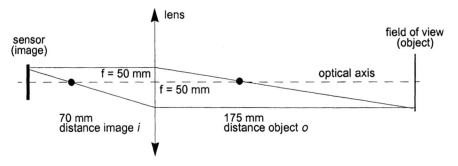

FIGURE 4.41   Figure for Example 4.10 (notice the double arrow symbolizing the lens).

The depth of field of an optical system is the interval distance along the optical axis where the object can be located while the corresponding image is still at the focus. The *confusion circle c* is the quantity of blur acceptable by the sensor without image degradation. In most vision systems, this corresponds to the pixel size. More specifically, the depth of field is given by (Figure 4.42)

$$\text{depth of field} = \text{far distance} + \text{near distance} = \frac{co}{A-c} + \frac{co}{A+c} \qquad (4.38)$$

A point on the object surface will be at focus if located in the interval far distance to near distance. Interestingly, as seen from eq. (4.38), the depth of field increases as the lens aperture A is reduced. At the limit, for a pinhole, the depth of field is infinite. The drawback is obviously that less light passes through the aperture, although this could be compensated by increasing the illumination of the scene. In photography, to get an image with both close and far objects in focus, the camera aperture is reduced and the scene is lighted with powerful flashes (e.g., the Balcar StarFlash 3 delivers 6.4 kJ of energy in about 15 ms; Section 8.1). These powerful flashes are also used in active thermography, as we will see later.

A practical rule says that the amount of acceptable blur is given approximately by

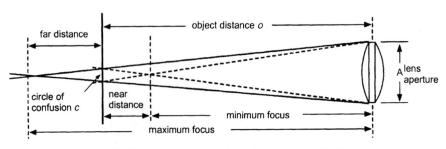

FIGURE 4.42   Computation of the depth of field.

$$\text{amount of blur} = \text{magnification} \times \text{pixel size} = \frac{\text{object distance}}{\text{image distance}} \times \text{pixel size}$$

$$(4.39)$$

Thus the depth of field can be given approximately by the lens F-number multiplied by this amount of blur.

Example 4.9 Compute the approximate depth of field using the data in Example 4.8, assuming a sensor of 512×512 pixels.

SOLUTION   The magnification was found to be $M = 1/0.4$. The F-number is $50/25 = 2$ (F/2 lens). The pixel size is 20 μm. The amount of blur is thus $(1/0.4)(20) = 50$ μm, and the depth of field is about 2×50 μm $= 100$ μm $= 0.1$ mm.

Finally, to avoid misalignment, the image plane, object plane, and lens plane have to be maintained parallel; otherwise, even if the image impinges correctly on the sensor, it will be unacceptable because of geometric deformations (*keystone effect*). Moreover, if misalignment is larger than the depth of field, the image is further degraded by blur (Figure 4.43).

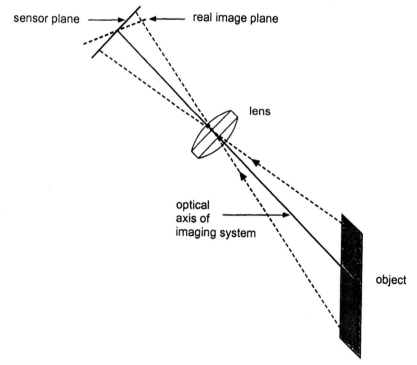

FIGURE 4.43 Imaging system with misaligned image plane with respect to sensor plane.

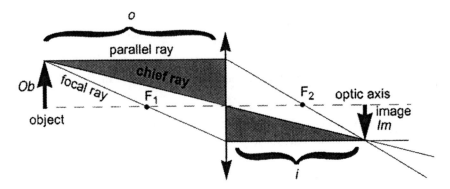

FIGURE 4.44   Ray tracing through a thin lens.

Ray tracing involves applying the reflection and refraction equations seen previously every time a ray intercepts a surface. Although specialized software is available for such a task, for simple systems, it is possible to get the position of the image using a few rules, as in the case of mirrors. Ray tracing in lenses is performed with the help of the chief, focal, and parallel rays, as shown in Figure 4.44. The chief ray is the one passing through the center of the lens. The focal ray is the one passing through the focal point F_1, striking the lens surface and emerging parallel to the optic axis. Finally, the parallel ray is the one originating from the object parallel to the optic axis. It strikes the lens surface and emerges by passing through focal point F_2. To locate images, it is sufficient to draw two of these rays.

Four cases are distinguished for images:

1. With a real object located outside the focal length of the lens, the image will be real and inverted (converging lens).
2. With a real object located inside the focal length of the lens, the image will be virtual, upright, and magnified (converging lens).
3. With a virtual object located at the right of a lens, the image will be real, upright, and reduced (converging lens).
4. For a diverging lens, the image will be virtual, upright, and reduced in size, no matter where the object is located.

More generally, to draw the propagation of a ray through a lens from point P, the procedure for a converging lens (Figure 4.45) is as follows:

1. If point P is outside focal plane F_1, draw the focal plane and the chief ray C from where the ray intersects the focal plane.
2. Continue the ray undisturbed up to the lens, and on the other side of the lens, make it parallel to the chief ray C (the ray after refraction).

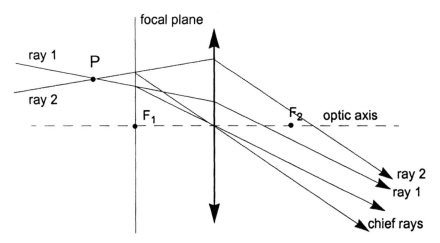

FIGURE 4.45 Arbitrary ray tracing through a thin converging lens (two cases illustrated).

More generally, to draw the propagation of a ray through a lens from point P, the procedure for a diverging lens (Figure 4.46) is as follows:

1. Draw the focal plane F_1, at the right of the lens.
2. From the ray, draw a line up to the focal plane, and from the intersection point of this line with the focal plane, construct the chief ray backward up to the lens center.
3. From the point on the lens where the ray emerges, draw a line parallel to the chief ray; this is the ray after refraction.

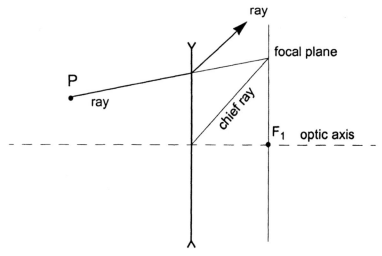

FIGURE 4.46 Arbitrary ray tracing through a thin diverging lens; notice the arrow shape symbolizing the diverging lens.

Example 4.10   We want an object to appear three times larger using a converging lens of focal length $f = 30$ cm. Where is the object to be located (real image)?

SOLUTION   With the similar triangles highlighted in Figure 4.44, we have immediately that $Im/i = Ob/o$ and thus $Im/Ob = i/o = M_{lat}$, which is eq. (4.37), where M_{lat} is the lateral magnification. Moreover, reorganizing eq. (4.36) and substituting $i = M_{lat}o$, we obtain

$$o = \frac{f}{M_{lat}}(M_{lat} + 1) \tag{4.40}$$

which is a general convenient relationship. Substituting $f = 30$ cm, $M_{lat} = 3$ in eq. (4.40) yields $o = 40$ cm, and the real inverted image will appear 1.2 m behind the lens.

4.10.3   Aberrations

Aberrations are all the effects degrading the image, caused either by the lens or by the way it is used. Most common aberrations are chromatic, spherical, coma, astigmatism, curvature of field, distortion, and vignetting. Different wavelengths will be in focus at different positions, due to variations in the refractive index of optical materials with wavelength; this is *chromatic aberration*. This is especially important in a high-resolution visible inspection system using white light (a mix of all visible wavelengths).

For a lens having curved surfaces, the light rays close and far from the optical axis are at the focus at different positions; this is known as *spherical aberration*. If a bundle of light passes through a lens obliquely, peripheral rays are at focus at a different longitudinal position with respect to the position of the light ray passing along the optical axis; this is the *coma*. The name refers to the cometlike appearance of this object point located off the optical axis.

Astigmatism occurs when the an off-axis object point does not result in a point image; this is due to the difference between the tangential and radial focus distance. Since a lens does not to transform a plane object to a plane image, when an image is projected on a flat screen, blur is observed at the image edges. This is the *curvature of field*, which can be corrected by projecting the image on a curved screen.

Distortion modifies the shape of the image; that is, it changes the location of image points even though the image is at the focus. It can be of two types: pincushion or barrel. There is no distortion in a thin lens; however, if a diaphragm is used to limit spherical aberration, distortion may appear.

Aberrations become more serious for large apertures. They can be reduced by juxtaposing lenses (such as in doublet configurations) or by using lenses with a high F-number (small aperture). Reducing lens aperture with a diaphragm, for example, may create other problems, such as *vignetting*. Vignetting is due to

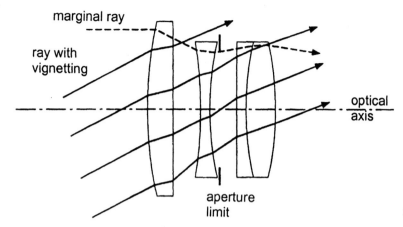

FIGURE 4.47   Vignetting in optical systems (left, vignetting principle in a Tessar objective; see also Figure 5.9).

an apparently different aperture for light rays along the optical axis with respect to oblique rays (Ballard and Brown, 1982). Observation of a uniformly lighted screen results in an image having darker edges with respect to the brighter image center. In infrared thermography, vignetting occurs with dependence on the temperature of the objectives, as seen in Figure 4.47. Section 4.8.1 presents an image-processing technique to correct for vignetting. Monochromatic lighting, work along the optical axis, and a combination of lenses having opposing aberrations contribute to reducing aberrations.

4.10.4   Filters

Filters are optical elements that permit the transmission of light at specific wavelengths while other wavelengths exhibit limited or null transmission. Filters find application in many different areas in TNDT. For example, a glass filter with the middle wavelength set at 5 μm and a half-width of 145 nm is required for temperature measurement of glass (Section 11.4.4). Another use is in observation of the welding pool, where it is necessary to restrict the spectral band of interest to reduce arc radiation interference: a 10,000-K high-temperature arc emits strongly up to 5 μm, while a lower-temperature 1800-K weld pool emits mostly at 0.5 to 20 μm (Nagarajan et al., 1994). Another application is in two-color pyrometry (Section 4.8.4).

A filter is made of a multilayer thin-film device bonded on a substrate so that particular transmission and reflectance properties are achieved. In some cavity configurations the thin film is bonded between two substrates. Since these desired transmission properties are obtained due to destructive interference [eq. (4.32)], filters are often referred to as *interference filters*. Substrates are selected for their homogeneity, uniformity of transmission.

Both edge and bandpass filters are available. Bandpass filters transmit a desired wavelength interval while rejecting simultaneously longer and shorter wavelengths. The interval can be either broad or narrow (down to a few nanometers). An important specification of bandpass filter is thus the FWHM (full width at half maximum). It is important to point out that since the interference phenomenon does occur at a specific thickness and thus at a specific angle of incidence, filters are generally specified for normal incidence, 45° incidence, or any other incidence.

Another type of filters is the *neutral density filter*, used to attenuate light at known relatively constant levels over a wide spectral range. Optical density (D) is derived from transmission Tr as

$$D = \log_{10} \frac{1}{\text{Tr}} \tag{4.41}$$

The reciprocal of Tr is known as *opacity*. There is no particular density unit; $D = 0.6$ refers simply to a density of 0.6 (or a transmission of 25%). This definition is similar to the definition of decibel in electronics, with the same advantage of logarithm summation when juxtaposing filters (provided that there are no multiple reflections between juxtaposed filters). Following eq. (4.41), transmissions are multiplicative and optical densities additive.

Example 4.11   Assuming no multiple reflections between two juxtaposed filters of density 0.5, find the corresponding transmission.

SOLUTION   Following eq. (4.41) and since densities are additive, we find that a density of $0.5 + 0.5 = 1$ corresponds to a transmission of $1/10^1 = 10\%$. A density of zero corresponds to a transmission of 100%.

PROBLEMS

4.1   **[detector parameters]** During design tests, alternative output voltage and associated current from an IR detector of capacitive type are measured as 5 mV and 5 µA, respectively. What is the detector impedance Z?

   Solution   Equation (4.1) is in vector form, and for capacitors it is known that the voltage lags the current by 90°; therefore, the answer is $Z = 5$ mV$/\underline{-90°}/5$ µA $= 1000/\underline{-90°}\Omega$. Such a purely capacitive component is called *reactance*. In fact, in its most general form, the impedance Z is made of two components, a resistive component R and a capacitive component X, with j being the imaginary number ($j = \sqrt{-1}, j^2 = -1$):

$$Z = R + jX$$

   In our case we thus have $Z = -1000j$ Ω [remembering the Euler relationship: $e^{j\theta} = \cos\theta + j\sin\theta = e^{j(-90°)} = \cos(-90°) + j\sin(-90°) = 0 - j = -j$].

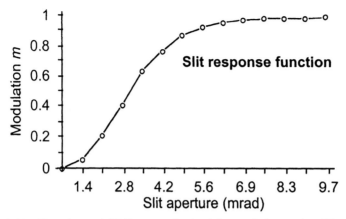

FIGURE 4.48   Experimental SRF curve obtained for an Inframetrics 600L using the procedure of Figure 4.2 (see Problem 4.2).

4.2   **[slit response function]** The SRF curve of an infrared camera Inframetrics 600L is measured and shown in Figure 4.48 with an image size of 256×256 pixels. If a modulation of 50% is accepted, what is the actual number of pixels per line?

Solution   Recalling the discussion of SRF in Section 4.1.7, we see in Figure 4.48 that the slit aperture for $m = 50\%$ is about 2.8 mrad. Assuming a field of view of width W at a distance L from the camera (Figure 4.49), the pixel size is thus given by $2L \, \mathrm{tg} \, \dfrac{(2.8 \text{ mrad})}{2} = L \, (2.8 \text{ mrad}) = 0.0028L$. The number pixels of available is thus $W/0.0028L$, that is, about $357(W/L)$, which can differ from the apparent number of pixels, which here is 256.

4.3   **[optics: infrared]** What is the reflectance at a water–glass interface?

Solution   Following eq. (4.31) and with $n_1 = 1.33$ and $n_2 = 1.5$, we obtain $R = 0.0035$, which means that 0.35% of the radiation is reflected and

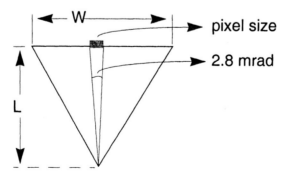

FIGURE 4.49   Field of view configuration for Problem 4.2 (not to scale).

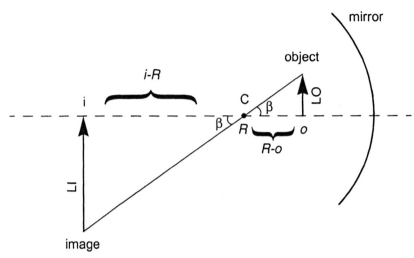

FIGURE 4.50   Configuration for Problem 4.4.

99.65% is transmitted through the glass. Such an interface is typical of observation windows in swimming pools used to evaluate drivers.

4.4   **[optics: mirror]** We want an object to appear three times larger using a concave mirror of radius of curvature $R = 30$ cm. Where is the object to be located? (Consider only case 2 in Section 4.10.1.)

Solution   In case 2 the image is real and is located between the center of curvature and the focal plane. The parameters of interest are shown in Figure 4.50. From eq. (4.34) we find that object distance $o = Ri/(2i - R)$. Another equation is found from angle β, which is common to both triangles in Figure 4.50, yielding $\tan \beta = LI/(i - R) = LO/(R - o)$. Solving for o gives $o = [(L + 1)/L]R - i/L$ with $L = LI/LO = 3$. Solving for i in these two equations leads to the following quadratic equation: $(2i^2/3) - 60i + 1200 = 0$, and this gives the values $i_1 = 30$ cm and $i_2 = 60$ cm. The value i_1 is rejected since it does not conform to case 2 (an moreover with $i_1 = 30$ cm, a value of $o = 30$ cm is found). The solution is thus $i_2 = 60$ cm, for which a corresponding value of $o = 20$ cm is found.

4.5   **[optics: mirror]** We want an object to appear two times larger using a concave mirror of radius of curvature $R = 50$ cm. Where is the object to be located?

Solution   As in Problem 4.4 and Example 4.6, leading to two quadratic equations: real image (case 2 in Section 4.10.1): $i^2 - 125i + 3750 = 0$, $i = 75$ cm, $o = 37.5$ cm; and virtual image (case 3 in Section 4.10.1): $i^2 - 25i - 1250 = 0$, $i = -25$ cm, $o = 12.5$ cm.

4.6   **[optics: lens]** What is the F-number of a lens having $f = 200$ mm and a diameter of 100 mm?

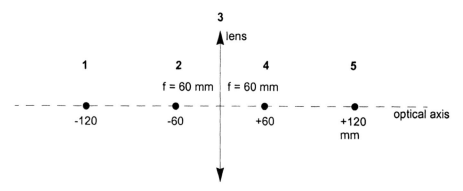

FIGURE 4.51 Configuration for Problem 4.7 (not to scale).

Solution Following the definition, the F-number is $200/100 = 2$. One talks about an F/2 lens.

4.7 [optics: lens] The focal length of the silicon lens of Figure 4.51 is 60 mm. Where is the image located if the object is located **(a)** in position 1; **(b)** at infinity on the right; **(c)** between positions 2 and 3; **(d)** in position 5; **(e)** in position 4?

Solution Equation (4.36) gives an image distance i of $1/(1/60 - 1/o)$. **(a)** 120 mm, thus in position 5; **(b)** 60 mm, thus in position 2; **(c)** let's say the object is at -50 mm and we obtained $i = -300$ mm, thus a virtual image left of position 2; **(d)** -120 mm, thus in position A; **(e)** ∞, thus infinite to the left.

4.8 [optics: lens] Using a converging lens of focal length $f = 10$ cm, we want an object to appear on a screen twice larger than its original size. Where is the object to be located in front of the screen (real image)?

Solution Substituting $f = 10$ cm, $M_{\text{lat}} = 2$ in eq. (4.40) yields $o = 15$ cm, and the real inverted image will appear 30 cm behind the lens. The object must then be located at a distance $i + o$ from the screen, that is, 45 cm in front of the screen.

4.9 [optics: filters] Assuming no multiple reflections between two juxtaposed filters of density 0.6 and 0.2, what is the corresponding transmission?

Solution Following eq. (4.41), and since densities are additive, we find that a $0.6 + 0.2 = 0.8$ density corresponds to a transmission value of $1/10^{0.8} = 16\%$.

CHAPTER FIVE

Images

In Chapter 2 we talked about thermal emission for TNDT, information about how energy is emitted from the surface inspected. In Chapter 3 the discussion was about how temperature differences are generated on these inspected surfaces through heat transfer. In Chapter 4 our attention turned to thermal sensors and optics in order to collect this emitted energy. The purpose of the present chapter is to discuss imaging: image fundamentals, data acquisition and recording, and basic image processing.

An inherent advantage of images is that "one obtains what one sees." Although true in visible applications such as in machine vision, this is not always the case in TNDT. However, infrared images can sometimes be related to their visible reality through a contour of inspected objects, for example. Two-dimensional (2D) infrared images are called *thermograms*. In some applications, one-dimensional (1D) imaging is more convenient, and this is discussed as well.

5.1 BASIC CONCEPTS

Many books are fully devoted to the topic of digital imaging, and we refer readers to Ballard and Brown, (1982), Gonzalez and Wintz (1977), Pratt (1991), and Nalwa (1993), among many other excellent books.

A *digital image* is an image $g(x, y)$ that has been discretized, that is, sampled in both spatial coordinates and intensity. This might be considered as a matrix whose row and column indices refer to a point in the image, while the matrix element corresponds to the intensity at that point. The particular value of g at a given location (x_i, y_i) is commonly referred to as a *picture element* or *pixel*. In this book, *image* refers to a thermogram generally 2D in nature, defined as a 2D radiance function $g(x, y)$, where x and y denote spatial coordinates and the value of g at any point (x, y) is proportional to the radiance or energy emitted from the scene at that location. Figure 5.1 illustrates the axis and image convention for an image having a maximum of Maxrow rows and Maxcol columns.

167

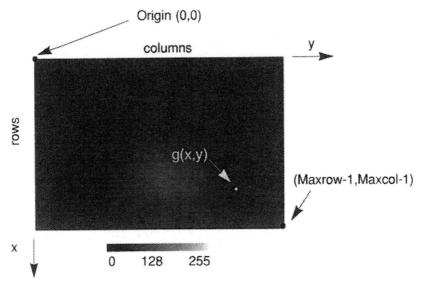

FIGURE 5.1 Axis convention used in digital thermogram representation.

The representation of Figure 5.1 is often referred to as the *gray scale representation*. The raw radiometric data obtained from an infrared camera are generally either of 8, 12, or 16 bits, giving $2^8 = 256$ (0 to 255), $2^{12} = 4096$ (0 to 4095), or $2^{14} = 16,384$ (0 to 16,384) intensity values. Traditionally, low intensities are represented by dark shades and high intensities by bright shades. Other possible representations are (Figure 5.2): 3D plot, false or pseudo color (Section 5.7), and negative image (with reversed gray scale: low intensity represented by bright shades and high intensity represented by dark shades). A 3D plot is an image function in perspective, with the third axis being the intensity value, higher peaks corresponding to greater intensities. Temperature (Section 4.8) or contrast (Section 5.6) representations are also common.

Image size is variable and depends on the infrared camera used. Common values vary from 64×64 to 512×512 and up. Rectangular arrays such as 68×105 or 120×160 are also common. In our discussion, the image size will be Maxrow × Maxcol:

$$g(x, y)$$

$$= \begin{bmatrix} g(0,0) & g(0,1) & \cdots & g(0, \text{Maxcol} - 1) \\ g(1,0) & g(1,1) & \cdots & g(1, \text{Maxcol} - 1) \\ \cdots & \cdots & \cdots & \cdots \\ g(\text{Maxrow} - 1, 0) & g(\text{Maxrow} - 1, 1) & \cdots & g(\text{Maxrow} - 1, \text{Maxcol} - 1) \end{bmatrix}$$

$$(5.1)$$

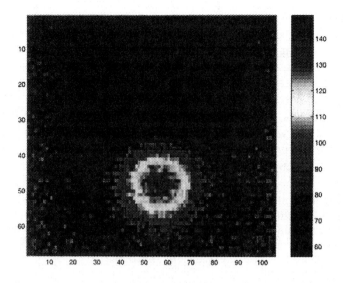

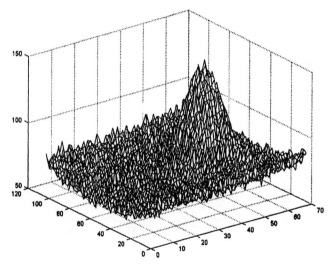

FIGURE 5.2   False color and 3D plot of Figure 5.1 ("jet" color map from Matlab).

Image size is defined as

$$\text{Maxrow} \times \text{Maxcol} \times b \tag{5.2}$$

where b is the number of bits needed to store 1 pixel.

Example 5.1   Determine the raw file size for an image having 480 columns by 320 rows, in 8 bits.

SOLUTION   Following eq. (5.2) we have $480 \times 320 \times 8 = 1{,}228{,}800$ bits or 150 kilobytes since 1 kilobyte is defined as 2^{10} or 1024 bytes and 1 byte = 8 bits. Often, an image file contains a header equal in length to one row, with image information such as the image size, date and time of acquisition, and image level and range (Section 4.8). If such an header is present, the total image size becomes $480 \times 320 \times 8 + 480 \times 8 = 1{,}232{,}640$ bits or 154,080 bytes. This image format has the typical 4:3 column/row ratio of TV standards. The file extension for such a file is *.raw*, with the image name *IMAGE1.RAW*, for instance. In Appendix F a Matlab script (read_uc.m) is provided to open and display such a *.raw* file with a gray or color scale or as a 3D plot.

The resolution, that is, the amount of detail in an image, depends strongly on the image format [Maxrow, Maxcol]. Greater values of Maxrow and Maxcol yield greater resolution for a given field of view. Unfortunately, greater values of Maxrow and Maxcol also increases the amount of storage [eq. (5.2)] and the computational time. Figure 5.3 shows the effect of changing the image sampling grid (the display area is kept constant by duplicating the pixels, thus causing a checkboard pattern). Since our brain has no natural common sense of infrared images, this effect is less dramatic than with visible images; it is, however, important in terms of the capability to detect a certain thermal entity within images. For instance, notice how the bright spot in a full-resolution image disappears in a low-resolution image. The number of pixels delivered by an in-

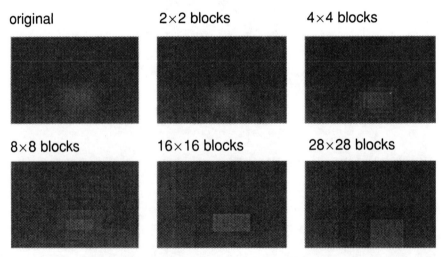

FIGURE 5.3   Effect of changing the image grid size on the image resolution for a given field of view (in all cases, the number of bits per pixels is kept to 8, corresponding to a possibility of 256 gray levels per bits, prepared with Matlab m-script: grid_sm.m (Appendix F).

original (8 bit coding)   7 bit coding   6 bit coding

5 bit coding   4 bit coding   3 bit coding

2 bit coding   1 bit coding

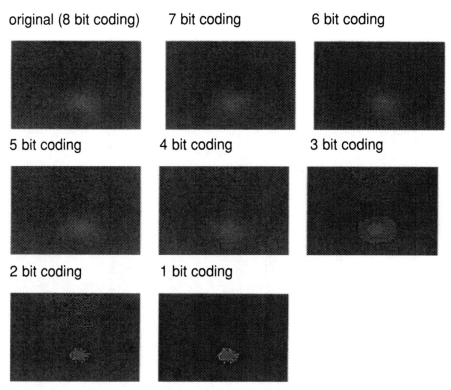

FIGURE 5.4 Effect obtained by changing the number of bits used to represent pixels in a thermogram from 8 bits to 1 bit per pixel, prepared with Matlab m-script: low_bit.m (Appendix F).

frared camera is thus an important issue to assess in order to resolve small thermal entities. This is, of course, also related to the line spread function (LSF) and slit response function (SRF) discussed in Section 4.1.7.

Figure 5.4 shows the effect obtained by changing the number of bits used to represent pixels in an image. Common values are 8 bits, corresponding to $2^8 = 256$ gray levels per pixel. Better-performing infrared systems provide images in 12 or 14 bits. An advantage of 14-bit coding is that the large span of data available ($2^{14} = 16,384$ gray levels per pixel) prevent the use of traditional range and level settings, which make absolute temperature computations trickier to handle (Section 4.8).

Before concluding this section, it is interesting to point out that Table 5.1 lists all Matlab m-scripts used to produce most of the figures in this chapter (and elsewhere as well). A list of these m-scripts is provided in Appendix F. In this way, readers are given the opportunity to make use on their own thermograms of the concepts developed.

TABLE 5.1 Matlab M-Scripts Available in Appendix F

Name of Matlab M-Script	Short Description, Purpose
read_uc.m mat = read_uc(file)	Opens *file* of unsigned characters and saves it as matrix *mat*. Size is specified in read_uc.m (including header). Example of use: **mat = read_uc('image1.raw');**
write_uc.m count = write_uc(file,mat)	Saves on disk the matrix *mat* of unsigned characters under name *file*. Returns value *count* to specify the number of bytes saved. Example of use: **count = write_uc('image1.raw', mat);**
plot_img.m plot_img(mat)	Display a 3D plot of the matrix *mat*. Example of use: **plot_img(mat);**
planck.m planck;	Displays Plancks law. No parameters to be specified. Example of use: **planck;**
grid_sm.m mat1 = grid_sm(mat,block);	Reduces the grid size of the matrix *mat* by block of size *block*. Returns the matrix *mat1*. See Figure 5.3 for an example. Example of use: **mat1 = grid_sm(mat, 4);**
low_bit.m mat1 = low_bit (mat,ratio);	Reduces the number of bit per pixel of the matrix *mat* by value *ratio*. Returns the matrix *mat1*. See Figure 5.4 for an example. Example of use: **mat1 = low_bit (mat,8);**
read_fl.m mat = read_fl(file)	Opens *file* of float numbers (32 bits) and saves it as the matrix *mat*. Size is specified in read_fl.m (including header). Example of use: **mat = read_fl('image1.raw');**
mat_fl_to_uc.m mat = mat_fl_to_uc(matfl) Uses internally function **calc_echel.fm**	Converts the floating-point matrix *matfl* as an unsigned character (1 byte per pixel) matrix *mat*. Size is found automatically. Example of use: **mat = mat_fl_to_uc (mat_fl);**
black.m color = black(num)	Changes the color map of Matlab to *num* levels of black values and returns the new *color* color map matrix. Useful to get all black 3D plots from *plot_img.m*. Example of use (for 64 levels): **black(64);**

matresul.m

mat = matresul(matf,t1,t2)

Uses internally function

black.m

Extracts relevant time zone (*t1* to *t2*) data from MATRESUL.DIS matrixes (*matf*) and returns a new matrix *mat*, plotted automatically.

Example of use: **mat2 = matresul (matf,57,62);**

poin_pro.m

mat1 = poin_pro(mat, a, b)

Point processing routine with parameters *a* and *b*. Returns a processed image, based on Figure 5.13. Images are of unsigned character type.

Example of use: **mat1 = poin_pro(mat,10,25);** with $b = 255$, *a* becomes the threshold [eq. (5.15)].

median_filter.m

mat1 = median (mat, size)

Noise suppression routine through median filtering. Returns a processed image. Images are of unsigned character type. Parameter *size* defines the size of the kernel of interest.

Example of use: **mat1 = median(mat,3);**

kernel.m

mat1 = kernel (mat,a_scale, array)

Image enhancement routine through the use of a kernel of values, eq. (5.16). Returns an enhanced image. Images are of unsigned character type.

Example of use: **mat1 = kernel(mat, 2, array);** with array defined previously as for a 2×2 Roberts gradient: array = [1 0; 0 −1].

stretch.m

mat1 = stretch (mat, gmin, gmax)

Histogram stretching through linear remapping of the gray level of the image mat within the available gray-level range *gmin* to *gmax* [eq. (5.17)]. Images are of unsigned character type.

Example of use: **mat1 = stretch(mat,0,255);**

enh_pro.m

mat1 = enh_pro(mat,k1,k2)

Image enhancement routine with parameters k1 and k2. Returns an enhanced image, based on eq. (5.19). Images are of unsigned character type.

Example of use: **mat1 = enh_pro(mat,5,0.5);**

snr.m

snr (mat1,mat2)

Routine to compute the signal-to-noise ratio between two images mat1 and mat2 recorded one after another, based on eq. (5.13). Images are of unsigned character type.

Example of use: **snr(mat1,mat2);**

smooth.m

mat1 = smooth(mat,B)

Routine to smooth out noise from the image mat with parameter *B*, based on eq. (5.29). Images are of unsigned character type.

Example of use: **mat1 = smooth(mat,4);**

(Continued)

173

TABLE 5.1 (*Continued*)

Name of Matlab M-Script	Short Description, Purpose
ctrst.m mat1 = ctrst (TYPE, x1, y1, x2, y2, mat, mat0)	Routine to compute the thermal contrast of images (of unsigned character type). Returns a floating-point contrast image. Parameter TYPE allows us to compute the various contrasts discussed depending on the value provided (Section 5.6): (1) absolute contrast, (2) running contrast, (3) normalized contrast, (4) standard contrast. Defect-free zone within the image is defined with a rectangle having upper-left and bottom-right coordinates (x1,y1) and (x2,y2), respectively. The thermal image before heating is mat0. Example of use: **mat_ctr = ctrst(4, 5, 5, 10, 80, mat, mat0);**
profil.m profile = profil(mat, nrow, value)	Routine to extract a profile located at row *nrow* from image matrix *mat*. The function returns the profile vector *profile*, which can be displayed with the command **plot(profile,'-k');** or **plot(profile,':k');** where 'k' stands for black color plot, ' – ' for solid line, ':' for dotted line. The function displays the original image *mat* with a superimposed line indicating the profile location. This line has a *value* value. Example of use: **profile = profil(mat, 50, 130);** Useful related commands: **hold on;** or **hold off;** maintains or releases the figure displayed (useful to superimpose several profiles on the same figure).
color_to_bw.m mat = color_to_bw (matr,matg,matb,x1,y1,x2,y2)	Routine to convert a color image into a black-and-white image based on a color scale included in the image and located in the rectangle x1, y1 and x2, y2 (x for row, y for column). Provided are the three split channels on 8 bits: red (matr), green (matg), blue (matb). The matrix mat returned is the black-and-white image coded on unsigned character (0 to 255). Example of use: **mat = color_to_bw (matr, matg, matb, x1,y1, x2, y2);** This routine is useful to convert color-scanned printed thermograms whose black-and-white versions are not available. Split channel images (to be saved in *.raw* format) are obtained easily from the original image in software such as Paint Shop Pro,[a] among others. Image size and color scale location are obtained easily with such programs. *.raw* split images can be read with read_uc.m script modified for appropriate image size.

[a] www.jasc.com.

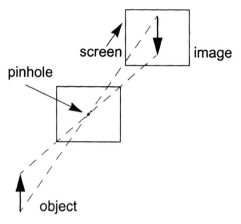

FIGURE 5.5 Pinhole camera model.

5.2 IMAGE FORMATION

We now discuss how images are formed by an infrared camera. Unless the scene being inspected is flat, a 3D-to-2D conversion takes place between the real world and the image [eq. (5.1)]. This conversion is performed by the camera optics. We briefly review three different camera models: pinhole, perspective projection, and orthographic projection.

5.2.1 Pinhole Camera Model and Perspective Projection

The pinhole camera model is made of an infinitesimal small aperture through which light rays (or infrared rays in our case) from the scene pass prior to hitting a screen (i.e., sensor surface) on which the image forms: straight rays pass through the pinhole and form an inverted image on the screen (Figure 5.5). The mapping from 3D to 2D is called the *perspective projection* and is defined as the projection of a 3D object or surface onto a 2D surface (sensor plane) by means of straight lines originating from a single point, as illustrated in Figure 5.6. The originating point is called the *center of projection*. As seen in the figure, since we are free to put the screen (image sensor) where we want, the image is no longer inverted as in the pinhole model.

The relationship between image coordinates and object coordinates is given with similar-triangle analysis by

$$x_i = \frac{f}{z_0}x_o \quad \text{and} \quad y_i = \frac{f}{z_0}y_o \tag{5.3}$$

where f is the distance between the center of projection and the image plane (screen) and the i and o subscripts denote image and object, respectively. Since eqs. (5.3) are not linear, they are not convenient to handle (a linear function is a

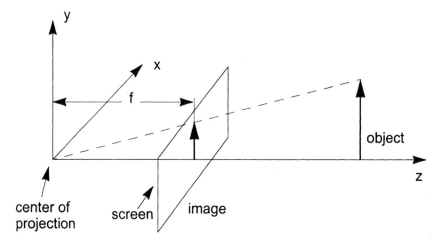

FIGURE 5.6   Perspective projection.

function whose variable appears only at the first power, excluding products between variables). One way to bypass this mathematical difficulty is to work in homogeneous coordinates. Homogeneous coordinates are defined in a four-dimensional (4D) space by mapping the 3D point (x, y, z) to the straight line passing through the origin of the 4D space such that $(X, Y, Z, W) = (wx, wy, wz, w)$.

Other important concepts related to perspective projection are the vanishing point and the spherical perspective projection. Two parallel lines in the 3D world will appear to merge together at infinity. The point where they seems to merge is called the *vanishing point.* A common example of that is when one looks at a railroad track: Although the distance between the two rails is constant, the rails appear to join far away. Obviously, such image distortion can be misleading in some applications. For these reasons, other camera models were developed. Strictly speaking, the perspective projection illustrated in Figure 5.6 is a planar perspective projection since the image is projected onto a plane. The problem with the planar perspective projection is that the image size depends on the position and orientation of the projection plane. One way to circumvent this difficulty is to project the image on a sphere of unity radius centered on the center of projection, called *spherical perspective projection.*

As discussed in Section 4.10, with a reduced aperture, a pinhole camera provides sharp images over a large depth of field. However, if the aperture is reduced further, blurring appears due to diffraction (diffraction is the bending of light rays intercepting an obstacle or an aperture of size close to their wavelength).

5.2.2   Orthographic and Parallel Projections

Figure 5.7 illustrates the principle of orthographic projection. In this case, 3D real world-to-2D image plane mapping is performed by parallel rays orthogo-

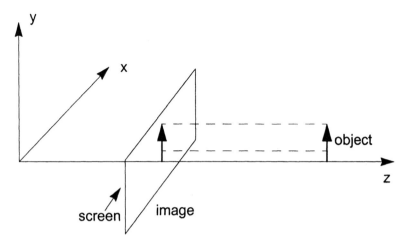

FIGURE 5.7   Orthographic projection.

nal (i.e., perpendicular) to the image plane, so that nonlinear eq. (5.3) reduces to the following linear equations:

$$x_i = x_o \quad \text{and} \quad y_i = y_o \tag{5.4}$$

Conveniently, orthographic projection approximates perspective projection to a scale factor if (1) the object is close to the optical axis (line orthogonal to the image plane and passing through the center of projection; Section 4.10) and (2) the size of the object is small with respect to the distance from the center of projection.

Orthographic projection can be generalized to parallel projection if the set of parallel rays projected from the object intercept the image plane at an angle other than 90°. As in orthographic projection, parallel projection is also a linear mapping of the 3D world and approximates the perspective projection to a scale factor.

5.3   DATA ACQUISITION AND RECORDING

5.3.1   Photographic Film Recording

Traditionally, IR images were recorded directly from a video monitor (color or black and white) on photographic film. Such an approach was good for archival purposes (to make a permanent record of the inspection) and rapid in the case of instant Polaroid cameras. IR images were used primarily for stationary thermal events, such as to pinpoint an abnormal condition found in a passive approach survey of an electric utility. However, in the case of the active TNDT approach, they suffer from many drawbacks. For instance, from a given experiment only a couple of images can be recorded. Transient thermal events are lost, and they are not practical for quantitative analysis (various distortions occur which are difficult to correct if the image is to be digitized, for instance on a flatbed digitizer).

Conventional photographic film is made of various layers. When the film is exposed to light, the emulsion (silver halide grains) absorbs optical energy so that areas of the film (strongly) exposed to light experience greater density of silver deposits after film development, while areas less exposed to light exhibit less. Exposure to light precipitates silver halide as metallic silver, and film development removes the remaining silver halide grains (i.e., those unexposed). Since silver deposits are opaque to light, a common negative image is obtained (areas of greater exposition appear darker). The process is repeated to get a positive image: The negative image is projected on photographic paper carrying a layer of silver halide, and after development the paper shows a positive image.

5.3.2   Videotape Recording

For a conventional North American black-and-white (B&W) video signal (RS-170), images are produced at a rate of 30 Hz (i.e., 30 images per second), 25 Hz in Europe. Suppose the image format is 512×512 pixels; this corresponds to a data rate of 7.5 megabytes per second (MB/s) to be recorded (1 megabyte $= 2^{20} \leftrightarrow 10^6$). Real-time recording will be achieved if the recording device can sustain a recording of 7.5 MB/s (*real time* means that the device is capable of recording all incoming data without loss). In this particular example, 2 min 16 s requires a storage capacity of 1 gigabyte (gigabyte $= 2^{30} \leftrightarrow 10^9$). Although hardware compression is possible, such values are quite demanding, and faster frame rates are still more demanding. This explains the popularity of the videotape recorder (VTR) to collect video signals, and for active approach experiments and during long surveys in the passive approach.

For example, a reported application involves airborne survey over high-powered electric lines in order to pinpoint abnormal overheating (e.g., static line arcing to ground through a damaged porcelain isolator). During such a survey, images from the IR camera are recorded on videotape, and analysis proceeds off-line. Special markers on the tape allow us to retrieve the location of interest on the electric line.

The good thing about VTRs is that they are affordable, portable (e.g., a camcorder can be used for such a purpose by setting the selection auxiliary switch so that the signal to record comes from the IR camera instead of from the camcorder itself), and convenient, since you can record many hours on a single \$2 cassette. On the other hand, problems with VTR recording are related to the limited video signal-to-noise ratio, which is around 40 dB typically. For 8-bit images, a dynamic range of at least 48 dB $[20 \log_{10}(256/1)]$ is needed (and much more for 12- or 14-bit systems).

Intuitively, the dynamic range can be viewed as the ratio between the smallest and largest observable signal expressed in decibels. If VTR specifications are not available, the following simple experimentation can be performed to get an estimate of the dynamic range figure. An image constituted of vertical bars is generated (i.e., at the output of a frame grabber; see below). The bars are alternatively values of 127 and 128, respectively (on an 8-bit frame grabber). The bar image is recorded and played back on the VTR while the VTR output is observed on a good-quality monitor. If the bar pattern is restored without *much distortion*, we consider the dynamic range as being about 48 dB (since the intensity difference in the bar pattern is $128/256 - 127/256 = 1/256$. Generally, commercial household VTRs have a dynamic range of about 40 dB, which thus slightly corrupts 8-bit infrared images and is unacceptable for 12-bit systems because of attenuation of high frequencies.

Another problem with VTR is automatic gain control (AGC), which, if not disabled, will displace the average image intensity value: dark images will be brighter, and bright images will be darker. This AGC is designed for comfort image viewing by the operator. However, it distorts the absolute pixel values. If it is not possible to disable the AGC, a gray scale has to be superimposed on the image prior to recording. Knowledge of the absolute values of this gray scale will allow us to restore the original pixel values after digitizing. This method is, however, time consuming.

As an example of the AGC effect, the following values have been obtained in the case of the intensity of a given white reference square always present in the infrared image area:

	All-White Image	All-Black Image
Direct transfer	181	180
Transfer via the VTR (AGC present)	167	145

Direct transfer implies digitizing of the image without prior recording on the VTR (so no AGC effect). The main sources of noise for a magnetic VTR recording unit are found especially at the level of recording and playback electronic circuitry rather than at the tape level, unless tape roughness hits the recording head, causing a burst of erroneous data.

As a last point concerning video signal recording, we should always be careful while propagating a video signal through the equipment (VTR, TV monitors, digitizer, etc.) to avoid signal degradation. Distribution video amplifiers (such as the Panasonic WJ3008) can be inserted in the video path to maintain a correct analog video signal amplitude, and in particular, to maintain the amplitude of synchronization pulses.

Example 5.2   Find the capacity needed to record a 5-s active thermography experiment if the IR camera produces 14-bit images of format 256×256 pixels at a frame rate of 400 Hz (no pixel packing).

SOLUTION Since images are of 14 bits and since the computer memory standard is 8 bits, the easiest solution (no pixel packing) is to store one pixel on 2 bytes (i.e., on $8 + 8 = 16$ bits). Thus one image requires $256 \times 256 \times 2 = 131{,}072$ bytes, and for 5-s duration this is $131{,}072 \times 400 \times 5 = 250$ MB.

5.3.3 High-Performance Hard Disk Storage

High-performance hard disk storage is possible on special recorders in which the video signal is stripped across multiple disk drives connected in parallel. In some of these recorders four or eight disks store the video signal in parallel, with one disk dedicated to a redundancy or parity check. Several companies, among them Storage Concepts, Inc.,* have products achieving up to 200 MB/s with a capacity of up to 1.6 terabytes (1 terabyte $= 2^{40} \leftrightarrow 10^{12}$). Smaller recorders with 80 or 100 MB/s and a capacity of 4, 9, and 18 GB are also available. Due to their high costs, these recorders are still reserved for high-end applications.

Example 5.3 How long can the 7.5-MB/s signal be recorded if a fast 1.6-TB capacity recorder is available?

SOLUTION It suffices to divide 1.6 TB by 7.5 MB: $1.6 \times 2^{40}/7.5 \times 2^{20} = 2.6$ days. This starts to be comfortable!

5.3.4 Direct Digital Recording

The first IR cameras had only a B&W video output and video signal was only available in analog format (e.g., the famous AGA 782 of the late 1980s; the company was later renamed Agema and is now known as Flir[†]). To digitize images, one had to rely on an external frame grabber board plugged into a computer (PC). Companies such as Data Translation (Marlboro, Massachusetts), Dipix (Ottawa, Canada), Matrox and Occulus (Montréal, Canada), and Sharp (Irvine, California), offer such products. Software for image manipulations was available from IR camera vendors or was home brewed.

More recently, thanks to progress in digital electronics, IR cameras have became available with various outputs, such as color NTSC (Networking and Telecommunication Standing Committee) compatible composite video and full or limited digital output in addition to the B&W analog video. This was, for instance, the case for the Inframetrics 600, featuring an RS-232 serial output, making it possible to extract an IR image one line (256 pixels) at a time on an 8-bit format (Anderson, 1990; Inframetrics is now known as Flir[†]).

Nowadays, current IR cameras have full real-time digital output. In such a case, a dedicated board is located in a computer to perform for interfacing.

*www.storageconcepts.com.
[†] www.flir.com.

Special software is required to control the interfacing board (generally provided by the IR camera vendor). Full-digital output is advantageous since analog-to-digital (A/D) conversion is performed by optimized electronics located close to the IR detector, thus helping to reduce the noise level. Moreover, since the frame rate is no longer tied to a video standard (e.g., 30 Hz or 25 Hz, as discussed above), it becomes possible to adjust it to particular needs. For example, one such IR camera from Raytheon* offers 140 Hz at full frame (256×256 pixels) up to 1800 Hz in a restricted subwindow (64×64). In such cases, the free memory of the host computer can be used to record the image sequence in real time (standard hard disks are too slow).

Example 5.4 How long can an active thermography experiment be recorded from a Raytheon IR camera operating in the 256×256 mode if the host PC has 128 MB of free RAM (random access memory)?

SOLUTION   IR images are generated at a rate of 140 Hz, thus requiring: $256 \times 256 \times 140 = 8.75$ MB/s. Since 128 MB is available, one can record for 14.6 s before the memory is full. Adding another 128 MB in that computer will boost the capacity to close to 30 s of full recording. Such an approach is particularly interesting since RAM memory is quite affordable [about $150 for 128 MB (2000 price)].

Although direct digital recording cannot compete with high-performance hard disk storage, it is a practical alternative. Adding more free RAM in the host computer obviously extends the recording period. Moreover, it can be lengthened by not recording all images. In Section 10.2, in the case of pulsed thermography, we discuss acquiring IR images on a logarithmic time scale. The idea is to modulate the recording cadence so that the acquisition rate is high shortly after the heating pulse, when thermal transients are likely to be relevant, and the acquisition rate is low later, when specimen temperature cools down slowly.

5.4   IMAGE DEGRADATIONS

Degradations are of three types: radiometric distortion, geometric distortion, and noise. In a formal manner, we can express the global acquisition process for position (i, j) in the image by

$$I_{CAM}(i, j) = S[h(i, j) * T(i, j)] + n(i, j) \qquad (5.5)$$

where $*$ is the convolution operator, $S[\cdot]$ is the radiometric distortion, $h[\cdot]$ a linear spatial operator that is concerned with geometric factors and with band-

* www.raytheon.com/rsc/ses/spr/spr_cir.

width, $n(\cdot)$ the random noise considered additive, $T(\cdot)$ the ideal temperature image, and $I_{CAM}(\cdot)$ the recorded signal corresponding to the radiance contribution on the calibration curve (Section 4.8). The objective of the restoration process is to recover $T(\cdot)$ starting with $I_{CAM}(\cdot)$. This process is complex since nothing a priori is known about $S[\cdot]$, $h(\cdot)$, and $n(\cdot)$, which thus must be found experimentally. The correction for radiometric effects in Chapter 4 allows to evaluate $S[\cdot]$, and we obtain

$$I'_{CAM}(i, j) \sim h(i, j) * T(i, j) + n(i, j) \tag{5.6}$$

Next, a filter can serve to *eliminate* the noise while maintaining edges in the image. The resulting image is given by

$$I''_{CAM}(i, j) \sim h(i, j) * T(i, j) + n(i, j) - n'(i, j) \tag{5.7}$$

This equation indicates that deconvolution is necessary to recover the original image $T(i, j)$ from $I''_{CAM}(i, j)$. In the frequency domain, this is expressed by an inverse filter. We rewrite eq. (5.7) as

$$IF''(u, v) \sim HF(u, v)TF(u, v) \tag{5.8}$$

where $IF''(u, v)$, $HF(u, v)$, and $TF(u, v)$ are the Fourier transform of $I''_{CAM}(i, j)$, $h(i, j)$, and $T(i, j)$, respectively. From $HF(u, v)$, an inverse filter $R(u, v)$ can be found:

$$R(u, v) = \frac{HF^*(u, v)}{|HF(u, v)|^2} \tag{5.9}$$

where $HF^*(u, v)$ denotes the conjugate of $HF(u, v)$. This elementary filter considers that $h(i, j)$ is known exactly and that no noise is present. It is also not very efficient near zeros of $HF(u, v)$, where $R(u, v)$ takes infinity values. We finally obtain

$$TF(u, v) \sim R(u, v)IF''(u, v) \tag{5.10}$$

All the restoration steps can be summarized as follows (Figure 5.8):

$$T(i, j) = S^{-1}[I_{CAM}(i, j)] * r(i, j) + n(i, j) - n'(i, j) \tag{5.11}$$

FIGURE 5.8   Signal restoration step. (From Maldague, 1993.)

It is interesting to consider the appropriateness of these operations, taking into account their *approximate* character and the need to satisfy in the context of TNDT:

1. *Radiometric distortions.* Consider this simple test. A thick brass (high-thermal-conductivity) plate is brought to above ambient temperature. It is observed with an infrared camera and an image is recorded. In most cases, the recorded image will not be uniform; moreover, we will observe that the nonuniformity of the image will depend on the plate temperature. Figure 5.9 presents an example of such a test. This simple test indicates that radiometric corrections are fundamental in performing quantitative measurements.

2. *Spatial geometry effects.* Consider this simple test. A black painted aluminum plate on which narrow grooves 1 cm apart are machined on the surface is observed with an infrared camera and an image is recorded.

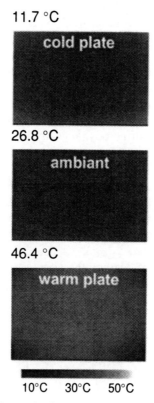

FIGURE 5.9   Effects of radiometric distortions: thermograms recorded on a thick brass plate of high thermal conductivity heated at a uniform temperature by immersion in a water tank at different temperatures (11.7, 26.8, and 46.4°C). The infrared camera objective was at ambient temperature. (From Maldague, 1993.)

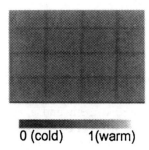

0 (cold)   1(warm)

FIGURE 5.10 Spatial geometry effects: thermogram of a plate with grooves (1 cm apart, 0.3 mm width, 0.1 mm depth). The plate is heated uniformly in a warm bath whose temperature is close to infrared camera temperature, to limit distortion effects; relative scale. (From Maldague, 1993.)

Figure 5.10 shows a typical image, and Figure 5.11 presents the image profile along a line. From these figures we notice the sharp transition of a groove span over 3 pixels instead of 1 (this is related to the apparent resolution discussed in Section 4.1.7). The temperature of the plate is close to the temperature of the camera casing, to limit the radiometric distortions, which are, however, visible. For TNDT applications, important image features (defects) appear as smooth transitions in the temperature images due to the diffusion process, as seen in Figures 5.1 to 5.4. For these applications, spatial geometry effects are not especially significant. In some cases where sharp temperature transitions must be observed, such as for the study of aerodynamically heated structures (e.g., missile atmospheric reentry), corrections based on LSF and SRF notions are possible (Wong, 1982).

3. *Noise effects.* Consider an infrared image recorded on a typical active TNDT scene with an infrared radiometer; a 3D plot is likely to resemble Figure 5.2. This figure reveals a strong high-frequency noise level which must be taken into account.

This study of thermogram degradation allows us to rewrite eq. (5.11):

$$T(i, j) \sim S^{-1}[I_{CAM}(i, j)] + n(i, j) - n'(i, j) \tag{5.12}$$

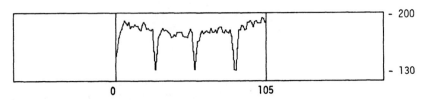

FIGURE 5.11 Profile along a row (row 38) of Figure 5.10. The four grooves are clearly distinguished. Horizontal scale, pixel coordinates along one row; vertical scale, thermogram intensity. (From Maldague, 1993.)

This relation is used to restore infrared images prior to their analysis. Radiometric effects (i.e., spectral response effects from the camera detector, optical path differences, entry pupil, etc.) were studied in Chapter 4, as was the calibration process (Section 4.8). The noise effect is studied next. Before we close this section, we would like to cite Fred E. Nicodemus (1967, p. 289), who advises those who are too confident in these *corrections* which can be applied to recorded signals: "Although often termed 'corrections,' they are unvoidable uncertainties in these transformations so that the raw measurements are usually more accurate or 'correct' (when rightly interpreted as conditions at the instrument) than are the final 'corrected results.'"

5.4.1 Noise (in Images)

In electronic vision systems, many different types of noise can be found and are depicted in Figure 4.5. To this *system* noise, particularly for infrared systems where signals are naturally weak (as discussed in Chapter 2), many adverse effects (e.g., parasitic emission, thermal reflections, random arrival of photons) contribute to further degrade the signal.

At the infrared detector level, we notice essentially Johnson or thermal noise (present in conductors and due to the random motion of charges in a solid), flicker or $1/f$ noise (present in semiconductors, it depends on the observation frequency), and shot noise (caused by the random and discrete photon arrivals in the incident radiation; Section 4.2). For infrared radiometers, at the end of the acquisition process, at the pixel level, the noise is additive, of Gaussian nature, and of high frequency with respect to the useful signals. An increase in the image exposure time, thus an averaging of consecutive images, will allow a reduction of these adverse effects. In this case the improvement of quality is a function of the square of the summed images. This averaging technique can be used whenever possible since it is simple and preserves edges, which is not the case when averaging is performed within the image itself (such as with local averaging in 3×3 kernels; Section 5.5).

5.4.2 Noise Evaluation

It is of interest to characterize the noise content present in infrared images using a technique proposed by D. J. Lee et al. (1987, p. 297) and Haddon (1988, p. 197). It is only necessary to record two images, A and B, of the same scene: If they are recorded in the same conditions, the only difference between them is noise, which can thus be evaluated. The signal-to-noise ratio (SNR) can be computed using the following formulas:

$$\text{SNR} = \frac{\text{average power image}}{\text{average power noise}}$$

$$= \frac{\sum_i \sum_j [A(i,j)]^2}{\frac{1}{2}[\text{standard deviation } \langle N(i,j) \rangle]^2 \cdot \text{Maxcol} \cdot \text{Maxrow}} \tag{5.13}$$

where N is the noise $\sim |A - B|$, and

$$\text{standard deviation } (N) = \sqrt{\frac{\sum_i \sum_j [N(i,j) - \eta]^2}{\text{Maxcol} \cdot \text{Maxrow}}}$$

where η is the average of N, $i = 0, 1, \ldots,$ Maxrow $- 1$ (Maxrow rows in images), and $j = 0, 1, \ldots,$ Maxcol $- 1$ (Maxcol columns in images). The $\frac{1}{2}$ factor is introduced in eq. (5.13) because the gray-level variance for the difference image is equal to twice the noise variance of the individual image, considering a Gaussian noise distribution. To add some scaling, a practical formulation is to multiply the result obtained from eq. (5.13) by $10 \cdot \log_{10}$.

As an example of this method, repeated tests gave an average SNR of around 30 for a given radiometer-based thermal imaging system and about 70 for an FPA-based system. As a comparison, the same procedure used with a CCD video camera (Panasonic WV CD-50) gave a value of 1330. As expected, noise is more important for the infrared radiometer. Mathlab m-File: snr.m (Table 5.1) can be used to compute the SNR using such a method (with a scaling factor of $10 \cdot \log_{10}$). In the next section, techniques are proposed to reduce noise effects.

5.5   IMAGE RECTIFICATION

Degradations from multiple sources might affect image quality as discussed in Section 5.4. Although in some instances it is possible to repeat a given measurement so that better images are obtained, this is not always possible, and sometimes we have to stick with what is available in the first place. In these situations image rectification is helpful. Entire chapters are devoted to such a topic in digital image processing books (e.g., in Pratt, 1991); in this section we present a few techniques found useful in TNDT. Often, image rectification is pursued for visualization purposes, for example to check whether or not a defect is present. In some other instances, image rectification precedes other manipulations, such as those required to extract quantitative information about specimens tested (estimation of defect depth, etc., Chapter 10). Interestingly, general image packages such as Paint Shop Pro or Corel Photo-Paint, among others, provide a variety of useful tools, while specialized programs offer more advanced features. Basically, there are two different methods for image enhancement: those based on the spatial domain and those based on the frequency domain.

5.5.1   Spatial Domain Image Enhancement

Spatial domain enhancement refers to processing directly on pixel values on the image plane, the general approach is based on the following equation:

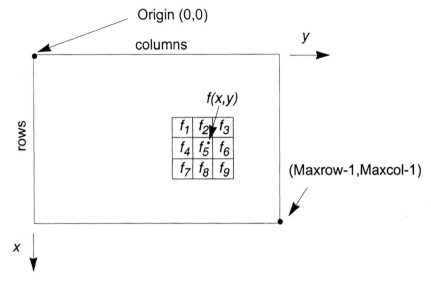

FIGURE 5.12   General principle of spatial domain enhancement, a 3 × 3 neighborhood (= kernel) centered on pixel of interest located in (x, y) is shown.

$$\hat{f} = g[f(x, y)] \qquad (5.14)$$

where f is the original image, $\hat{f}$ the processed image, g the processing operator, and (x, y) the pixel location. Since spatial enhancement techniques operate directly on the image plane, the operator g will be defined on a specific neighborhood, centered on a pixel of interest (x, y). Most common neighborhoods are square or rectangular; other geometries, such as circles, are possible but less common (Figure 5.12). At the limit, the neighborhood is restricted to the pixel itself (size 1 × 1) and this is called *point processing*. In such a case, depending on operator g contrast, stretching or binarization is obtained. For binarization, g is only a threshold (Figure 5.13a):

$$f(x, y) < \text{threshold} \rightarrow f(x, y) = 0 \quad \text{(black)}$$
$$f(x, y) < \text{threshold} \rightarrow f(x, y) = 1 \quad \text{(white)} \qquad (5.15)$$

For contrast stretching, the range of available values from black to white is divided unequally and devoted to a specific range of input values. Figure 5.13 illustrates some effects to emphasize low or high values with straight lines crossing at (a, b). Obviously, more complex functions can be devised.

For neighborhood different than 1, one speaks in term of mask processing or filtering with various effects. In such cases, the central pixel $f(x, y)$ is replaced by the value computed from the following equation, assuming the n kernel

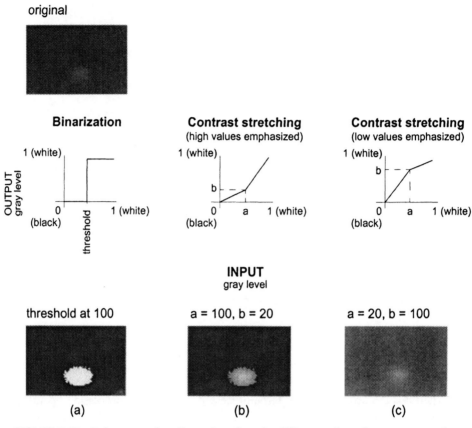

FIGURE 5.13   Point processing. Examples of use for different values of operators a and b, prepared with Matlab m-script: poin_pro.m (Appendix F).

values are $a_1, a_2, a_3, \ldots a_n$ (Figure 5.12):

$$f'(x, y) = a_{\text{scale}}(a_1 f_1 + a_2 f_2 + \cdots + a_n f_n) \tag{5.16}$$

The scaling factor a_{scale} is necessary to keep the processed image within the same range of values as the original image. For instance, smoothing or averaging (low-pass filtering) is obtained if all filter values are set to 1 with $a_{\text{scale}} = 1/n$. The averaging effect obtained reduces high-frequency components within the image, often affected by the noise (Figure 5.14a). High-frequency components of an image are zones of rapidly changing values such as sudden transition from cold (black or 0 value) to hot (white or 1 value). A more efficient noise filter that tends to preserve edges is the median filter described below.

When noise present in the image is spiky in nature, and when it is necessary to preserve edges, median filtering is useful. Using this method, the central pixel value of the kernel is replaced by its median value. For example, if the 3×3

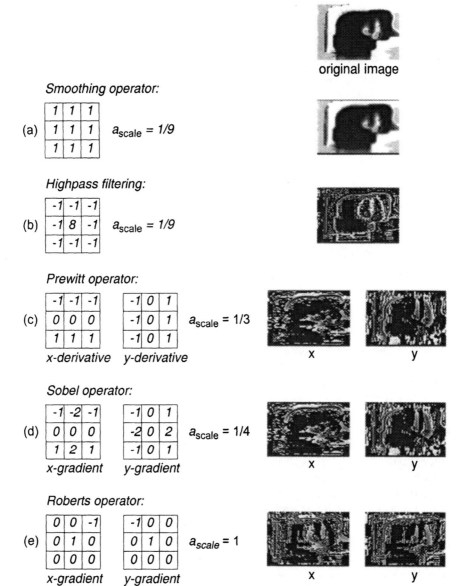

FIGURE 5.14 Various 3×3 kernels used with eq. (5.16) and their effects for image smoothing, sharpening or gradient and derivative computation, prepared with Matlab m-script: kernel.m (Appendix F). Original thermogram recorded during pipe inspection (Section 11.2.1).

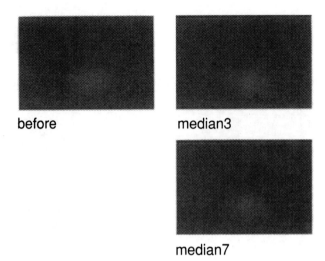

before   median3

median7

FIGURE 5.15 Median filtering and its effects for noise suppression, prepared with Matlab m-script: median_filter.m (Appendix F). Original thermogram of Figure 5.12.

kernel has the following values (sorted in ascending order: 63, 65, 66, 67, 68, 71, 72, 73, 140), the central pixel will be replaced by 68. This method has the advantage of preserving edges while rejecting high-frequency noise; it makes pixels appear more like their neighbors (Figure 5.15).

Equation (5.16) is also used for image sharpening (high-pass filtering). A 3×3 filter is obtained with all coefficients set to -1 but the central value, which is set at $+8$, and $a_{scale} = \frac{1}{9}$. It is noted that the sum of all coefficients is zero; moreover, such a filter reduces the image average to zero, and thus negative values might appear (that could be clipped to zero). This filter enhances image edges (Figure 5.14b).

Differentiation is also a convenient way to sharpen images. For this, eq. (5.16) is used to compute a gradient. In the case of a 3×3 kernel, derivative along the x direction is obtained when a difference between the first and last rows is performed while the derivative along the y direction is obtained when a difference between the first and last columns is performed (Figure 5.14c). Coefficients sum to zero since the derivative of a set of uniform values is zero. Such kernels are known as *Prewitt operators*, and they can be extended to a larger size. For instance, a 7×7 kernel for a y derivative is obtained with the central column set to zero, left columns to $+1$, right columns to -1, and a_{scale} to $1/21$ (Figure 5.14c). *Roberts and Sobel gradient operators* are defined similarly, still based on eq. (5.16) and particular mask values (Figure 5.14d,e). See also Section 10.2.4.

Other enhancing methods involve histogram manipulation; they have been reviewed by Pratt (1991) and by Gonzalez and Wintz, (1977). A gray-level histogram of an image is a graphic representation indicating the occurrence of each possible gray level within that image. For example, gray-level stretching

original image

stretched image: levels 0-128

stretched image: levels 128-255

FIGURE 5.16   Histogram stretching and its effects to improve image contrast, eq. (5.17). Prepared with Matlab m-script: stretch.m (Appendix F).

involves employing all possible gray levels through mapping. For a low-contrast image, the effect is a better visualization of information lost in too narrow a span of gray levels (Figure 5.16). For example, convenient linear mapping of an image having minimum and maximum values min and max to a stretched image with minimum and maximum available gray levels G_{min} and G_{max} is obtained with the following formula [$g(x, y)$ is the processed pixel and $f(x, y)$ is the original pixel value]:

$$g(x, y) = m \cdot f(x, y) + b \qquad (5.17)$$

where

$$m = \frac{G_{max} - G_{min}}{max - min}$$

$$b = G_{max} - m \cdot max = G_{min} - m \cdot min$$

Example 5.5   Suppose that in an 8-bit thermogram the minimum and maximum values observed are 40 and 234, respectively. To what values is a pixel of value 100 mapped after linear stretching?

SOLUTION   An 8-bit thermogram involves an available range of values spanning from 0 to 255 ($2^8 - 1$). Following eq. (5.17) we have $m = 1.31$ and

$b = -52.6$, and then the new mapped value turns out to be 78 and the global effect of the contrast stretching will not be dramatic. It is good practice to smooth out noise before performing such processing since a single high- or low-value spiky pixel might affect such processing completely. For example, assuming that 234 is in fact a spiky isolated pixel and that after smoothing the maximum observed is 158, our pixel of value 100 will be mapped to 130.

5.5.2   Detail Enhancement of Thermograms Using Local Statistics

Generally, infrared images tend to exhibit limited spatial contrast. Consequently, procedures are needed to enhance subtle details which are hardly noticed on the original thermal image. Techniques based on local statistics within the image are simple to implement (J. S. Lee, 1980). Pixel $p(i, j)$ of the thermogram is recalculated based on its local mean $m(i, j)$ obtained over a squared window 3 pixels wide centered in (i, j):

$$p'(i, j) = (1 - K_1)m(i, j) + K_1 p(i, j) \qquad (5.18)$$

where K_1 is the gain factor. With $K_1 = 1$, there is no effect. With $K_1 = 0$, the recalculated pixel $p'(i, j)$ is equal to the local mean and a smoothing effect is obtained, as with $0 < K_1 \leq 1$. With $K_1 > 1$, the image is sharpened through an increase in high frequencies, the result being similar to a high-pass filter. Such a method is advantageous since it does not require computation of the variance, as in the case of some other local statistics enhancement approaches (variance computation is time consuming). To increase flexibility, eq. (5.18) is modified as follows:

$$p'(i, j) = (K_2 - K_1)m(i, j) + K_1 p(i, j) \qquad (5.19)$$

This modification spreads values of $m(i, j)$. Figure 5.17 shows a typical result obtained in the case of the thermographic inspection of a turbine blade through a transmission technique by water circulation. A sudden temperature transition of circulating water (20 to 40°C) reveals the internal structure; details on this inspection technique are exposed in Section 11.2. In this case, interest is to show

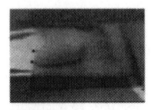 

before   after

FIGURE 5.17   Detail enhancement of thermograms using local statistics, for a turbine blade inspected through water circulation at different temperatures (Section 11.2.2), prepared with Matlab m-script: enh_pro.m (Appendix F).

more clearly the internal structure (deflectors) of the cooling ducts within the blade. Equation (5.19) was applied to the left image with $K_1 = 5$ and $K_2 = 0.5$, resulting in the right enhanced image having a larger distribution of pixel values: minimum 0 and maximum 236 out of 256, versus minimum 0 and maximum 92 in the case of the unretouched image.

5.5.3 Frequency-Domain Image Enhancement

Frequency-domain processing is based on the convolution theorem which allows to go back and forth between the spatial and frequency domains while processing the images:

$$g(x, y) = h(x, y) * f(x, y) \leftrightarrow G(u, v) = H(u, v)F(u, v) \qquad (5.20)$$

where G, H, and F are the Fourier transforms of g, h, and f. The spatial domain corresponds to the image plane (Figure 5.1). The frequency corresponds to the speed at which pixels change intensities within the image. For example, on an 8-bit (0 to 255) image, pixel values changing one after the other from 0 to 255 correspond to a high frequency, while a low frequency would appear as a slowly varying trend within the image. For example, thermogram distortion by vignetting is an example of low frequency (Sections 4.8.1 and 4.10.3). Trend removal is an image-processing technique that removes low frequencies within an image f by subtracting or dividing f with $\hat{f}$, where $\hat{f}$ is the first-, second-, or higher-degree polynomial fit of f.

In eq. (5.20), particular selection of $H(u, v)$ allows us to obtain specific characteristics in the processed image g, such as high- or low-frequency enhancement with

$$g(x, y) = F^{-1}[H(u, v)F(u, v)] \qquad (5.21)$$

where F^{-1} denotes the inverse Fourier transform and f is the unprocessed image. Interest in the convolution theorem is that multiplication and inverse Fourier transform are less computation intensive than is direct convolution. For image processing purposes, in eq. (5.20) h is often considered as the global transfer function of the system, f as the original image (the one that is sought), and g as the image obtained. The transfer function agglomerates all transformations undergone by the original image, such as optical effects (such as distortions), atmosphere degradation, quantification, and electronic noise. Obviously, if h is known, it becomes possible to recover f [Section 5.4, eq. (5.4)].

5.5.4 Spatial and Temporal Reference Thermogram Enhancement Technique

Special image processing techniques by spatial or temporal references are also useful to improve defect visibility. Spatial reference techniques are useful to

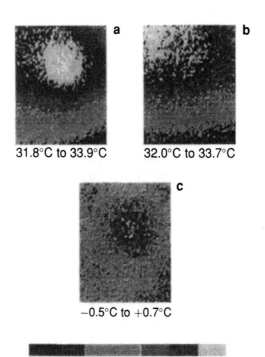

a   31.8°C to 33.9°C

b   32.0°C to 33.7°C

c   −0.5°C to +0.7°C

FIGURE 5.18 Spatial reference technique in pulsed thermography: (a) thermogram over a defect (Teflon implant) of 20 × 20 mm, 2.25 mm beneath the surface; (b) thermogram over a sound area; (c) enhanced defect visibility after subtraction: (a)–(b). Graphite-epoxy plate. (From Maldague, 1993.)

compensate for repetitive variations of nonhomogeneity of the thermal perturbation source. This spatial reference technique consists of comparing the thermal behavior of the specimen inspected with results obtained from a sound reference specimen. An example of such an approach is illustrated in Figure 5.18. Figure 5.18a shows a thermogram obtained using pulsed thermography above a square defect, 20 × 20 mm, located 2.25 mm beneath the surface. In this image, the thermal contrast is poor, especially because of the nonhomogeneity of the heat source. These effects can be eliminated by subtraction with the reference image shown in Figure 5.18b, which is obtained under similar conditions above a sound area of the sample. The difference image (Figure 5.18c) shows more distinctly the position of the defect. Notice the blurring effect around the edge in the subtracted image, which is due to the "large" defect depth in this highly anisotropic material (Section 9.1).

When image degradations are nonrepetitive, for example when surface emissivity of the workpiece changes in an unpredictable manner or when unknown thermal reflections are present, temporal methods of image processing can be applied to identify defects. An example of such an approach is depicted on Figure 5.19. In this case, an artificial Teflon defect of 10 × 10 mm was

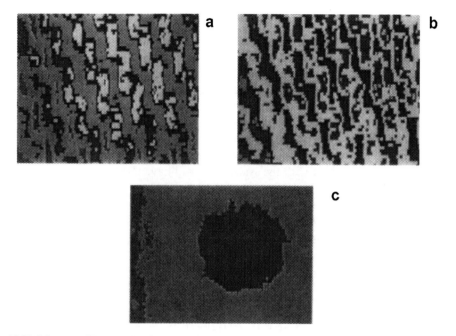

FIGURE 5.19 Temporal reference technique in pulsed thermography. Thermograms (a) and (b) were recorded over the same area of the specimen in extreme conditions of thermal noise (parasitic reflections), 3 s and 5 s, respectively, after beginning of heating. The subtracted image in (c) offers improved defect visibility. Graphite-epoxy plate. (From Maldague, 1993.)

inserted, before curing, 0.5 mm beneath the surface of a graphite epoxy plate of 3 mm thickness. Images were recorded in pulsed thermography 3 and 5 s after heating was begun. The thermal pulse was applied in transmission with a 1000-W projector placed on the back side while an IR camera was pointed at the front surface. During the experiment, a strong thermal noise caused by a nearby welding station perturbed the measurement. The diagonal pattern visible on the images is caused by differences in reflectivity on the surface of this woven composite. Since the texture is the same on both images, reflection noise can be suppressed by subtraction of both images. This is shown in Figure 5.19c. The image subtracted represents thermal evolution of the heated zone between the two observation times and is thus related to the thermal propagation properties of the sample.

An interesting aspect of these two basic techniques is that due to their simplicity, they can be implemented in real time. For the *spatial* reference technique, this can be done as a condition to have recorded the reference image of the sound workpiece prior to inspection, while for the *temporal* technique, subtraction can proceed just after the second image is acquired.

5.5.5   Alternative Smoothing Routine for Noise Processing

In addition to the spatial enhancements techniques described previously, another efficient method to smooth images consists of employing a sliding Gaussian window. To simplify, we first present the unidimensional case. Let $v(\Omega)$ be the raw signal delivered by the system (vector of values to smooth; Ω is the index variable) and $f(\Omega)$ the transfer function of the system (comprising global effects from the camera, etc.). What is important to notice here is that the system cannot *itself* produce frequencies greater than those specified by its function $f(\Omega)$. In this sense, any signal whose frequency is greater with respect to $f(\Omega)$ is said to be *inconsistent* with the system and will have to be eliminated.

The signal $v(\Omega)$ can be expressed as follows: $v(\Omega) = v_n + n(\Omega)$, meaning that in the nth iteration, the signal is equal to the summation of the ideal (smoothed) signal $v_n(\Omega)$ with the noise $n(\Omega)$. Since the noise is considered statistically uncorrelated with the system function, we have

$$f(\Omega) * n(\Omega) = 0 \tag{5.22}$$

For the first approximation, we suppose that

$$v_1(\Omega) = f(\Omega) * v(\Omega) \tag{5.23}$$

is equal to the ideal smoothed signal. The noise is thus given by the difference between the raw signal and the ideal signal as a first approximation:

$$n(\Omega) = v(\Omega) - v_1(\Omega) \tag{5.24}$$

If we suppose an iterating convergent process toward the ideal (smoothed) signal, we can thus assume that at the nth approximation, we will have an ideal signal made up of the *ideal signal of the previous approximation* v_{n-1} plus a *residual noise*:

$$v_n(\Omega) = v_{n-1}(\Omega) + [v(\Omega) - v_{n-1}(\Omega)] * f(\Omega) \tag{5.25}$$

Residual noise

Since the number of iterating steps is large and following eq. (5.22), the *residual noise* is reduced to zero. In addition to eliminating the signal's *incompatible* content, which is eliminated within the first few iterations, a larger number of iterations will not degrade the signal. Practically, two iterations are generally sufficient.

This smoothing process is fast and efficient. For images it is applied sequentially along rows and columns. The $f(\Omega)$ function is given by the normal curve; expressed in continuous form, it is given by

$$f(\Omega) = \frac{1}{\sigma\sqrt{2\pi}} \exp\left[-\frac{(\Omega - \mu)^2}{2\sigma}\right] \tag{5.26}$$

with μ the average and σ the standard deviation of the distribution, and for the discrete case,

$$f(i) = \frac{C}{B} \exp^{-1/2} \left[\frac{i - (i_{max}/2)}{B} \right]^2 \tag{5.27}$$

with i_{max}, the number of elements for the Gaussian $= 10R + 1$, the value typically established through testing; $i = 0, 1, 2, \ldots, i_{max}$; and $B = 5$ (for a a 512×512 image, typically), $= 4$ (for a 100×100 image, typically), and $= 2$ (for a 70×70 image, typically). The C factor in eq. (5.27) is adjusted so that summation of all the Gaussian elements yields 1:

$$C = \frac{B}{\sum\limits_{i=0}^{i_{max}} \exp - \frac{1}{2} \left[\frac{i - (i_{max}/2)}{B} \right]^2} \tag{5.28}$$

This avoids the presence of an undesirable gain: A uniform signal will pass through the process without being affected. The discrete convolution operation $v'(\Omega) = v(\Omega) * f(\Omega)$ of Gaussian $f(\Omega)$ with vector $v(\Omega)$ of N elements gives, for each element Ω of $v(\Omega)$, the new element $v'(\Omega)$:

$$v'(\Omega) = \frac{1}{i_{max} + 1} \sum\limits_{i=0}^{i_{max}} \left[v\left(\Omega - \frac{i_{max}}{2} + 1 + i \right) f(i) \right]$$

$$= \frac{1}{i_{max} + 1} \sum\limits_{i=0}^{i_{max}} [v(\text{index}) f(i)] \tag{5.29}$$

where index $= \Omega - (i_{max}/2) + 1 + i$, and for the limit cases, if index $< 0 \to$ index $= 0$ and if index $> N \to$ index $= N$.

Typically, the Gaussian curve is first initialized with eqs. (5.27) and (5.28); then the smoothing is done on two passes using eqs. (5.23), (5.25), and (5.29). The parameter B is defined to obtain optimal smoothing: This is the classic trade-off between the maximum reduction of the random noise and the minimum deterioration of data. This technique leads to an improved SNR of 152 in the case of infrared images of the inspection station originally having an SNR of 32 for raw data (as stated at the beginning of Section 5.4). This five-fold improvement is due to the high-frequency noise content (1 cycle $\sim$ 4 pixels) of these raw infrared images, for which the smoothing process is found to be particularly efficient: Figure 5.20 shows a smoothed image (the raw image is pictured on Figure 5.2). The Matlab m-file smooth.m is provided in Appendix F for such purposes (Table 5.1) and as a C program language listing in Appendix B.

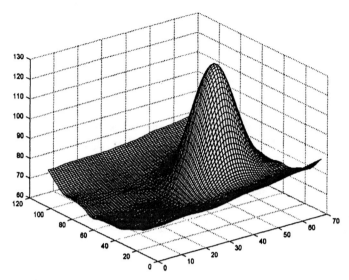

FIGURE 5.20   Three-dimensional plot of the smoothed infrared thermogram of Figure 5.2, prepared with Matlab m-script: smooth.m (Appendix F) and parameter B set to 4.

5.6   THERMAL CONTRAST COMPUTATIONS

In Section 4.8 we saw how to compute temperature from the radiometric signal of the infrared camera. The next step performed is often to compute the thermal contrast, which is of interest to evaluate defect visibility, enhance image quality, and also for quantitative purposes, as discussed in Chapter 10. Many definitions are used to compute the thermal contrast; they are reviewed below.

Absolute contrast $C^a(t)$ corresponds to *increased* or *excess temperature* with respect to a reference region at a given time t. It is computed as follows:

$$C^a(t) = \Delta T(t) = T_{\text{def}}(t) - T_s(t) \tag{5.30}$$

where T is the temperature signal, t the time variable, and the indexes def and s refer to the signal over a suspected defective location (in fact, any pixel in the image) and over a sound area, respectively. Typically, a sound region in the image is identified either automatically or by an operator, and then C^a is computed. The advantage of computing C^a is better visualization of defects with respect to the background. On the other hand, C^a is linearly related to the energy absorbed, and this limits comparisons among experiments. This formulation makes use of the well-known spatial reference technique, in which sound values are subtracted from suspected values to improve defect visibility (Section 5.5.4).

Running contrast $C^r(t)$ is less affected by surface optical properties because suspected and nondefect points are located over areas of the same emissivity/ absorptivity. It also depends less on the energy absorbed. It is computed as follows:

$$C^r(t) = \frac{C^a(t)}{T_s(t)} = \frac{\Delta T(t)}{T_s(t)} \tag{5.31}$$

with parameters defined as before.

Normalized contrast $C^n(t)$ is computed with respect to the values of temperature at the instant t_m when the excess temperature is maximal (this parameter depends on heating stimulation). Normalization may also be done with values at the end of the thermal process (i.e., at time t_e). With t_m, it is computed as follows:

$$C^n(t) = \frac{T_{\text{def}}(t)}{T_{\text{def}}(t_m)} - \frac{T_s(t)}{T_s(t_m)} \tag{5.32}$$

with parameters defined as before. The problem with this definition is that t_m must first be identified; moreover, $C^n(t_m) = 0$.

A final definition of contrast is *standard contrast* C^s, which is computed before heating temperature distribution at time t_0 (to suppress the adverse contributions from the surrounding environment) and normalized by the behavior of a sound area. A unit value is obtained over a nondefect area. It is computed as follows:

$$C^s(t) = \frac{T_{\text{def}}(t) - T_{\text{def}}(t_0)}{T_s(t) - T_s(t_0)} \tag{5.33}$$

C^a and C^r can be computed on a single thermogram regardless of the inspection procedure (active or passive thermography). In such a case, the variable t is not relevant. C^a and C^r (as well as C^n and C^s) can also be computed as part of the active pulsed thermography procedure (Chapter 9), either on a single thermogram or for all the thermograms in the temporal sequence (thus the reference to the time variable t in the afore-mentioned definitions). Figure 5.21 shows thermograms computed using all these definitions. As seen in Figure 5.21, thermal contrast images are generally "cleaner" than raw thermograms. Contrast is then often computed as a first step in the analysis of the thermogram sequence (discussed in detail in Section 10.1).

The main problem with thermal contrast computations in all these contrast definitions is that they require a priori knowledge of a nondefect area within the field of view. Unless some sort of automatic defect detection algorithm is used (Chapter 6), this requires operator intervention, which is impractical. In Section 10.4 we discuss an alternative to contrast-based computations that does not require us to identify a sound area—pulsed phase thermography processing.

(a) raw thermogram (best defect visibility at t = 20.9 s)

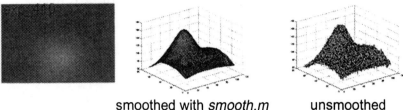

smoothed with *smooth.m*   unsmoothed

(b) absolute contrast

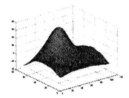

(c) running contrast

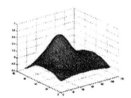

(d) normalized contrast with t_e = 119.95 s (end of of thermal process)

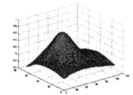

(e) standard contrast

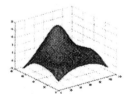

FIGURE 5.21 Thermal contrast definitions: (a) raw thermogram; (b) absolute contrast; (c) running contrast; (d) normalized contrast; (e) standard contrast (computed at $t =$ 20.9 s). Case of the pulsed thermography on a graphite-epoxy specimen with 12-mm-diameter drilled hole, 2 mm beneath the surface, pulse duration: 1.5 s, 2500 W. Prepared with Matlab m-script: ctrst.m (Appendix F).

5.7 MANUAL INTERPRETATION OF COLOR SCALES AND PROFILES

5.7.1 Color Scales

Before any processing attempt, it is a good practice to have a look at the images. Although raw images can be improved by processing techniques such as the ones described in this chapter, if nothing is seen upon acquiring images, it is a bad sign. One of the simplest ways to look at incoming images is to use the various color maps provided by the acquisition system (including various shades of color and black and white). Our eyes are made up of two kinds of sensitive cells: cones, which are highly sensitive to color, and rods, which are not sensitive to color. Cones are located primarily in the central portion of the retina (the area called the fovea), and each cone is connected to its own nerve ending. On the other end, rods are spread all over the retina and several of them connect to a nerve ending. This explains why we can resolve subtle shades of color much better than we can shades of gray. For example, in Figure 5.4 we do not perceive much difference between image coding at 5, 6, 7, and 8 bits, although the number of gray levels changes from 32 at 5 bits to 256 at 8 bits, an eightfold difference. In fact, physiological studies reveal that our eyes can resolve about 30 gray levels. This also explains why color thermograms provide great reports. Color coding of IR images reveals some details, and since color mapping allows real-time visualization with simple hardware, such a feature is offered by most IR systems. Moreover, since IR images are not directly related to the reality we perceive directly with our eyes, the color mapping can be adjusted for best visibility: What is green in one image can be turned to red or blue without changing the signification we associate with that particular feature.

As we noted earlier, since our eyes are strongly sensitive to color, it is better to use a continuous color palette such as the *jet* color map available in Matlab (Figure 5.2) to avoid a disturbing mosaic effect (as seen in Figure 5.22 with the *prism* color map, for instance). Reduction in the number of shades enlarges the individual color zones, and such integration of many different values within one color reduces the noise effect without processing (see, e.g., the 3-bit-coded image in Figure 5.4). However, if differences between color bands are too large (e.g., $>1°C$), some features can go unseen. On the other hand, occasionally, a judicious choice of the color palette significatively enhances the visibility of a particular feature. For example, in Figure 5.22, with an *hsv* color map a strong nonuniformity appears clearly, although it becomes less visible with other color maps and much less in the black-and-white version of this image (Figure 5.1), due to the limited sensitivity of our eyes in black and white shades. Interestingly, when it is necessary to print out a color image, the *pink* color map available in Matlab provides a continuous scale from black to white once printed on a noncolor printer (Figure 5.22). As a rule, it is advised always to look first at a black-and-white thermogram, to check the presence of particular features. Next, a color palette can be selected to enhance features seen in the black-and-

COLORCUBE colormap PRISM colormap

HSV colormap PINK colormap

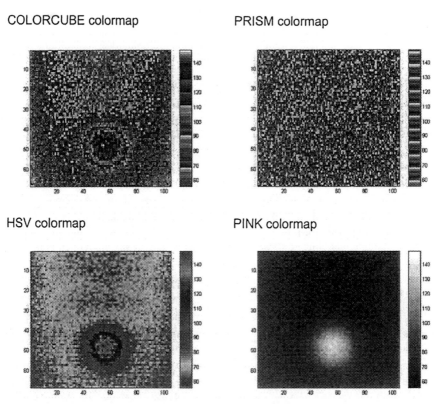

FIGURE 5.22 Effects of changing different color maps on thermogram appearance (color maps shown are available in Matlab). Image of Figure 5.1; see also Figure 5.2.

white image. Techniques described in Section 5.5 are useful to intensify the visibility of features in thermograms.

Another interesting feature of some IR cameras is the display of isotherm data (i.e., the highlighting of an image area having the same temperature). Some Agema and Inframetrics IR cameras offer such capability, which is somehow similar to color coding since a given temperature span is associated with a specific color shade. The difference is that in the isotherm mode, only the temperature span of interest is highlighted. Figure 5.23, which illustrates the isotherm function, was prepared using the Matlab command **colormap (gray(64))**, and then changing position 25 with **gray(25,1) = 1; gray(25,2) = 1; gray(25,3) = 1; colormap(gray)** on the thermogram of Figure 5.1.

5.7.2 Profiles

Profiles are useful when comparing certain zones within thermograms, especially when they are not displayed in color. Both 3D (Figure 5.2) and 2D

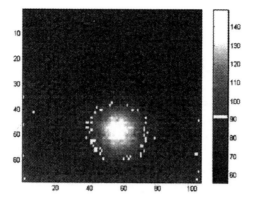

FIGURE 5.23   Isotherm display higlighting raw thermal value 92; original image of Figure 5.1.

(Figure 5.24) plots are useful for such purposes. In fact, a profile feature is available in real time on several IR cameras, including the Inframetrics 600. Superimposition of several 2D profiles or a full 3D plot indicates subtle image trends (notice, for instance, the nonuniform background due to vignetting on Figure 5.2). In many instances, such 3D plots are a useful alternative for displaying results when color printing is not available or when black-and-white printouts fail to render the information properly.

PROBLEMS

5.1 **[image format]** What is the raw file size for an image of 68 rows and 105 columns, with pixels on 12 bits and with an extra row for the header?

Solution   Following eq. (5.2), we have $68 \times 105 \times 16 + 105 \times 8 = 115{,}080$ bits, or about 14 kilobytes. Notice here that since the memory standard for computers is based on bytes, pixels are stored on 2 bytes = 16 bits even if 12 bits would be enough. This is a common practice which makes storage easier even though it wastes some place.

5.2 **[image sequence storage]** What capacity is needed to record a 5-s active thermography experiment if the IR camera produces 14-bit images of a format of 256×256 pixels at a frame rate of 400 Hz (with pixel packing)?

Solution   Since the images consist of 14 bits, and since pixel packing is activated, 1 pixel packing requires $14/8 = 1.75$ bytes (1 byte = 8 bits). Thus one image requires $256 \times 256 \times 1.75 = 114{,}688$ bytes and for a 5-s duration this is $114{,}688 \times 400 \times 5 = 219$ MB, a gain of 12% with respect to a no-packing approach (Example 5.1).

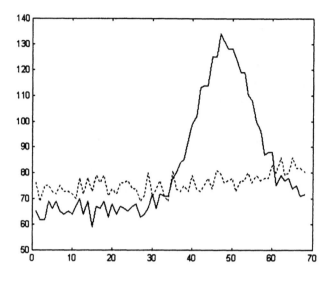

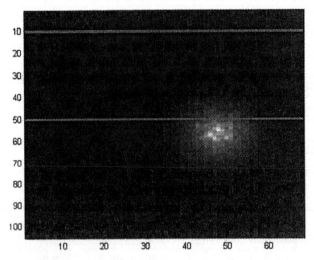

FIGURE 5.24 Profiles for rows 10 and 50 as indicated; thermogram of Figure 5.1 (horizontal scale: column number, vertical scale: raw thermal values). Prepared with Matlab m-script: profil.m (Appendix F).

5.3 **[gray-level stretching]** The minimum and maximum values observed in a low-contrast 14-bit thermogram are 40 and 234, respectively. To what value is a pixel of value 100 mapped after linear stretching?

Solution A 14-bit thermogram involves an available range of values spanning from 0 to 16,383 ($2^{14} - 1$). Following eq. (5.17), we have $m = 84.45$, $b = -3378$, and then the newly mapped value turns out to be 5066.

CHAPTER SIX

Automated Image Analysis

In Chapter 5 we discussed imaging and basic image processing applied to TNDT. The purpose of this chapter is to go deeper into image processing for automatic image analysis. Two techniques are presented: a robust algorithm for automatic defect detection and application of a neural network to TNDT.

6.1 BASIC CONCEPTS

The basic idea is to identify automatically where defects (if any) are located in a thermogram representing the field of view inspected. This technique could be applied in the case of repetitive TNDT inspection (e.g., on the production line). The goal is then to produce a complete map of the component inspected, where defect locations and gross shapes are depicted. Since a list of defects with their approximate size is available, automatic inspection becomes possible: The components inspected can be sorted as accept or reject, based on a probability detection curve (Figure 7.8). Obviously, the inspection method must be able to identify all critical defects with reliability, which means that the critical threshold must be greater than the threshold for false alarms. Rejected components must be eliminated from the production line or fixed if the cost of refurbishing is acceptable. In this case, a more sensitive NDE technique such as ultrasonics can be used to assess defective zones with greater resolution (Section 7.5).

For example, in the case of bonded Al–foam laminates (Section 7.4.1), it is noted that the threshold for detection (say, 0.2°C in a given system) corresponds to a radius/depth ratio on the order of 40 (Figure 11.9c), and defects detected (Figure 6.14) have a ratio on the order of this value. This is at the limit of detectability for TNDT in this case. It is important to notice that for such Al–foam components, defects that must be detected absolutely, in real life, have a much higher ratio. Typical defect sizes are on the order of 30×30 cm (i.e., about 1 ft^2) with a corresponding radius/depth ratio of about 150. Undetected defects of this size can be really dangerous, due to reduced mechanical properties for such damaged components. TNDT is thus particularly well suited as an inspection technique for this kind of material.

6.1.1 Defect Detection Algorithms

Concerning detection algorithms, the conference series on image processing algorithms of the Society of Photo-Interpretive Engineers (SPIE*) is a good source of information on this subject, as are the PAMI (Pattern Analysis and Machine Intelligence) and SMC (Systems, Man and Cybernetics) transactions of the Institute of Electrical and Electronics Engineers (IEEE[†]). Many other relevant journals are noted in the reference section of this book.

There exist a wide variety of algorithms whose purpose is to perform *image segmentation* to separate regions of interest in images. Various techniques exist for such purposes, such as those based on edge detection, growth around key points called *seeds*, histogram analysis (a widely used technique that does not require complex computations), and symbolic modeling. These utilize either discontinuity or similarity of characteristic attributes (e.g., pixel intensity, gradients) to label all the pixels within the image (i.e., to associate them with a particular region). Considering the wide variety of image analysis problems, the variety of algorithms is extreme. Generally, algorithms are *ad hoc*, and if applied in another context, they can fail pitifully. This is because an image represents an enormous number of possibilities: In the case of small infrared images of size 68 (Maxrow) $\times$ 105 (Maxcol), for example, we have 7140 pixels on 8 bits, or $(2^8)^{7140}$ possibilities!

Another important aspect is the image background. In TNDT, because of heating effects (nonuniformity), image borders tend sometimes to be hotter than the other parts of images. This has inspired *trend removal* elimination procedures. The basic idea is to produce a *synthetic image* from the original image by making use of a polynomial function to obtain, after subtraction or division of the synthetic image by the original image, a uniform background on which defects appear more clearly. Next, a threshold segmentation algorithm based on valley detection in the image histogram is used to identify a threshold between the two modes (defects and background) at either the global (whole image) or local level (using a small running window over the image). However, this method does not work very well if separation between the two modes is not sharp. Although the trend removal approach is well suited for some types of images, such as radiographic images, it generally reveals deception in TNDT images (Figure 6.1).

6.1.2 Validity of Defect Detection Procedures

A difficult question is to determine whether or not the segmentation is valid. One approach to this question is to consider the degree of agreement with the

---

* www.spie.org/home.html.
[†] www.ieee.org.

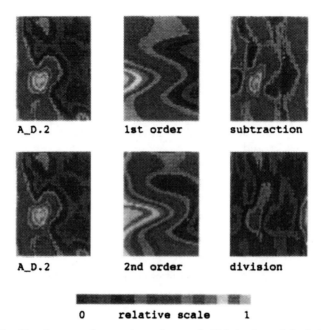

A_D.2   1st order   subtraction

A_D.2   2nd order   division

0   relative scale   1

FIGURE 6.1 Trend removal procedure. Image A_D.2 is the original thermogram. Middle images are synthetic images of first and second order. Right images show result after subtraction and division. (From Maldague, 1993.)

human interpretation. In fact, the eye–brain combination is extraordinary powerful, as evidenced by the fact that about 50% of cells in the cortex are dedicated to visual tasks. This is why, in most cases, an experienced operator is perfectly able to segment TNDT images as they appear originally. In fact, the human nervous system seems analyze images using threshold techniques that make it possible to separate objects based on their relative intensity. Obviously, a *cultural* aspect is also necessary to perform such tasks. This is why, for instance, a newborn baby has to learn all about its surrounding environment before being able to recognize things and objects. This *cultural* aspect corresponds to *heuristic* rules in the case of machine vision algorithms. Finally, if comparisons are to be done with human vision, it is also important to realize that the eye response is logarithmic, which provides an extremely wide dynamic range.

6.2 IMAGE FORMATION

To obtain a high inspection rate, the number of computations to be performed on the images obtained from the inspection station should be restricted to a minimum. For this reason, the number of images to be analyzed following the inspection procedure must be restricted as much as possible. Obviously, this contradicts the requirement to record the entire thermal history curve of the part inspected in order to catch abnormal thermal events.

Generally, in the pulsed TNDT technique, images can be obtained from either a static or a mobile configuration. In a static configuration, the camera observes the same surface continuously, and a succession of images is recorded beginning at the instant $t = 0$, corresponding to the firing of the thermal pulse (e.g., the moment when heating lamps are turned on). Notice that in some instances, it may be of interest to start image acquisition slightly before the thermal pulse in order to obtain a reference "cold" image, which allows us to reduce the spurious effects of thermal reflections by subtraction with other images (Sections 5.6 and 10.2).

For the static configuration, the *moment method* proposed initially by Balageas et al. (1987b) can be used. The temporal moment, of order M, for a temperature T_0 on a sound area is defined by

$$M = \int_0^\infty \Delta T_0(t) \, dt \tag{6.1}$$

The Δ operator means that we are interested in increasing the temperature with respect to the ambient room temperature T_a, $\Delta T = T_0 - T_a$. This moment M tends to infinity. If a defect is present, we can form $[\Delta T_{\text{def}}(t) - \Delta T_0(t)]$, where T_{def} corresponds to the temperature above the defect zone. Consequently, we can evaluate the temporal moment ΔM of order zero. It can be demonstrated that this moment has a finite value equal to

$$\Delta M = \int_0^\infty [\Delta T_{\text{def}}(t) - \Delta T_0(t)] \, dt = Q \, R_{\text{def}} \left(1 - \frac{z_{\text{def}}}{L} \right)^2 \tag{6.2}$$

where Q corresponds to the absorbed energy by the sample of thickness L, while R_{def} is the thermal resistance of the defect and z_{def} is its depth. If the sample is very thick, ΔM becomes equal to $Q \, R_{\text{def}}$ and defects will appear with the same contrast whatever their depth. Equation (6.2) can be applied simply by adding together all the images recorded in the time domain. This also has the advantage of improving the SNR ratio by the square of the number of summed images. Consequently, if this summation process is applied, it is not necessary to use the noise reduction techniques described in Chapter 5. However, caution should be exercised in this summation process; not "all" images have to be added, because for a given thermal event of finite duration, the thermal contrast tends to vanish as images in which it is absent are summed together.

Such a summation method can be applied to reduce the number of images to be processed following a TNDT pulsed inspection procedure. After (or during) the experiment, it is only necessary to add together all the images acquired in a given time window. When selecting this window it is necessary to have it sufficiently wide so that thermal contrasts of potentially present defects have had the opportunity to develop sufficiently, taking into account the thermal diffusivity α of the material analyzed [eq. (9.7)]. However, in practice, this constraint is quite flexible. If we call I the image obtained in this fashion during time window $[t_a, t_b]$ corresponding to individual images $G(t_i)$, we have

$$I = \sum_{t_a}^{t_b} G(t_i) \tag{6.3}$$

In the case of a mobile configuration where the infrared camera records the complete motion of the component passing in front of it, such a direct summation process cannot be applied directly because the field of view is constantly changing. In this case a special technique can be used in which specific columns of pixels are extracted from every recorded image in order to *reconstruct* the entire component as seen at a particular time. Next, the reconstructed images are summed following the method of eq. (6.3), to obtain the I' image (the prime indicating a reconstructed image).

In Figure 6.2 we illustrate the principle of inspection for components in motion in a field of view of width L and at time $t_0, t_1, \ldots, t_i, \ldots t_N$. This corresponds to acquisition of N images during the inspection experiment. In Figure 6.3 we show the reconstruction process for the image $G'k$ obtained through juxtaposition (operator J) of columns $C_k(t_i)$ extracted from images $t_0, t_1, \ldots, t_i, \ldots t_N$:

$$G'_k = \mathrm{J}_{i=0}^{N} C_k(t_i) \tag{6.4}$$

with $k = 0, 1, 2, \ldots, \text{Maxcol} - 1$, where Maxcol = number of columns in one image.

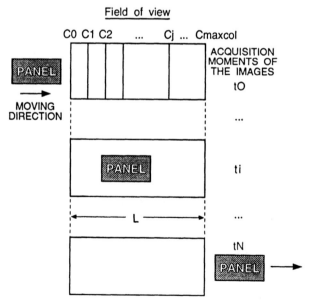

FIGURE 6.2 Motion of the panel in the field of view (mobile configuration in reflection). (From Maldague, 1993.)

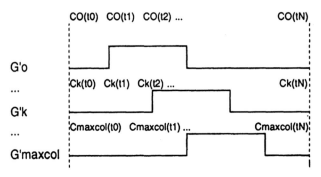

FIGURE 6.3   Reconstruction process for the images. (From Maldague, 1993.)

Due to the motion of the component within the field of view, all the reconstructed images G'_k correspond to total observation of the component when it is in the thermal state $t(k)$ since the component is observed at the same distance from the heating unit (the same extracted column k in the field of view) and since the lateral speed v is constant, from Figure 6.4 we obtain

$$t(k) = \frac{L'}{v} + \frac{LC_k}{vC_{\text{Maxcol}}} = \frac{1}{v}\left(L' + \frac{LC_k}{C_{\text{Maxcol}}}\right) \qquad (6.5)$$

From this, we see that images G' correspond somewhat to images G of eq. (6.3), since

Image	Corresponds to component at time	N_{col}
G'_0	$\left[\dfrac{1}{v}\right]L'$	0
G'_k	$\left[\dfrac{1}{v}\right]L' + L(C_k/C_{\text{Maxcol}})$	k
$G'_{\text{Maxcol}-1}$	$\left[\dfrac{1}{v}\right]L' + L$	$\text{Maxcol} - 1$

$$(6.6)$$

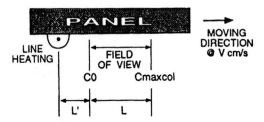

FIGURE 6.4   Studied geometry. (From Maldague, 1993.)

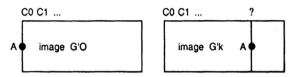

FIGURE 6.5 Study of the shifting in the reconstructed images. (From Maldague, 1993.)

where N_{col} is the column number of the original infrared images having Maxcol columns.

In Figure 6.3, the horizontal scale is a temporal scale related directly to the acquisition time of images $t_0, t_1, \ldots, t_i, \ldots t_N$, and every reconstructed image corresponds to a particular thermal state of the component [eqs. (6.5) and (6.6)]. However, we notice on this figure that the component is not present at the same positions in the sequence of reconstructed images $G'_0, \ldots, G'_{\text{Maxcol}}$. This shifting of the image G'_k obtained by *juxtaposition* of the kth column of the N acquired images can be evaluated. The question is as follows: *On which column of image G'_k will point A that appears on column 0 of image G'_0 appear?* This situation is depicted in Figure 6.5. Since the point is moving at speed v, it will move from column C_0 to C_k within the field of view in a time interval given by (Figure 6.6)

$$t = \frac{L}{v} \frac{C_k}{C_{\text{Maxcol}}} \tag{6.7}$$

We obtain N images in time t_N, one image is thus acquired in $[t_N/N]$ seconds. Consequently, point A will appear in the "column"; notice here that we talk about the columns of the reconstructed images which correspond to images 0 to N:

$$\text{Sh\_col}(k) = \frac{L}{v} \frac{C_k}{C_{\text{Maxcol}}} \bigg/ \frac{t_N}{N} = \frac{L}{v} \frac{NC_k}{t_N C_{\text{Maxcol}}} = \frac{L}{v} \frac{RC_k}{C_{\text{Maxcol}}} \tag{6.8}$$

where R is the acquisition rate (number of images acquired per second). Every $G'k$ image will thus have to be shifted by $\text{Sh\_col}(k)$ columns before we proceed

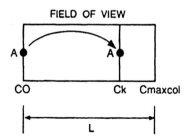

FIGURE 6.6 Study of the shifting in the reconstructed images. (From Maldague, 1993.)

to the summation following eq. (6.3), as in the static case:

$$I' = G_0' + \text{SHIFT}[G_1', \text{Sh\_col}(1)] + \cdots + \text{SHIFT}[G_i', \text{Sh\_col}(i)] + \cdots$$
$$+ \text{SHIFT}[G_{\text{Maxcol}}', \text{Sh\_col}(\text{Maxcol})]$$
$$= \sum_{i=0}^{\text{Maxcol}} \text{SHIFT}[G_i', \text{Sh\_col}(i)] \tag{6.9}$$

where SHIFT $[G_i', \text{Sh\_col}(i)]$ corresponds to the shifting operation of Sh_col(i) columns on image G_i'.

This study shows that it is possible to obtain a *reconstructed* image corresponding to the entire width of the component inspected. Due to the lateral motion of the component, its full width is inspected. It is important to mention that this study [computations of eq. (6.9)] is an approximation, since we take into account the *apparent separation* of the image columns instead of the *real separation*, which takes into consideration the slit response function (Section 4.1.7). However, this way to proceed reveals adequate detection analysis, at least for the defect.

Since the temporal information is lost in the image formation process [for the static (eq. 6.3) and dynamic (eq. 6.9) configurations], the depth of the defects detected cannot be computed unless it is known due to the geometry of the components inspected, as in the case of known depths in bonded laminates. This is also the case for defect size: No *corrective factor* can be computed to recover the real shape from the apparent size (see also Section 10.2.5). However, these limitations do not limit automatic defect detection, as mentioned earlier.

In Chapter 4 we studied which corrections need to be applied to raw images for temperature conversion. In the case of static configuration in pulsed TNDT, the entire temperature history curve of defects detected is available. After the defect detection step it is only necessary to reprocess and correct the images to obtain a quantitative information, as discussed in Chapter 10. In the case of the mobile configuration, this study can also be done based on the individual reconstructed shifted SHIFT[$G'k$, Sh_col(k)] images [eq. (6.9)]. However, due to the approximate nature of the reconstruction process, the margin of error on such quantitative computations may be unacceptable.

To maximize the computational speed and taking into account previous limitations on defect size and depth, the automatic detection algorithms (described in the next sections) can be applied directly on raw thermal value images. This saves the time needed to execute the temperature computation step (Section 4.8). The fact that the algorithms work even in the absence of these corrections is a positive aspect. Obviously, the algorithms can be applied on temperature-converted images. Basically, the main difference between defect detection on raw and corrected temperature images will be the time of execution (needed for temperature computations).

Moreover, with the same desire to maximize execution speed, images I or I' can be converted to an *unsigned character* type, which takes less space in computer memory. In this way one pixel is coded on one byte (8-bit system) in computer memory instead of four or eight bytes required for the coding of floating-point variables. This also allows us to make use of *integer arithmetic*, which is faster than *floating-point arithmetic* in most computer implementations. Conversion of an image in which pixels are expressed in floating-point values (I_{float}) to an image in which pixels are expressed in characters (I_{char}) is performed as follows [note the similarity to eq. (5.17)]:

$$I_{\text{char}}(i, j) = mI_{\text{float}}(i, j) + b \qquad (6.10)$$

where

$$m = \frac{B_{\max} - B_{\min}}{F_{\max} - F_{\min}} \quad \text{and} \quad b = B_{\max} - mF_{\max}$$

Here $B_{\max}$ and $B_{\min}$ are maximum and minimum values of image I_{char}. On 8-bit implementations, they are 255 and 0, respectively.

$F_{\max}$ and $F_{\min}$ are maximum and minimum values of image I_{float}, and i and j are all the pixel positions in the images. All subsequent image-processing steps will be performed on the image I_{char} or I'_{char}, which will be denoted by I, for the sake of simplicity. Mathlab m-files mat_fl_to_uc.m (Table 5.1) performs such a conversion.

In the following examples, the thermal perturbation source deposits energy on the specimens and the inspection proceeds in reflection (pulsed thermography approach); consequently, potential defects will be represented in image I by areas of higher temperature with respect to their immediate surroundings. Defect edges are often represented by ramps of temperature that span over a few pixels because of the 3D spreading of the heat flow (Section 9.1). Because of this ramp aspect, the use of edge detector operators is less attractive.

Example 6.1 Panels are to be inspected in a mobile configuration as depicted in Figure 6.4. Suppose a field of view of 20×20 cm, an image size of 256×256 pixels, and a moving speed of 10 cm/s. The acquisition rate is 20 ms per image. What is the shifting factor needed to make up the moment image?

SOLUTION   The frame rate is 1/20 ms = 50 Hz (50 images/s). The panel crosses the field of view in 2 s (20/10 s), corresponding to a total acquisition of 100 images (2×50). Each reconstructed image will have 100 columns. Following eq. (6.8), it is

$$\text{shifting} = \frac{LRC_k}{vC_{\text{Maxcol}}} = \frac{(20 \times 50)C_k}{10 \times 256} = 0.391 C_k$$

This means that the first reconstructed image will not be shifted, while the last reconstructed image will be shifted by 100 pixels ($0.391 \times 256 = 100$). The reconstructed image size is height 256 pixels and width 100 pixels. To compensate for the small horizontal size, columns are often duplicated, just for visualization (Figure 6.11).

6.3 AUTOMATIC SEGMENTATION ALGORITHM

It is interesting to determine both the location and the gross size of defects that may be present in TNDT images. In chapter 10 we present an approach for the quantitative evaluation of defects once they have been detected. The detection procedure we discuss now is nevertheless informative. For example, for aluminum-bonded laminates, the defect depth is known since it is given by the thickness of the aluminum sheet above the bonding epoxy layer (Section 7.4.1). Moreover, in many situations, the important point is whether or not a defect is present in a given part.

The algorithm described next is based on the fact that TNDT images have a limited number of spatial features. This is a very different situation with respect to visible images characterized by complex edge structures. Moreover, this algorithm makes use of some heuristics common in machine vision, as discussed previously. In a first step, defect localization is performed, and next, specific thresholds are found in the image to estimate the defect border. The originality of this algorithm comes from the fact that *seeds* are first located reliably based on a global sorting process within the image. (A *seed* is defined in the literature as the center point of a defect, hottest for hot thermal perturbation schemes in TNDT.) One threshold will be established per defect detected. Each threshold is found by means of a *region growing* approach, which starts at the central point of a defect and stops when either an image border is hit or the number of pixels agglomerated together around the seed increases abruptly (meaning that the *image background* is reached).

6.3.1 Part I: Defect Localization

For each defect found in image I, the first part of the algorithm produces localization of the corresponding hottest pixel (i.e., the seed). To limit computations, it is supposed that defects have at least one pixel at an intensity greater than the image average Avg. In this respect, many background pixels will be neglected since only pixels greater than Avg will be processed. Of course, all the pixels of the image can be processed, the computational time will be greater. This way to proceed is a short cut rather that a limitation, it allows, grossly, to cut in half the processing time. Since the algorithm proposes an estimation of defect characteristics as opposed to a quantitative value, raw thermal images can be used, this speeds up the process since correction and temperature computation stages are skipped.

All the pixels $I(i, j)$ of the image are compared with Avg. If larger $[I(i, j) > \text{Avg}]$, they are loaded in a four-vector structure (first initialized to zero):

$$vx(r) = j \text{ [i.e., position of the pixel } I(i, j) \text{ along the column]}$$

$$vy(r) = i \text{ [i.e., position of the pixel } I(i, j) \text{ along the row]}$$

$$gl(r) = I(i, j) \text{ [i.e., gray level of the pixel } I(i, j)]$$

$$lb(r) = \text{label associated with the pixel } I(i, j)$$

where r is the rank in the vector structure: $0, 1, 2, \ldots$ [Maxcol × Maxrow].

Initially, vector lb is set to zero, so that no label is associated to any pixel. After the loading phase, all the nonzero elements of vector gl are sorted in decreasing order. For all r values, we obtain

$$gl(r) \geq gl(r+1) \tag{6.11}$$

The content of the two other vectors vx and vy is also processed to avoid mixing the information related to individual pixels. After this *gray-level sorting*, sorting is performed *spatially*, so that for every position r of pixels having the same gray-level value, we have

$$vx(r) \geq vx(r+1) \tag{6.12}$$

As mentioned previously, the content of the other vectors is updated to avoid mixing pixel information. These sorting operations ensure that neighboring pixels belonging to the same class (see below) will be close in the vector structure. Once the two sorting procedures have been accomplished, the absolute values of pixels (the gl values) are no longer important compared to the relative rank of the pixels with respect to the others in the vector structure. In this sense we can say that the algorithm adapts itself to the histogram distribution of image I.

The final step consists of labeling all the pixels, that is, assigning them to a given class. Class 1 is assigned to the first pixel of the structure ($r = 0$). All subsequent pixels are tested with all previously labeled pixels, starting with the last that was tested (i.e., its closest neighbor). This speeds up the labeling process (the likelihood of having neighboring pixels with the same class being high). The criterion for assigning a new label to a pixel r_i is

$$\text{if } vx(r_i) - vx(r) > \text{MND} \quad \text{or} \quad \text{if } vy(r_i) - vy(r) > \text{MND} \tag{6.13}$$

The constant MND, the *minimum neighbor distance*, is established through trial and error. Since MND represents a distance between pixels rather than an absolute value, it is independent of the image. Rather, it depends on the size of the field of view and on the minimum defect size necessary for detection.

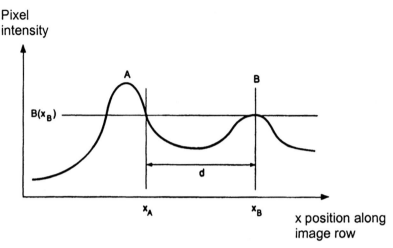

FIGURE 6.7   Principle of the algorithm for defect detection (see the text). (From Maldague, 1993.)

This part of the algorithm can be understood more easily by reference to Figure 6.7, where the one-dimensional case is shown. Two defects (blobs) A and B are represented. Suppose that the value of a hot-test pixel of blob B is denoted by $B(x_B)$, with x_B referring to the x position along one image row and $A(x)$ the gray level of any pixel of A. The algorithm will proceed by assigning label A to pixels sorted in decreasing values up to the moment it has to label pixel $B(x_B)$. At this point, $B(x_B) = A(x_A)$, and a different label will be assigned to blob B (label B) at the condition that

$$d > \mathrm{MND}$$

(where $d = x_B - x_A$), independent of the absolute intensity of blobs A and B. In the case of image I, this process is extended to both axes (rows and columns).

The labeling process stops whenever all the pixels present in the vector structure are labeled or when a predefined number of labels have been posted (i.e., a predefined number of defects has been found in the image I). In fact, as pixels of low gray-level values are labeled, the risk of false alarms grows. Moreover, if the condition to reject an inspected part is that at least a pre-defined number of defects is detected, the computational time can be reduced if the labeling process is stopped as soon as this predefined number of detected defect is reached (in the case of a bad component).

To validate this procedure, the procedure can be redone with

$$\mathrm{MND}' \rightarrow \mathrm{MND} + 1 \tag{6.14}$$

until the same defect set is obtained in two successive tests. Each test is executed rapidly, since sorting within the vector structure is already available. It is only necessary to reinitialize the label vector lb to zero and to redo the labeling process as MND'.

6.3.2 Part II: Defect Edge Estimation

At this stage of the algorithm, all the seeds corresponding to the defects are available. In this section, the gross defect shape will be estimated by growing a region around the seeds. Each seed is processed individually and one threshold per defect is established. If many defects are present in an image, many thresholds (one per defect) are obtained. The purpose of the technique discussed in this section is to give an estimation of defect shape *rapidly*. In Chapter 10 methods are discussed to determine defect shape quantitatively.

In the image I, for the defect (i.e., the seed) located in (i_d, j_d), the threshold Th is first set:

$$\text{Th} = \text{Th}_{\max} = I(i_d, j_d) \tag{6.15}$$

where $I(i_d, j_d)$ is the value of the pixel located in (i_d, j_d) in image I. The number of neighbors $\text{Nr}(\text{Th}_{\max})$ having the same gray-level value as $I(i_d, j_d)$ is computed using a recursive procedure and assuming an eight connectivity (i.e., neighbor pixels at four edges plus four corners). The search is then redone with

$$\text{Th}' \leftarrow \text{Th} + 1 \tag{6.16}$$

until, using a recursive procedure, an image border is hit. At this moment, the vector Nr holds, for all the potential thresholds Th, the number of pixels agglomerated around the seed located in (i_d, j_d). Figure 6.8 shows a schematic example of the Nr content (Figure 6.9 shows typical real values.) If present in Nr, a sudden increase in the number of agglomerated pixels indicates that the image background intensity is reached. This is especially the case for reconstructed images I' in the mobile configuration (Section 8.1.2, Figure 8.8). Using simple *heuristic* rules, this threshold is located at Th_{est} if the corresponding difference $|\text{Nr}(\text{Th}_{\text{est}}) - \text{Nr}(\text{Th}_{\text{est}} - 1)|$ is found greater than three times the average step computed from the differences $|\text{Nr}(\text{Th}_i) - \text{Nr}(\text{Th}_{i-1})|$ between adjacent positions in the vector Nr from $i = \text{Th}_{\max}$ to 0. In fact, this corresponds to computing a derivative of vector Nr and to applying rules to locate and validate the sharp background transition. Notice that part I of the algorithm (Section 6.3.1) could be applied on a vector Nr derivative to locate the sudden increase in agglomerated pixels, which corresponds to a large derivative value.

If no sharp transition is found within Nr, the maximum is kept:

$$\text{Th}_{\text{est}} = \text{Th}_{\max} \tag{6.17}$$

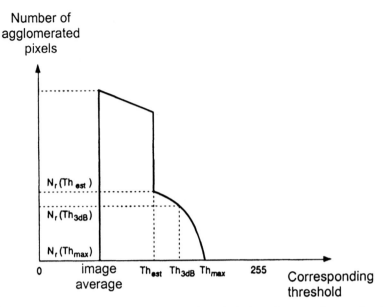

FIGURE 6.8 Schematic variation of the number of pixels agglomerated around a defect (seed) as a function of the threshold. The sudden increase corresponds to the background value, thus allowing us to establish the threshold TH_{3dB}. Note that computation was done only for pixels with values above image average (see the text); this explains the abrupt transition seen on the left for lower values. (From Maldague, 1993.)

Many tests indicated that the threshold Th_{est} obtained using this procedure does not correspond "perfectly" to manual segmentation. A corrective 3-dB factor allows us to obtain better agreement. The defect threshold is thus set at

$$Th_{3dB} \leftrightarrow \frac{1}{\sqrt{2}} Nr(Th_{est}) \tag{6.18}$$

The threshold specified by eq. (6.18) was found adequate in most cases. This use of *heuristic rules* can be understood since it provides interpretation program knowledge not available otherwise.

6.3.3 Results and Discussion

Let's now present some typical results obtained with the segmentation algorithm just described, with images obtained on a variety of samples tested in either a static or a mobile configuration. The mobile configuration is of interest since it allows us to inspect large surfaces at a rapid pace. The segmentation algorithm makes it possible to obtain a *map* of the surface inspected, where positions of the defects detected are indicated. If only defect detection is neces-

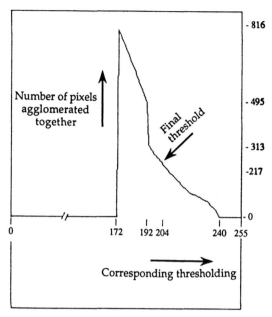

FIGURE 6.9 Typical example of vector N_r, case of central defect of Figure 6.12. (From Maldague, 1993.)

sary, part II of the algorithm is not required. It may, however, be of interest to evaluate the severity of the damaged areas in order to reject false alarms based on probability criteria of size, relative intensity, or other characteristic features (perimeter, surface, etc.). False-alarm detection is related to the notion of *detection probability curve* (Figure 7.8), in particular the question of small defects, taking into account the amount of noise present in the thermograms.

Generally, an interesting approach to validating segmentation and the results of NDE techniques consists of establishing a statistical database using, for instance, the Tanimoto detection criterion, which can be defined as follows in inspection situations (Yanisov et al., 1984):

$$R = \frac{N_R - N_M}{N_R - N_F} \tag{6.19}$$

where N_R is the number of real defects, N_M the number of missed defects, and N_F the number of false alarms. Obviously, to use this statistical method, a great number of components must be considered and destructed after the inspection in order to determine N_R, N_M, and N_F. If component destruction is not practical, it may be valuable to go through production and maintenance records in order to establish the Tanimoto criterion value. Alternatively, the sample of interest can be inspected nondestructively with another pertinent NDE technique (ultrasonics, x-ray, etc.) to assess the parameters of eq. (6.19) (see Section 7.5).

Figure 6.10 shows few segmentation results in the case of the static configuration in reflection. The images are organized by pairs as follow: on the left, raw images, and on the right, segmented images. The components tested are typical of the aerospace industry. In the mobile configuration (Section 8.1.2), the field of view depends on the size of the component inspected as well as of the moving speed. A smaller *minimum neighbor distance* (MND) is selected in order to pick up smaller defects, which are less visible since they fill less space in the *reconstructed* images. Figure 6.11 shows some results; in all cases, MND = 9. Among these images (all but part *b*) it is worthwhile to mention the noncontinuous texture, which is due to the low image transfer rate of the equipment used (3.8 images per second in that case). Figure 6.11b was obtained with image transfer performed from an experiment recorded on a VTR tape. Since more images are available in this case (30 images per second), the texture of the reconstructed image is smoother.

As stated previously, the temporal information is lost in both the *reconstruction* and *summation* processes, eqs. (6.3) and (6.9). Consequently, it is not possible at this stage to evaluate defect depth and size quantitatively. This was not the purpose of this study. However, for this reason, it is important to note that identical defects located at two different depths will appear in the segmented image as having different diameters.

Back to the *Tanimoto criteria* explained previously, the performance of this segmentation algorithm was checked and yields to a 90% figure. Of course, unless a component is cut in pieces after the inspection, it is difficult to estimate the quality of the inspection/detection/interpretation procedure, especially in the case of real defects. For example, in Figure 6.12 we show the image obtained in the case of static inspection of an impact-damaged graphite epoxy component. Two defects show up in the segmented image, and the *butterfly* shape (Section 11.1.1), is easily distinguished. To determine either the second defect is present on its own or is only an outgrow of the main defect (this is likely to be the case for impact damages), the part would have to be cut in pieces and/ or inspected with another NDE technique, such as a water-immersed ultrasonic C-scan (Section 7.5).

This segmentation algorithm is well suited for automatic defect extraction in low-spatial-content images such as in the case of TNDT images, while the *defect maps* produced by the algorithm are easily interpreted. The algorithm uses information at the *global* level to ease the decision-making process at the *local* level. Since complex operation is not required, it can be implemented easily. This is an important aspect in the case of real-time inspection/interpretation procedures, where hardware implementation becomes mandatory in achieving high throughput. Current digital signal processors (DSPs) are particularly well suited for such task.

The MND constant, which is based on spatial distance rather than on absolute intensity, is robust and relatively insensitive to noise, background uniformity, defect size (provided that defects are not too small, of course), number

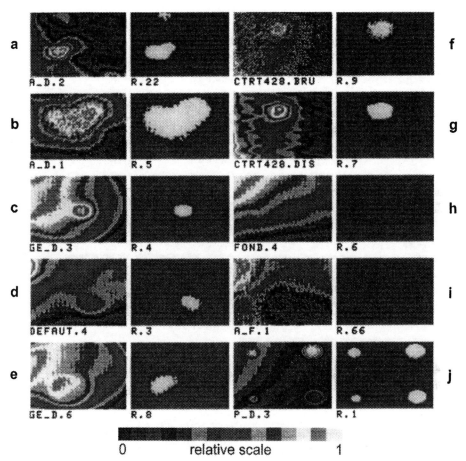

FIGURE 6.10 Segmentation results, static configuration, pulsed thermography. MND = 23. The field of view is 5 × 5 cm for all cases except (j), where it is 12 × 12 cm. (a,b) Aluminum honeycomb (aluminum sheet 0.5 mm thick, epoxy bonded on a honeycomb core). A bonding defect (lack of epoxy) is present at the bonding interface. Two images having a different field of view are shown with the defect detected having, of course, the same shape. A false reading is detected in (a), at the top of the image. (c–e) Graphite-epoxy panel with artificial defects (Teflon implant, 10 mm in diameter) inserted at different depths under the surface: (c,d) 1 mm; (e) 0.5 mm. Note the strong radiometric distortions on raw images (d) and (e). (f,g) Graphite epoxy panel as in (c) and (d). These images are obtained differently; they are, in fact, thermal contrast images which serve for the quantitative analysis (Chapter 10). (f) is the raw image and (g) is the smoothed image (using the technique presented in Chapter 5). This demonstrates the good noise immunity of the algorithm. (h,i) Graphite-epoxy panel without defect. (j) Plexiglas plate in which holes were drilled at different depths from the back surface (1.12 mm top row and 2.25 mm bottom row, with diameters 10 mm on left and 20 mm on right). These are raw images where the effects of the apparent spatial resolution are not corrected, thus explaining the elliptical rather than circular shapes observed. These defects are truly academic; this plate is tested because it has many known defects. Matlab m-script: defect.m performs such computations (see Appendix F). (From Maldague, 1993.)

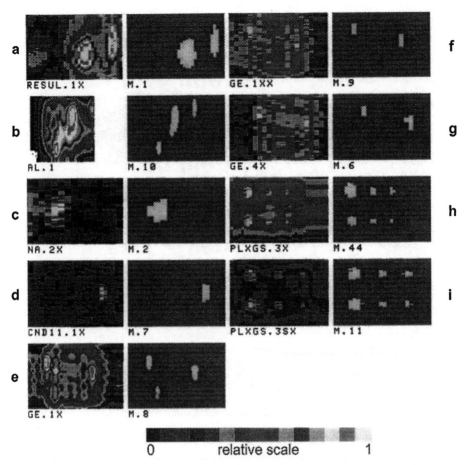

FIGURE 6.11 Segmentation results, mobile configuration, reconstructed images, pulsed thermography. (a,b) Aluminum laminate (aluminum sheet 0.5 mm thick, epoxy bonded on a foam core). Two bonding defects (lack of epoxy) are present at the interface. Field of view, 12 cm height on the complete panel length (40 cm). These panels are used, for instance, in the transport industry, in the construction of refrigerated vehicles. (c) Aluminum honeycomb panel (a,b), field of view: 12 cm height on the complete panel length (15 cm). (d) Graphite epoxy panel with artificial defect (Teflon implant, 10 mm in diameter inserted 0.5 mm under the surface). (e) Graphite-epoxy panel with two artificial defects (Teflon implant, 10 mm in diameter, inserted 1 mm under the surface). A trend removal procedure was used (Section 6.1.1) before calling the segmentation algorithm with the goal of increasing the visibility of the defects. On the contrary, a false reading appears at the center of the segmented image. Field of view: 12 cm height (complete panel length 17 cm). (f,g) Same panel as in (e) for two different runs (reconstructed images without previous trend removal procedure). (h,i) Plexiglas plate of Figure 6.10j. (h) Reconstructed image as is. (i) Spatial subtraction technique studied previously: the image shown was obtained after subtraction from a reconstructed image of a Plexiglas plate without defect. As expected, (i) is "cleaner." Nevertheless the algorithm performed well in both cases and reveals the six drilled holes of 20, 10, and 5 mm diameter, re-

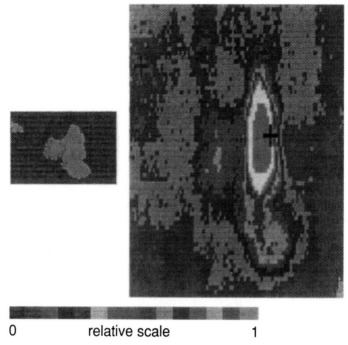

0   relative scale   1

FIGURE 6.12   First part of the algorithm and segmented image of a graphite-epoxy panel damaged by impacts. Crosses indicate the position of the defects (seeds) as located by the algorithm. What seems to be a false reading is visible on the left, although the delamination can stretch over large distances. Static configuration, field of view 4 × 4 cm. Matlab m-script: defect.m performs such computations (see Appendix F). (From Maldague, 1993.)

of defects, defect intensity, and defect orientation. In Chapter 10 we will see how quantitative analysis completes results presented in this section. Note, however, that it is estimated that the detection/localization approach is sufficient in 80% of inspection situations.

The concept of automatic inspection through image *reconstruction* and segmentation is well suited for repetitive inspection of large components whose surfaces are relatively planar (nonplanar surfaces are discussed in Section 10.6). For cases where the inspection must be performed manually by an operator,

---

spectively, from left to right, located at 1.12 mm beneath the surface (top row) and 2.25 mm beneath the surface (bottom row). Note that the smallest defect (5 mm diameter, 2.25 mm depth) is at the detectability limit since its radius/depth ratio is about 1.1 in this isotropic material. Lateral speed was different for (h) and (i), explaining the different positions of detected defects in the images. Matlab m-scripts: defect.m and juxta.m performs such computations (see Appendix F). (From Maldague, 1993.)

it is important to maximize defect visibility in the thermograms. Techniques reviewed in Chapter 5 are pertinent for these tasks.

Example 6.2 In a static pulsed TNDT inspection procedure, the field of view is 20×20 cm and the image size is 256×256 pixels. What value of MND would be recommended if it was established that a 1-cm circular defect is the minimum size the procedure can detect with reliability?

SOLUTION MND is not a very sensitive parameter, but that also makes it difficult to establish. Without taking into account the SRF (slit response function), a 20-cm field of view, corresponding to 256 pixels, means that a 1-cm defect correspond to 13 pixels. A rule is to take about half the defect size of interest as *minimum* defect separation, thus making a MND value of 7 a start here. For example, in Figure 6.10, the field of view is 5 cm with 105 pixels per row. The size of defects of interest is about 2 cm, corresponding to 42 pixels $(42/2 = 21)$. The MND specified is 23. Here, MND = 7 can result in spurious defect detections, so that the user will probably want to increase that value somehow.

6.4 LATERAL SCANNING DEFECT DETECTION ALGORITHM

We now introduce an algorithm especially suited for defect extraction in images obtained in a mobile configuration where workpieces are scanned laterally following the pulsed thermographic approach (Figure 6.13) (see also Figure 8.8).

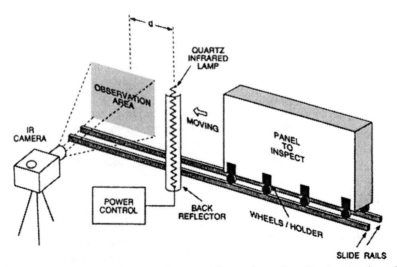

FIGURE 6.13 Schematic diagram of the mobile configuration: line heating in pulsed active thermography for inspection of large components mounted on a slide. (From Maldague, 1994b, with permission from Gordon and Breach Publishers.)

The principle is as follows. Observation of the image sequence during motion of a panel in front of a heating lamp reveals a very interesting fact: Even in the presence of noise, defects are easily seen because of their lateral motion. Noise is random, while defects follow panel motion. If only one image is observed, this dynamic effect is lost and the distinction among defect, noise, and sound area is more delicate. This is because the eye–brain entity integrates the dynamic noise but cannot proceed in such manner when the noise pattern is fixed. This phenomenon can also be observed by watching an image sequence recorded from a VTR and comparing the perceived quality obtained with a particular image while the VTR is in the "pause" mode: The fixed image looks *terrible* but the image sequence is acceptable. In fact, such an effect is used with profit in some low-rate image transmission systems where dynamic noise is superposed on images to make them more appealing to the viewer by reducing the *contour effect* due to the quantification performed on a limited number of gray levels.

The algorithm is based on analysis of many images (two, typically) recorded at different moments. In this algorithm, subtraction by spatial reference is first performed as described in Section 5.5. Subtracted images are next binarized (Figure 6.14d and e) through application of a threshold function. Binary images are coded with only two levels: high and low. Notice that for this application what is of interest is defect detection—in particular, defect localization and gross size estimation. Consequently, the relative intensity of defects is not relevant in this study.

In a first step, knowing the lateral speed of the panels (expressed in pixels/s), it is possible to compare two (or more) images by taking their lateral motion into account. Each binary image is divided in a succession of small matrixes (3×3). The matrix of an image located in (a, b) is compared with the corresponding matrix in the other image (recorded a little later), which is located at the position $(a + \Delta a, b)$, where Δa is the interval due to the lateral motion (along the x axis). The passage rule to obtain the result is simple: If at least one element is at a high level in both matrixes considered, the resulting matrix (output image O_1) is set to high; otherwise, it is set to low. This simple technique allows us to reduce significantly the amount of noise present in the image (Figure 6.14f).

This first step is, however, not sufficient to *purify* image O_1 sufficiently. For this reason, another noise-suppressing technique is applied, this time by working only on image O_1. At this stage, a matrix of bigger size is used. Its size corresponds grossly to the minimum defect size we are interested to put in evidence. In our example, this corresponds to a matrix of 32×32 pixels. This large matrix is scanned over image O_1 from left to right and top to bottom, at every location, and the ratio of the number of pixels at a high level over the number of pixels at a low level is computed. If this ratio is smaller than a preestablished value (20%), all the windows in O_1 are set to a low level; otherwise, the window content remains unaffected. This type of processing, by *erosion*, is quite radical and permits us to eliminate completely the residual noise present in O_1. The only structures subsisting in the new output image O_2 are the defects, if any (Figure 6.14g).

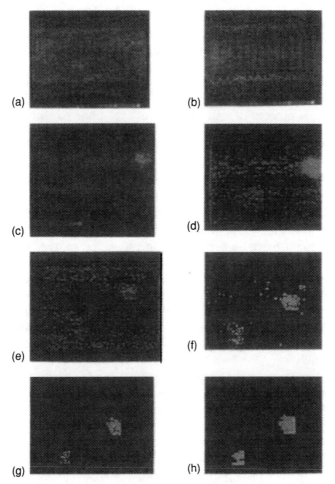

FIGURE 6.14 Steps needed for defect extraction from images recorded over an aluminum laminate (bonding defect). Raw image recorded at $t = 3.04$ s after beginning of heating: (a) above a defect; (b) above a sound area. (c) [= (a)–(b)] Two defects are visible despite the presence of strong noise. (d) Binary image of (c). (e) Binary image recorded at $t = 3.33$ s. The same steps as for d have been applied on this image. (f) Result obtained from (d) and (e), gross localization of two defects thanks to their motion in the field of view (see the text). (g) Noise suppression through erosion. (h) Defect reconstruction through dilation. Note that a 530-W 18-cm-long lamp is used for this test; configuration of Figure 6.13. (From Maldague, 1993.)

An *erosion* process is generally followed by a *dilatation* process in order to recover somewhat the original shape of the defects, which went through two successive erosive phases. The third stage, the dilatation process, is similar to the second stage since it also works with a matrix scanned over the image, as before. However, this time, the matrix size is fixed at 5×5 pixels, a sufficiently small size to track defect edges. The rule is as follows: If at least one element of the window considered is at the high level, the entire window is set to high; otherwise, nothing happens. This dilatation method has the advantage of filling the "holes" in the defects and of partially reconstructing their edges; the third image output O_3 is shown in Figure 6.14h.

As a possible improvement in the algorithm, a flexible matrix grid would be more appropriate than a fixed grid, to avoid edge truncation. Drawbacks of these modifications are increased programming complexity and slower execution time, however. The algorithm is very powerful for the segmentation of thermograms recorded during lateral inspection. It requires two input parameters: the threshold (to binarize the images) and also the lateral motion speed (expressed in pixels per second). In the case of repetitive inspections, these values are set in advance. It is, however, important to check the validity of these parameters periodically. In fact, the use of an absolute threshold is not always reliable. In the preceding section, another algorithm which allows us to relax this constraint was presented.

Before closing this section, we would to point out that even if binary images are used in the algorithm, by performing an AND operation between the output image O_3 and the first analyzed image, the original gray levels of the defects are restored, thus enabling quantitative analysis of the defects detected (Chapter 10).

6.5 NEURAL NETWORK DETECTION

Neural networks are powerful data analysis methods based on calibration with known data, as opposed to methods based on an analytical formulation of the physical system under study, such as some of those presented in Chapter 10. Neural network methods are introduced in this section.

6.5.1 Neural Network Fundamentals

As is well known, neural network (NN) architecture is inspired by biology and they offer interesting properties for thermographic analysis and other fields as well, since they are adaptive and robust. Interested readers will find additional information in the following references: C. Lau (1991), Kröse and Van Der Smargt, (1995), Ben et al. (1995), Hagan et al. (1996), Sarle (2000). The present section is based on the following references: Largouët (1997), Maldague et al. (1998), Vallerand and Maldague, (2000).

A *neural network* (NN) is a highly interconnected set of small processors, each of which has a small memorization capability they learned from information submitted to them in the first place. Generally, the network is divided into layers and the parallel nature of NNs make them well suited to work on multiparameter problems. Figure 6.15 shows a simple perceptron with two inputs and one output. The basic idea is really simple. A neuron is like a cell with various inputs $x_1, x_2, \ldots, x_n$ and one scalar output, y. The neuron multiplies the

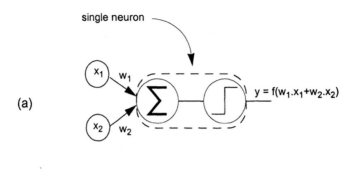

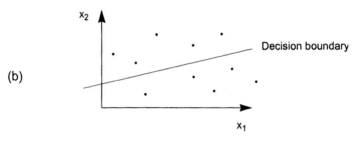

FIGURE 6.15   (a) Simple perceptron with two inputs and one output; (b) corresponding simple decision boundary (the neuron is fired in for the points located above the border).

inputs by synaptic weights w_1 to w_n and combines these products linearly to form

$$S = x_1 w_1 + x_2 w_2 + \cdots + x_n w_n. \tag{6.20}$$

The output signal is

$$y = f(S) \tag{6.21}$$

In a simple case, the function $f(\cdot)$ is a simple threshold function. In such a case, y is not significant unless S reaches the threshold Th, which fires the neuron. From this single building block neuron, various architectures are possible, as, for instance, the Hopfield–Kohonen growing cell structure (GCS), and growing neural gas (GNG), among others. This allows us to establish more complex decision boundaries than the straight line shown in Figure 6.15b. A common activation function is the log-sigmoid function, which allows us to mimic the delay present in biological neurons, in that case f (Figure 6.16):

$$f = \frac{1}{1 + e^{-n}} \tag{6.22}$$

Moreover, the nonlinearity of this function is particularly convenient for han-

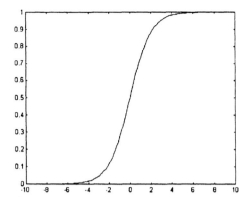

FIGURE 6.16 Shape of the NN logsig activation function f, eq. (6.22). Matlab m-script: logsig.m performs such computations (see Appendix F).

dling complex problems. Such a function is available in the Mathlab neural network toolbox (function logsig; Table 6.1), which is useful in implementing and testing NNs without the burden of programming.

Learning proceeds by modification of synaptic weights w_i. In supervised learning, for each entry vector an output vector with the value desired or expected is presented. In the case of unsupervised learning, no imposed output is presented to the network, and the output classification is done by the NN itself.

6.5.2 Neural Network Design

Establishing the architecture of an NN is not a trivial operation. In fact, it is necessary to specify the: number of layers, number of neurons per layer, activation function, and learning algorithm. Unfortunately, no theory is available to answer these questions. People do rely on experimental expertise, knowledge, and trial and error. At this point a thermal model (Section 3.4.2) of the inspection task is convenient to enable rapid investigation of different NN topologies without the necessity to build and test real samples. To investigate the NN architectures of interest with more reliability, such simulated data are often corrupted by added (Gaussian) noise to resemble more closely experimental data.

The number of neurons must be high to represent complex functions; on the other hand, the number should be kept reasonable, to avoid overlearning, which is characterized by multiple diverging values between correct outputs for learned inputs, thus preventing the NN from generalizing. Another question concerns the number of layers. Complex problems with several parameters to sort and many inflection points in the decision border require more neuron layers to enable a more complex decision hyperplane. In fact, processing capability increases with the number of layers used. However, the Kolmogorov theorem stipulates that a problem of any complexity can be modeled by a three-

TABLE 6.1 Matlab M-Scripts Available in Appendix F

Name of Matlab M-Script	Short Description, Purpose
defect.m [mat1, mat_out] = defect (mat, MND,Max_def, cross, varargin);	Implementation of the automatic defect detection procedure on an unsigned character image mat with parameter MND. Max_def is the maximum number of defects that can be found in an image. If the parameter cross is set to 1, the image mat1 returned is equal to mat with only crosses indicated at the defect center. Any other value of cross returns ma1 as a map of defect(s) detected in mat (value 0 at background and 1 at the defect location). The matrix mat_out contains as the first column, the row of the defect center; second column, the column of the defect center; third column, the value of the defect center. Example of use: **[mat1, mat_out] = defect(mat,50,1,1);** *Note:* If you have the image processing toolbox (functions *bwla-bel* and *bwmorph* are then used), you may add a parameter (any number will do) after the cross to enable the use of functions from this toolbox. You may obtain better results with these functions. Example of use with the image processing toolbox: **[mat1, mat_out] = defect(mat,50,1,1,1);**

moment.m
mat = moment (dir, pre, ext, ta, tb, incre, option);
	Return the moment image from the image sequence present on the disk in the directory dir. Computation proceeds in the time period *ta* to *tb* (coded in the name of the file having the prefix *pre* and extension *ext*). *incre* is the time increment between images. Option parameters are (1) add all images, (2) average all images, (3) add all images and convert to unsigned character. A header is accounted for. Adjustable parameters are:
	   Header: number of bytes before the image
	   Maxrow: number of pixels per row
	   Maxcol: number of pixels per column
	Example of use: **mat = moment('G:\images\','d_','uc', 200,2000,10,2);**

logsig.m logsig;	Plot the logsig NN activation function [eq. (6.22)]. Example of use: **logsig;**

layer network. Although theoretically true, practically it is not always possible to restrict the problem to three layers. A three-layer NN such as that shown in Figure 6.17 is a common architecture in the field of TNDT (Prabhu et al., 1992; Bison et al., 1994a; Tretout et al., 1995). It belongs to the multilayer perceptron (MLP) family. A hidden layer is a layer that is not connected directly to the output; the NN of Figure 6.17 has two hidden layers.

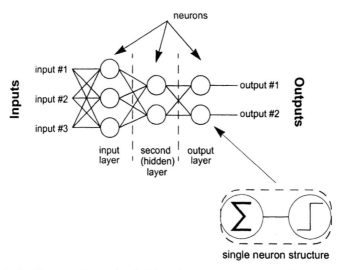

FIGURE 6.17   Example of a three-layer perceptron neural network.

6.5.3   Learning

Hidden layers complicate the learning process since it becomes difficult to reflect back on each weight the error computed on the output layer, especially with nonlinear activation functions. The back propagation algorithm of the error difference handles such problems. First, weights are initialized to random values and an input–output combination is presented to the network. This is called the *propagation* or *relaxation phase*. The error is next computed as the quadratic summation of differences for each output neuron and back-propagated through the network. At this step, weights are possibly modified. This process is repeated with a new input–output combination until the quadratic error is within the desired limits or until a certain number of iterations is reached. An interesting property of this logsig function [eq. (6.22)] is that it can be derived; this is essential in the back-propagation algorithm (BPL).

A property of the MLP NN is its poor ability to extrapolate. For TNDT this implies training for use of the NN over the entire depth range and for all material of interest. The simulated data set should cover the same depth range as the specimens of interest. During the learning phase, to avoid network sensitivity to depth ranking instead of to absolute values, data should be presented to the network randomly. To train the NN independently of the material of interest, it is possible to normalize the data using the thermal diffusivity α of the material.

6.5.4   Example of Neural Network Use

Lets now illustrate the NN use in TNDT. The task is to determine automatically the percentage of corrosion in aluminum specimens using the pulsed

Principle of inspection:

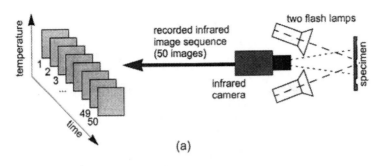

(a)

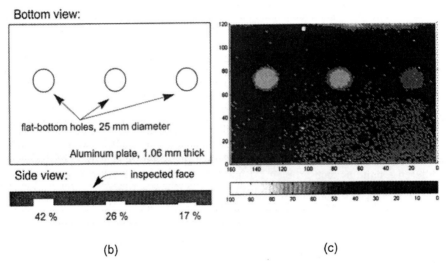

(b)   (c)

FIGURE 6.18 Application of neural network to TNDT: (a) pulsed thermography configuration and recorded thermogram sequence; (b) specimen geometry; (c) neural network output calibrated in percent of corrosion detected in the specimen. [(c) From Vallerand and Maldague, 2000, reprinted with permission from Elsevier Science.]

thermography approach. The processing approach relies on an NN. A MLP is created having a 50–10–7 structure (each number represents the number of neurons for each respective layer: first, second, and third). Such a structure is implemented in Matlab. The 50 neurons in the first layer correspond to the 50 thermograms recorded after pulsed heating of the specimen (Figure 6.18a). The NN is thus presented with the vector of 50 temperature data for a given pixel in the image. Its output is calibrated as a percentage of aluminum corrosion for that pixel. In fact, the seven outputs correspond to the seven corrosion grades of interest. The 10 neurons in the hidden layer are determined through trial and error. The three-layer NN is classic in many applications.

Before using the NN, all weighting factors are adjusted according to the BPL algorithm. This means that the NN needs to learn how to class input vectors into different corrosion classes: 0% (no defect), 17%, 26%, 42%, 52%, 62%, and 83%. In this case, the NN training is supervised because training samples are provided. A training sample is an input vector with a known output, generated using samples with known flaws. Once trained, the NN is ready to assign unknown input data to a specific corrosion value (i.e., to a specific class).

Figure 6.18 illustrates the application of this NN on a specimen whose geometry is shown in part (b). The NN output is shown in part (c). Several pixels uncorrectly classified appear in the output image, especially at the bottom right. This is due to strong nonuniform heating in that area. The edges of the flat-bottomed holes are not characterized correctly. The corrosion computed for the middle flaw is reported as 42% instead of 26% corrosion (real value). The two other holes are characterized correctly.

To assess the result correctness, the percentage of correct characterization (PCC) is computed. This value is established by computing the ratio of all correctly categorized pixels over all pixels in the region of interest. In Figure 6.18c, the PCC measured for this specimen over the entire image is 86%. In Section 10.4.4, a two-step statistical method is presented featuring a 98% PCC (whole image). The two-step method consists, first, of detecting the presence of a defect and then assessing its characteristics (such as its depth), Bison et al., 1994a. Nevertheless, the NN described here gives fair results in terms of flaw detection and localization.

Notice that in this example, the limited acquisition rate of the IR camera used (20 ms between IR images or 50 Hz) might explain the discrepancies observed, especially in the case of high-thermal-conductivity aluminum inspection. Reported use of NN in the case of lower-thermal-conductivity specimens (such as plastic) with the same acquisition rate yields to better agreement.

6.5.5 Remarks on Neural Networks

NNs are now commonly used in various applications, especially since the physical process involved can be considered as a blackbox: You do not need complex inverse mathematical equations to extract relevant data from your experiments, such as defect location and depth, as in the case of the example just discussed. Moreover, common software tools such as Matlab allow for easy configuration of the NN. The drawback of this blackbox approach is that you do not really know what is happening, and this is especially annoying when the results obtained are not as expected.

Example 6.3 Design an NN for defect detection in TNDT. Thermogram sequences are 64 images long, starting when pulsed heating (duration 100 ms) fires, at $t = 0$ ms. The acquisition rate is 20 ms.

SOLUTION   As discussed earlier, NN design is often defined through operator experience and trials and error. Nevertheless, based on past achievement in

TNDT, an MLP with one hidden layer would be a good choice. Since only defect detection is requested, a one-output neuron would be sufficient. As for the input layer, we have 64 images to work with. However, with flash pulse duration of five images ($100 \text{ ms} = 5 \times 20 \text{ ms}$), it is likely that at least the first five thermograms will be useless, due to saturation through reflection of infrared emitted radiation from the flash pulse. Then it would be recommended to work with the $64 - 5 = 59$ thermograms ($t \geq 120 \text{ ms}$). Thus for every pixel in the field of view, a vector v of 59 values would be extracted and presented to the NN, yielding a defect/nondefect decision for that pixel location. Extending this processing to all pixel location produces a defect map. The number of hidden neurons is less trivial to establishe. The case study presented in Section 6.5.4 had a 50–10–7 configuration. It should be noted, however, that 50–4–7 and 50–7–7 were tested, with similar results obtained. Increasing the number of neurons makes the NN more rigid in the sense that it tends to stick to learned values, with less ability to generalize when noisy data sets are presented. If follows that perhaps seven would be a good starting guess, thus with a configuration of 59–7–1 (a ratio of 10 with respect to the input layer plus 1, rounded).

PROBLEMS

6.1 **[image formation]** Panels are to be inspected in a mobile configuration as depicted in Figure 6.4. Assume a field of view of 18×15 cm, an image size of 128×128 pixels, and a moving speed of 1.3 cm/s. The acquisition rate is 270 ms per image. What is the shifting factor needed to form a moment image?

Solution   The frame rate is 1/270 ms = 3.7 Hz (3.7 images/s). The panel crosses the field of view in 13.9 s (18/1.3 s), corresponding to a total acquisition of 52 images (13.9×3.7). Each reconstructed image will have 52 columns. Following eq. (6.8), it is

$$\text{shifting} = \frac{LRC_k}{vC_{\text{Maxcol}}} = \frac{(18 \times 3.7)C_k}{1.3 \times 128} = 0.4C_k$$

This means that the first reconstructed image will not be shifted, whereas the last reconstructed image will be shifted by 52 pixels.

6.2 **[defect detection]** Assume a field of view of 10×10 cm in a static pulsed TNDT inspection procedure. The image size is 128×128 pixels. What value of MND would be recommended if it were established that 3-cm circular defects are the minimum size the procedure can detect with reliability?

Solution   The 3-cm defect corresponds grossly to 39 pixels. As in Example 6.2, a good starting value would be about half of that value (i.e., 20). Such a large value with respect to the field of view should not result in too many false detections.

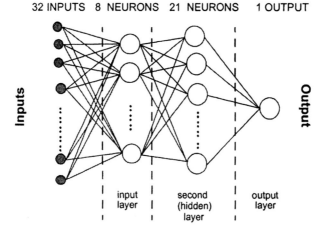

FIGURE 6.19   Three-layer neural network with 8–21–1 configuration.

6.3   **[neural network]** It is asked to design an NN for defect assessment in TNDT. Thermogram sequences are 100 images long, starting when the pulsed heating (duration 1 s) fires at $t = 0$ s. The acquisition rate is 250 ms. The range of defect depth of interest is up to 2 mm in the material inspected.

Solution   As before, a one-hidden-layer MLP could be a solution. To the difference of Example 6.3, it is now asked to characterize the defects so the output layer goes up to 9 neurons to sort defect in 0.25 mm depth increments plus a nondefect class. The first four images (1 s = 250 ms × 4) are likely to be useless, due to saturation, so they will not interest us here. A choice of $100 - 4 = 96$ neuron input layers is surely not good because of possible convergence problems in the learning phase with the BPL algorithm. As in Figure 6.19, it is not necessary for the number of neurons on the input layer to equal the number of inputs. Thus with an acceptable ratio of 4, $96/4 = 24$ neurons are selected. As for the hidden layer, using the same approach as before (a ratio of 10 with respect to the input layer plus 1, rounded) changes the configuration to 24–4–9. For reference, some configurations reported to have been used in TNDT are 50–10–7, 50–7–7, 50–4–7, 8–21–1, 9–17–1, and 50–12–13.

CHAPTER SEVEN

Materials

The aim of this chapter is to cover materials fundamentals in a very concise way. Writing this chapter was inspired by many sources. Among them, G. S. Brady and Clauser, (1986), Dorlot et al. (1991), Moore and McIntire, (1996), and the Matweb Web site* were found particularly helpful, and we invite interested readers to refer to these documents, which provide more information than it is possible to cover in a single chapter. Some topics are not relevant to the present discussion and are not included, such as acidity and alkalinity, biotic materials, flavor, viscosity, and colors. The chapter starts with a review of materials properties, followed by a presentation of common industrial materials. A brief overview of other NDE methods follows, together with a general discussion of TNDT applicability. The chapter concludes with a discussion of the probability of defect detection.

7.1 NEW MATERIALS

In our modern world, thanks to their specific advantages, numerous new materials (as contrasted with traditional materials) are now employed in particular application niches, displacing in these markets the traditional supremacy of, say, iron (5% of Earth's crust) and aluminum (8% of Earth's crust) even at higher cost: approximately $6 per kilogram instead of $1.50 per kilogram typically for traditional materials. Applications of these new materials are in the automobile, leisure boat, bicycle, shoe, and building industries. There are many reasons for the growing use of these new materials. Typically, they are less prone to corrosion, require fewer finishing operations, and sometimes also make possible a reduction in the number of parts needed to complete an assembly. For instance, fabrication of the trunk lid of the Fiat Tipo necessitated 20 parts using steel, whereas the plastic version required only 2 parts, resulting

---

*www.matweb.com.

in a substantial reduction of assembly time. The new materials include multi-layer laminates, CFRP (carbon fiber–reinforced plastic) composites (also called graphite epoxy composites), honeycomb panels (Figure 11.18), ceramic coatings, thermal barriers, metallic matrix composites, new alloys (such as Al–Cu–Li, Al–Li–Cu–Mg NiTi), and so on. They often come from the aerospace and military industries, such as ceramic coatings which enable weight reduction, increased efficiency, and substantial fuel consumption savings for jet engines operating at higher temperatures. In the following sections we discuss about both traditional and new materials.

7.2 PHYSICAL AND MECHANICAL PROPERTIES OF MATERIALS

The following list is adapted from (G. S. Brady and Clauser, (1986). Table 3.7 and Appendix E list property values for some materials.

- *Brittleness:* property to break without visible deformation or observable sign.
- *Bursting strength:* capacity of a material (in sheet form) to support hydrostatic pressure without rupture.
- *Compressibility:* extent to which a material is compressed when submitted to a given load (e.g., a gasket). *Permanent set:* percentage of original thickness that the material does not return. *Recovery:* percentage of original thickness that the material returns in a given time (generally, this diminishes in time as the loading lasts).
- *Conductivity:* relative rate to which the material conducts heat or electricity at normal temperature (16°C).
- *Creep rate:* rate at which strain or deformation occurs in a material under loading. *Creep strength:* maximum force that a material can sustain without rupture in a given period of time at a given temperature. *Creep recovery:* percentage of decrease in deformation once loading is removed.
- *Ductibility:* material property of being permanently deformed by tension without rupture (i.e., property of being drawn thin without breaking).
- *Elasticity:* capacity for a material to return to its original shape once the force causing the deformation is removed.
- *Elastic limit:* the largest stress unity that a material can sustain without permanent deformation.
- *Elongation:* increase in length of a bar under test as expressed as a percentage of length when breaking occurs in the original length.
- *Factor of safety:* for a material, ratio of the maximum stress with respect to working stress.
- *Fatigue strength:* measurement [in megapascal (MPa)] of the capacity to sustain loading without failure when material is submitted to such loading

a number of times. Fatigue strength is generally higher than prolonged tensile strength. *Fatigue life* is a measurement of the useful life of a material through a number of cycles of loading of given magnitude without failure.

- *Flash point:* minimum temperature to which a material or its vapor ignites or explodes.
- *Flow (creep):* gradual and continuous distortion of a material under constant loading, generally at high temperature.
- *Fusibility:* ease to which a material is melted.
- *Hardness:* property of a solid or a very viscous liquid to be firm or express solidity. Different methods measure hardness such as the *Brinell method* (a hard ball is pressed into the material, hardness is obtained from measurement of the indentation mark with respect to tabulated values), *scleroscope* (hardness is read on a graduated scale at the rebound height of a hammer with a diamond tip, from a fixed height), *Rockwell* (hardness related to the depth of penetration under loading of a steel ball or a diamond cone), *Vickers* (same as Brinell or Rockwell methods but with a diamond pyramid for hard materials), *Bierbaum* [hardness is measured as the width (measured with an microscope) of a scratch performed with a cubic-shaped 3-g diamond load], *Mohs hardness* scale is a scratch comparison method for abrasive and minerals, with 1 for talc and 10 for diamond. Table 7.1 presents hardness for some materials, as measured on the Knoop scale with a Knoop indentor.
- *Impact strength:* force [in joules (J)] required to break a material upon sudden application.
- *Malleability:* capacity to undergo permanent deformation without rupture; capacity of being rolled or hammered into thin sheet.
- *Modulus of elasticity (Young's modulus):* constant designated E, ratio of stress to corresponding strain in elastic limits of the material, without breaking. E is a measure of the stiffness [units of pascal (Pa)].

TABLE 7.1   Knoop Indentor Hardness of Hard Materials

Material	Hardness
Diamond	6000–6500
Boron carbide	2255
Silicon carbide	2135
Topaz	1250
Quartz	750
Hardened steel	400–800
glass	300–600

Source: Adapted from S. P. Parker (1984).

- *Modulus of rigidity:* ratio of unit shear stress to displacement for a unit length of material when submitted to shear stress causing displacement.
- *Plasticity:* property of a material to be deformed when submitted to a force and not to return to its original shape once the force is removed.
- *Porosity:* for a material, ratio of the volume of interstices to the volume of its mass.
- *Reduction of area:* percentage between area of a bar before stress and after rupture.
- *Refractive index:* indicates the ability of the material to transmit light; a larger value indicates less transmission. Vacuum has an refractive index of 1.00 (light transmission is 100%, reflection is 0%); ice, 1.30; window glass, 1.52; and polished diamond with parallel faces, 2.42 (light transmission is 83%) (see also Section 4.9).
- *Resilience:* energy of elasticity, energy stored in a material under stress which allows the material to return to its original shape when the stress is removed, within the elastic limit.
- *Softening point* (Vicat): temperature at which a flat-tip needle of 1 mm^2 penetrates 1 mm below a material surface under a 1-kg load when the material temperature is increased at a constant rate of 50°C/h.
- *Specific gravity:* ratio of weight of a given volume of material to the weight of pure 4°C water of the same volume.
- *Specific heat:* for 1 g of material, the number of calories required to increase its temperature by 1°C (also expressed as the number of joules for 1 kg of material). See Table 3.7 for values of common materials.
- *Stiffness:* material property measured by the rate at which stress in the material increases under strain.
- *Strain:* distortion taking place in a material submitted to an external force.
- *Strength:* capacity of a material to resist an imposed physical force.
- *Stress:* internal forces in a material following application of an external force.
- *Tensile strength:* maximum tensile loading per square unit of original cross section that the material can support. Most common measurement of ductibility for metals.
- *Thermal conductivity:* number of calories transmitted per second between two opposite faces of a cube 1 cm × 1 cm × 1 cm when the temperature difference between opposite faces is maintained at 1°C (also expressed as J s^{-1} m^{-1} °C^{-1} = W m^{-1} °C^{-1}).
- *Thermal expansion:* for a material, the coefficient of linear thermal expansion is equal to the length increase for a 1°C temperature difference.
- *Thermoplastic:* for a material, capacity of being molded and remolded at a temperature higher than normal without rupture under pressure and heat.

Thermosetting means that the material *sets* as a hard solid under heat and pressure without the possibility of being remolded.

- *Toughness:* relative degree of impact resistance without breaking; capacity of a material to absorb energy when stressed at a value greater than the elastic limit but without breaking.
- *Ultimate strength:* stress computed following the original cross section of the material when the applied force is maximum and causes material to rupture.
- *Yield point:* minimum tensile stress required to produce continuous deformation of a solid material.

7.3 INDUSTRIAL MATERIALS

Atoms are the basic building blocks of elements forming all material (Table 7.2). There are 92 stable elements organized in the periodic table, starting with hydrogen, with an atomic number of 1, up to uranium with an atomic number of 92. Materials having atomic number higher than 92 do exist but are unstable and decay in time (this is measured as the *half-life*). Elements are used alone or in a mixture with others elements, with an infinite number of combinations and proportions. Alloys, which are blends of metals in different percentages of constitutive elements, are a common example of such a mixture. Elements combining into molecules with distinctive geometrical shapes are another example of such a mixture (e.g., a carbon-based organic compound).

Depending on the mobility of the molecules, solid, liquid, and gas states are observed. For some materials, the liquid phase is very short. In fact, this is as if such a material changes directly from a solid state to a gaseous state. Such a process is called *sublimation*. For example, iron has free electrons and disintegrates into contact with air and moisture to form iron oxide. On the other hand, gold has no free electrons, and due to its high-energy state, is not affected by other elements. It is thus said to be noncorrosive. In the case of aluminum, oxygen cross-links with free electrons, and this process prevents subsequent corrosion.

7.3.1 Iron-Based Alloys

Steels and products derived therefrom make up about 90% of the world production of metallic materials (in weight), because of low cost, ease of mass manufacturing, and the possibility of obtaining specific properties thanks to specific alloys and particular thermal processes. Generally, steels are divided into the following families: general-purpose carbon steels, steel alloys with thermal processing (allied or not), tool steels, stainless steels, and cast iron. Most steels are alloys of iron, carbon (less than 1.5%), and some other elements

TABLE 7.2   Natural Elements Discussed

Name	Symbol	Atomic Number	Atomic Weight[a]	Melting Point (°C)
Aluminum	Al	13	26.97	660
Carbon	C	6	12	3700
Chromium	Cr	24	52.01	1800
Cobalt	Co	27	58.94	1490
Copper	Cu	29	63.57	1083
Gold	Au	79	197.2	1063
Hydrogen	H	1	1.0078	−259.2
Iron	Fe	26	55.84	1535
Lead	Pb	82	207.22	327.4
Magnesium	Mg	12	24.32	650
Manganese	Mn	25	54.93	1260
Molybdenum	Mo	42	96.0	2625
Nickel	Ni	28	58.69	1455
Niobium	Nb	41	92.906	2468
Nitrogen	N	7	14.008	−210
Oxygen	O	8	16.0000	−218.8
Silicon	Si	14	28.06	1430
Tantalum	Ta	73	180.88	3000
Tin	Sn	50	118.7	231.9
Titanium	Ti	22	47.9	1820
Tungsten	W	74	184.0	3410
Uranium	U	92	238.14	1850
Vanadium	V	23	50.95	1735
Zinc	Zn	30	65.38	419.5
Zirconium	Zr	40	91.22	1700

Source: Adapted from S. P. Parker (1984) and G. S. Brady and Clauser (1986).

[a]Number assigned to each chemical element which specifies the average mass of its atom (taking into account the relative proportion of its isotopes, if any). The reference is carbon (^{12}C).

in small quantities (such as silicon and manganese). Carbon steels without further processing are of general use and constitute 85% of the total steel production. These alloys have an elasticity limit below 350 MPa. Improved properties (elasticity limit up to 500 MPa) are possible with specific thermomechanical processing (such as rolling and quenching) and in addition, in small quantities (0.1 to 0.5%) of other elements, such as niobium, vanadium, and titanium for microallied alloys, and chromium and molybdenum for dual-phase steels used in the automobile industry. Without external coatings such as paints and galvanization, steels are prone to atmospheric corrosion. However, the addition of elements such as copper (0.2 to 0.55%), sometimes supplemented by nickel (0.5%) and chromium (0.5%), improves resistance to atmospheric corrosion, thanks to the formation of a thin protective oxide layer. The atomic steel structure is austenitic (i.e., cubical face-centered lattice) over 910°C and exhibits a ferrite matrix structure (i.e., centered cubical) below that temperature. Effects of such structures on steel properties are important; for instance, austenitic stainless steel is a nonmagnetic material.

Tool steels makes only 0.1% of the world production, but they are of great importance. They have properties similar to those of carbon steels but their final use implies improvement of some of their properties, such as hardness and wear resistance. These are obtained with the addition of elements such as carbon, manganese, vanadium, tungsten, molybdenum, and cobalt in various proportions depending on the application. Some thermal processing completes the manufacturing, such as quenching at room temperature followed by subsequent cooling at temperatures between −50 to −196°C. In the United States, more than 75 different alloys are produced. For example, steel used for plastic molds contains 0.1% carbon, 2.6% chromium, and 1.25% nickel.

Various kinds of stainless steels are available with different degrees of resistance to corrosion. The key additive element is chromium, which added in proportions above 12% and up to 30% makes the steel corrosion-resistant, thanks to the formation of a passive protective surface film. The second important additive element in some alloys is nickel (up to 17%). Other additives appear in smaller proportions, such as carbon (<0.2%), aluminum (<1.2%), nitrogen (<0.25%), molybdenum (<2.5%), and tantalum (<1%). As for the other steel families, some thermal processing are also necessary (such as quenching, air and/or water cooling, aging at high temperature for a given period of time (such as 1 h at 500°C). A common stainless steel alloy is the 18–8 (18% chromium, 8% nickel) alloy used to make tanks for storage and transportation of liquefied gas at low temperature.

Cast iron is a low-cost material obtained only by casting since it cannot sustain deformation at any temperature. Cast irons are generally Fe–C–Si alloys with 2 to 4% carbon and some silicon (0.4 to 3%). Depending on the thermal processes (slow heating, aging, low cooling, quenching), different properties are obtained. Cast irons are used in various products, such as crushing balls, armor shields, pump casings, cogwheels, and crankshafts.

7.3.2 Nonferrous Alloys

Although the cost of nonferrous alloys is superior to the cost of ferrous alloys just described, for some applications such selection is made thanks to advantages derived from specific properties such as weight; resistance to extreme temperature, corrosion, and oxidation; ease to manufacture and to shape; and electrical, magnetic, and thermal properties. The most common nonferrous alloys are those based on copper, aluminum, magnesium, zinc, and titanium.

Although aluminum is only 2% of iron-based production, aluminum alloys are second-best in production, owing to their interesting properties. The mass density is three times less than for steel (2700 kg m^{-3}). The corrosion resistance is high, due to the thin protective oxide layer (Al_2O_3) forming on the surface.

However, under some circumstances, such as fatigue, severe corrosion is experienced. The electrical conductivity is 62% with respect to copper (a copper wire weighs double and heats more than its aluminum counterpart when carrying the same electrical current). Pure aluminum has weak mechanical properties; however, this can be improved with the addition of other elements (such as copper, magnesium, silicon, zinc, and tantalum) and special processing. Ductibility at all temperature makes it ideal for shaping in thin forms (e.g., for soda cans). With a relatively low fusion temperature (660°C, 1500°C for steel), aluminum is more easily manipulated and put in shape than iron alloys. Table 7.3 presents common aluminum alloys and their use.

Copper was the first metal used by humans and is still indispensable in numerous applications. Its main quality is electrical conductivity, responsible for half the world's production of electrical wires, motors, and transformers. Mechanical properties are weak but improvable with particular processing. Copper is very ductile, is easy to shape, has excellent thermal conductivity, and good corrosion resistance (applications such as water piping, plumbing, pumps, and valves). Brass alloys are made of copper and zinc (with a copper content of up to 40%) and with small additions of other elements, such as lead. As the zinc content increases, the alloy color changes from red to yellow with increased mechanical resistance and ductibility. Brass alloys are suitable for precision milling, are nice looking, can be plated (with nickel, chromium, and gold), offer resistance to corrosion, and are less expensive than copper, due to the lower price of zinc. Bronzes are alloys of copper and tin (up to 16%) but are more expensive than brass since tin is more expensive than copper. Bronzes are used in electrical contact, springs, cogwheels, pistons, steam plumbing, and gear pinions, thanks to their good corrosion and wear resistance. Other copper alloys are obtained with the addition of aluminum (less than 9%), nickel (8 to 18%), and zinc (17 to 27%). These last alloys, known as nickel–silver, are used primarily for their good resistance to corrosion and attractive pale color (due to the nickel), for example, in jewelry, decorative objects, clocks, and optical parts.

Most uses of magnesium are derived from its low mass density (1740 kg m^{-3}, 66% of the mass density of aluminum). Applications are in the aircraft and automobile industries, ladders, portable tools, and housings of machinery submitted to high rotation speed (printing, textile). Magnesium alloys have good sound-damping qualities, are resistant to alkalies (caustic hydroxides capable of neutralizing acids) but have to be used at low temperature ($<150°C$ typically) and under light dynamic stress. Atmospheric corrosion resistance is better than for steel but less than for aluminum alloys. In the presence of humidity, they must be protected (e.g., by painting).

Pure zinc offers poor mechanical properties but excellent resistance to corrosion; it serves, for instance, as roofing sheets. Half of the production is dedicated to steel protection through galvanization. Cast zinc alloys (with the addition of magnesium, copper, or aluminum) with their low fusion temperature (around 380°C) and excellent shaping ability permit the manufacture of complex, thin

TABLE 7.3 Aluminum Alloys (Aluminum Association)

Alloy	Series[a]	Use
Aluminum casting alloys	100–800	Pistons in internal combustion engines, etc.
Al ≥ 99%	1000	Electrical wires, tubes for evaporators and radiators, protective layer against corrosion (through veneering), building industry (weatherboarding, etc.)
Al–Cu and Al–Cu–Mg	2000	Transportation material (airplane, train, automobile, truck, and naval industries), sporting goods, screws; excellent mechanical properties
Al–Mn	3000	Sheet, bars, tubes; same use as 1000 series but with improved mechanical properties
Al–Si	4000	Complex shapes in automobile industry; alloy highly fluid in sand molds
Al–Mg	5000	Pipes, building industry (weatherboarding, etc.), soda cans; good corrosion resistance (use in marine atmospheres)
Al–Mg–Si	6000	Wires, tubes, welded structures, transportation material [automobile frame (e.g., Audi), truck, naval industries]; average mechanical properties, ease of shaping and welding, excellent corrosion resistance
Al–Zn–Mg and Al–Zn–Mg–Cu	7000	Transportation material (automobile, truck, train industries), bolts; best mechanical properties among Al alloys; low resistance at temperatures above 120°C
Al plus other elements	8000	High iron content
Unused series	9000	Commercial grade, not much added value

Source: Adapted from Dorlot et al. (1991). Special thanks to Jacques Boutin from ALCAN, Jonquière, Quebec, Canada.

[a] The last two digits indicate the minimum Al percentage (e.g., 1030 contains a minimum of 99.30% Al, 1085 a minimum of 99.85%). The second digit indicates the modification from the original alloy (0 for the original).

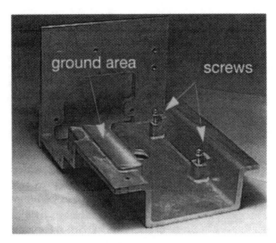

FIGURE 7.1   Example of complex part made from zinc alloy. After injection molding this part went through some finishing operations such as drilling (for screws) and grinding.

(down to 400 μm) shapes at a cost competitive with that of plastic. Zinc alloys are widely used in automobile and consumer goods (camera casings, faucets, locks, etc.; Figure 7.1).

Although titanium is fairly abundant in Earth's crust, occurring in a great variety of minerals, the difficulty of reducing it from the oxide, and production in protected atmospheres, led to its high price and limited uses, especially in the aerospace and chemical industries (e.g., the white pigment of titanium oxide), due to the mass density between aluminum and steel and excellent mechanical and corrosion resistance (higher than for stainless steel). Interestingly, NiTi is an alloy much used in shape-memory alloys, which have numerous applications as actuators or sensors (these alloys return to their original shape after experiencing a specific temperature cycle). The shape-memory effect is related to structural transformation from austenite to martensite (this can be provoked by temperature changes or by tensile stresses). Transient heat injection and transport are thus key factors for parts made with shape-memory alloys. In fact, thermography is reported to characterize such NiTi alloys (in particular, measurement of thermal effusivity and diffusivity; Gibkes et al., 2000).

7.3.3 Plastics

Nowadays, plastics compete everywhere with traditional materials. The main constituent of plastic is carbon from petroleum, which connects with other elements (carbon and others) to form giant molecules called *polymers*. Depending on how the polymerization occurred, various types of plastic with different properties are obtained. Numerous manufacturing processes are involved, using basic constituents as pellets, granules, powders, sheets, and fluids (S. H. Goodman, 1984):

- *Injection molding.* Plastic materials are heated and the resulting fluid is injected in a colder mold where solidification takes place. Advantages: speed of production, minimum postmolding operation, and simultaneous multipart molding.
- *Extrusion.* Continuous forming of parts is obtained when liquefied plastic material is forced through an opening or a die having the form of the cross section desired, such as for pipes, profiles, and wires. Advantages: production of long shapes which can be cut into smaller segments as required.
- *Blow molding.* In this process a tube called a *parison* is first formed and then blown with air or other gases against the wall of a mold. Advantage: making of hollow shapes.
- *Thermoforming.* Plastic sheets are formed into objects of various shapes through heat and pressure. Advantages: most inexpensive process, suitable for large and small parts.
- *Rotational forming.* Plastic material such as powder is rotated in a heated mold until melted; at that point the liquid material covers the inner mold surface uniformly; after solidification a hollow part is obtained. Advantage: a complete hollow part is obtained.
- *Compression and transfer molding.* Plastic material (powder) is introduced in a mold; a mating mold is than pressed against it, and pressure and heat forms the shape after solidification.
- *Foam processes.* Various processes make it possible to create a range of foams, from soft and flexible to hard and rigid. These include the addition of a chemical blowing agent generating gas in the polymer liquid, and direct gas injection into the melt.
- *Casting and encapsulation.* Plastic is poured into a container of the desired shape and liquid plastic material is poured, polymerized with heat, and solidified into the desired shape.
- *Calendaring.* The plastic material is rolled as a thin film which subsequently impregnates paper or fabric through an additional rolling stage.

The mass density of plastics is small, and this is advantageous when mechanical properties are not critical, this is due primarily to their basic elements (carbon and hydrogen have low atomic weight). The addition of heavier atoms such as chlorine increases the mass density. For example, polyethylene is 960 kg/m^3, polyvinyl chloride (PVC) is 1200 to 1500 kg/m^3. Thermal expansion is much higher for plastic with respect to metals, and this should be taken into consideration when assembling metal and plastic parts. For example, the coefficient of linear expansion is 50 to 180 ($\times 10^{-6}$ m $°C^{-1}$) for PVC and only 11 for iron. Thermal conductivities are low, making plastic materials attractive for thermal insulation purposes (e.g., polystyrene foams have thermal conductivities close to 0.035 W m^{-1} °C^{-1}). Plastic materials are also good electrical insulators, with high electrical resistivity (10^{15} to 10^{18} $\Omega \cdot$ cm) since they do not have electrical charge carriers. They are widely used as insulators on electrical wires, to manufacture electrically insulated casings, as dielectric in capacitors, and so on. Good light transmission of plastics (such as for polycarbonate) also make them attractive as window panes, optical lenses, safety goggles, motorcycle windshields, and headlight casings. Rigidity in polymers is not related to cohesion between atoms as in metals or ceramics, but rather, is based on secondary interactions between macromolecules; this explains their lower rigidity figures. The useful temperature span for polymers is much smaller than for metals and ceramics. Finally, all polymers do absorb water in various proportion (up to a few percentage points), and this affects their mechanical properties.

7.3.4 Ceramics

The first ceramics, potteries, and bricks made by humans consisted of fired or baked clay. Progress in ceramic technology has significatively enlarged the spectrum of fabrication processes, constituent materials, and the range of applications. In addition to traditional ceramics originating from clay, we can distinguish glasses, pure oxides, carbide, nitride, boride, and concrete. Not all ceramic materials are obtained through high-temperature processes. For example, concrete and plaster set through a chemical reaction. Mechanical properties of ceramics depend on the raw material used and manufacturing processes. In traditional processes based on unprocessed raw materials, ending products are heterogeneous and contain numerous impurities and porosities, leading to internal constraints, thus explaining why such ceramics are so fragile. Traditional ceramics made of easily available materials (clay, sand, feldspar) are produced at low cost. These include pottery (bricks, tiles, smoke ducts, drainpipes), stoneware, sandstone, and porcelain (chinaware), with fabrication temperatures from 950 to 1400°C (for porcelain).

Desirable microstructure (such as very low porosity) is obtained if particular care is paid to the selection and preparation of raw materials (Figure 7.2). Moreover, since ceramics offer high wear resistance, sustain high temperature, and are chemically inert, they are particularly attractive in various contexts, such as mechanics, electrotechnology (power line insulators, pulse and power transformer cores, permanent-magnet ferrites in small electrical motors and loudspeakers), electronics (capacitors, inductors, antennas, piezoeletric elements), surgery (such as artificial hips, based on the good biocompatibility of ceramics), optics, and abrasive and cutting tools. For the latter products, their high hardness Knopp values make them competitive with natural or synthetic diamonds (Knopp hardness of 7000). For example, cubic boron nitride with a Knopp hardness of 5000 is the second-hardest material; it is used in cutting tools (such as circular blades). Ceramics are also used in composite materials

FIGURE 7.2 Examples of ceramics: silicon nitride and floor commercial grade. Notice the finer grain of the silicon nitride ceramics.

(see the next paragraph) and as thermal-barrier coatings in engines (e.g., ZrO_2 layer on piston valves).

An important class of ceramics are the refractories employed where resistance to very high temperature is required, as in furnace linings and melting pots (inspected by TNDT; Kauppinen et al., 1999; Sahr, 1999). Materials with a melting point higher than 1580°C are called *refractories*; those with a melting point higher than 1790°C are called *high refractories*. Other factors important for refractories are mechanical resistance at high and low temperatures and to thermal shocks, limited size variation (to limit brick and binder dislocation following repeated cycles of heating and cooling), heat transfer, and electrical resistivity. Refractories are acid (e.g., silica, SiO_2; alumina, Al_2O_3), neutral (e.g., graphite, chromite), or basic (e.g., magnesite, MgO; bauxite). Acid processes require acid refractories to avoid rapid degradation of the refractory.

Glass is an amorphous (i.e., not made up of crystals) solid because during cooling, crystals do not have time to crystallize. The primary properties of glass are transparency, hardness, rigidity at ordinary temperatures, and the capacity for plastic working at high temperatures, resistance to weathering, and resistance to most chemicals (except hydrofluoric acid). Glass is made by fusing silica (SiO_2) with a basic oxide. The production steps are melting and refining, forming and shaping, and heat treating and refining. Depending on the composition, various properties are achieved. For example, glass ceramics are obtained through controlled crystallization of the structure. This leads to better mechanical properties, since for example, in a crystalline structure, propagation of a crack is deviated against grain faces, whereas it is not stopped in an amorphous solid. Corning 9606, with 56% SiO_2, 15% MgO, and 9% TiO_2, is an example of glass ceramics (by comparison, regular glass contains more than 99.5% of SiO_2). TNDT is important in the glass industry (Section 11.4.4; Pajani, 1987c).

Cement is a construction material made up of clay and limestone mixed with water. A chemical reaction turns the resulting paste into a rocklike mass (it takes up to one month to achieve full mechanical resistance). In one reported application, monitoring of cement cooling in a dam was performed through embedded fibers through Brillouin-distributed temperature measurement (Thévenaz, 2000). The most common cement is portland cement. Different adjuvants are used to speed up or slow down the chemical reaction. Aggregates such as sand and gravel are often added to cement to reduce its price; the resulting product is called concrete. In reinforced concrete a steel internal structure composed of rods or mesh is added to improve the low resistance of concrete to traction (concrete works well in compression).

7.3.5 Composite Materials

The materials reviewed so far have specific advantageous properties but also drawbacks. The idea of composite materials is to obtain interesting properties from a combination of materials that naturally do not mix. Often, the goal behind the replacement of a traditional material by a composite is to reduce structure weight while retaining rigidity. This explains the growing interest for composites in the aircraft industry. Composites are generally made of a matrix and of a reinforcing material, often a fiber (of finite dimension). In unidirectional composites, all fibers are oriented in the same direction, generally the constraint direction. To prevent the resulting anisotropy, plies of unidirectional composites can be stacked on each other in different directions (such as 0–90° or 0–45–90°). In multidirectional composites, fibers appear as a woven pattern or are dispersed randomly (Figure 5.19); mechanical properties are isotropic.

Reinforcing fibers are the key to the final properties of the material. Glass fibers were used first in the 1940s; they are obtained at low cost (about $1 per kilogram) from extrusion with a diameter of 5 to 15 µm. Associated with a light matrix of weak rigidity (e.g., a polymer), they provide improved rigidity with a good reduction in weight with respect to, say, aluminum. Other fibers include polymer fibers (e.g., Kevlar), carbon fibers, metallic fibers (e.g., boron), ceramic fibers (e.g., silicon carbide), and natural low-cost material such as asbestos (used in asphalt shingles) or mica.

Matrixes have two roles: to transmit constraints to the fibers and to incorporate them at low cost. For mass-market production, average properties are obtained with thermoplastic polymer (polyethylene, polystyrene, etc.) matrixes reinforced with low-cost fibers (asbestos, mica, cellulose, etc.). Better properties are obtained with thermoset polymers, called *resins* in the industry; these include polyester resins (in combination with glass fibers to obtain low-cost composites), melamin resins (colorable, used to make decorative panels), and epoxy resins (much used in the aircraft industry, with good mechanical properties up to 200°C, in combination with carbon fibers).

Since plastic matrixes cannot sustain high temperature and require particular assembly procedures, metal matrixes (aluminum, silver) have been developed, although mass density and cost are higher than for plastic matrixes, while possible reactions between metal matrix and fibers can lead to problems, such as cracks growth and rupture. Ceramics are good candidates as matrixes, due to their rigidity, chemical inertness, refractory properties, and mechanical resistance. In combination with fibers, improved properties are achieved. This is, for instance, the case for cement; addition of asbestos, Kevlar, or glass fibers to the initial mix improves the mechanical resistance substantially (e.g., for use as undulated panels and pipes).

Wood is a natural composite material made of cellulose fibers incorporated in a lignite or hemicellulose matrix. Properties are strongly anisotropic and depend on species (more than a 1000 worldwide), growing rate and conditions, moisture content, and other factors. To obtain large size, or uniform and enhanced properties, wood is often processed as veneer, fiber and particle boards, layered–bonded beams, and so on. Moreover, such processing allows us to use smaller trees and to recycle wood residues from sawmills and veneer industries.

7.4 MATERIALS INSPECTED BY TNDT

TNDT can be deployed to inspect parts made from most of the materials described in Section 7.3. In fact, there are three key aspects to consider. First, TNDT is a border technique, so it is mostly sensitive to defects close to the surface. This is, however, a vague statement, since it depends on the material of interest. Studies indicate that it is possible to reveal defects buried 2 cm under the surface in concrete but only 2 mm in the case of strongly anisotropic CFRP composites. The second aspect to consider is the defect itself, thermal properties of which have to be different from the surroundings to enable detection. As a general rule, TNDT is sensitive to voids, inclusions, cracks, and delaminations, and to the presence of water. In addition to defect detection, thermal properties of material can be obtained, an important field for TNDT (Section 11.5). Finally, the problem of TNDT is surface emissivity (Section 2.9.1). In fact, with emissivity below 0.2, direct detection of relevant temperature differences is not possible. Provided that these three factors are under control, TNDT is possible.

Metal alloy (iron-based and nonferrous) components are inspected routinely by TNDT (after low emissivity has been taken care of through painting, for instance; Section 8.2). Some examples of successful TNDT inspection are lap joints and rivets in aluminum aircraft fuselages (Figure 10.49), corrosion in steels, bonding in aluminum laminates, blockage in turbine blades (Chapter 11), and so on. Moreover, passive thermography is deployed routinely in continuous steel and aluminum casting, aluminum electrolysis (Pajani, 1991), and so on.

For plastic material, active TNDT has to be deployed carefully to avoid damaging the inspected part through overheating. It is sometimes interesting to rely on a cooling approach. For example, TNDT enables detection of internal bubbles in extruded plastic material through IR observation of the extruded material after cooler air blowing.

Ceramics are a big field of application for TNDT, as it is possible to measure layer thicknesses and disbondings, for instance. Active TNDT is also suitable for inspection of specific materials in the nuclear industry, such as for the study of thermal barriers projected by plasma (Lewak, 1994). Civil engineering applications have also emerged recently: for example, to detect steel reinforcing bars, air, water inclusions, and cracks (de Halleux et al., 1998; Sakagami et al., 1999).

TNDT is also well adapted to detect various anomalies in various composites, such as in CFRP (see below). Due to mass-market production, wood products are an increasing field of interest for TNDT, with applications in bonded decorative panels and veneers, for example (Wu et al., 1997b).

Detailed case studies are presented in subsequent chapters, but we are ready to look at some major TNDT applications.

7.4.1   Bonded Assemblies

Sandwich structural panels are made through the assembly of two plates, S_1 and S_2 (Figure 7.3), made of high mechanical resistance strength (e.g., sheets of Al, Ti, steel, etc.) and of a core A of low density (foam, honeycomb, balsa

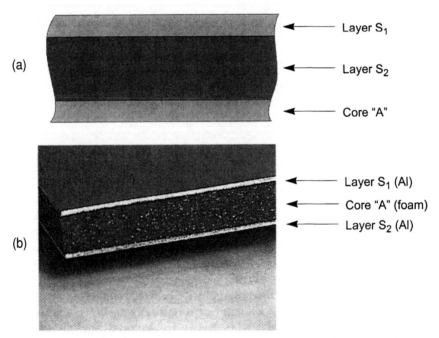

FIGURE 7.3   (a) Cutaway view of a sandwich structural panel; (b) example of an Al–foam–Al panel.

FIGURE 7.4 Example of an aluminum bonded assemblies. TNDT is useful here, for example, to assess epoxy bondline presence.

wood) used as a spacer element between the plates. These materials are employed to build structures having great strength, low mass, and eventually, good thermal insulation capability as well as good fire protection.

Bonded assemblies offer interesting advantages: better distribution of mechanical loading, superior resistance to fatigue with respect to the more traditional point-by-point welding and riveting anchorages, lengthened life span, airtightness and watertightness, improved resistance to corrosion, and the ability to ensure electrical or thermal insulation (Figure 7.4). The quality (strength and durability) on a bond depends mainly on the interaction of the adhesive with the adherent (surfaces to bond). The quality of preparation is thus essential. Bonding is affected particularly by ambient temperature and humidity at the time of bonding.

If all the fabrication parameters are not perfectly controlled, different types of defect are likely to occur, such as (1) lack of adhesive (bubbles, air layers, foreign materials), (2) cohesion defects (breaking within the adhesive), and (3) bonding defects (breaking at the surface–bond interface). Defects of types 2 and 3 are very difficult to detect whatever the NDE method is used. TNDT offers good possibilities for defects of type 1. For example, the Lockheed C-5 transport aircraft is made of some 3400 m^2 of bonded structures, inspected completely using infrared thermography (Figure 7.5). In the aerospace industry where inspection is of prime importance at all the fabrication stages, mainly for security reasons, there is a need for fast and economical NDE inspection techniques such as TNDT.

7.4.2 Graphite-Epoxy Structures

Graphite-epoxy structures are constituted of a matrix made of carbon fibers drowned in a epoxy resin bath (up to 32% resin). They are widely used due to their excellent mechanical properties. These properties authorize molding in a

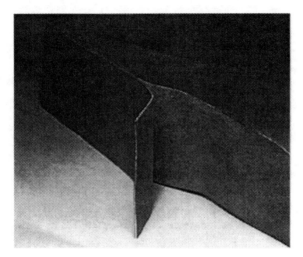

FIGURE 7.5 A common use of aluminum bonded assemblies is for an aircraft fuselage where two pieces of aluminum are bonded together. The picture is a fuselage sample with sheet disbonded (destructive peeling test). TNDT is useful here, for example, to assess presence of foreign material such as internal corrosion.

wide variety of shapes, thus reducing the quantity of parts to assemble and consequently, lowering the fabrication costs. Moreover, since the anisotropy is easily controllable (it depends on the orientation of the individual plies), the designer is free to adjust the strength and mechanical resistance with respect to the envisaged mechanical loading of the part. Typical defects include voids, inclusions, areas with unbalanced resin content, badly cured areas, surface cracks, and broken fibers. Saw cutting and hole drilling tend to induce delaminations, which occur on the edges and which can alter fiber orientation, causing a loss of rigidity. Of importance also is damage caused by impact. They induce delaminations and cracks propagating parallel to the plies. These delaminations take place especially within the more disoriented plies. Often, as we will see in Section 11.1.2, the damage is barely visible on the impact side but considerable on the other side. Often called *blind-side impact damage*, it sometimes has a butterfly shape (Figure 11.1). In the case of graphite-epoxy structures, TNDT offers good possibilities for the detection of delaminations, inclusions, and impact damage, as well as for the evaluation of fiber content and orientation.

An example of the generalized use of graphite-epoxy composite materials is the Boeing 777, in which 10% of the total weight of the structure is made of such a product, as compared to a mere 3% in the case of other Boeing airplanes. In fact, TNDT has always been quite active in the aerospace field as the numerous papers on the subject demonstrate. One reason for its good acceptance in the aerospace industry is that TNDT is adapted to the inspection of both low-thermal-conductivity materials such as CFRP or ceramic, and also, with the advent of fast IR cameras, to the inspection of high-thermal-conductivity workpieces such as aluminum fuselages.

7.4.3 Thin Coatings

Although the focus of this book is on thermal bidimensional imaging, it is important to note that point heating (Figure 9.3a) offers very attractive characteristics for specific applications, such as in advanced ceramics. Since heating can be performed with a laser beam, short heating pulses that approximate the delta (impulse) function can be achieved (e.g., in the range of 15 ns with a 1.06-μm Nd:YAG laser in the Q-switch mode). Detection with a focused single detector is much faster than the bidimensional IR camera image rate. Consequently, fast thermal events can be caught while a particular wavelength selection is achieved with a broadband single detector and an appropriate filter. This makes it possible to select a particular thermal emission wavelength. Such a point thermal wave method can be used, for example, to measure clearcoat thickness in metallic paint systems of car bodies. The surface is pulsed-laser heated (at 532 nm) and the peak delay time of the thermal radiation emitted chosen by the single detector is related to the clearcoat thickness (in the range 20 to 50 μm). In this application it is important to select an excitation wavelength that is transmitted by the coating and absorbed by the substrate, so that from the substrate–film interface to the top surface, the thermal wave reflected diffuses through the film thickness: The thicker the coating, the longer the travel time, thus delaying the temperature peak observed.

Thermographic imaging of surface finish defects in coatings on metal substrates is also possible by heating the coating system from the back slightly above ambient temperature (up to 20 to 30°C) with a heating pad and viewing it with an IR camera. Surface finish defects (craters or protrusions) are visible due to coating thickness variation between the defect and sound areas, since coating thickness affects the thermal radiation emission/reflection/transmission properties of a coating system (typical resolution about 20 μm). The step heating approach (Section 9.4) is commonly used for coating thickness measurement.

7.5 OTHER NDE TECHNIQUES

In addition to TNDT, several other NDE methods are available, among them the so-called *big five*: ultrasonics, x-ray radiography, eddy current testing, penetrant and magnetic flux leakage (see, e.g., Moore and McIntire, 1996, for a complete discussion of NDE methods).

Ultrasonics is probably the most widely used NDE method. Ultrasonics offers interesting possibilities but is often characterized by (slow) point-by-point inspection rates. The principle of ultrasonic NDE is based on a pulse-echo scheme in which the defect reflects (or, in transmission, absorbs) ultrasonic waves of high frequency (1 to 25 MHz). The necessity of contact with the zone being inspected in order to ensure good signal propagation between the transducer and the workpiece is a major disadvantage of this method. This contact can be achieved through the use of water jets or immersion in a water tank (squirt method). These two methods, which are more easily automated than the manual procedure using wedge transducers and gel couplant, have, however, the drawback that the workpiece is exposed to water, which in some cases is undesirable. Recently, techniques of generation and detection of ultrasound without contact by laser have become available commercially.

The principle is as follow. The thermal stress induced by the generation laser initiates ultrasound propagation. Reflected or transmitted mechanical ultrasonic waves locally deform the workpiece surface, and these mechanical deformations (a few nanometers in amplitude) are picked up by the detection laser through interferometric methods (Figure 7.6). Finally, it is worthwhile to mention that ultrasound imaging cameras with imaging arrays made of a piezoelectric layer and several thousand pixels are now commercially available (such

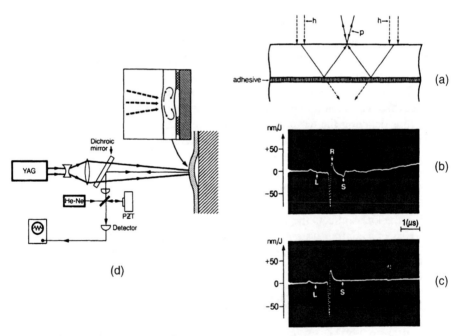

FIGURE 7.6 Inspection of adhesively bonded structures by laser-generated ultrasonic bulk waves: (a) experimental configuration, where h is the cross section of the annular YAG laser beam (of 1.5 mm diameter) and p is the focused probing beam (Michelson laser interferometer); (b) signal detected in an unbonded area, where L and S indicate the longitudinal and shear reflected echoes, respectively, and R indicates the surface-wave pulse; (c) signal detected in a well-bonded area (the vertical scale is in nanometers per unit energy of the incident laser pulse). (d) Schematic diagram of the experimental apparatus. (From Cielo et al., 1985a. Copyright © 1985 The American Society for Nondestructive Testing, Inc.; reprinted with permission.)

as the 128×128 detector element Acoustocam I100 of Imperium*). In this case, an acoustic lens replaces the beam forming or mechanical scanning often used for traditional ultrasonic imaging (the array needs to be in intimate contact with a water column, which makes it possible for the ultrasounds to propagate).

X-ray radiography or *gamma radiography* (for deeper penetration) allows us to detect some types of defects (voids, air layer, presence of foreign materials) because of the differential absorption of radiation based on density differences. This NDE method has some problems related to the health security of nearby personnel and is not efficient to detect thin air layers (since the x-ray beam is left unaffected in this case). Moreover, real-time radiography still experiences limited resolution compared with film radiography.

In *eddy current* (EC) *testing*, eddy currents induced in a part (a metallic material, nonmagnetic or ferromagnetic) generate their own magnetic field. Magnitudes, time lags, phase angles, and flow patterns of the eddy currents within the part are detected by measuring resultant magnetic fields with another set of sensing coil or solid-state magnetic field detectors (Hall effect devices; Moore and McIntire, 1996). Analysis of the signal collected from the sensor allows for subsurface defect detection and quantification. The generating coil and sensor can be placed together in a compact head or can be positioned on both sides of the workpiece. Variations in a part, such as in geometry, discontinuities, alloy composition, hardness, corrosion, cracking, or thermal or mechanical processing that modifies the magnetic permeability, change the eddy current, leading to possible detection. Point-by-point measurement is not the strongest feature of EC testing. However, arrays of EC detectors have been reported and EC testing is very efficient for the inspection fast-moving sheets and wires. The Hall effect is the development of a transverse electric potential gradient in a current-carrying conductor placed in a magnetic field when the conductor is positioned with the magnetic field perpendicular to the direction of current flow (Blatt, 1984).

In *penetrant testing*, a penetrant liquid is applied on the surface being inspected. In penetrant leak testing, observation on the opposite surface can reveal leaks and thus through-cracks and porosities. An alternative method uses a penetrant liquid capable of entering open discontinuities on a surface. Often, the surface is then wiped and subsequently illuminated with ultraviolet or visible light, depending on the penetrant used. Since penetrants are highly visible even in small traces under such illumination, it becomes possible to detect open discontinuities. Obviously, application and removal of penetrant is not always practical. However, penetrants are cost-effective and can reveal difficult-to-detect cracks (such as on gear tooth roots).

Magnetic flux leakage (MLF) is a method for the detection and analysis of a surface discontinuity using the flux that leaves a magnetically saturated or nearly saturated test object at a discontinuity (Moore and McIntire, 1996). It is suitable for ferromagnetic materials. First, the test piece can be magnetized

---

*www.imperiuminc.com.

Magnetization coil

FIGURE 7.7 Example of deployment of magnetic flux leakage used in inspection of steel cable mounted on a moving stage.

by means of an encircling coil. Magnetic flux leakage caused by discontinuities in a part is detected by various sensors, such as coils, Hall elements, and magnetodiodes. Processed signal indicates the presence of a discontinuity. An important application of MLF is for wire rope inspection, such as those used in cable cars (Figure 7.7). In magnetic particle testing (MT), material discontinuities show up due to magnetic particles sensitive to magnetic leakage fields. Magnetic particles are made of fine ferromagnetic material capable of individual magnetization and sensitive to flux leakage fields. Surface and subsurface discontinuities can be revealed by MLF and MT. Since a magnetic field is directional, orientation of the discontinuity with respect to the magnetic field is a problem. Best results are obtained when the discontinuity is perpendicular to the field. Moreover, MLF and MF work only on ferromagnetic material; they cannot be used on aluminum, copper, or austenitic stainless steel.

Another type of NDE method is *visual testing*, which relies on the visible light reflected on a test piece and observed by an operator or through machine vision. Surface changes can affect the reflected light pattern somehow, leading to the detection of some defects. Obviously, this is one of the cheapest methods available. An example of its use on aircraft aluminum fuselages is for detection of corrosion detection around rivets (the Diffracto D-sight method*). Corroded material expands (this is known as the *pillowing effect*), the surface orientation changes, and projected grazing light is then reflected in a different manner over corroded areas. The surface image is projected magnified on a screen to maximize observation by an operator. All types of visual aids (e.g., borescopes) are available to carry out visible inspection inside machinery.

* www.diffracto.com.

A simple NDE technique is *tapping*. In this case an operator hits a surface with a hammerlike object and listens to the sound propagation (the method is similar to ultrasonics but uses audible sound waves of frequency below about 20 kHz). Cracks and discontinuities within the object affect sound propagation, and this is perceived by the operator (everybody has experienced detection of a cracked dish, glass, or cup in this way) (Figure 11.19). Although primitive, it is possible to automate tapping. A reported application is for the detection of cracks in steel balls used in ore crushing. The balls are launched in a tube from a given height and fall to a block. A mike is attached to the block, and simple signal processing identifies cracked balls in real time.

Other NDE methods are used on new materials with variable success, these include holographic interferometry, positron annihilation, acoustic emission, and the mirage technique (Lu et al., 1998; Pao et al., 1998). The *mirage technique* makes possible extremely sensitive measurements. The sample surface is heated periodically by a modulated light beam, and deviation measurement of a probe beam propagating in a parallel direction just above the heated area is related to sample periodic temperature. In turn, sample temperature will reveal specific sample characteristics. For example, it is possible to determine optical losses in mirrors coating with a resolution of a few tens per million or to detect very thin air slices (with a resolution of ~0.1 μm) in metallic components. Drawbacks of the mirage technique include its sensitivity to mechanical vibrations, beam misalignment, and beam optical noise, which may interfere with careful design on the inspection setup.

7.6 PROBABILITY OF DEFECT DETECTION, RELIABILITY, AND INSPECTION PROGRAMS

The physics of the damage and mechanical theory of failure bring us to the concept of *tolerance of the damage*. The idea is to be able to predict part behavior as a function of local constraints, it thus becomes possible to define *acceptable defects*. In this respect, Figure 7.8 shows the probability of defect detection as a function of defect size. This is an experimental curve valid for a specific method of inspection. On this curve, we can notice that the ratio between the threshold of detection and a reliable detection value can be as great as 10. It is thus important to have an *inspection strategy* that allows deferentiation between superficial tolerable defects and unacceptable defects.

Important considerations concern the reliability of a given part or system (see also Kraft and Wing, 1981). In fact, it is well known that reliability can be expressed as a *bathtub curve* (Figure 7.9). During the early life of a component or system, the failure rate is high. As early failures (primary failures) are replaced, the component or system settles down to a long, relatively steady period at a lower and constant failure rate. This period is often referred to as the *useful life*. Later, in the *wear-out period*, the components or system deteriorates rapidly and the failure rate rises again. The wear-out period can be avoided if

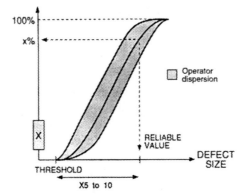

FIGURE 7.8 Probability of detection curve. The region marked with a cross indicates where false readings occur. (From Maldague, 1993.)

components are replaced before they reach this period (inspection is important to determine when to replace components). The following relationship expresses the reliability R:

$$R(t) = \exp(-\lambda t) = \frac{N_{\text{working}}}{N_{\text{total}}} \tag{7.1}$$

with λ the failure rate expressed in FITS (1 failure in 10^{-9} h), N_{total} the total number of units put into operation in the first place, and N_{working} the remaining working units at time t (among the N_{total} original units). The last formulation is useful to establish reliability values based on experimental measurements: for example, from data gathered in maintenance reports.

Example 7.1 Find the reliability at $t = 500$ h if for a reliability test, 98 of the original 100 engines inspected are still operating. Compute the failure rate λ in FITS.

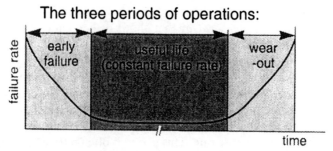

FIGURE 7.9 Typical bathtub curve of failure rate as function of time. (Inspired from Kraft and Toy, 1981.)

SOLUTION   Following eq. (7.1), we have $R = 98/100 = 0.98$, and rewriting that equation yields

$$\lambda = -\frac{1}{t} \ln R = -\frac{1}{500} \ln(0.98) = 4.04 \times 10^{-5} = 40{,}400 \text{ FITS}$$

In a system made up of two components of individual reliability, R_1 and R_2, the equivalent reliability is obtained from serial, parallel, or serial–parallel arrangement:

$$R_{\text{total–series}} = R_1 + R_2 \tag{7.2}$$

$$R_{\text{total–parallel}} = 1 - (1 - R_1) \times (1 - R_2) \tag{7.3}$$

These relationships can be generalized to n components.

Finally, an interesting parameter is the mean time between failures (MTBF) given by

$$\text{MTBF}_{\text{series}} = \frac{1}{\lambda_0} \tag{7.4}$$

$$\text{MTBF}_{\text{parallel}} = \frac{3}{2} \times \frac{1}{\lambda_0} \tag{7.5}$$

λ_0 being the total failure rate [for a serial system made up of two components, $\lambda_0 = \lambda_1 + \lambda_2$; for a parallel system, eqs. (7.1) and (7.3) are used to compute λ_0]. In most cases a serial arrangement is assumed since failure of one subsystem causes failure of the entire system (see Problem 7.4).

Example 7.2   Calculate the reliability after 10 years of a small electric substation made up of three components having the following MTBF values (available from manufacturers' data sheets): $\text{MTBF}_1 = 130{,}000$ h, $\text{MTBF}_2 = 430{,}000$ h, and $\text{MTBF}_3 = 330{,}000$ h. The arrangement is shown in Figure 7.10.

SOLUTION   Let's first compute the individual failure rate λ_i for the three components from eq. (7.4): $\lambda_i = 1/\text{MTBF}_i$, giving $\lambda_1 = 7.69 \times 10^{-6}$, $\lambda_2 = 2.33 \times 10^{-6}$, and $\lambda_3 = 3.03 \times 10^{-6}$ failure/h, respectively. From this we can compute the total failure rate for the branch with components 1 and 2: $\lambda_0 = \lambda_1 + \lambda_2 = 10 \times 10^{-6}$ failure/h, and then the corresponding reliability after 10 years ($t = 10$ years $\times$ 365 days/year $\times$ 24 h/day $= 87{,}600$ h) is given by eq. (7.1):

$$R_{12} = \exp(-\lambda t) = \exp(-10 \times 10^{-6} \times 87{,}600) = 0.42$$

The reliability of the third component is

$$R_3 = \exp(-\lambda t) = \exp(-3.03 \times 10^{-6} \times 87{,}600) = 0.77$$

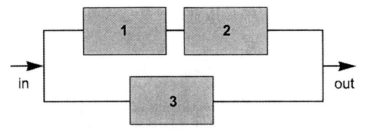

FIGURE 7.10 Figure for Example 7.2 (reliability computations of a small electrical substation).

and then the total reliability R_{123} comes from eq. (7.3):

$$R_{123} = 1 - (1 - R_{12}) \times (1 - R_3) = 1 - (1 - 0.42) \times (1 - 0.77)$$
$$= 0.87 \quad \text{or} \quad 87\%$$

Interestingly, the total reliability is greater. This is because a parallel arrangement can be seen as an arrangement whose one branch is like a spare tire. Following the failure of one branch, the system can still operate on the second branch, at least from a theoretical point of view.

The human life cycle follows a similar pattern to that described above, with an high infant mortality rate which lasts up to about the tenth year, followed by a period in which the majority who survive live to an old age. In the seventieth year or so, the curve rises, as people begin to die of natural causes. The death rate increases until all in the group are dead.

Normally, the failure rate λ is constant during the useful life period. It can, however, increase if the system or component is disturbed by accident, unusual operations, external conditions, or bad maintenance. This discussion implies than if a thermographic inspection program is set up as part of a maintenance program, after some time it should not continue to produce results, that is, to report findings (detection of defects). This situation does not mean, however, that the inspection program needs to be stopped; on the contrary, it means that maintenance is being done well and inspection is still needed to double-check.

PROBLEMS

7.1 **[NDE methods]** Discuss the TNDT procedure for specimens such as the one shown in Figure 7.3b.

Solution TNDT would be a good candidate since the diameter-to-depth ratio is likely to be large ($\gg 1$) and thus favorable to TNDT (defects of interest in these structural panels are large). Moreover, since the defect depth is known (defects are in the bond line, about 1.5 to 2 mm under the surface of the aluminum sheet), it is possible to deploy pulsed thermography in a mobile configuration (Sections 6.4 and 8.1) with proper spacing between the heat source and infrared imaging or lock-in thermography with heating at a frequency related to defect depth. Lock-in thermography is used for inspection of large areas at a time (a few square meters). Static pulsed thermography is also possible. Inspection examples of such panels are presented in this book.

7.2 **[NDE methods]** Discuss the TNDT procedure for specimens such as the one shown in Figure 7.1.

Solution   For critical applications such as molding of a casing holding in place moving parts such as gears, defects of interest are likely to be cracks close to mounted shafts and possibly also some inclusions under the surface that affect structure rigidity. When parts are complex in shape, uniform radiative heating is tricky. Better approaches are likely to be vibrothermography [e.g., putting the specimen on a shaker and visualizing the hot spots that develope at weak points around cracks (Sections 1.2.4 and 9.5)] or loss-angle lock-in thermography [with mechanical excitation of the specimen obtained at ultrasonic frequencies (Section 9.5)]. Point thermography with, say, laser heating could possibly resolve subtle subsurface defects on critical areas of the sample (Section 9.2). In any case, it is likely that the specimen or the IR camera/heating head will have to be moved to inspect from different points of view. The specimen can be mounted on a rotating computer-monitored stage or the IR camera/heating head installed on a robotic arm (Section 8.1.2).

7.3 **[NDE methods]** Discuss the TNDT procedure for specimens such as the one shown in Figure 7.11.

Solution   TNDT would be a good candidate, along other methods, such as ultrasonics, neutron or x-ray radiography, computerized tomography, or manual investigation. The advantage of using thermography here is to exploit the fact that such a specimen is hollow. The specimen could, for example, by heating from the inside by injecting cool/hot gas while observing thermal contrasts developing on the surface. Blocked cooling channels will affect the internal flow, modifying the thermal signature recorded on the surface, enabling possible detection. Moreover, the metal thickness involved is relatively small. Such investigations are discussed in Section 11.2.

7.4 **[reliability]** An electrical system is made up of 10 components, each having $\lambda_i = 10{,}000$ FITS. Assuming a serial configuration, compute the MTBF for the entire system and the reliability at the beginning of its useful life and at $t = 10{,}000$ h.

Solution   We first compute the total failure rate, which is given by $\lambda_0 = \lambda_1 + \lambda_2 + \cdots$. For 10 components we have $\lambda_0 = 10 \times 10{,}000$

cooling channel exhausts

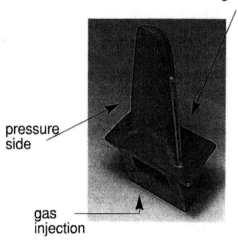

pressure
side

gas
injection

FIGURE 7.11   Typical turbine blade used in jet engine (Problem 7.3).

FITS $= 100{,}000$ FITS $= 1 \times 10^{-4}$ failure/h. The MTBF is given by eq. (7.4), MTBF $= 1/\lambda_0 = 10{,}000$ hours (or 1 failure per about 1.1 years, statistically). (*Be careful:* The next failure could be tomorrow; that's the problem with statistics!) The reliability is given by eq. (7.1). At the beginning of its useful life (at $t = 0$), we have $R = 1$ (or 100% reliability, normally, as the system is new). At $t = 10{,}000$ h, we have $R = \exp(-100{,}000 \text{ FITS} \times 10{,}000) = 0.37$ (or 37%).

CHAPTER EIGHT

Experimental Concepts

8.1 INSTRUMENTATION FOR PASSIVE AND ACTIVE INFRARED THERMOGRAPHY

A system used for the TNDT inspection based on principles described in this book is composed of many different subsystems, such as an infrared (IR) camera, an image acquisition and analysis system, and a thermal stimulation system (Figure 8.1). In this chapter we focus our attention on the main elements of the thermographic inspection system illustrated in Figure 8.1 in addition to the IR detectors and cameras that were studied in Chapter 4. In this book we concentrate especially on imaging systems, which are of more interest for fast TNDT inspection procedures rather than on point or line systems (Chapter 1). Obviously, this two-dimensional study can also be applied (partly) to line or point systems. In this section we review the elements of a typical TNDT apparatus, and to illustrate this study, we refer to a real TNDT inspection station.

There are different ways that the infrared equipment can be deployed for TNDT depending on the specific application that is envisaged. As specified in Chapter 1, thermography can be deployed in either a passive or an active fashion. In the *passive* configuration (Figure 8.2) the infrared imaging camera is pointed on the scene, looking at objects to inspect while no external thermal perturbation is applied. Applications such as building inspection, control of industrial processes, maintenance in power plants, military surveys, welding, fire (forest) detection, and medicine make use of this mode of operation, where surface temperature distribution, as is, contains relevant information concerning possible disorders, the presence or absence of targets, and so on. Some of these applications are discussed in this book in Chapter 11.

In the *active* configuration an externally applied thermal stimulation is needed to generate meaningful contrasts that will yield to the detection of subsurface abnormalities (Chapter 1). Without such a foreign applied thermal perturbation, no information can be drawn from the inspection since initial surface temperature distribution is not related to subsurface structure. Appli-

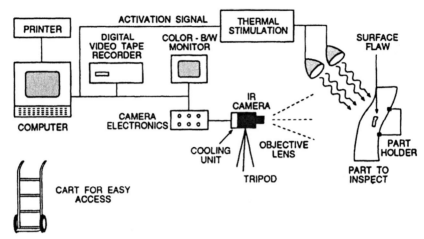

FIGURE 8.1 Typical active infrared TNDT experimental setup. (Adapted from Maldague, 1993.)

cations of active thermography include material analysis, aerospace NDE, printed circuit boards (PCBs), and evaluation, among many others. Some of these applications are discussed in this book (Chapters 9 to 11).

In this chapter we review the use of infrared equipment in both *active* and *passive* configurations.

8.1.1   Instrumentation for the Passive Approach

In the passive configuration, inspection is done by a qualified operator who generally proceeds to the interpretation of results directly in the field. In some cases, such as for air-lifted power line infrared inspection or fire forest detection performed over long distances (hot-spot detection; Section 11.4), the interpretation of results may be done (automatically) off-line at a later time.

For the passive approach, the equipment needed may be summarized as follows:

- Acquisition head (Figure 8.2)
  - *Infrared camera.* For aerial investigations, dedicated sensors are recommended (Figure 8.3); see the discussion of atmospheric and remote sensing later in this chapter.
  - *Appropriate objective* (including zoom, close-up, telescope, expander tubes which are needed depending on the application). Since these are made for a small market due to the particular material required for good transmission in the infrared, they are quite expensive, and thus the objective has to be selected carefully (Sections 4.9 and 4.10).

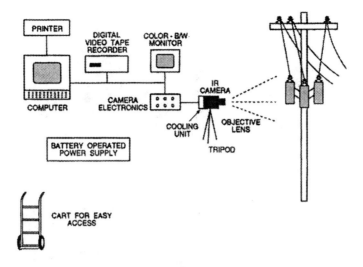

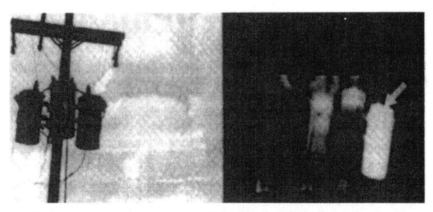

FIGURE 8.2 Typical passive infrared TNDT experimental configuration, here for inspection of three-phase transformers. Application is shown below: no anomaly is apparent in this visible image (left) while the IR image indicates that one of the three transformers is overheating). (Adapted from Maldague, 1994b and from Stovicek, 1987, with permission from Gordon and Breach Publishers.)

- *Tripod, cart, or any convenient way to hold in place and move the infrared camera and equipment.* Some manufacturers offers integrated systems limiting the number of pieces of equipment that must be carried (see the next item).
- Associated electronics and the analysis equipment
  - A videotape recorder (VTR) might be useful for measurements (since a tape cassette is probably the cheapest available recording medium: $2 for about 20 G on a VHS tape cassette compared to 650 M for a recordable

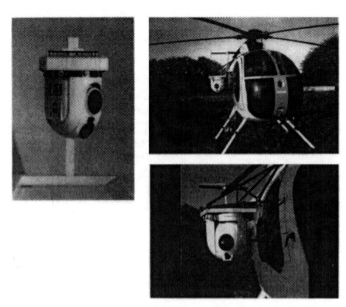

FIGURE 8.3 For airborne surveys, dedicated sensors are recommended with long-wavelength response, zoom capability, visible and infrared cameras onboard, and stabilized platform among desirable features. For example, the unit shown on the left is particularly dedicated for airborne surveys of transmission lines and power distribution sites. It incorporates a 320 × 240 uncooled microbolometer FPA and a 12° field-of-view optic. Right images show a typical setup with the sensor mounted on the side of an helicopter (Left: courtesy of Flir Systems; right: courtesy of Centre for Built Environment, Division of Information Technology for Land, Building and Infrastructure, Sweden, Dr. S. Å. Ljungberg.)

5-inch CD*). However, some modern IR camera systems come with on-board memory and memory cartridges [such as PC-MCIA (Personal Computer Memory Card International Association†) cards], allowing to record infrared images as well as related comments (voice recording) in a compact one-unit system (Figure 8.4). A 35-mm instant Polaroid or digital still camera or printer can be used instead or in combination to have a permanent record of the inspection. A portable (digital) VTR is a better alternative to hardcopy printouts since it allows off-line image processing, printing, and editing (problems and alternatives to analog VTR recording were discussed in Section 5.3). In either case, a portable (12 V dc) unit is more convenient, facilitating work in the field. A *digital* camcorder is an interesting alternative to conventional analog VTR. Although more expensive, they do not further degrade the signal, which can later be downloaded into a computer through various interfaces, such as I.link

---

*1 K $= 2^{10}$, 1 M $= 2^{20}$, 1 G $= 2^{30}$.
† www.pcmcia.org.

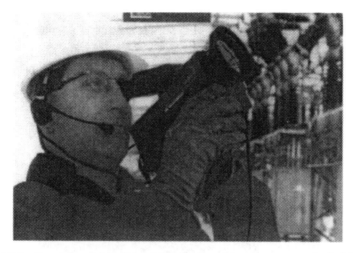

FIGURE 8.4 Recording infrared images as well as related comments (voice) in a compact one-unit system. (Courtesy of Flir Systems.)

for some Sony products [= IEEE 1394 FireWire (on Apple Macintosh computers) protocols*]. These protocols currently permit the transfer of up to 400 megabits per second. Moreover, the camcorder can be used to grab a few visible pictures of the setup, helping significatively to document later experimentations.

- A computer is generally not needed on the field since image processing (such as enhancement) will be done off-line at a later time. However, a laptop computer might serve to display (and preprocess) ingoing images that are next recorded on a hard disk (or on PC-MCIA cards; Figure 8.5). Commercial image processing software can be purchased from infrared camera vendors for such purposes. Thanks to their crunching ability, portable computer can assist interpretation by the operator through specific processing (such as those reviewed in Chapters 5 and 6).

- For remote operation, a power generator can be used, or more conveniently, if possible, 12-V dc from a car/van battery or solar power can be employed as well (major items such as an IR camera, digital camcorder, or laptop computer can be powered from a car/van 12-V battery (depending on the power drain, it might be necessary to turn on the engine to recharge the battery, etc.). Moreover, battery-operated power supplies may be needed to power sensitive electronic equipment, especially in electromagnetically contaminated environments such as in areas close to high-powered machinery. In such cases RF shield casings may be required as well. An advantage of dc-operated power supplies is that power fluctuations (such as the ripples in ac–dc conversion) are elimi-

*www.1394ta.org.

FIGURE 8.5 Compact portable IR system, including an IR sensor, tripod, laptop computer, batteries, and case. (From Rencz, 1998. Copyright © 1998, reprinted by permission of John Wiley & Sons, Inc.)

nated. For example, photoconductive detectors require an external current to measure conductivity change. Since biasing current is needed for photoconductor-type detectors, high-charge-capacity alkaline batteries can be used to induce minimum noise and ripple, which is essential to achieve stable results. Finally, inverters (e.g., Enerwatt model EW-2500) up to 2500 W allow us to get ac (e.g., 115 V 60 Hz) power from 12 V dc. (In such a case it is important to add appropriate fuses with proper wiring between the battery and the inverter since current on the dc side can reach 375 A in case of a short allowable 4500-W power surge; power = current × voltage.)

- Unless the infrared camera is thermoelectrically cooled or does not require cooling, special procedures will have to be taken to ensure proper cooling operation (Section 4.6), if needed: for example, sufficient supply of liquid nitrogen to allow periodic refilling. A well-insulated 10-L dewar tank typically keeps up to 10% of its content over a 1-month period (without liquid withdrawal; Figure 8.6).

8.1.2 Instrumentation for the Active Approach

In this section we discuss equipment configuration in the active approach. This discussion is completed with more formal information, which can be found in Chapter 9, devoted entirely to material inspection using the active scheme. The

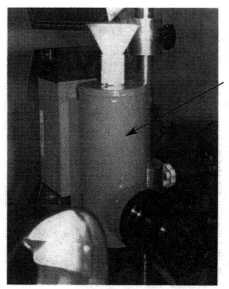

Dewar

FIGURE 8.6 Liquid nitrogen vapor coming out from the funnel used to fill out the IR camera dewar.

setup configuration for the active approach is quite similar to the equipment needed for the passive approach (discussed in preceding section). A typical setup is shown in Figures 8.1 and 8.7.

- Acquisition head
  - Infrared camera
  - Appropriate objective lens
  - Cooling fluid (if required)
  - Thermal stimulating device
  - Tripod, cart
  - Battery-operated power supply
- Sample holder
  - Part holder
  - Moving slide
  - Robotic arm
  - $x–y–z$ translation stage (or $x–y$ wide table)
  - Appropriate objective lens
  - Tripod, cart
- Associated electronics and analysis equipment
  - Computer and associated software [more sophisticated algorithms are likely to be used; stronger computing capabilities are needed, especially

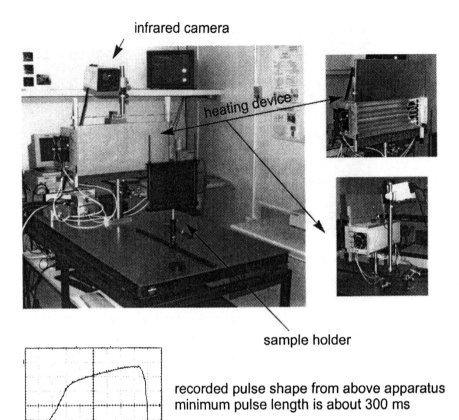

FIGURE 8.7 Active TNDT experimental setup (the heating device is made up of six quartz infrared tubes of 1.2 kW each; the computer-controlled shutter allows for sharp thermal pulse as short as 0.3 s (shown open and closed). The IR camera is mounted on an $X-Y$ translation stage. Typical pulse shape is also provided.

to solve inverse problems (Chapter 10) or to perform thermal modeling (Chapter 3)]

- Cart for moving the equipment around
- 35-mm Polaroid, printer, VTR, digital recording system, digital camcorder (These are needed especially to keep a record of the experiment if measurements are performed in the field and if several thermogram sequences are recorded. The 35-mm camera makes it possible to take pictures of the experiment going on, useful for later reference.)
- Battery-operated power supply

They are, however, differences with the passive configuration, as discussed below.

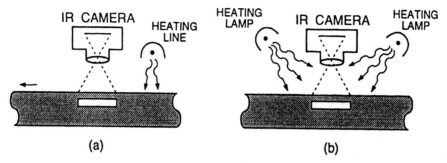

FIGURE 8.8 Configuration for thermal inspection: (a) mobile; (b) static. (Adapted from Maldague, 1993.)

Acquisition Head. The acquisition head is composed basically of two elements, the infrared camera and the thermal stimulation device. Infrared cameras (radiometers) and infrared detectors were reviewed in Chapter 4. The acquisition head can be either static (e.g., mounted on a tripod) in front of the part to be inspected, or mobile, for example mounted on a robotic arm in the inspection of large complexly shaped workpieces (see Section 8.1.4).

Hot and Cold Thermal Stimulation Devices. A difference with respect to the passive approach concerns the thermal stimulating device that is needed to generate a thermal response from the part. This is a crucial aspect since, in fact, lack of success in TNDT is often related to problems with the thermal stimulation device, such as the lack of homogeneity of the heating source. We now concentrate on this subject.

In pulsed thermography, when step heating is used to reveal subsurface defects, it is necessary to generate a thermal transient inside the part (Chapter 1). Such a thermal transient can be generated as either a static or a mobile configuration (Figure 8.8). Such a thermal transient can be either hot (hotter than the part) or cold (cooler than the part) since a thermal front propagates in the same fashion, being either cold or hot. The cold approach has, under certain circumstances, several advantages over the warm approach: for example, if the part to be inspected is already hot, such as at the output of the curing oven, in the case of graphite–epoxy panels. In these cases, it is uneconomical to add extra heat to the part; cooling is then more economical. Another benefit is that a cool thermal stimulating device will not generate any thermal noise (which can reflect on the part; Section 4.8) since it is at a reduced temperature with respect to the component to be inspected. Such a *cool* thermal perturbation device can be deployed using:

- Cool water jets
- Cool air/gas jets (Figure 8.9)
- Snow or ice (with contact)
- Cool water bag (with contact)

FIGURE 8.9 TNDT using liquid nitrogen spray to find defects in a filament-wound graphite-epoxy tank. (From Burleigh in Maldague, 1994b, with permission from Gordon and Breach Publishers.)

More traditionally, a *hot* thermal perturbation is used, which can be:

- High-powered cinematographic lamps (Figure 8.10)
- Bank of quartz infrared line lamps
- High-powered photographic flashes (Figure 8.11)
- (Scanned) laser beam
- Heat gun, hot water jets, hot air jets, hot water bag (Figure 8.12)

The main desirable characteristics of the thermal perturbation (either cool or warm) is repeatability, uniformity, timeliness, and duration. Repeatability is needed in order to compare among results obtained on many identical parts. Notice than lamp aging can reduce power deposit significantly (up to 40% of the light intensity can be lost through deposition of filament evaporated tungsten on the inner lamp glass envelope). Uniformity is necessary when inspection is performed surface wide. Uneven thermal stimulation will produce spurious hot (cold) spots, which may be interpreted as subsurface flaws (especially in the case of the pulsed scheme; Chapters 1 and 9). This uniformity characteristic is hard to achieve, however. Special deconvolution algorithms, special calibration, and contrast computations (Chapter 5) can allow for stimulation correction after the fact.

Laser scanning of the surface inspected is an attractive possibility, although it may complicate the setup design; however, it provides high, uniform energy

FIGURE 8.10 High-powered (1000 W) cinematographic lamp used to induce thermal stimulation in active lock-in thermography. (Courtesy of IKP, Universität Stuttgart, Germany, Prof. G. Busse.)

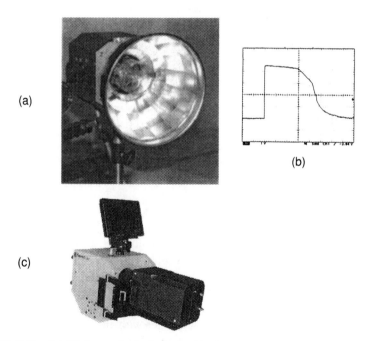

FIGURE 8.11 (a) High-powered photographic flashes (Star flash 3 from Balcar with two flash tubes mounted in a cross configuration 6.4 kJ, 15 ms). Care should be exercised to protect one's eyes during firing of such high-powered photographic flashes, for example with shrouds or screens. (b) Recorded pulse shape for flash of (a), at a distance of 57 cm. (c) Some vendors offer integrated heads with the shroud, the flashes, and the IR camera. (Image courtesy of Thermal Wave Imaging, Inc.)

with warm water bag (thermocouple reading: 54.9°C)

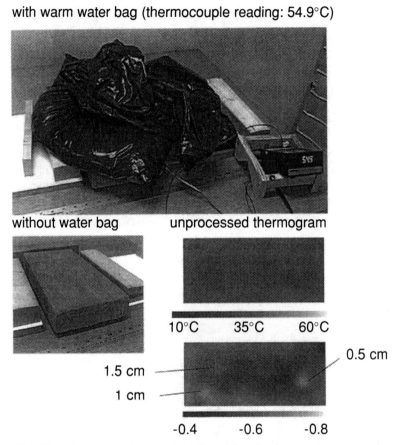

without water bag   unprocessed thermogram

10°C   35°C   60°C

1.5 cm   0.5 cm

1 cm

-0.4   -0.6   -0.8

FIGURE 8.12 Warm-water bag used to thermally stimulate a concrete specimen to find subsurface delaminations. Raw and processed (slope) images shown with three subsurface defects (flat-bottomed holes of 2.5 cm diameter at indicated location and depth).

deposit over the surface. A CO_2 laser emitting at 1.06 μm is often used as a stimulation device since the heating beam is not seen by the infrared camera at either 3 to 5 or 8 to 12 μm. In this discussion we refer more to imaging, although point-by-point inspection is also possible, to determine fiber orientation in composites, for example, or to evaluate material physical properties such as thermal diffusively. The latter application is discussed in Section 11.5.

For the pulsed scheme, timeliness and duration are other important issues for thermal stimulation. For better agreement with theoretical models and to enable data inversion (Chapter 10), a square thermal pulse is needed. In that case a mechanical shutter may be used in front of the thermal perturbation device. In this configuration the shutter is constituted of one (Figure 8.7) or two screens (Figure 8.13). The first intercepts the radiation of a quartz infrared line lamp bank. At $t = 0$ it is released and partial thermal stimulation starts. At $t = t_f$ the second screen is released, blocking the radiation. Pulse duration of

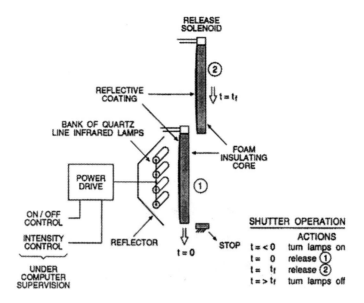

FIGURE 8.13   Principle of operation of a two-screen shutter for sharp and short heat pulse generation in active pulsed configuration. The picture shows the details of the shutter–solenoid configuration (case of Figure 8.7). (Top: from Maldague, 1994b, with permission from Gordon and Breach Publishers.)

tens of milliseconds can be achieved in this way. Depending on the partial thermal characteristics of the specimen, mainly the thermal conductivity k, different pulse durations have to be selected, short for high-conductivity samples and long otherwise [refer to eq. (8.1) below]. Photographic flash (xenon) lamps or mechanical shutters make it possible to realize short high-powered pulses, which are desirable for inspection of high-conductivity parts such as those made of aluminum.

A rosette of, say, six unfocalized projectors mounted with a parabolic back reflector around an infrared camera and connected to a timer allows us to obtain thermal pulses of variable length (minimum ~ 0.5 s). Careful orientation of the individual projectors within the rosette helps to maximize the uniformity of the heat deposited on the surface inspected. Experimental trials have shown that for a six-1000-W-projector rosette, the maximum area inspected is about 20×20 cm (Figure 8.14a).

Figure 8.14b shows the typical shape of a heating pulse obtained with such incandescent projectors. It is characterized by a fast rise time (power on) and

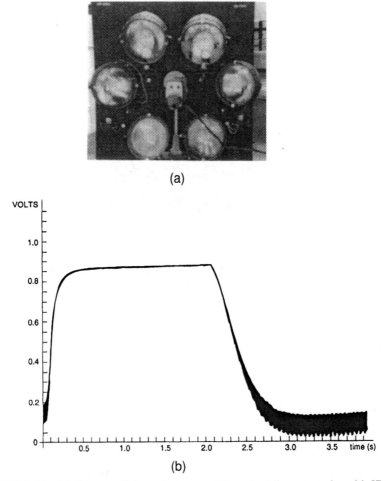

(a)

(b)

FIGURE 8.14 (a) Rosette of six projectors used in pulsed thermography with IR camera mounted at the center; (b) shape of the heating pulse obtained with one of these 1000-W incandescent projector. [(b) From Maldague, 1993.]

slow cooling of the filament (power off). Compare this with the pulse shape obtained with the mechanical shutter apparatus of Figure 8.7: The shutter reduces the pulse tail from 0.5 s down to about 0.1 s. One problem with using incandescent sources as heat stimulation devices is precisely the warm-up of the mechanical structure which occurs as inspection tasks are repeated. This warm-up causes the lamp structure to emit parasitic thermal radiation reflected on the sample surface, picked up by the infrared camera and superposed with the useful signal, thus causing measurement perturbation. Fans or cooled-water circulation help reduce this effect. Lamp aging is another drawback which affects measurement reproducibility.

To improve the uniformity of heat deposition in the static configuration, an attractive possibility is to mount four line heating xenon flash tubes in a square configuration with the infrared camera placed at the center (Figures 8.15 and 8.11c): pulse lengths of less than 15 ms (Figure 8.11b) are possible with a homogeneous energy deposit of up to 20 kJ over a 30 × 30 cm area (such a configu-

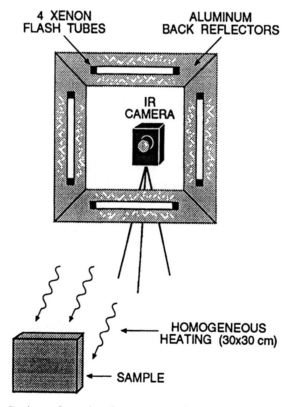

FIGURE 8.15 Static configuration: homogeneous heat deposition is obtained by using four xenon tubes with back reflector mounted in a square with the camera at the center. Short heat pulses of about 15 ms are possible. (From Maldague, 1993.)

FIGURE 8.16   Radiant heater mounted on a moving slide. An aluminum honeycomb specimen is positioned on the sample holder (on left). (From Maldague, 1993.)

ration is commercially available from Mecica, Reims, France, among other vendors).

Another heating technique consists of using a heating radiator such as the one pictured in Figure 8.16 (Watlow, St. Louis, Missouri, model 8745C, 2500 W, heating surface: 30 cm height × 20 cm width), which is mounted on a moving slide such that it heat-stimulates a workpiece while passing in front of it. This kind of stimulation yields better uniformity of heat stimulation than with the rosette discussed previously. As an example, given the unit of Figure 8.16, a deposit of 20 J cm^{-2} is obtained at a speed of 4 cm s^{-1}, 8 J cm^{-2} at 10 cm s^{-1}, and 1.6 J cm^{-2} at 50 cm s^{-1}. Even if this configuration is not really static as for the rosette, it is an interesting approximation for experimental studies.

For the detection of surface cracks, a practical way to proceed consists of have the thermal front propagating along the surface, called *lateral surface heating*. The stimulation can be achieved with the specimen suddenly being brought into contact with a thermal mass (Figure 8.17). In this case a surface crack of higher thermal resistance (Chapter 3) opposes passage of the thermal front, leading to surface temperature differentials detectable by the IR camera. Surface propagation of a thermal front can also be achieved by vertical immersion of one section of the specimen in a hot (or cold) water tank while observation is conducted on the nonimmersed section with the thermal front propagating bottom up. Although this approach does not lead directly to repetitive inspection on the plant floor, it can be of great interest for specific applications.

A convenient way to exploit such lateral surface heating to detect the presence of surface cracks (e.g., in concrete) is as follows. Once deposited on a sur-

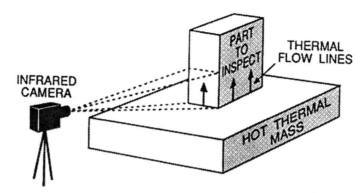

FIGURE 8.17   Heat front propagation along the surface after sudden contact of the sample with a thermal mass of different temperature. (From Maldague, 1993.)

face, the energy propagates both under the surface and laterally. The presence of a crack on the surface acts as thermal resistance, reducing the lateral thermal propagation and causing heat to build up. An IR camera pointed on the surface registers areas of abnormal temperature, making it possible to locate surface cracks (Lesniak et al., 1996–8). Such an approach, used to locate surface cracks on concrete, was tested successfully (Figure 8.18). In that case, a slide projector provided the heating strips, which are spaced apart a distance based on the

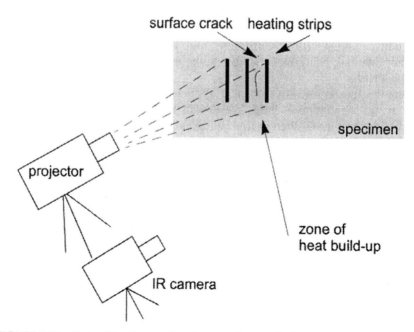

FIGURE 8.18   Lateral heating used to detect surface cracks (on concrete, a total heating surface of about 20 × 20 cm is possible with a standard slide projector of 150 W).

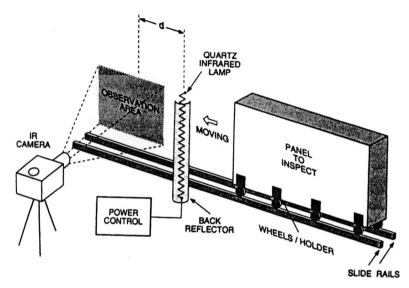

FIGURE 8.19 TNDT configuration for the inspection of large components mounted on a moving slide. (From Maldague, 1993.)

thermal time of propagation of the material of interest and on vanishing of thermal contrast over the surface. For example, for concrete, it was demonstrated (e.g., Figure 8.12) that the maximum detection depth for delamination is about 1.5 cm. A strip spacing of 1.5 cm would thus be appropriated. In some instances, such as for inspection of low-thermal-conductivity specimens (or for qualitative TNDT), time constraints and heating uniformity requirements are not essential; heat gun or lamps can be used for thermal stimulation (e.g., Figure 11.19).

Line heating is also an attractive approach (mobile configuration; Figure 8.8). In this configuration large panels can be inspected. They move first in front of the heater and than in front of the infrared camera. The surface is next imaged by the infrared camera (Figure 8.19). Line heating can be deployed using a quartz infrared lamp (e.g., the General Electrics QH 1600T 1600-W lamp, 30 cm long) mounted with a back aluminum reflector. Line cooling using a column of air/water jets is an attractive alternative for thermal stimulation, if possible, as mentioned before. However, the contact of water with the workpiece may not always be acceptable.

This configuration allows inspection to proceed at a fast pace, up to 1 m^2 s^{-1} in some instances. This may be convenient for inspection of laminates, to look for bonding integrity between a core and a skin (Section 7.4.1). Since the distance d between the thermal stimulation area and the observation area is related to potential subsurface flaw depth, it is an important parameter which may be evaluated for homogeneous materials using a practical rule:

$$z = \sqrt{\alpha t} \qquad (8.1)$$

where z is the thickness of material to be inspected (meters), t the required time of observation (seconds), α the thermal diffusivity $= k/\rho c$ m^2 s^{-1}, k the thermal conductivity (W m^{-1} °C^{-1}), ρ the density (kg m^{-3}), and c the specific heat J kg^{-1} °C^{-1}. For example, a 1-mm-thick aluminum coating ($\alpha = 95 \times 10^{-6}$ m^2 s^{-1}) requires an observation time on the order of 10 ms per spot inspected, whereas a similar material 10 mm thick requires a relatively long inspection time of 1 s per point.

This configuration may also be inverted; in this, the line heater is mounted on the slide and the part is not moving. The inspection can proceed in transmission (IR camera on one side and heater on the other) or in reflection (IR camera and heater on the same side). Such lateral motion is advantageous since heating uniformity is obtained along the moving direction, which may not be the case for a static configuration. Also, lateral motion allows interesting image-processing possibilities by integrating multiple images for noise reduction (see also Sections 6.4 and 9.2.2). Contact thermal stimulation, such as with ice, snow, or a cold/hot water bag, is possible, although it does not allow for easy automation or repeatability (Figure 8.12).

When a part is hollow, it can be of interest to use internal stimulation with liquid (water) or gas (air) flow. In this configuration, change in flow temperature (hot to cold, or the reverse) allows us to discover abnormal variations of wall thickness or blocking passages because of the delayed arrival of the thermal perturbation. Following eq. (8.1), a variation of twice the wall thickness will produce a fourfold variation in thermal perturbation on the outer surface after internal perturbation is initiated. For example, such a method has been used for jet turbine blade inspection and for wall thickness evaluation in case of pipe cavitation (Figure 8.20). One advantage of this method is that since the thermal perturbation resides inside the part, it does not generate any spurious thermal reflection, which may contaminate the measurement, as in the case of external sources of radiation heating (more about this in Section 11.2).

Microwave heating is another interesting heating mode. In microwave heating, a time-gated microwave source introduces heat into the specimen, and for homogeneous specimens this allows detection of subsurface microwave absorbing features (such as water-filled areas, metal wires, or fibers) through infrared imaging. Quantitative information on these features including their depth and thermal diffusivity, is possible by analyzing the time and spatial dependence of the surface temperature (Osiander et al., 1994; d'Ambrosio et al., 1995). If the host is transparent to microwaves, the microwave source heats these features directly. In this case, a higher spatial resolution is noted than in the surface heating case since the transit time of the thermal front is shorter, and consequently, three-dimensional spreading losses are reduced. The thermal front only has to diffuse to the surface instead of being of round-trip nature. Among reported applications is purity assessment in composite fabrication (Osiander et al., 1995a,b). In such a case, the quality of the structures is related directly to the purity of the various layers. Step heating is used (Chapters 1 and 9): A 9-GHz oscillator heats the specimen, while a single flare horn antenna

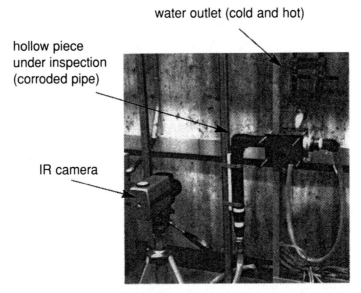

water outlet (cold and hot)

hollow piece
under inspection
(corroded pipe)

IR camera

FIGURE 8.20   Internal thermal stimulation with flow of water to assess pipe cavitation. (Adapted from Maldague, 1994b, with permission from Gordon and Breach Publishers.)

sends out a beam at about $50°$ on a specimen located 15 cm away, thus achieving a power density of 20 mW cm^{-2}. Carbon fiber contaminants in fiberglass epoxy have a particular thermal signature, allowing both detection and depth evaluation (depth: up to 0.75 mm). Another application is for detection of water-filled cracks in concrete to assess the safety of aging concrete structures, and in particular for highway and railroad bridges (Sakagami et al., 1999). Since water is heated favorably by the microwave with respect to sound, dry concrete, cracks in which water (rain) penetrates achieve a particular heated pattern of 6 to $8°C$ higher than that of sound concrete pulse-heated under the same conditions (e.g., 2.45 GHz, 1400-W domestic microwave oven for 5 s), thus enabling crack detection (e.g., 4 cm long, 1 cm deep, and 0.2 and 0.4 mm width). Figure 8.21 presents an application of microwave heating to detect rotten wood.

In the case of lock-in thermography, sinusoidal heating is used to stimulate the part. This can be accomplished using incandescent projectors (such as those of Figure 8.10) combined with appropriate electronics. The system is first calibrated by having the IR camera looking at the scene; this allows the electronics to adjust itself to the lamps so that real sinusoidal heating is obtained.

Sample Holder. A part holder may be needed to hold a specimen if it does not stay in place by itself or if a particular surface area must be presented to the IR camera. This is especially important to get repetitive measurements if many identical parts are to be inspected. If the number of objects to be inspected is

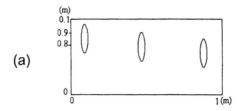

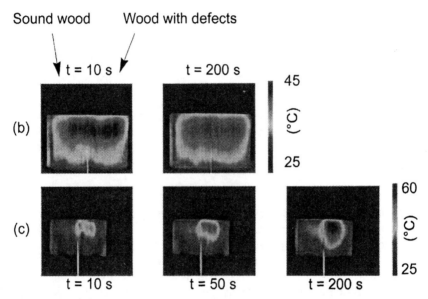

FIGURE 8.21 Microwave heating to investigate rotten wood. For all tests, comparison is always made with the heating of a dry, sound wood piece (at left of the line shown). A 1250-W domestic oven heats the specimens for the indicated period of time. (a) Geometry of the wood piece with three simulated defects (holes) of 2 cm diameter located 0.5, 1, and 1.5 cm below the surface; (b) defect (holes) filled with wood chips of the same humidity content as that of sound wood. (c) The wet case is more interesting (holes filled with wet wood chips) since defect geometry becomes visible.

large, they may hold themselves in place, or they can be mounted on a moving slide (e.g., Unislide from Velmex, East Bloomfield, New York*) for fast inspection of large flat surfaces (Figure 8.19). For smaller parts, a special holder may be necessary (refer, e.g., to Figure 8.7).

The sample holder is also important for maintaining a specimen over an experimental bench surface, thus preventing generation of spurious thermal reflections caused by the cavity created by the combination of bench surface

*www.velmex.com.

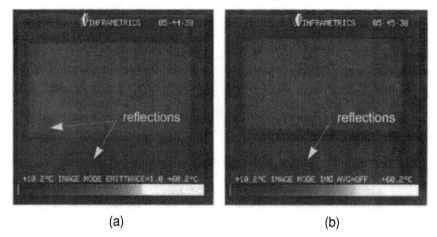

(a)   (b)

FIGURE 8.22   A sample holder maintaining the specimen over the experimental bench surface helps prevent the generation of spurious thermal reflections: (a) specimen directly on the bench surface; (b) specimen mounted on spacers. The effect of the thermal reflections caused by the cavitylike structure created with the combination of bench surface and specimen surface is visible.

and specimen surface when a specimen is deposited directly on a bench surface (Figure 8.22).

A potential difficulty may arise when inspection must proceed horizontally, and if the IR camera is of the liquid nitrogen cooling type, since the internal camera dewar (Figure 8.6) must be held vertically (to avoid spilling liquid coolant). In this case, a highly reflective mirror set in place at 45° in front of the camera allows horizontal viewing (Figure 8.23). First-surface gold or aluminum mirrors are almost 100% reflective for wavelengths from 3 μm and up and are well suited for these applications (Section 4.9). When using such a mirror, it is important to recall that recorded images will be inverted (Section 4.10.1).

Robotic Thermographic Inspection.   To maintain a fast inspection pace during repetitive inspection tasks on the same parts or when inspecting large structures, it may be of interest to mount the thermographic inspection head on a robotic manipulator. In one reported case, such a thermographic robotics scheme made it possible to reduce inspection costs by 30 to 60% over those of a traditional radiographic or point-by-point ultrasonic method (inspection of diffusion-bonded structures used for the F-15E fuselage; Theilen et al., 1993).

Two approaches are possible for such a task: The robot arm can be moved to preprogrammed locations, or the robot can proceed with more autonomy by itself mapping the shape curvature of the part to inspect with a 3D range sensor. In this approach, the self-navigation capability of the system reduces the need for exact positioning of the structure to be inspected. Figure 8.24 shows such an inspection system; 1.2-kW back-reflector line heating is used.

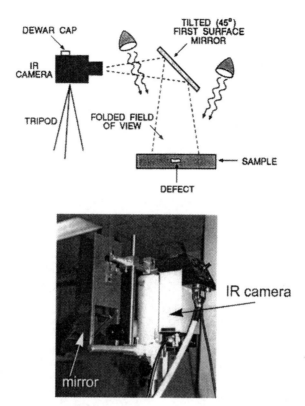

FIGURE 8.23 TNDT configuration making use of a tilted mirror for down-view inspection and picture of a typical setup for such purposes (with first-surface gold mirror). (Top: from Maldague, 1993.)

This system has a two-step procedure. In the first step the surface being inspected is modeled. This step is necessary in order to record the shape of the surface. In the system depicted in Figure 8.24, to localize the boundaries of the specimen it is assumed that the specimen is always positioned over a flat surface. Of course, for large components such as an aircraft fuselage, this question is of no concern since the surface inspected covers the full workspace. If available, the presence of a flat background also has practical concerns since it allows us to calibrate range data expressed in the 3D camera reference system into world 3D coordinates. Moreover, it allows us to set the first location for the scanning process: The lower left corner is given the origin of the world coordinate system, and the workspace is then located in the positive x–y quadrant (z-axis normal to the surface; Figure 8.25a).

In the modeling step, the scanning proceeds as follows. From the origin, a ranging scan is taken along the x-axis; then the robot arm moves a little along the y-axis and another ranging scan is taken. A set of two parallel lines of

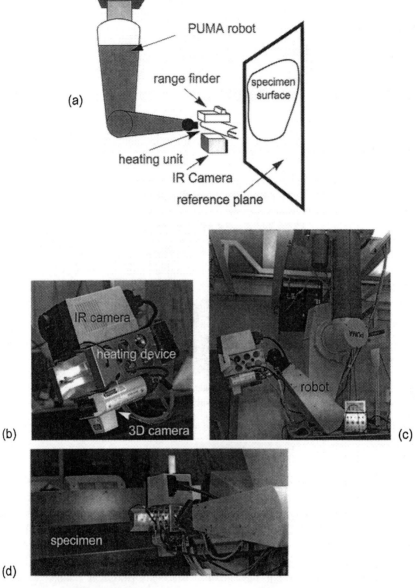

FIGURE 8.24 TNDT robotic inspection with self-navigation capability: (a) schematic diagram; (b) inspection head; (c) complete configuration with the robot; (d) inspection of a complex shape specimen.

points (256 in total) is thus obtained and serves to build a small triangular patch whose vertices are the three neighboring points (Figure 8.25a). The vector normal to every triangle is calculated, and the corresponding surface orientation is used to decide if the current robot head orientation and height above the sur-

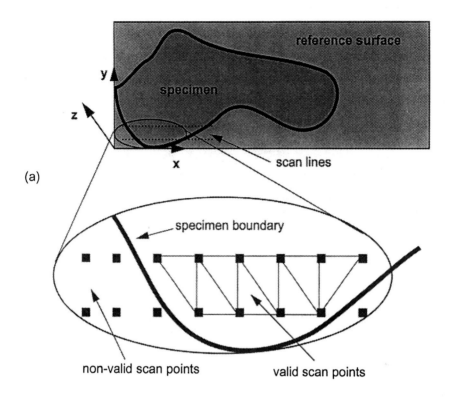

(a)

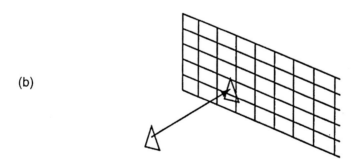

(b)

FIGURE 8.25 Three-dimensional modeling of the inspected surface: (a) 3D scanning of the surface inspected; (b) projection of a triangular element onto the parametric grid.

face are adequate. If all is correct, the system pursues the next triangular patch in the same fashion. If a new point of view is required, the system will adjust its height and orientation before proceeding to the next patch.

After covering the entire workspace, collected range data are integrated in the surface model, which is based on a parametric grid oriented parallel to the

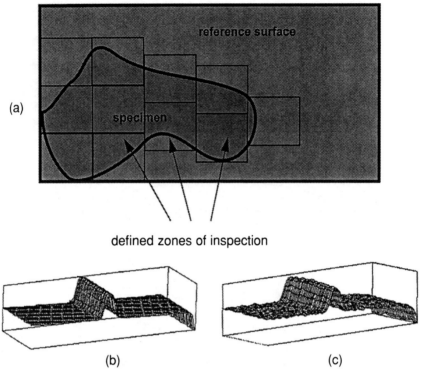

defined zones of inspection

(b)   (c)

FIGURE 8.26 TNDT robotic inspection: (a) strategy to perform the inspection; (b) thermal line modeling; (c) corresponding actual temperature profile measured (vertical scale 21 to 27°C, horizontal scales: positions).

world coordinate system onto which each triangular patch is projected (Figure 8.25b). In general, the patches obtained during the scanning process overlap, and some areas of the inspected surface are represented by many patches; the parametric grid allows to suppress such redundancy.

At this point, the object is modeled and ready for TNDT inspection. A scanning strategy is required to inspect the surface so that disturbing heating from one field to the next is minimized (the thermal front spreads in all directions; Figure 8.26a). To perform this and to set the scanning speed of the inspection head over the surface, thermal modeling of the experiment is needed (Figure 8.26b and c). The complete system is controlled from a workstation.

For specific applications, in addition to the cost reduction discussed previously, such a robotic thermographic inspection procedure has several advantages in terms of reliability, repetitivity, automatization of the inspection procedure, and data acquisition.

Associated Electronics: Timer Operation. In addition to camera electronics, some additional equipment may be needed, such as a video timer (to superimpose acquisition time on images) or special equipment to trigger thermal stimulation devices. It is not really possible to describe such equipment in great detail since it is very installation dependent.

Concerning the video timer (e.g., some models are manufactured by FOR-A, Japan*), it may be convenient to generate timing information on images in a machine-readable fashion. This is essential if the sequence of images (e.g., temperature evolution over the part inspected) have to be digitized and processed. One way to achieve this is through the use of a bar code superimposed on infrared images. Some IR camera manufacturers (such as Inframetrics 600L, Flir[†] Systems, among others) already use this scheme to code in real time pertinent information on the top first lines of thermograms (this is called the *header* of the image). This way to proceed is convenient since it is only necessary to digitize the infrared image to obtain all related information (image itself, thermal range and level, time and date of acquisition, atmospheric coefficient factor, etc.).

A video timer is controlled through a three-input connector for the *reset*, *start*, and *stop* functions. In the case of a sample moving on a slide, an optodetector (e.g., the TIL-138, Texas Instruments, or equivalent) can be actuated by a blade mounted on the sample holder, which breaks the emitter-detector beam while passing by. Figure 8.27 shows a typical drive circuit for such an optodetector. The *start* signal obtained with the circuit of Figure 8.27 can also be used to trigger an image acquisition sequence.

Recording Other Signals. If absolute temperature is to be computed from infrared images, it is necessary to record the video signal and all the infrared camera settings simultaneously. As mentioned in Section 4.8, on some IR radiometers, there are generally two settings of interest: the thermal range and the thermal level (refer to Figure 4.19). During a given infrared test, the thermal range is generally not changed (to keep the same temperature resolution during all the experiments), and consequently, only the thermal level is to be recorded along with infrared images. Some manufacturers (such as Flir) combine this information with the video signal. In this case, recording the video signal suffices to compute the temperature (Section 4.8). Otherwise, the thermal level needs to be recorded separately.

Such a thermal-level signal can be derived from the infrared camera electronic unit as an analog signal which can be digitized by dedicated acquisition plug-in boards such as the A/D DAS-8 or DAS-800 board from Metrabyte-Keithley[‡] (Cleveland, Ohio). When doing so, it is important to match the dynamic range of the signal to record and the board. The circuit of Figure 8.28 allows us to perform such matching. For example, in a typical setup, direct digitizing without using such a conditioning circuit would result in the use of only 84 of the 4096 possible values available with the DAS-8 board (to cover

---

* www.for-a.com/index.htm.
† www.flir.com.
‡ www.keithley.com.

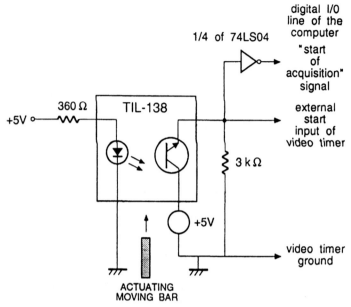

FIGURE 8.27   Schematic diagram of the interface circuit needed to interface a TIL-138 optocoupler to the video timer. (From Maldague, 1993.)

the thermal-level span of interest). The circuit of Figure 8.28 is formed of three parts: an *RC* filter to cut off high frequencies ($f_{3dB} = 1/2\pi RC \sim 3.4$ Hz, high enough to follow the slow rotating action of the operator's fingers on the level knob); gain and offset to bring possible signal variations in the 0 to 10 V input span of the DAS-8 board, and a differential amplifier which make it possible to suppress undesirable *ground loop* effects.

Thermal-level signal is digitized in fixed time increments (such as 50 ms) on a disk file. Since the acquisition time is short, it is possible during one time increment to average multiple values in order to reduce the noise level (noise is reduced by a factor of 3 dB every time we double the number of averaged values). It is also possible to smooth this disk file to reduce the noise further (Section 5.5). When temperature processing takes place, it is necessary to extract a level value corresponding to the acquisition time of the image.

This double recording process (images and level) is necessary on some infrared radiometers. Notice, however, that recent infrared systems with a 12-bit (or even 14-bit) digital output have a greater dynamic range (12 bits: 72 dB, 16 times greater than that for more conventional 8-bit systems), so that separate recording of thermal range and level is no longer necessary: 14-bit availability makes it possible to cover the entire span of the calibration curve directly without requiring the *artificial* introduction of level and range.

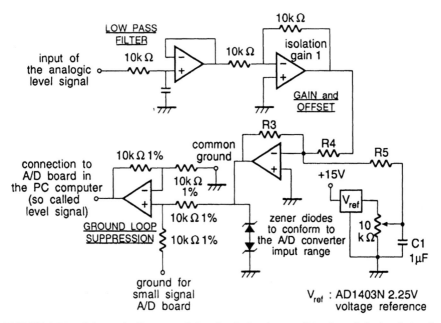

FIGURE 8.28 Schematic diagram of the circuit for the conditioning of the level signal; the amplifiers are OP-7 8215 (type 741). The output V_o is given by (R_3, R_4, and R_5 are 1% resistors): $V_o = V_{in}(R_3/R_4) - V_{ref}(R_3/R_5)$. (From Maldague, 1993.)

8.1.3 Measurement Reproducibility

It is interesting to address the question of measurement reproducibility. The question is to determine if we are susceptible to obtaining the same absolute values for a given parameter from experiments that took place at different moments. Of course, the experimental conditions and the work piece must all be identical. All the instrumentation should also have reached a permanent operating regime after a warm-up period of about 3 h. This is particularly important for an infrared camera whose casing itself acts as a parasitic emitter. This emission has to be taken into account in the calibration process for complete correction of the radiometric response (Section 4.8). Consequently, temperature variations of the infrared camera casing will affect the measurement if not yet stabilized. This warm-up period is thus necessary to keep signal drift at a minimum.

As an example, Figure 8.29 shows two typical data sets obtained on two different samples (Plexiglas and graphite epoxy plates). For each sample, the same thermographic investigation was repeated seven times over a one-month period. The parameter measured was the time of occurrence for maximum contrast over an artificial subsurface defect (refer to Section 5.6 for details on contrast computations). For the Plexiglas sample the maximum deviation is 4.6%, with an average deviation of 1.8%, while for the graphite epoxy, we

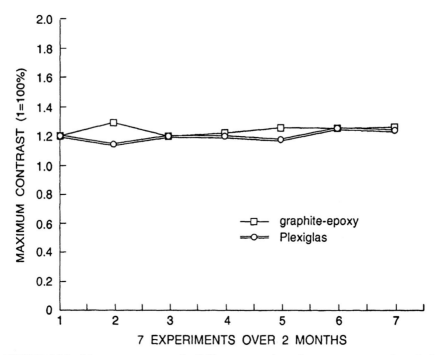

FIGURE 8.29   Measurement reproducibility test conducted over a two-month period. The horizontal scale corresponds to the reproduction of one test. Graph illustrates an important parameter needed for quantitative evaluation: the value of the maximum thermal contrast in pulsed thermography (Section 10.3). (From Maldague, 1993.)

obtain 5.8% and 1.4%, respectively. These deviations are in agreement with what is generally expected of infrared systems (as discussed in Sections 4.8 and 10.2.5): agreement of a few percent.

8.1.4   Analysis for an Experimental Setup Configuration (Active Pulsed TNDT)

Before closing this section, we study an example that allows us to introduce some interesting considerations about the experimental setup. When deciding to perform an inspection task by thermography, one of the first concerns is to evaluate the required spatial resolution. A probability detection curve such as the one in Figure 7.8 may help us find the minimum size of defect detectable. This determination allows to specify the field of view for the inspection task and decide on the experimental configuration.

For example, suppose that we want to detect lack of adhesive in bonded aluminum panels. Suppose that the aluminum panels are made of sheets 2 mm thick that are epoxy-bonded on a 5-cm foam core (these types of panels are used to build thermally insulated cold boxes for trucks; Figure 7.3). Pulsed active TNDT is retained as the NDE method of interest. The panels are flat and of large size, the minimum size of disbondings to be detected is 25 mm in diameter. A practical *rule of thumb* concerning the minimum size for defect detection by infrared thermography is the following (Section 9.2): The defect radius-to-depth ratio much be greater than 1; we see that for this example, the criterion is satisfied: $12.5/2 = 6.25$, although in this case the insulating properties of the foam strengthen this constraint. With respect to Figure 7.8, we notice that we are over the minimum detection threshold. This brief study indicates that thermography appears a good candidate for detection of these defects, provided that problems of low aluminum emissivity and high thermal conductivity are solved: for example, using black painting (Section 8.2) and short thermal perturbations (such types of samples are studied further in Section 11.1.2).

We can next evaluate the field of view. In our example we make use of the data of Figures 4.3 and 4.4. For a given 12° lens, without an expansion ring, the manufacturer specifies a minimum distance for focus of about 0.7 m. Suppose that we install the infrared camera at a 1-m distance; this gives us a field of view of about 2×1000 tg 6° ~ 210 mm. If the image size is 105 pixels wide, a 25-mm-wide defect will appear as a blob about 13 pixels wide (probably less, in fact, due to 3D heat flow effects) or, if expressed as an angular width, $25/1000 \sim 25$ mrad. Recalling the SRF curve of Figure 4.3, it appears that such a defect corresponds to a modulation factor of 1 (close to 6 mrad and above is needed for $m = 1$). Consequently, contrast detection for this kind of defect should not cause problems.

For this example a field of view of size about 21 cm is acceptable (of course, larger fields of view can be envisaged). It is, however, important to consider the spatial resolution of the camera as we did here (will the defect contrast be sufficient?) and whether the thermal stimulation device will allow a uniform perturbation to be applied on the entire field of view. In the case of inspection of large flat panels, the inspection can proceed on an area-by-area basis. They are many ways to do this. The straightforward method is to divide the panel surface into 21×21 cm overlapping areas and then inspect them one by one manually using thermographic inspection equipment (e.g., IR camera and associated thermal stimulation device mounted on tripods, with associated electronics and analysis equipment carried on a cart). Another possibility is to mount the acquisition head with the thermal stimulation device on a robotic arm which can be programmed to perform the inspection task on a 21×21 cm area basis (as in Figure 8.26). This method is convenient if many identical panels have to be inspected. Another attractive possibility is to mount the sample on a $x-y-z$ translation stage. This is similar to the robotic arm configuration but with the sample moving instead. The advantage of the robotic arm approach is that larger (curved) surfaces can be inspected, as discussed previously.

Perhaps a faster inspection method is to mount the flat panel on a moving slide with the inspection proceeding on a line fashion (Figure 8.19). A fast inspection rate of up to 1 m^2 s^{-1} can be achieved in this way. This line method can be deployed since defects are always located at the same depth (2 mm, the sheet thickness). This imposes a specific distance between the line heater and the observation area on the surface (Section 9.2.2). It is important to note that this method is impractical if the surface is not flat, in which case other approaches have to be selected (Section 10.6).

8.2 SOLUTIONS TO EMISSIVITY PROBLEMS

The use of TNDT is difficult, if not impossible, for materials of low emissivity. It can be shown that radiation measurement cannot be performed successfully if material emissivity is lower than 20% of blackbody emissivity. This is, for example, the case for materials having highly reflective surfaces, such as bare metal sheets. Bonded aluminum laminate samples (Chapter 7) used in the aeronautic and transport industries are good examples of such materials. Table 8.1 lists common materials and their emissivities.

Polished metal surfaces have a small absorptivity and thus a weak emissivity in the infrared, about 5% of the emissivity of the blackbody (Section 2.2). Consequently, radiance emitted by such a metallic surface is weak and produces only faint infrared images. Worse, the presence of grease patches or slightly oxidized zones can change surface emissivity by factors of 5, 10, or 20% of the blackbody emissivity (an oxidized surface exhibits a higher emissivity; Table 8.1). This yields *apparent hot spots* in the infrared images, hot spots that may be interpreted falsely as damaged zones.

Moreover, high reflectivities of metal surfaces introduce a problem of parasitic reflection when emitted radiations reflected by warm surrounding bodies are reflected on the metal surface, which acts as a mirror in the infrared spectrum. This reflected image is superimposed on the image emitted by the metal surface inspected, creating parasitic hot spots which further complicate thermogram interpretation; eq. (4.14) introduces this question into evidence. An illustration of the problem is shown in Figure 8.30. An infrared camera is pointed directly on a shiny aluminum sheet, and instead of revealing a uniform temperature distribution over the sheet, the image reveals the isotherm distribution of the infrared camera operator, which reflects on the sheet, acting as a mirror at theses wavelengths (3 to 5 μm). Consequently, extreme care is required when TNDT is to be deployed on materials that have reflective surfaces. Notice that in the case of natural outdoor scenes, emissivity is generally high, except for water or ice surfaces observed under grazing incidence ($<40°$).

Another concern is that emissivity varies with viewing angle. Figure 8.31 displays polar emissivity diagrams for common industrial materials. In this figure we notice that emissivity is rather constant for nonmetallic materials for wide variation in the observation angle from normal incidence. For smooth metallic surfaces, emissivity tends to be lower at normal incidence than at grazing incidence. For outside material (organic soil, silt, vegetation, gravel), reports indicate a variation of less than 1% of the emissivity for an angle be-

TABLE 8.1 Emissivity of Common Materials

Material	Temperature (°C)	Emissivity
Metallic Materials		
Aluminum		
Polished	50–100	0.04–0.06
Rough surface	20–50	0.06–0.07
Oxidized alumina	50–500	0.2–0.3
Anodized	100	0.55
Brass		
Polished	100	0.03
Oxidized	200–600	0.60
Bronze		
Polished	50	0.10
Rough	50–150	0.55
Powder	—	0.80
Cast iron		
Molded	50	0.81
Liquid	1300	0.28
Polished	200	0.21
Chromium	50	0.10
Polished	500–1000	0.28–0.38
Copper		
Polished	100	0.03
Oxidized	50	0.70
At fusion temperature	1100–1300	0.13–0.15
Iron		
Rough, nonoxidized	20	0.24
Rusted	20	0.61–0.85
Oxidized	100	0.74
Galvanized	30	0.25
Polished	400–1000	0.14–0.38
Lead		
Gray, oxidized	20	0.28
Shiny	250	0.08
Magnesium		
Polished	20	0.07
Powder	—	0.86
Mercury	0–100	0.09–0.12
Nickel		
Polished	20	0.05
Oxidized	200–600	0.37–0.48
OR		
Polished	100	0.02
Platinum	200–600	0.05–0.10
Polished	1000–1500	0.14–0.18
Silver		
Polished	200–600	0.02–0.03

(Continued)

TABLE 8.1 *(Continued)*

Material	Temperature (°C)	Emissivity
Steel		
Alloy, 8% Ni–18% Cr	500	0.35
Mild, at fusion temperature	1600–1800	0.28
Galvanized	20	0.28
Oxidized	200–600	0.80
Rusted	20	0.69
Polished	100	0.07
Stainless	20–700	0.16–0.45
Buffed	—	0.16
Tin		
Polished	20–50	0.04–0.06
	—	0.40
Tungsten	200	0.05
	600–1000	0.10–0.16
	1500–2200	0.24–0.31
	3300	0.39
Zinc		
Polished	200–300	0.04–0.05
	400	0.11
Oxidized	1000–1200	0.5–0.6
Powder	—	0.82
Sheet	50	0.20
Nonmetallic Materials		
Asphalt	—	0.96
Basalt		
Rough	—	0.93
Blackbodies		
Commercially available		0.99
Carbon		
Filament	1000–1400	0.53
Graphite	20	0.98
Lampblack	20–400	0.95–0.97
Cement		0.54
Clothes, dark	20	0.98
Concrete	20	0.92
Dolomite		
Polished	—	0.93
Rough	—	0.96
Durite	—	0.86
Enamel	—	0.89
Feldspar	—	0.87
Fused silica	20	0.93
Window glass	—	0.94
Glass, polished	20	0.94
	20–100	0.94–0.91
	250–1000	0.87–0.72
	1100–1500	0.7–0.67

TABLE 8.1 *(Continued)*

Material	Temperature (°C)	Emissivity
Granite		
Granite	—	0.82
Rough	—	0.90
Lime	—	0.3–0.4
Marble (gray, polished)	20	0.93
Obsidian	—	0.86
Oil (lubrification)	20	0.82
Paint (oil-based)		
Average, 16 colors	—	0.94
Various colors	100	0.92–0.96
Matt black (3M Velvet)	—	0.98
Paper		
White	20	0.7–0.9
Yellow	—	0.72
Red	—	0.76
Dark blue	—	0.84
Green	—	0.85
Black	—	0.9
Matt	—	0.93
Plaster	20	0.91
Porcelain		
Glassy	20	0.92
Shiny white	—	0.7–0.75
Red brick	20	0.93
Brick	—	0.95
Rubber		
Hard	20	0.95
Soft	20	0.86
Sand	20	0.6–0.9
Quartz sand	—	0.91
Silica sandstone	—	0.91
Skin (human)	32	0.98
Soil		
Dry	20	0.9
Wet	20	0.95
Stucco	—	0.93
Talc		
Powder	—	0.24
Tanned leather	—	0.75–0.80
Tar	—	0.79–0.84
Terra cotta	70	0.91
Water		
Water	—	0.99
Distilled	20	0.96
Smooth ice	−10	0.95
Frost	−10	0.98

(Continued)

TABLE 8.1 *(Continued)*

Material	Temperature (°C)	Emissivity
Snow	−10	0.85
With thin petroleum film	—	0.97
Wood		
Tree	—	0.5–0.7
Plank	20	0.8–0.9
Wood	—	0.95
Wood coal	—	0.96

Source: Data from Gaussorgures (1989, p. 59) (metallic materials) and Colwell (1983, Vol. 2, p. 1583).

tween 10 and 53°; for sand, the change is up to 4% (band at 8 to 10 μm) and 2% (bands at 3 to 5 and 10 to 14 μm) (Hook et al., 1998).

8.2.1 Blackpainting

A common approach to solving the emissivity problem consists of covering metallic surfaces with high-emissivity paint before inspection. Such *black* paints of uniform and high emissivity (ε: 0.9 to 0.98, Table 8.1) in the infrared bands have a low reflectivity [eq. (2.28)]. Consequently, if it is applied on low-

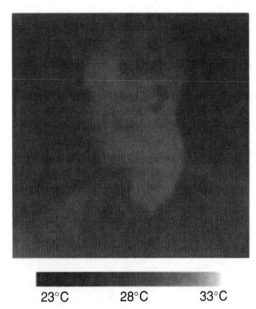

23°C   28°C   33°C

FIGURE 8.30 The problem of thermal reflections: The thermal image of the operator standing behind the IR camera is reflected on the aluminum sheet, which acts as a mirror. The IR camera points at the sheet (notice how the image is distorted).

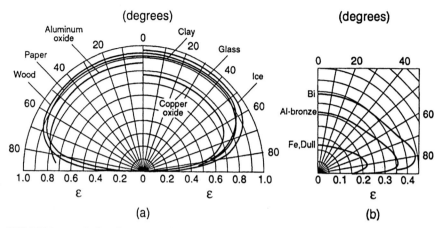

FIGURE 8.31 Polar diagrams of the infrared emissivity of a number of (a) nonmetallic and (b) metallic materials. (From Maldague, 1993.)

emissivity surfaces, interferences caused by neighboring warm bodies are reduced and thermographic inspection becomes possible. The need to proceed to a painting step prior to inspection and to a paint-removing step after inspection is not, however, attractive for an NDE technique which claims that a fast inspection pace is one of its strongest advantages with respect to other NDE techniques (Section 1.4).

Moreover, this practice of "blackpainting" components to increase surface emissivity up to an acceptable level prior to TNDT inspection can also modify radiometric readings. For example, if such painting is applied in order to proceed to passive inspection of PCBs, temperature distribution over the components inspected can be altered due to the enhanced heat dissipation that such paintings cause (see heat sink computation, Chapter 3). Emissivity depends on the paint coating thickness. Tests showed that if three or more paint applications are deposited on the surface, the emissivity becomes constant.

Interestingly, the paint required to increase the emissivity does not need to be "black" (in the visible spectrum). Following Table 8.1, we see that various colors of oil-based paint have a high emissivity (an average of 16 paint colors gives a value of 0.94). In TNDT of low-emissivity surfaces, it is thus possible to paint a surface permanently with such a paint, and since the color of the paint selected is attractive, it does not require removal.

8.2.2 Techniques Applied to Solve the Emissivity Problem

Besides blackpainting, different techniques have been reported to solve the emissivity problem (see, e.g., Ono, p. 565 in DeWitt and Nutter, 1988). An original technique well suited for small specimens whose emissivity is unknown or uneven over object surfaces was suggested by Vleck (cited by Öhman, 1981). His technique consists in keeping the specimen between clasped hands long

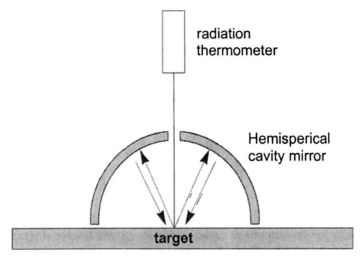

FIGURE 8.32 Hemisperical mirror cavity used to increase the apparent emissivity of the target through summation of multiple reflections.

enough to ensure that the specimen reaches skin temperature (this can be checked with a small temperature probe inserted between the sample and one of the palms). Once specimen temperature has stabilized, the hands are opened, and both palms and specimen are immediately snapshotted with the infrared equipment. Since skin emissivity is high ($\varepsilon \sim 0.98$), emissivity distribution over the specimen surface can be determined by comparing temperature distributions recorded over both the palms and the specimen, which due to this primitive thermal transfer imaging technique are the same (if the exposure proceeds quickly enough). Of course, object emissivity must be sufficiently high to limit the problem of thermal reflection.

Another possible approach to solving emissivity problems is to use a reflecting cavity. Hemispherical reflectors positioned in contact with the surface of interest are well known to create near-blackbody conditions. For example, if a 5-cm-diameter hemispherical mirror of 0.95 reflectivity at a temperature of 350 K covers a lambertian surface at 500 K with an emissivity of 0.3, the apparent emissivity increases up to 0.83 at 2.5 μm and up to 0.87 at 10 μm (for a 1-cm-diameter aperture in the mirror, a nearly monochromatic radiation thermometer, with a surrounding temperature of 300 K, Ono, p. 579 in DeWitt and Nutter, 1988; Figure 8.32). However, the performance of such devices drops rapidly as the gap between the surface and the hemisphere increases. This technique is appropriate when it is necessary to obtain an average temperature reading above the surface. It is, however, not suitable for thermal imagery, since multiple reflections tend to average the radiation emitted by the surface covered by the cavity. Such an approach is, however, appropriate when it is needed to measure temperature over, say, a moving surface such as in contin-

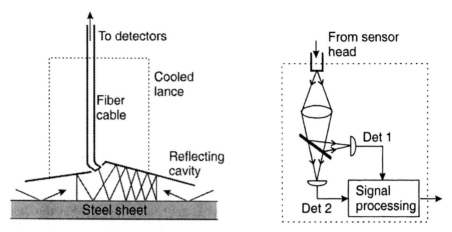

FIGURE 8.33 Schematic diagram of the conical-fiber sensor head (left) and detection module based on two-color pyrometry. (From Cielo et al., 1992.)

uous casting and when it is acceptable to obtain an average reading over the sensor head (Figure 8.33; Cielo et al., 1988a, 1992; L. Chen et al., 1990; Krapez et al., 1990).

Other techniques are possible, but they require a more complex experimental apparatus. For example, it is possible to measure the surface reflectivity directly using a laser reflectometer prior to take temperature readings (D. R. Green, 1968; R. G. Peacock, 1988). Other techniques involve two temperature sensors. For example, in the steel industry, to determine oven temperature, one sensor is pointed at the steel surface while the other is pointed at the oven wall in order to take into account spurious thermal reflections (R. B. Johnson et al., 1988; Ramelot et al., 1988). Notice that these methods are not useful if contributions from spurious reflections are more important than those of interest. Another drawback is that theses techniques generally have a limited spatial resolution (point measurement only).

8.2.3 Multiwavelength Pyrometry

The techniques of multiwavelength pyrometric are other popular approaches to this problem. However, these techniques are generally restricted to high temperature. Multiwavelength pyrometry is based on the assumption that a simple relationship between emissivity and wavelength exists so that it becomes possible to calculate the temperature from measurements done at different wavelengths (it is like solving a set of N equations having N unknowns, with temperature and emissivity being unknowns). For two-color pyrometers, temperature is computed from the ratio R of the two signals recorded at both wavelengths, since for a given set of wavelengths λ_1 and λ_2 and temperature T, the spectral radiance ratio is unique. In the case of two-color pyrometry, the hypothesis is that the emissivity ratio is constant over the wavelength span of interest.

Experimental measurement of the signal ratio R corresponds to a temperature value $T = f(R)$ after proper calibration, where $f(\cdot)$ is the calibration polynomial. The assumption of a constant emissivity ratio holds for *gray bodies* (for which emissivity is constant with wavelength) and to some extent for *colored bodies* (for which emissivity varies with wavelength); consequently, the calibration function $f(\cdot)$ of the pyrometer is specific for a given material. Temperature computation for two-color pyrometer was reviewed in Section 4.8.3.

8.2.4 Thermal Transfer Imaging

We now review some techniques that solve this emissivity problem through thermal transfer imaging. The principle of thermal transfer imaging is based on a high-emissivity material brought into contact with the structure inspected. Since the thermal observation is performed on the high-emissivity material which picks up the temperature distribution of the low-emissivity material with which it is in contact, TNDT inspection becomes possible. Such a technique is suitable for low-temperature inspection and flat surface components such as aluminum sandwich panels (Figure 7.3).

Using this transfer technique, the emissivity problem is solved by placing along the inspection line a high-emissivity freely rotating flexible roller. The roller is positioned such that it is continuously in good thermal contact with the (low-emissivity) surface. The thermal front propagates from the surface to the roller, so that after the contact, the temperature distribution of the roller surface reproduces the temperature distribution of the surface. This transferred temperature distribution can be observed by pointing the infrared camera directly on the flexible roller surface, as shown in Figure 8.34. Since a roller

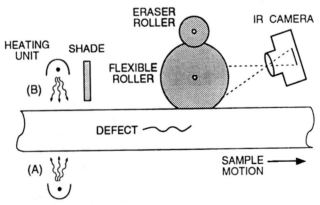

FIGURE 8.34 Thermal transfer imaging using a compliant roller and shown in (A) a transmissive and (B) a reflective configuration for the detection of defects of known depth in a laminar specimen. (From Maldague, 1993.)

surface is of high emissivity ε and thus of low reflectivity ρ, because of the link between those parameters (Section 2.9.5: $\varepsilon = 1 - \rho$), the thermal image picked up on the roller has a relatively good thermal contrast and is not affected by parasitic reflections.

The flexible roller can be made of polymer foam painted with high-emissivity paint: The large number of gas bubbles ensures good contact and low thermal conductivity. In the region where the roller is in contact with the surface inspected, the gap bubbles are collapsed due to the pressure: This ensures good transfer of the thermal front. After the contact, the bubbles are reshaped back to their initial volume. Since foam thermal conductivity is weak, the image transferred, observed with the infrared camera, offers a good thermal contrast.

An auxiliary "eraser" roller, preferably made of high-thermal-conductivity metal and cooled by means of internal water circulation or by Peltier elements is also brought into contact with the flexible roller. It acts as an eraser for the thermal image transferred. This approach is suitable for the pulsed thermography approach only (Chapter 9) since signal pickup is possible only at a given thermal state of the specimen and since the thermal stimulation might also be challenging: Delivering a thermal pulse is achievable, but not more! The thermal perturbation can be deployed either in reflection (inspection of thick components) or in transmission (inspection of thin components). Since the emissivity of the surface inspected is small, instead of using a radiative type of thermal perturbation source (such as a line heating), a contact thermal stimulation source such as water or air jets should be preferred. Notice, however, that if a line heating is deployed in conjunction with a back reflector, due to multiple thermal reflections between the parabolic reflector and the surface inspected, the effective emissivity increases, thus making possible (under some instances) energy deposit on the surface.

Alternatively, in the case of shallow defects, the thermal perturbation unit can be suppressed by maintaining the auxiliary eraser roller at a sufficiently different temperature with respect to the surface. In this case, the thermal perturbation is introduced on the surface during contact with the flexible roller, and the response to this perturbation is collected immediately by the roller. Notice also that the infrared camera can be replaced by an infrared line scanner.

Another configuration based on the principle of the setup of Figure 8.34 is shown in Figure 8.35. In this case a closed-loop membrane connects both rollers. The encumbrance of the transfer head is reduced. This makes it possible to collect short thermal contrasts which would be lost if thermal transfer occured later (case of shallow subsurface flaws or inspected material with high thermal conductivity). An additional benefit of this configuration is the ease with which the infrared image can be observed and subsequently erased due to the membrane length.

For the configurations shown in Figures 8.34 and 8.35, subsurface defects must have a depth on the order of the thermal propagation length corresponding to the period of time between injection of the perturbation and transfer of

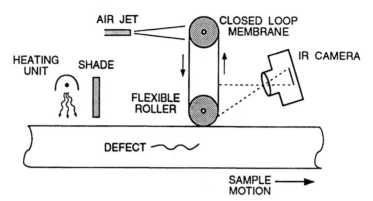

FIGURE 8.35   Thermal transfer imaging using the same basic principles as the configuration of Figure 8.34 and shown for the reflection approach with a closed-loop membrane for reduced head circumference and more convenient erasure of the images transferred. (From Maldague, 1993.)

the temperature distribution by the roller. Consequently, the distance between the thermal source and the roller must be selected according to the depths of the possible defects (e.g., the thickness of the bonded sheet in sandwich assemblies; Section 7.4.1).

For specimens having a wide range of possible defect depths, the modified configuration of Figure 8.36 can be deployed. In this case, a high-emissivity membrane made of a thin film is brought into contact with the panel surface by means of stretching rollers. Tiny water sprays are used to produce, by capillar-

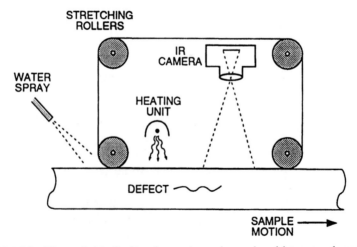

FIGURE 8.36   Thermal transfer imaging system using a closed-loop membrane applied on the workpiece surface for the detection of defects of unknown depth. (From Maldague, 1993.)

ity, a thin film of liquid between the membrane and the surface in order to have good thermal contact and to maintain the membrane in place. Pulsed thermography is deployed following methods discussed in Chapters 1 and 9. The thermal front is eventually affected by the presence of subsurface flaws, thus causing hot spots or cold spots (depending on the nature of the thermal perturbation) to appear on the membrane surface observed by the infrared camera. Modifications of the configuration of Figure 8.36 can be introduced. For example, the thermal perturbation can be applied by the water sprays or by the first stretching roller if a sufficiently large temperature differential is maintained between this roller and the surface.

For unplanar surfaces, it is possible to modify the configuration of Figure 8.36. A high-emissivity flexible membrane can be forced to follow the specimen contour either by pressing it on the surface (a thin film of water helps to maintain the membrane in place) or by mean of a vacuum sealing machine. Following this operation, TNDT inspection can proceed as if the surface would have been blackpainted using techniques described in Chapters 1 and 9.

Finally, notice that thermal contrasts observed by these transfer methods are smaller than contrasts obtained by direct observation (using, e.g., blackpainting; see the next section for details).

Physical Behavior. It is possible to simulate the behavior of the thermal transfer process numerically in order to better understand the underlying physical principles. An adapted version of the model presented in Section 3.4.2 can be used for this purpose. In this section we restrict this discussion to some meaningful results.

The thermal process can be divided into three phases (Figure 8.37). In the first phase the surface of the sample is heated and cools down. In the second phase, from time t_1 to t_2, the flexible roller is compressed and is in contact with the surface inspected. In the third phase, at time t_2, both materials are separated and the flexible roller reestablished. Notice than instead of using a hot thermal perturbation source, cold thermal perturbation can be used. In this case the sample surface starts to warm just after the cold thermal perturbation is over.

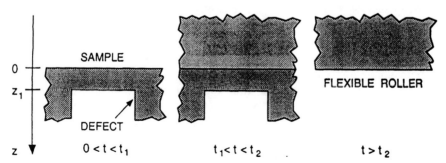

FIGURE 8.37 The three phases of the thermal transfer process. (From Maldague, 1993.)

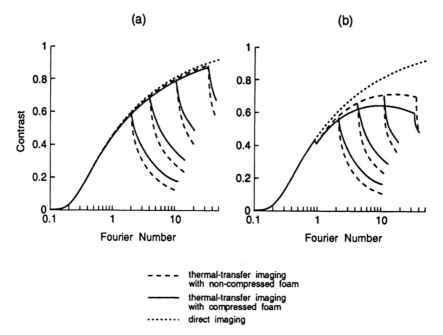

FIGURE 8.38 Contrasts computed with different imaging methods for materials of (a) high and (b) low thermal conductivity. (From Maldague, 1993.)

Figure 8.38 presents some interesting results. Notice that the normalized time is expressed on the abscissa in Fourier numbers [eq. (10.11)], which makes it possible to derive more general results. On the figure, the second phase starts at Fourier number $F_{phase2} = 1$, this time corresponds to a thermal propagation time of the thermal front to the depth z_{def} of the defect on the order of $t = z_{def}^2/\alpha$. The various results considered in Figure 8.38 are for various durations of the second phase (period of contact foam specimen), $F_{phase2} = 1$, 3.2, 10, and 32 with $F_{phase3} = 10$. Two types of materials are investigated: those of high thermal conductivity such as aluminum and those of low thermal conductivity such as graphite epoxy. The contrast obtained by observing the surface directly is drawn with a dotted line (in this case the thermal model assumes that the surface is covered with high-emissivity paint prior to inspection). Other cases shown correspond to thermal transfer imaging observation in the case of both uncompressed and compressed foam. Notice that the results of this modeling are less valid for longer time values (Fourier number >10) since the model used to draw Figure 8.38 is unidimensional and does not consider thermal losses. This explains why, in this figure, the contrast increases steadily. In a real situation, the contrast in direct observation would reach its maximum around Fo ∼ 1 to 10 before decreasing because of thermal losses (e.g., Figure 10.10).

Nevertheless, study of Figure 8.38 reveals interesting aspects: Thermal transfer imaging is less efficient than direct observation in all but in the case of low-emissivity surfaces. In this case, images must be transferred as soon as possible after separation since the temperature then drops abruptly. Consequently, the configuration of Figure 8.35 is more interesting than the configuration of Figure 8.34 because of its reduced head encumbrance. Uncompressed foam also seems to give higher contrasts, but this advantage disappears rapidly after separation. In fact, at longer times, more suitable for acquisition of the image sequence, compressed foam is more advantageous since it offers higher contrasts.

Experimental Results. We now illustrate the operation principle of the thermal transfer imaging technique by a few experimental examples. Materials of high and low thermal conductivity are tested. First, an aluminum–foam bonded panel is tested; this specimen is similar to specimens studied in Chapter 7. In the case of this component, it is decided to detect an artificial disbonding defect, a lack of adhesive over a 4-cm-diameter circular area, 1 mm below the surface (Al sheet thickness = 1 mm). A Plexiglas plate in which a hole is drilled under the surface is also studied. Samples are inspected in reflection with a 1600-W line heater.

Figure 8.39 shows the experimental apparatus with a flexible foam roller. Figure 8.40 presents a few results obtained on both the aluminum panel and the

FIGURE 8.39 Experimental apparatus using the configuration of Figure 8.34: the IR camera, flexible blackpainted roller, and lamp heater, as well as the aluminum laminate mounted on a slide (extreme right) are visible. (From Maldague, 1993.)

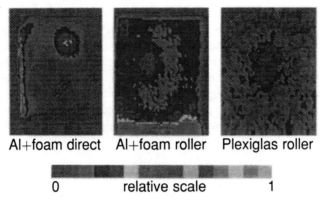

Al+foam direct   Al+foam roller   Plexiglas roller

0   relative scale   1

FIGURE 8.40   Experimental results for the aluminum panel and Plexiglas plate (see the text). (From Maldague, 1993.)

Plexiglas plate (hole: 1.6 mm depth, 1 cm in diameter). Three images are shown; on the left the blackpainted aluminum panel is tested in reflection with direct observation (camera pointed directly at the painted surface) while the panel mounted on a slide moves at about 10 cm/s in the field of view. Due to the high emissivity of the paint ($\varepsilon \sim 0.9$), the disbonding defect is easily visible in the middle of the thermogram. A test performed under the same conditions, on the same unpainted panel, did not reveal the defect because of the low emissivity of the surface.

Figure 8.40 (middle) shows the same panel, but observed using the thermal transfer imaging setup of Figure 8.39, with the camera pointed at the surface of the flexible foam roller. Although the thermal contrast is lower than for direct observation, the defect is still visible. Notice that the aluminum surface was blackpainted to stimulate the specimen sufficiently with the heat source, which is of radiative type. The hotter band at the bottom of the thermogram corresponds to observation of the blackpainted panel surface, hotter than the roller and observed by the camera.

The right image of Figure 8.40 shows a thermogram obtained on the Plexiglas plate. The image transferred reveals clearly the presence of the subsurface hole (infrared camera pointed at the roller surface). The plate surface was blackpainted in order to absorb enough energy from the radiative stimulation device. For low-emissivity surfaces, radiative heating is not very efficient. Contact thermal stimulation devices using either water or air jets are more appropriate, as discussed in Section 8.1.

Figure 8.41 shows the experimental apparatus for the configuration of Figure 8.35, in which a rubber membrane 1.5 mm thick is used to catch the thermal images. Figure 8.42 shows a reconstructed image (using techniques described in Section 6.2). The total encumbrance of the inspection head is on the order of 10 cm; this is much less than the 30 or so centimeters of the setup shown in Figure 8.39. This space saving allows us to catch thermal contrasts

FIGURE 8.41 Experimental apparatus using the configuration of Figure 8.35: the IR camera, flexible blackpainted roller, closed-loop membrane, and lamp heater are visible as well as an aluminum laminate with foam core mounted on a slide. Notice the unpainted aluminum surface being inspected. (From Maldague, 1993.)

earlier, thus making it possible to observe subsurface bonding defects on this shiny aluminum panel. In fact, the flexible rollers of large diameter catch thermal contrasts later, at a moment when defect visibility is already reduced (Figure 10.10). A pressurized air curtain is used to erase thermal imprints on the membrane (box visible on the left of Figure 8.41).

Figure 8.43 shows results obtained for the configuration shown in Figure 8.36. A thin high-emissivity membrane is brought into contact with the high-reflectivity aluminum surface to increase its emissivity. For the test of Figure 8.43, a 50-μm film of black polyurethane is brought into contact through capillarity with the metal surface, as explained previously. The segmented image (Section 6.3) is shown in Figure 8.43a. Figure 8.43b presents a segmented image

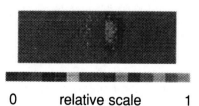

0   relative scale   1

FIGURE 8.42 Reconstructed image of a defect (see the text). (From Maldague, 1993.)

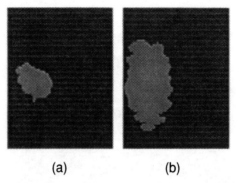

(a)   (b)

FIGURE 8.43   (a) Segmented image of a defect obtained with thermal transfer imaging apparatus of Figure 8.36; (b) segmented image of the same defect obtained in direct observation with blackpainted sample surface and no membrane. Same sample as for Figure 8.41. Uncorrected size of field of view along rows and columns. (From Maldague, 1993.)

of the same bonding defect but in direct observation (with the aluminum surface blackpainted and without using the polyurethane film). Comparison of the two images reveals a loss of contrast [in (a)]; the defect is, however, visible in both cases. For this configuration, since all the temporal evolution of temperature recorded above the defect is available, characterization techniques described in Chapter 10 can be employed (with adjustments to compensate for the loss of contrast). This is not the case for the other two configurations, where observation is done only once.

8.3   ATMOSPHERE

Atmosphere is a complex mix of various gases, particles, and aerosols in different concentrations. A description of the transmittance of electromagnetic energy through this mix is thus a complicated task. In this section we describe briefly the processes involved; interested readers can consult Gaussorgues (1989) and Rencz (1998), among others, for more information on this topic.

The atmosphere has a few annoying characteristics with regard to TNDT. The main problem concerns the transmittance of electromagnetic energy, which is less than 100%. Self-absorption by gas making up the atmosphere (e.g., H_2O, CO_2, O_3) and diffusion due to particles such as molecules and aerosols are responsible for this. Moreover, various factors, such as the presence of thermal gradients and turbulences, make the index of refraction inhomogeneous and contribute further to degradation of the IR measurement through the atmosphere. Finally, the atmosphere itself emits IR energy which is captured by the IR sensor. Most of these effects depend on the distance, wavelength, and meteorological conditions; they are thus not easily taken into account.

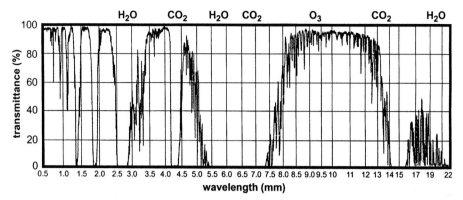

FIGURE 8.44 Atmospheric transmittance for the 1976 U.S. standard atmosphere, 15°C, 5.9 mm of precipitable water, 46% relative humidity, 101.3 kPa atmospheric pressure at sea level for an horizontal path of 1 km. Also shown are the main gases causing absorption in the infrared bands.

In order of importance, main gases that absorb the radiation transmitted are first water vapor (H_2O), then carbon dioxide (CO_2) and ozone (O_3). In combination with the other gases (e.g., CO, O_2) they make up the following transmission windows (in μm): 0.4 to 1 (this includes the visible spectrum), 1.2 to 1.3, 1.5 to 1.8, 2.1 to 2.5, 3 to 5, and 8 to 13 (Figure 8.44). From these we recognize the two main bands of interest for TNDT, discussed in Section 4.7. Following Planck's law (Figure 2.6), the longwave band (8 to 13 μm) is of particular interest for measuring radiation from objects at room temperature (an example is for the detection of intruders by law enforcement agencies) The shortwave band (3 to 5 μm) is best suited for warmer objects (an example is any process releasing CO_2, such as combustion engines, e.g., target tracking by the military). Absorption by gases is a complex phenomenon and, due to the scale, Figure 8.44 does not reveal the imbricate pattern of individual absorption lines.

It is, nevertheless, possible to describe the absorption mathematically. One way is through the exponential absorption law, which stands in the case of a given monochromatic wavelength (it can also be used in the case of spectral bands if absorption does not depend on the wavelength. It is expressed by

$$F = F_0 \exp(-k_F x) \tag{8.2}$$

where k_F is the medium absorption coefficient which depends on the wavelength λ and x is the distance. At null distance, $x = 0$ and thus $F = F_0$ (this can be useful for calibration purposes). The coefficient of transmittance Tr is then expressed by

$$\mathrm{Tr} = \frac{F}{F_0} = \exp(-k_F x) \tag{8.3}$$

Sometimes, it is preferred to characterize the medium in term of optical density D, which is obtained as

$$D = \log \frac{1}{\text{Tr}} = -\log[\exp(-k_F x)] = 0.43 k_F x \tag{8.4}$$

As mentioned before, diffusion of the radiation by particles is another concern. This process changes the spatial distribution of the transmitted energy, whose intensity is also affected. Large particles (with respect to the wavelength of interest) do not contribute significatively to the diffusion process, whereas small ones, those whose size is similar to the wavelength of interest, affect the diffusion following the theory of diffraction. Interestingly, the diffusion process of solar radiation by particles explains why the sky is blue at noon (sun at the zenith) and turns red at dusk (sun at the horizon).

In the presence of humidity, particles (around 0.5 μm) suspended in the air agglomerate water molecules. This creates mist, which turns to fog as water continues to agglomerate, making droplets or ice crystals (in this case the diameter rises to a few micrometers). Finally, when droplet size reaches about 0.25 mm, the droplets become too heavy and it rains. Water vapor absorption occurs particularly at 2.6 μm in the range 5.5 to 7.5 μm and above 20 μm.

As the sun heats up atmospheric layers, turbulences are created by air convection due to gap densities, which are inversely proportional to gas temperature. These turbulences affect the refraction index n_{air} of the atmosphere following Gladstone's law, expressed as

$$n_{\text{air}} = 1 + k_n \rho_{\text{air}} \tag{8.5}$$

where k_n is a constant and ρ_{air} is the air mass density. Normally, $n_{\text{air}} = 1$ as an average. It is difficult to take turbulence into account, since air mass density is affected by local temperature variations but also by winds, humidity, and other factors. In general, computation of atmospheric transmittance assumes that turbulence is homogeneous and isotropic. In the presence of turbulence, radiation is transmitted but the propagation path no longer follows a straight line. Various phenomena are observed. In the lower atmosphere, due to the stratification of air layers, radiation rays curve (this explains mirages, for example). Points are also displaced with respect to their normal positions on the image plane, causing images to fluctuate. This is due to rapid displacement of large zones (with respect to rays transmitted) with inhomogeneous n_{air} values (if these zones are small, the image scintillates, i.e., spots appear). Finally, defocalization of images and degradation of spatial cohesion are also observed.

It is not an easy task to take into account all these effects, especially since one has first to measure the parameters involved, such as the wind speed, temperatures along the path of interest, and the concentration of various gases (primarily H_2O, CO_2, and ozone). Moreover, these parameters are not constant in time, due to meteorological conditions and solar heating through the day. In fact, it is probable that measurements turn obsolete as soon as they become available! For these reasons, models of the atmosphere have been developed.

A variety of radiative transfer models have been developed. The discrepancies among models vary from up to about 10%, depending on, for example, cloud optical thickness. Among common models are the band model Modtran and the line-by-line radiative transfer model FASCOD3P (Fast Atmospheric Signature Code). These were originally developed by the U.S. Air Force (USAF) Phillips Laboratory, Hanscom AFB, Massachusetts. Modtran 3.7 evolves from Lowtran, whose Fortran code has been patented by the USAF (Lowtran-6 is available). Commercial and freeware versions of these programs are available from various vendors.* For example, PcLnTran/PcLnWin for Windows is a commercial PC version of FASCOD3P that predicts atmospheric transmittance and radiance at high (line-by-line) spectral resolution. An on-line version of Modtran 3 is also available from the University of Chicago, Department of Geophysical Studies.[†] In the IR, the situation is more complex than in the shortwave range because of assumptions made regarding line shape and line cutoff, for example.[‡]

8.3.1 Simple Computation of Atmospheric Transmittance

We now present a simple method for computations of atmospheric transmittance which is fully detailed in Gaussorgues (1989). The method has a few inputs: air temperature, relative humidity, visibility, distance of interest. Although limited, this method enables us, nevertheless, to study various scenarios.

The global transmittance Tr_λ is expressed by the following formula (for a given wavelength λ and distance x):

$$Tr_\lambda = Tr_{\lambda, H_2O} Tr_{\lambda, CO_2} Tr_{\lambda, d} Tr_{rain} \qquad (8.6)$$

where Tr_{λ, H_2O} is the transmittance due to water vapor in the atmosphere, Tr_{λ, CO_2} that due to carbon dioxide, and $Tr_{\lambda, d}$ that due to particle diffusion; Tr_{rain} takes into account the effect of rain, if any. Tr_{λ, H_2O} are Tr_{λ, CO_2} are obtained from Passman and Larmore tables (Tables 8.2 and 8.3). Water vapor in the atmosphere is expressed as the height h_w of precipitable water vapor expressed in millimeters as seen in Figure 8.45 for different temperatures and

---

*See, for example, the following Internet references:

www.ontar.com.

www.aer.com/groups/rs/fase.html.

www.members.bellatlantic.net/~smrogers/index.html.

www.lidar.ssec.wisc.edu/papers/dhd_thes/thesis.htm.

[†] www.geosci.uchicago.edu/~archer/cgimodels/radiation.html.
[‡] www.arm.gov/docs/documents/project/er_0585/gautier/radiative_2.html.

TABLE 8.2 Passman and Larmore Tables for H$_2$O: Spectral Transmission for Water Vapor (Horizontal Path, Sea Level)

Wavelength (µm)	Precipitable Water (mm)												
	0.10	0.20	0.50	1	2	5	10	20	50	100	200	500	1000
0.3	0.980	0.972	0.955	0.937	0.911	0.880	0.802	0.723	0.574	0.428	0.263	0.076	0.012
0.4	0.980	0.972	0.955	0.937	0.911	0.860	0.802	0.722	0.574	0.428	0.263	0.076	0.012
0.5	0.986	0.980	0.968	0.956	0.937	0.901	0.861	0.804	0.695	0.579	0.433	0.215	0.079
0.6	0.990	0.986	0.977	0.968	0.955	0.929	0.900	0.860	0.779	0.692	0.575	0.375	0.210
0.7	0.991	0.987	0.980	0.972	0.960	0.937	0.910	0.873	0.800	0.722	0.615	0.425	0.260
0.8	0.989	0.984	0.975	0.965	0.950	0.922	0.891	0.845	0.758	0.663	0.539	0.330	0.168
0.9	0.965	0.951	0.922	0.890	0.844	0.757	0.661	0.535	0.326	0.165	0.050	0.002	0
1.0	0.990	0.986	0.977	0.968	0.955	0.929	0.900	0.860	0.779	0.692	0.575	0.375	0.210
1.1	0.970	0.958	0.932	0.905	0.866	0.790	0.707	0.595	0.406	0.235	0.093	0.008	0
1.2	0.980	0.972	0.955	0.937	0.911	0.860	0.802	0.723	0.574	0.428	0.263	0.076	0.012
1.3	0.726	0.611	0.432	0.268	0.116	0.013	0	0	0	0	0	0	0
1.4	0.930	0.216	0.844	0.782	0.695	0.536	0.381	0.216	0.064	0.005	0	0	0
1.5	0.997	0.994	0.991	0.988	0.982	0.972	0.960	0.944	0.911	0.874	0.823	0.724	0.616
1.6	0.998	0.997	0.996	0.994	0.991	0.986	0.980	0.972	0.956	0.937	0.911	0.860	0.802
1.7	0.998	0.997	0.996	0.994	0.991	0.986	0.980	0.972	0.956	0.937	0.911	0.860	0.802
1.8	0.792	0.707	0.555	0.406	0.239	0.062	0.008	0	0	0	0	0	0
1.9	0.960	0.943	0.911	0.874	0.822	0.723	0.617	0.478	0.262	0.113	0.024	0	0
2.0	0.985	0.979	0.966	0.953	0.933	0.894	0.851	0.790	0.674	0.552	0.401	0.184	0.006
2.1	0.997	0.994	0.991	0.988	0.982	0.972	0.960	0.944	0.911	0.874	0.823	0.724	0.616
2.2	0.998	0.997	0.996	0.994	0.991	0.986	0.980	0.972	0.956	0.937	0.911	0.860	0.802
2.3	0.997	0.994	0.991	0.988	0.982	0.972	0.960	0.944	0.911	0.874	0.823	0.724	0.616
2.4	0.980	0.972	0.955	0.937	0.911	0.860	0.802	0.723	0.574	0.428	0.263	0.076	0.012
2.5	0.930	0.902	0.844	0.782	0.695	0.536	0.381	0.216	0.064	0.005	0	0	0
2.6	0.617	0.479	0.261	0.110	0.002	0	0	0	0	0	0	0	0
2.7	0.361	0.196	0.040	0.004	0	0	0	0	0	0	0	0	0
2.8	0.453	0.289	0.092	0.017	0.001	0	0	0	0	0	0	0	0

2.9	0.689	0.571	0.369	0.205	0.073	0.005	0	0	0	0	0	0	0
3.0	0.851	0.790	0.673	0.552	0.401	0.184	0.060	0.008	0.	0	0	0	0
3.1	0.900	0.860	0.779	0.692	0.574	0.375	0.210	0.076	0.005	0	0	0	0
3.2	0.925	0.894	0.833	0.766	0.674	0.506	0.347	0.184	0.035	0.003	0	0	0
3.3	0.950	0.930	0.888	0.843	0.779	0.658	0.531	0.377	0.161	0.048	0.005	0	0
3.4	0.973	0.962	0.939	0.914	0.880	0.811	0.735	0.633	0.448	0.285	0.130	0.017	0.001
3.5	0.988	0.983	0.973	0.962	0.946	0.915	0.881	0.832	0.736	0.635	0.502	0.287	0.133
3.6	0.994	0.992	0.987	0.982	0.973	0.958	0.947	0.916	0.866	0.812	0.738	0.596	0.452
3.7	0.997	0.994	0.991	0.988	0.982	0.972	0.960	0.944	0.911	0.874	0.823	0.724	0.616
3.8	0.998	0.997	0.995	0.994	0.991	0.986	0.980	0.972	0.956	0.937	0.911	0.860	0.802
3.9	0.998	0.997	0.995	0.994	0.991	0.986	0.980	0.972	0.956	0.937	0.911	0.860	0.802
4.0	0.997	0.995	0.993	0.990	0.987	0.977	0.970	0.960	0.930	0.900	0.870	0.790	0.700
4.1	0.977	0.994	9.991	0.988	0.982	0.972	0.960	0.944	0.911	0.874	0.823	0.724	0.616
4.2	0.994	0.992	0.987	0.982	0.973	0.958	0.947	0.916	0.866	0.812	0.738	0.596	0.452
4.3	0.991	0.984	0.975	0.972	0.950	0.937	0.910	0.873	0.800	0.722	0.615	0.425	0.260
4.4	0.980	0.972	0.955	0.937	0.911	0.860	0.802	0.723	0.574	0.428	0.263	0.076	0.012
4.5	0.970	0.958	0.932	0.905	0.866	0.790	0.707	0.595	0.400	0.235	0.093	0.008	0
4.6	0.960	0.943	0.911	0.874	0.822	0.723	0.617	0.478	0.262	0.113	0.024	0	0
4.7	0.950	0.930	0.888	0.843	0.779	0.658	0.531	0.377	0.161	0.048	0.005	0	0
4.8	0.940	0.915	0.866	0.812	0.736	0.595	0.452	0.289	0.117	0.018	0.001	0	0
4.9	0.930	0.902	0.844	0.782	0.695	0.536	0.381	0.216	0.064	0.005	0	0	0
5.0	0.915	0.880	0.811	0.736	0.634	0.451	0.286	0.132	0.017	0	0	0	0
5.1	0.885	0.839	0.747	0.649	0.519	0.308	0.149	0.041	0.001	0	0	0	0
5.2	0.846	0.784	0.664	0.539	0.385	0.169	0.052	0.006	0	0	0	0	0
5.3	0.792	0.707	0.555	0.406	0.239	0.062	0.008	0	0	0	0	0	0
5.4	0.726	0.611	0.432	0.268	0.116	0.013	0	0	0	0	0	0	0
5.5	0.617	0.479	0.261	0.010	0.035	0	0	0	0	0	0	0	0
5.6	0.491	0.331	0.121	0.029	0.002	0	0	0	0	0	0	0	0
5.7	0.361	0.196	0.040	0.004	0	0	0	0	0	0	0	0	0
5.8	0.141	0.044	0.001	0	0	0	0	0	0	0	0	0	0
5.9	0.141	0.044	0.001	0	0	0	0	0	0	0	0	0	0

(Continued)

TABLE 8.2 *(Continued)*

Wavelength (μm)	Precipitable Water (mm)												
	0.10	0.20	0.50	1	2	5	10	20	50	100	200	500	1000
6.0	0.180	0.058	0.003	0	0	0	0	0	0	0	0	0	0
6.1	0.260	0.112	0.012	0	0	0	0	0	0	0	0	0	0
6.2	0.652	0.524	0.313	0.153	0.043	0.001	0	0	0	0	0	0	0
6.3	0.552	0.401	0.182	0.060	0.008	0	0	0	0	0	0	0	0
6.4	0.317	0.157	0.025	0.002	0	0	0	0	0	0	0	0	0
6.5	0.164	0.049	0.002	0	0	0	0	0	0	0	0	0	0
6.6	0.138	0.042	0.001	0	0	0	0	0	0	0	0	0	0
6.7	0.322	0.162	0.037	0.002	0	0	0	0	0	0	0	0	0
6.8	0.361	0.196	0.040	0.004	0	0	0	0	0	0	0	0	0
6.9	0.416	0.250	0.068	0.010	0	0	0	0	0	0	0	0	0
7.0		0.569	0.245	0.060	0.004	0	0	0	0	0	0		
7.1		0.716	0.433	0.188	0.035	0	0	0	0	0	0		
7.2		0.782	0.540	0.292	0.085	0.002	0	0	0	0	0		
7.3		0.849	0.664	0.441	0.194	0.017	0	0	0	0	0		
7.4		0.922	0.817	0.666	0.444	0.132	0.018	0	0	0	0		
7.5		0.947	0.874	0.762	0.582	0.258	0.066	0	0	0	0		
7.6		0.922	0.817	0.666	0.444	0.132	0.018	0	0	0	0		
7.7		0.978	0.944	0.884	0.796	0.564	0.328	0.102	0.003	0	0		
7.8		0.974	0.937	0.878	0.771	0.523	0.273	0.074	0.002	0	0		
7.9		0.982	0.959	0.920	0.842	0.658	0.433	0.187	0.015	0	0		
8.0		0.990	0.975	0.951	0.904	0.777	0.603	0.365	0.080	0.006	0		
8.1		0.994	0.986	0.972	0.945	0.869	0.754	0.568	0.244	0.059	0.003		
8.2		0.993	0.982	0.964	0.930	0.834	0.696	0.484	0.163	0.027	0		
8.3		0.995	0.988	0.976	0.953	0.887	0.786	0.618	0.300	0.090	0.008		
8.4		0.995	0.987	0.975	0.950	0.880	0.774	0.599	0.278	0.077	0.006		
8.5		0.994	0.986	0.972	0.944	0.866	0.750	0.562	0.237	0.056	0.003		

8.6	0.029	0.169	0.411	0.702	0.837	0.915	0.965	0.982	0.992	0.996
8.7	0.030	0.173	0.416	0.704	0.839	0.916	0.966	0.983	0.992	0.996
8.8	0.031	0.177	0.421	0.707	0.841	0.917	0.966	0.983	0.993	0.997
8.9	0.032	0.180	0.425	0.709	0.843	0.918	0.966	0.983	0.992	0.997
9.0	0.037	0.193	0.440	0.719	0.848	0.921	0.968	0.984	0.992	0.997
9.1	0.046	0.215	0.464	0.735	0.858	0.926	0.970	0.985	0.992	0.997
9.2	0.052	0.228	0.478	0.744	0.863	0.929	0.971	0.985	0.993	0.997
9.3	0.057	0.239	0.489	0.750	0.867	0.930	0.972	0.986	0.993	0.997
9.4	0.061	0.248	0.498	0.756	0.870	0.933	0.973	0.986	0.993	0.997
9.5	0.066	0.257	0.507	0.762	0.873	0.934	0.973	0.987	0.993	0.997
9.6	0.070	0.265	0.516	0.766	0.876	0.936	0.974	0.987	0.993	0.997
9.7	0.073	0.270	0.521	0.770	0.878	0.937	0.974	0.987	0.993	0.997
9.8	0.077	0.277	0.526	0.773	0.880	0.938	0.975	0.987	0.994	0.997
9.9	0.080	0.283	0.532	0.777	0.882	0.939	0.975	0.987	0.994	0.997
10.0	0.083	0.289	0.538	0.780	0.883	0.940	0.975	0.988	0.994	0.998
10.1	0.083	0.289	0.538	0.780	0.883	0.940	0.975	0.988	0.994	0.998
10.2	0.083	0.289	0.538	0.780	0.883	0.940	0.975	0.988	0.994	0.998
10.3	0.085	0.292	0.540	0.781	0.884	0.940	0.976	0.988	0.994	0.998
10.4	0.086	0.294	0.542	0.782	0.885	0.941	0.976	0.988	0.994	0.998
10.5	0.087	0.295	0.544	0.784	0.886	0.941	0.976	0.988	0.994	0.998
10.6	0.089	0.300	0.548	0.786	0.887	0.942	0.976	0.988	0.994	0.998
10.7	0.091	0.302	0.550	0.787	0.887	0.942	0.976	0.988	0.994	0.998
10.8	0.087	0.295	0.544	0.784	0.886	0.941	0.976	0.988	0.994	0.998
10.9	0.085	0.292	0.540	0.781	0.884	0.940	0.976	0.988	0.994	0.998
11.0	0.082	0.287	0.536	0.779	0.883	0.940	0.975	0.988	0.994	0.998
11.1	0.080	0.283	0.532	0.777	0.882	0.939	0.975	0.987	0.994	0.998
11.2	0.056	0.237	0.487	0.750	0.867	0.931	0.972	0.986	0.993	0.997
11.3	0.048	0.218	0.467	0.738	0.859	0.927	0.970	0.985	0.992	0.997
11.4	0.055	0.235	0.485	0.748	0.865	0.930	0.971	0.986	0.993	0.997
11.5	0.059	0.243	0.493	0.753	0.868	0.932	0.972	0.986	0.993	0.997
11.6	0.069	0.262	0.513	0.765	0.875	0.935	0.974	0.987	0.993	0.997

(Continued)

319

TABLE 8.2 (Continued)

Wavelength (μm)	Precipitable Water (mm)												
	0.10	0.20	0.50	1	2	5	10	20	50	100	200	500	1000
11.7		0.996	0.990	0.980	0.961	0.906	0.820	0.673	0.372	0.138	0.019		
11.8		0.997	0.992	0.982	0.989	0.925	0.863	0.733	0.460	0.212	0.045		
11.9		0.997	0.993	0.986	0.972	0.932	0.869	0.755	0.495	0.245	0.060		
12.0		0.997	0.993	0.987	0.974	0.937	0.878	0.770	0.521	0.270	0.073		
12.1		0.997	0.994	0.987	0.975	0.938	0.880	0.773	0.526	0.277	0.077		
12.2		0.997	0.994	0.987	0.975	0.938	0.880	0.775	0.528	0.279	0.078		
12.3		0.997	0.993	0.987	0.974	0.937	0.878	0.770	0.521	0.270	0.073		
12.4		0.997	0.993	0.987	0.974	0.935	0.874	0.764	0.511	0.261	0.068		
12.5		0.997	0.993	0.986	0.973	0.933	0.871	0.759	0.502	0.252	0.063		
12.6		0.997	0.993	0.986	0.972	0.931	0.868	0.752	0.491	0.241	0.058		
12.7		0.997	0.993	0.985	0.971	0.929	0.863	0.744	0.478	0.228	0.052		
12.8		0.997	0.992	0.985	0.970	0.926	0.858	0.736	0.466	0.217	0.047		
12.9		0.997	0.992	0.984	0.969	0.924	0.853	0.728	0.452	0.204	0.041		
13.0		0.997	0.992	0.984	0.967	0.921	0.846	0.718	0.437	0.191	0.036		
13.1		0.996	0.991	0.983	0.966	0.918	0.843	0.709	0.424	0.180	0.032		
13.2		0.996	0.991	0.982	0.965	0.915	0.837	0.703	0.411	0.169	0.028		
13.3		0.996	0.991	0.982	0.964	0.912	0.831	0.690	0.397	0.153	0.025		
13.4		0.996	0.990	0.981	0.962	0.908	0.825	0.681	0.382	0.146	0.021		
13.5		0.996	0.990	0.980	0.961	0.905	0.819	0.670	0.368	0.136	0.019		
13.6		0.996	0.990	0.979	0.959	0.902	0.813	0.661	0.355	0.126	0.016		
13.7		0.996	0.989	0.979	0.958	0.898	0.807	0.651	0.342	0.117	0.014		
13.8		0.996	0.989	0.978	0.956	0.894	0.800	0.640	0.328	0.107	0.011		
13.9		0.995	0.988	0.977	0.955	0.891	0.793	0.629	0.313	0.098	0.010		

Source: Data from Gaussorgues (1989, pp. 98–103).

320

TABLE 8.3 Passman and Larmore Tables for CO₂: Spectral Transmission of Carbon Dioxide (Horizontal Path, Sea Level)

Wavelength (μm)	0.10	0.20	0.50	1	2	5	10	20	50	100	200	500	1000
								Distance (km)					
0.3	1	1	1	1	1	1	1	1	1	1	1	1	1
0.4	1	1	1	1	1	1	1	1	1	1	1	1	1
0.5	1	1	1	1	1	1	1	1	1	1	1	1	1
0.6	1	1	1	1	1	1	1	1	1	1	1	1	1
0.7	1	1	1	1	1	1	1	1	1	1	1	1	1
0.8	1	1	1	1	1	1	1	1	1	1	1	1	1
0.9	1	1	1	1	1	1	1	1	1	1	1	1	1
1.0	1	1	1	1	1	1	1	1	1	1	1	1	1
1.1	1	1	1	1	1	1	1	1	1	1	1	1	1
1.2	1	1	1	1	1	1	1	1	1	1	1	1	1
1.3	1	1	1	0.999	0.999	0.999	0.998	0.997	0.996	0.994	0.992	0.987	0.982
1.4	0.996	0.995	0.992	0.988	0.984	0.975	0.964	0.949	0.919	0.885	0.838	0.747	0.649
1.5	0.999	0.999	0.998	0.998	0.997	0.995	0.993	0.990	0.984	0.976	0.987	0.949	0.927
1.6	0.996	0.995	0.992	0.988	0.984	0.975	0.964	0.949	0.919	0.885	0.838	0.747	0.649
1.7	1	1	1	0.999	0.999	0.999	0.998	0.997	0.996	0.994	0.992	0.987	0.982
1.8	1	1	1	1	1	1	1	1	1	1	1	1	1
1.9	1	1	1	0.999	0.999	0.999	0.998	0.997	0.996	0.994	0.992	0.987	0.982
2.0	0.978	0.969	0.951	0.931	0.903	0.847	0.785	0.699	0.541	0.387	0.221	0.053	0.006
2.1	0.998	0.997	0.996	0.994	0.992	0.987	0.982	0.974	0.959	0.942	0.919	0.872	0.820
2.2	1	1	1	1	1	1	1	1	1	1	1	1	1
2.3	1	1	1	1	1	1	1	1	1	1	1	1	1
2.4	1	1	1	1	1	1	1	1	1	1	1	1	1
2.5	1	1	1	1	1	1	1	1	1	1	1	1	1
2.6	1	1	1	1	1	1	1	1	1	1	1	1	1
2.7	0.799	0.718	0.569	0.419	0.253	0.071	0.011	0	0	0	0	0	0
2.8	0.871	0.804	0.695	0.578	0.432	0.215	0.079	0.013	0	0	0	0	0

(Continued)

TABLE 8.3 (Continued)

Wavelength (μm)	Distance (km)												
	0.10	0.20	0.50	1	2	5	10	20	50	100	200	500	1000
2.9	0.997	0.995	0.993	0.990	0.985	0.977	0.968	0.954	0.927	0.898	0.855	0.772	0.683
3.0	1	1	1	1	1	1	1	1	1	—	—	—	—
3.1	1	1	1	1	1	1	1	1	1	—	—	—	—
3.2	1	1	1	1	1	1	1	1	1	—	—	—	—
3.3	1	1	1	1	1	1	1	1	1	—	—	—	—
3.4	1	1	1	1	1	1	1	1	1	—	—	—	—
3.5	1	1	1	1	1	1	1	1	1	—	—	—	—
3.6	1	1	1	1	1	1	1	1	1	—	—	—	—
3.7	1	1	1	1	1	1	1	1	1	—	—	—	—
3.8	1	1	1	1	1	1	1	1	1	—	—	—	—
3.9	1	1	1	1	1	1	1	1	1	—	—	—	—
4.0	0.998	0.997	0.996	0.994	0.991	0.986	0.980	0.971	0.955	0.937	0.911	0.859	0.802
4.1	0.983	0.975	0.961	0.944	0.921	0.876	0.825	0.755	0.622	0.485	0.322	0.118	0.027
4.2	0.673	0.551	0.445	0.182	0.059	0.003	0	0	0	0	0	0	0
4.3	0.098	0.016	0	0	0	0	0	0	0	0	0	0	0
4.4	0.481	0.319	0.115	0.026	0.002	0	0	0	0	0	0	0	0
4.5	0.957	0.949	0.903	0.863	0.807	0.699	0.585	0.439	0.222	0.084	0.014	0	0
4.6	0.995	0.993	0.989	0.985	0.978	0.966	0.951	0.931	0.891	0.845	0.783	0.663	0.539
4.7	0.995	0.993	0.989	0.985	0.978	0.966	0.951	0.931	0.891	0.845	0.783	0.663	0.539
4.8	0.976	0.966	0.945	0.922	0.891	0.828	0.759	0.664	0.492	0.331	0.169	0.030	0.002
4.9	0.975	0.964	0.943	0.920	0.886	0.822	0.750	0.652	0.468	0.313	0.153	0.024	0.001
5.00	0.999	0.998	0.997	0.995	0.994	0.990	0.986	0.979	0.968	0.954	0.935	0.897	0.855
5.10	1	0.999	0.999	0.998	0.998	0.996	0.994	0.992	0.988	0.984	0.976	0.961	0.946
5.20	0.986	0.980	0.968	0.955	0.936	0.899	0.857	0.799	0.687	0.569	0.420	0.203	0.072
5.30	0.997	0.995	0.993	0.989	0.984	0.976	0.966	0.951	0.923	0.891	0.846	0.760	0.666
5.40	1	1	1	1	1	1	1	1	1	1	1	1	1

(Continued)

5.50	—	—	—	—	—	—	—	—	—	—	—	—
5.60	—	—	—	—	—	—	—	—	—	—	—	—
5.70	—	—	—	—	—	—	—	—	—	—	—	—
5.80	—	—	—	—	—	—	—	—	—	—	—	—
5.90	—	—	—	—	—	—	—	—	—	—	—	—
6.00	—	—	—	—	—	—	—	—	—	—	—	—
6.10	—	—	—	—	—	—	—	—	—	—	—	—
6.20	—	—	—	—	—	—	—	—	—	—	—	—
6.30	—	—	—	—	—	—	—	—	—	—	—	—
6.40	—	—	—	—	—	—	—	—	—	—	—	—
6.50	—	—	—	—	—	—	—	—	—	—	—	—
6.60	—	—	—	—	—	—	—	—	—	—	—	—
6.70	—	—	—	—	—	—	—	—	—	—	—	—
6.80	—	—	—	—	—	—	—	—	—	—	—	—
6.90	—	—	—	—	—	—	—	—	—	—	—	—
7.00		—	—	—	—	—	—	—	—	—	—	—
7.10		—	—	—	—	—	—	—	—	—	—	—
7.20		—	—	—	—	—	—	—	—	—	—	—
7.30		—	—	—	—	—	—	—	—	—	—	—
7.40		—	—	—	—	—	—	—	—	—	—	—
7.50		—	—	—	—	—	—	—	—	—	—	—
7.60		—	—	—	—	—	—	—	—	—	—	—
7.70		—	—	—	—	—	—	—	—	—	—	—
7.80		—	—	—	—	—	—	—	—	—	—	—
7.90		—	—	—	—	—	—	—	—	—	—	—
8.00			—	—	—	—	—	—	—	—	—	—
8.10			—	—	—	—	—	—	—	—	—	—
8.20			—	—	—	—	—	—	—	—	—	—
8.30			—	—	—	—	—	—	—	—	—	—
8.40			—	—	—	—	—	—	—	—	—	—
8.50			—	—	—	—	—	—	—	—	—	—

TABLE 8.3 (Continued)

Wavelength (μm)	Distance (km)												
	0.10	0.20	0.50	1	2	5	10	20	50	100	200	500	1000
8.60	—	—	—	—	—	—	—	—	—	—	—		—
8.70	—	—	—	—	—	—	—	—	—	—	—		—
8.80	—	—	—	—	—	—	—	—	—	—	—		—
8.90	—	—	—	—	—	—	—	—	—	—	—		—
9.00	—	—	—	—	—	—	—	—	—	—	—		—
9.10	—	—	—	0.999	0.999	0.998	0.995	0.991	0.978	0.955	0.914		
9.20	—	—	—	0.999	0.998	0.995	0.991	0.982	0.955	0.913	0.834		
9.30	—	0.999	0.997	0.995	0.990	0.975	0.951	0.904	0.776	0.605	0.363		
9.40	—	0.993	0.982	0.965	0.931	0.837	0.700	0.491	0.168	0.028	0.001		
9.5	—	0.993	0.983	0.007	0.935	0.842	0.715	0.512	0.187	0.035	0.001		
9.6	—	0.996	0.990	0.980	0.961	0.906	0.821	0.675	0.363	0.140	0.029		
9.7	—	0.995	0.986	0.973	0.947	0.873	0.761	0.580	0.256	0.065	0.004		
9.8	—	0.997	0.992	0.984	0.969	0.924	0.858	0.730	0.455	0.206	0.043		
9.9	—	0.998	0.995	0.989	0.979	0.948	0.897	0.811	0.585	0.342	0.123		
10.0	—	1	1	0.999	0.997	0.994	0.989	0.978	0.945	0.892	0.797		
10.1	—	0.997	0.999	0.998	0.996	0.990	0.980	0.960	0.902	0.814	0.663		
10.2	—	0.997	0.994	0.988	0.977	0.943	0.890	0.792	0.558	0.312	0.097		
10.3	—	1	0.994	0.987	0.975	0.939	0.881	0.777	0.532	0.283	0.080		
10.4	—	1	1	0.999	0.998	0.995	0.991	0.982	0.955	0.913	0.834		
10.5	—	1	1	0.999	0.998	0.998	0.995	0.991	0.978	0.955	0.914		
10.6	—	1	1	0.999	0.999	0.998	0.995	0.991	0.978	0.955	0.914		
10.7	—	1	1	1	0.999	0.999	0.997	0.995	0.986	0.973	0.947		
10.8	—	1	0.999	0.999	0.998	0.998	0.995	0.991	0.978	0.955	0.914		
10.9	—	1	0.999	0.999	0.997	0.993	0.986	0.973	0.934	0.872	0.761		
11.0	—	1	0.999	0.999	0.997	0.993	0.986	0.973	0.934	0.872	0.761		
11.1	—	1	0.999	0.998	0.997	0.992	0.984	0.969	0.923	0.855	0.726		

11.2	1	0.999	0.998	0.995	0.989	0.978	0.955	0.892	0.796	0.633
11.3	0.999	0.999	0.997	0.994	0.985	0.971	0.942	0.862	0.742	0.552
11.4	0.999	0.998	0.997	0.993	0.983	0.966	0.934	0.842	0.709	0.503
11.5	0.999	0.998	0.996	0.992	0.980	0.960	0.921	0.814	0.661	0.438
11.6	0.999	0.998	0.995	0.991	0.977	0.955	0.912	0.794	0.632	0.399
11.7	0.999	0.998	0.995	0.991	0.977	0.955	0.912	0.794	0.632	0.399
11.8	0.999	0.998	0.997	0.993	0.983	0.966	0.934	0.842	0.709	0.503
11.9	1	0.999	0.998	0.995	0.989	0.978	0.955	0.892	0.796	0.633
12.0	1	1	0.999	0.999	0.997	0.993	0.986	0.966	0.934	0.872
12.1	1	1	0.999	0.998	0.998	0.995	0.991	0.978	0.955	0.914
12.2	1	1	0.999	0.998	0.998	0.995	0.991	0.978	0.955	0.914
12.3	0.998	0.995	0.990	0.981	0.952	0.907	0.823	0.614	0.376	0.142
12.4	0.994	0.985	0.970	0.941	0.859	0.738	0.545	0.218	0.048	0.002
12.5	0.987	0.968	0.936	0.877	0.719	0.517	0.268	0.037	0.001	0
12.6	0.980	0.950	0.903	0.815	0.599	0.358	0.129	0.006	0	0
12.7	0.996	0.989	0.979	0.959	0.899	0.809	0.654	0.346	0.120	0.015
12.8	0.990	0.974	0.949	0.901	0.770	0.592	0.351	0.072	0.005	0
12.9	0.985	0.962	0.925	0.856	0.677	0.458	0.210	0.020	0	0
13.0	0.991	0.977	0.955	0.912	0.794	0.630	0.397	0.099	0.010	0
13.1	0.990	0.974	0.949	0.900	0.768	0.592	0.348	0.071	0.005	0
13.2	0.978	0.946	0.895	0.803	0.575	0.330	0.109	0.004	0	0
13.3	0.952	0.884	0.782	0.611	0.292	0.085	0.007	0	0	0
13.4	0.935	0.846	0.715	0.512	0.187	0.035	0.001	0	0	0
13.5	0.901	0.767	0.593	0.352	0.070	0.005	0	0	0	0
13.6	0.901	0.792	0.627	0.351	0.097	0.009	0	0	0	0
13.7	0.916	0.803	0.644	0.415	0.110	0.012	0	0	0	0
13.8	0.858	0.681	0.464	0.215	0.021	0	0	0	0	0
13.9	0.778	0.534	0.286	0.082	0.002	0	0	0	0	0

Source: Data from Gaussorgues (1989, pp. 98–103).

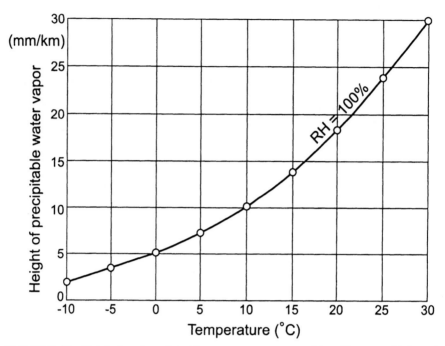

FIGURE 8.45   Height h_w of precipitable water vapor expressed in millimeters. (Adapted from Gaussorgues, 1989.)

for 100% relative humidity RH and a distance d of 1 km. If other values of RH and d are of interest, h_w is easily obtained from that curve.

Example 8.1   Compute h_w if the temperature of interest is 10°C, RH = 60%, and the distance is 2 km.

SOLUTION   From Figure 8.45 we find that h_w is 10 mm for 1 km and RH = 100%. For RH = 60% and d = 2 km, h_w is computed as follow:

$$h_w = 10 \text{ mm/km} \times 0.6 \times 2 = 12 \text{ mm}$$

Once h_w is established, Tr_{λ, H_2O} is obtained directly from Table 8.2 and Tr_{λ, CO_2} from Table 8.3.

Example 8.2   Compute Tr_{λ, H_2O} and Tr_{λ, CO_2} under the following conditions: the temperature of interest is 10°C, RH = 60%, the distance is 2 km, and the wavelength is 10 μm.

SOLUTION   From Example 8.1 we have h_w = 12 mm directly. Looking at Table 8.2 we get (by interpolation) $Tr_{10 \text{ μm}, H_2O}$ = 0.862 and $Tr_{10 \text{ μm}, CO_2}$ = 0.997.

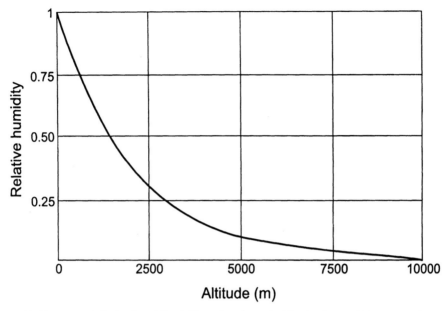

FIGURE 8.46   Relative humidity RH as function of altitude; RH = 100% at sea level.
(Adapted from Gaussorgues, 1989.)

If RH is known at sea level, Figure 8.46 allows to compute its value in altitude. The table is given for RH = 100% at sea level. Multiplication of the curve
value by the RH value of interest yields the corrected RH at a given altitude.

The next term to be computed, $Tr_{\lambda, d}$, is related to diffusion by particles. It is
computed by

$$Tr_{\lambda, d} = \exp(-\gamma_\lambda x) \tag{8.7}$$

where x is the distance and γ_λ is the coefficient of diffusion at wavelength λ.
Practically, γ_λ is computed with respect to the visibility at $\lambda_0 = 0.6$ μm. The coefficient of diffusion in the visible spectrum $\gamma_{0.6}$ is first computed as (x expressed
in kilometers):

$$\gamma_{0.6} = \frac{3.92}{x} \tag{8.8}$$

and for a particular wavelength γ_λ,

$$\gamma_\lambda = \gamma_{0.6} \left(\frac{0.6}{\lambda}\right)^{1.3} \tag{8.9}$$

Combining eqs. (8.7), (8.8), and (8.9) gives $Tr_{\lambda, d}$, as illustrated below.

Example 8.3 Compute $Tr_{\lambda,d}$ under the following conditions: the distance of interest is 2 km, the wavelength is 10 μm, and the visibility is 12 km.

SOLUTION $\gamma_{0.6}$ is first computed with eq. (8.8):

$$\gamma_{0.6} = \frac{3.92}{x} = \frac{3.92}{12} = 0.327$$

Next, γ_λ is obtained with eq. (8.9):

$$\gamma_{10\,\mu m} = \gamma_{0.6}\left(\frac{0.6}{\lambda}\right)^{1.3} = 0.327\left(\frac{0.6}{10}\right)^{1.3} = 0.00844$$

and $Tr_{10\,\mu m,d}$ is then obtained from eq. (8.7):

$$Tr_{10\,\mu m,d} = \exp(-\gamma_{10\,\mu m}x) = \exp(-0.00844 \times 2) = 0.983$$

As said before, diffusion of particles affects radiation propagation such as spatial coherence but not greatly affect the transmittance.

The last parameter in eq. (8.6) concerns the effect of rain, which can be evaluated taking into account the coefficient of diffusion for rain γ_{rain}, as listed in Table 8.4 combined with eq. (8.7).

Example 8.4   For Example 8.3, compute the global transmittance Tr of the atmosphere both in the absence of rain and when rain is very strong.

SOLUTION   We first compute Tr_{rain}: γ_{rain} for strong rain is 0.52 (Table 8.4), with distance $x = 2$ km and eq. (8.7):

$$Tr_{rain} = \exp(-\gamma_{rain}x) = \exp(-0.52 \times 2) = 0.353 \qquad (8.10)$$

In the absence of rain, $Tr_{rain} = 1$ ($\gamma_{rain} = 0$). Finally, $Tr_{10\,\mu m}$ is obtained from

TABLE 8.4   Coefficient of Diffusion for Rain Under Various Conditions

Condition	Coefficient of Diffusion for Rain, γ (km^{-1})
Light rain	0.07
Average rain	0.16
Strong rain	0.23
Very strong rain	0.52

Source: Adapted from Gaussorgues (1989, p. 109).

eq. (8.6) with values computed in Example 8.3:

$$Tr_{10\,\mu m} = Tr_{10\,\mu m,H_2O}Tr_{10\,\mu m,CO_2}Tr_{10\,\mu m,d}Tr_{rain}$$

$$= 0.862 \times 0.997 \times 0.983 \times 0.353 = 0.298 \quad \text{(in the case of rain)}$$

$$= 0.862 \times 0.997 \times 0.983 \times 1 = 0.845 \quad \text{(in the absence of rain)}$$

As seen, most of the effects come from the presence of water. Finally, it is worthwhile to mention that atmospheric transmittance computation applets for various standard atmospheres are available interactively on the Web.* Moreover, for short distances in TNDT experiments (below 1 to 2 m), the effect of atmosphere is usually neglected with an atmospheric transmittance very close to 100%.

8.4 REMOTE SENSING

Writing of this section was inspired by many sources. The *Manuals of Remote Sensing* of the American Society for Photogrammetry and Remote Sensing (Colwell, 1983; Rencz, 1998) were found particularly helpful, and we invite interested readers to refer to these documents, which bring more additional information than it is possible to cover in a single section. Interestingly, three successive editions of the *Manuals of Remote Sensing* covering that field in detail have been published since the 1940s. A list of other national remote sensing societies is available on the Web.[†]

Increased pressure on resources due to growing human and economical development has pushed the need for the noncontact acquisition of various kinds of environmental data. This field of knowledge is known as *remote sensing*. Remote sensing (RS) is a precious tool for obtaining cultural, human, and mineral resources data. Generally, the information is obtained from airplanes, satellites [such as the Canadian *Radarsat* I and II, ESA (European Space Agency) *SPOT* with 10-m resolution or *Helios A* (1995) and *B* (2000) with 1-m resolution], and the NASA (U.S. National Aeronautics and Space Administration) Space Shuttle [such as the February 2000 *Endeavour* C/X-band synthetic aperture radar topography mission that mapped 80% of Earth's terrain (119 million km^2) with data stored on 330 digital cassettes—equivalent to 20,000 CDs]. Spatial, spectral, and temporal domains are important considerations in RS. In some surveys, for example, to maximize the spatial resolution, a more 'local' device such as a truck with a hydraulic mast or lift platform permits more detailed measurements (Figure 8.47).

Another example of RS use is in thermal archaeology (Weil and Graf, 1992). Salisbury, North Carolina, is a U.S. Civil War site where up to 10,000 im-

---

*See, for example, www.ciks.cbt.nist.gov/nef/dbm/envdb.html.
† www.casi.ca/listof.htm.

FIGURE 8.47   Mobile sensing unit for building, technical infrastructure, and environmental applications. The infrared camera is mounted on a hydraulic mast with an image analysis system inside the mobile unit. (Courtesy of Centre for Built Environment, Division of Information Technology for Land, Building and Infrastructure, Sweden, Dr. S. Å. Ljungberg.)

prisoned Union soldiers crowded a prison camp located on a 2.4-ha compound. Many people died and are now buried there. In 1998,* an IR survey discovered that the camp's latrines were about 20 m away from the well water supply, suggesting that many of those buried there died from dysentery. Another IR survey found more than 100 unmarked graves for blacks outside a cemetery for whites.

In remote sensing, three scale factors are generally considered: (1) detection (a feature is present or absent, such as a crop is there), (2) identification (determining the nature, such as type of crops), and (3) analysis (studies to identify and explain particular aspects from the whole, such as are any infested trees from, say, an orange grove). Going from scale factor 1 to 2 involves a threefold improvement in spatial resolution (Section 4.1), while from 2 to 3 a tenfold factor is required. Greater resolution sensing is generally restricted to smaller areas (such as to survey ground use in urban areas; Figure 8.48) and the trade-off is then to lose the capability to establish regional patterns.

Multispectral gathering of data provides additional information not available if only the visible band is used. For example, observation of crop growth and development is seen in visible and also in reflective IR (smaller than 2.5 μm), thermal IR bands suits volcanism studies and forest fire detection, and microwave allows to assess the state of the sea (Figure 8.49).

Some features change over time; others do not. Temporal evolution of remote sensing data brings an additional dimension, so that these surveys need to be repeated periodically with a cycle that depends on the application (obviously,

*www.infobeat.com/stories/cgi/story.cgi?id=2565069010-037.

FIGURE 8.48   SPOT Satellite image (combined multispectral and panchromatic) from Quebec City, Quebec (Canada) taken October 25, 1989. Resolution: 10 m, image area: 4.5 × 4.5 km. The fine resolution shows much detail in the historic part of the city, as well as allowing interpretation of some of the industrial features of this port on the St. Lawrence River (e.g., shadows of the storage tanks give a clue to their height, A). The multispectral or colored nature of this merged image permits interpretation of some of the residential and recreational portions of the city (e.g., street pattern, B and location where trees are more prominent, such as in parks, C). (Image courtesy of the Canada Centre for Remote Sensing, Natural Resources Canada.)

this implies that the sensor must be able to follow the cycle of interest). Relevant changes spans from a few minutes (e.g., winds), to hours (e.g., floods), to days and weeks (e.g., crop growth), to years (e.g., urban evolution, glaciers).

8.4.1   Infrared (Thermal) Remote Sensing

Human activity involves the loss of energy in the environment. In winter there is the heating of buildings; in summer refrigeration and insulation deficiencies cause energy losses (this explains why urban areas are called *heat islands*). Many industrial processes also release massive amounts of energy in water or in the air (e.g., thermal power plants). Thermal remote sensing is of prime importance to detecting, identifying, and analyzing these thermal manifestations.

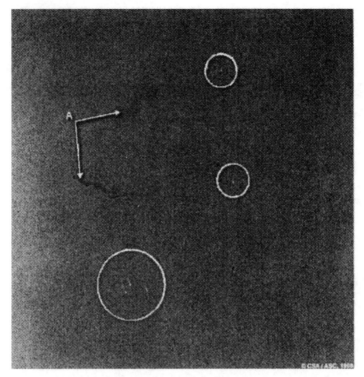

FIGURE 8.49   Very small portion of a standard (S7) RADARSAT image acquired over the Flemish Cap, an area in the Atlantic Ocean southeast of the coast of Newfoundland on August 2, 1996 (47°N, 46°30'W). Sensor resolution: 20 m, image area: 33 × 36 km approximately. The monitoring of ocean features and detection of ships over large expanses of coastal waters would be impractical and prohibitively expensive if we could not rely on remotely sensed data and automated image analysis techniques. This image illustrates the ability of a new system called the Ocean Monitoring Workstation (OMW), an automated system that uses RADARSAT data to locate ships to produce data to be ingested into wave forecast models and to identify various mesoscale ocean features. On this subimage OMW identified two natural slicks (A) and five ships (two identified east of the slicks and three clustered to the south). Wakes are clearly visible behind the three ships at the bottom of the image. This information can be used to determine their speed and direction of travel. (Image courtesy of the Canada Centre for Remote Sensing, Natural Resources Canada.)

In the shortwave band (also called *middle infrared* in remote sensing: 3.5 to 5.5 μm, in contrast to the near infrared, 0.7 to 1.1 μm), radiation is both reflected and emitted, while in the longwave band (also called *far infrared* in remote sensing: 8 to 14 μm), radiation is emitted. As discussed in Section 8.2, emissivity is an important parameter affecting the emission of radiation. In thermal remote sensing, the variability of building materials, with specific emissivities, makes quantitative analysis more complex (e.g., emissivity values for most roofing materials are in the range 0.88 to 0.94 but the emissivity of an unpainted metallic surface is below 0.2; Table 8.1).

Thermal inertia P is a measure of the thermal response of a material to a temperature change. It is defined as (in J m^{-2} °C^{-1} s$^{-1/2}$)

$$P = \sqrt{k\rho C_p} \tag{8.11}$$

where k is the thermal conductivity (W m^{-1} °C^{-1}), ρ the mass density (kg m^{-3}), and c the specific heat (J kg^{-1} °C^{-1}). Materials with higher thermal inertia exhibit a more uniform temperature day and night than that of material with a lower P value. For example, the thermal inertia of water is greater than the thermal inertia for soils and rocks, and thus water temperature is higher at night and cooler during the day with respect to soils. Rocks heat up and cool down more rapidly; because of the thickness of the material, the effect of solar heating is small, and thus absorption and release of energy are fast (see Problem 8.5). The high thermal inertia of plants and trees (water content) also explains why trees are cooler than their surroundings during the day and warmer at night, especially in urban areas: Leaves evaporate water during the day, thus maintaining a cooler temperature.

Rooftop thermal sensing is an important application for evaluating, for example, the quality of housing from both the structural and environmental (energy management) points of views. Reasonable cost and a fast acquisition procedure (overflight) allow us to deploy such surveys over entire cities. Deficiencies detected during such surveys include insufficient attic insulation, structural leaks, and damaged pipelines (steam transport for heating, etc.). Ideally, thermal remote sensing is coupled to ground-based (thermal) measurements to complete the survey (e.g., detection of uninsulated doors and windows; Figure 8.50).

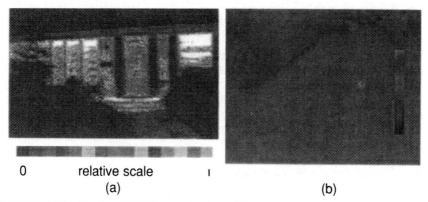

0   relative scale   ı
        (a)   (b)

FIGURE 8.50 Passive TNDT investigation of housing thermal insulation efficiency. (a) Old house: severe heat losses through the front door are clearly visible on this thermogram (notice spurious reflections on the icy path in front of the stairs). (b) New house which has windows with improved efficiency (notice the fan vent on the facade, between left two windows and thermal reflections on trees). Exact temperature can be compared only if emissivities are known. Thermograms recorded on a cold (-20°C) winter night with a clear sky. (Left: from Maldague, 1993.)

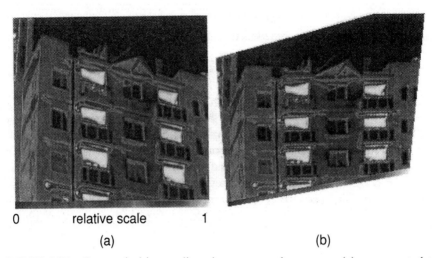

0   relative scale   1

(a)   (b)

FIGURE 8.51   Geometrical image distortions seen on thermogram (a) are corrected to set image back to a plane perpendicular to the optical axis in (b); windows are now parallel to image borders. (Courtesy of Institute of Electronics, Technical University of Lodz, Poland, Prof. B. Wiecek.)

Qualitative analysis are relatively easy to perform. Quantitative analysis is more challenging since it requires knowledge of building structure, thermal conductivities, and emissivities of involved materials, local weather at the time of scanning, and possible geometrical image distortions. Provided that this is available (with the help of residents), computations of heat flow described in Chapter 3 can be performed to estimate heat loss and insulation factors (especially since it is often possible to assume a permanent regime). Madding (1979) presents such an analysis (see also Section 11.4.2). Geometrical image distortions can be corrected to set images back to a plane perpendicular to the optical axis [see Zwolenik and Wiecek, (1998) on 2D perspective transformation (Figure 8.51); see also Section 10.6.3 on image registration].

In addition to housing quality, thermal remote sensing is performed for a variety of purposes: for example, to monitor the effect of warm water discharge (permitted or prohibited) of a plant in a lake or river. Such analysis is cost-effective since the temperature distribution of the plume is instantaneously accessible; however, no depth profile is available (Figure 8.52).

8.4.2   Conditions for Thermal Surveys

The following factors allow us to maximize the quality of data obtained during surveys:

- Perform the survey late at night to minimize the disturbing effects of the sun. Ideally, surveys are performed just before dawn; it is when maximum thermal contrasts are generally observed.

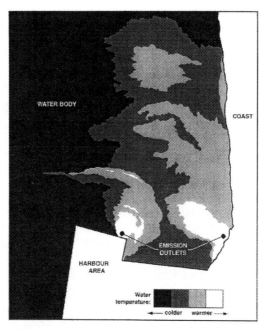

FIGURE 8.52 Dual discharge of effluents from a pulp and paper mill. This level sliced image shows the effluent plume in discrete tones. Each tone represents a 2°C change in temperature. Aerial thermography provides a complete overview of the size and distribution of the plume. The cooling to ambient temperature is clearly visible. Centerline decay can be calculated from the thermography. (Data from Colwell, 1983.)

- The air temperature should be cool, preferably below 2°C.
- The sky should be clear to avoid the spurious thermal reflections from an overcast sky on roofs and surfaces.
- Winds must be calm (below about 10 km h^{-1}).
- The best period is generally just after a cold front, since the air is then calm and dry and the sky is clear with limited turbulences.
- Water, frost, ice, and snow on roofs and surfaces is to be avoided, as they create thermal anomalies (they are sometimes informative, however; Figure 8.53). If inspection is to proceed under the rain (with the observed surfaces dry: for example, walls protected by a roof), a common trick to protect an IR camera is to cover it with a clean trash bag held in place with a rubber band (Hurley, 1994). It is important to extend the film fully in front of the lens. Such a trick is good for qualitative thermography, since such a bag is quite transparent in the IR bands (Figure 4.32); a limited effect on thermograms will be noted.

In warm and dry regions (e.g., in the southern United States), roofs are generally flat, whereas in colder and wetter areas (e.g., in Canada), roofs are

FIGURE 8.53 In the springtime, particular combinations of light snow during the night followed by sunshine sometimes provide astonishing results. For example, on these pictures, the internal roof structure become clearly visible thanks to such visible thermography. Obviously, however, such situations occur only under exceptional circumstances and for a short period of time, making such an approach difficult (since as snow melts later in the morning, all disappear).

steeply pitched or gabled to evacuate rainwater and minimize the snow load. Such roof orientation is an important problem in thermal remote sensing. In gabled roofs, warm air tends to be trapped at ridge lines, causing local heat losses in these areas unrelated to structural problems, thus making thermal data interpretation more difficult.

Another concern is the orientation of the scanner with respect to the roof. Lambert law's of cosines specifies that radiation intensity is related to the angle between the normal to the surface and the line of observation. In the case of a flat roof, both directions are parallel (0° interval) and no effect is observed (cos 0° = 1). Observation of gabled roofs makes things different. For example, emitted radiation is reduced by 50% if angle between the normal and the line of observation is 60° (cos 60° = 50); thus the side facing the scanner will appear warmer than the opposite side, although it may be assumed that both sides radiate the same amount of heat. Moreover, at an angle, observation is performed over a larger areas, and this may compensate for orientation. For example, for the geometry of Figure 8.54, the roof has a slope of 0.33 [slope = tan (corner angle) = tan 18.5°] and although on the sensor plane the distances a and b are similar, this corresponds to half the distance on the roof ($A = 2 \times B$). On the other hand, with an angle of 77° to the surface, only 22% of the energy is emitted in the sensor direction (cos 77° = 0.22). In this case the compensation for a larger sensed area is not enough to counterbalance the low viewing angle. Unless roofs are flat or slightly slooping or their orientation can be accounted for, thermal remote sensing is less accurate.

Finally, in forested areas, trees can cover rooftops completely or partially with a warmer signature that affects measurements. Nevertheless, thermal remote sensing is a precious tool, especially if additional information is available, such as maps and ground measurements performed at the same time as the overflight.

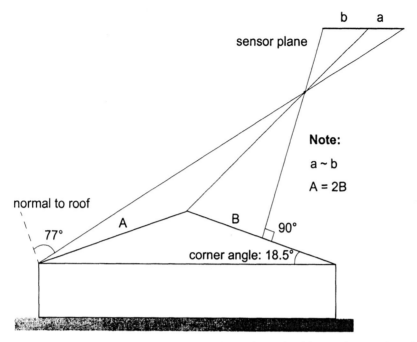

FIGURE 8.54 The problem of the thermal inspection of gabled roofs: actual surface area viewed and amount of radiation received by the sensor varies with observation angle.

8.4.3 Air Surveys

More information about air surveys can be found in Hurley (1994; see also Section 11.4). Flight altitude is generally about 300 m above ground, often referred to as the *altitude above target*, while flight speed is about 100 km/h. To maintain a sufficient spatial resolution (Section 4.1) in such conditions, an IR zoom (such as 4×) is recommended. In the case of aerial flight restrictions or obstacles (trees, buildings, towers), greater distances are imposed, further challenging the system's spatial resolution.

Simultaneous recording of visible and IR imagery enables us to remove possible ambiguities about what was observed at flight time. Additionally, a distance logger allows for later retrieval of particular locations of interest (detected hot spots) over the flight path (e.g., distance traveled superimposed on the images visible). The negative effect of moisture in air was discussed in Section 8.3 and should be taken into consideration since it reduces atmospheric transmission in IR bands.

Although aerial surveys using regular IR equipment are possible with reduced performance, dedicated equipment is preferred (Figure 8.3). Regular IR equipment can be pointed through open airplane windows or doors but with inconvenience for operators: noise, wind, uncomfortable temperature in cold weather. Moreover, if the IR equipment has to be refilled periodically with coolant (such as liquid nitrogen), precious flight time is lost and refilling is neither easy nor safe in a tight space (helicopter flight duration is about 3 h between tank refills). On the other hand, regular IR equipment is less costly and does not require aerial certification if not attached on the aircraft.

Dedicated aerial IR equipment can be operated year-round when controlled remotely from inside aircraft (orientation and zoom capabilities): for example, 180° along the flight path and 270° across the flight path at a speed of 60° s^{-1}. Moreover, options include vibration compensation and self-stabilization. This is particularly important for maintaining the target continuously within the field of view.

PROBLEMS

8.1 **[precipitable water vapor]** Compute h_w if the temperature of interest is 20°C, RH = 90%, and the distance is 0.5 km.

Solution From Figure 8.45 we find that h_w is 17 mm for 1 km and RH = 100%. For RH = 90% and $d = 0.5$ km, h_r is computed as follows:

$$h_w = 17 \text{ mm km}^{-1} \times 0.9 \times 0.5 = 8 \text{ mm}$$

8.2 **[atmosphere transmittance: gas absorption]** Compute Tr_{λ, H_2O} and Tr_{λ, CO_2} in the following conditions: the temperature of interest is 20°C, RH = 90%, the distance is 0.5 km, and the wavelength is 4 μm.

Solution From Problem 8.1 we have directly that $h_w = 8$ mm. Looking in Table 8.2, we get (by interpolation) $Tr_{4 \text{ μm}, H_2O} = 0.973$ and for $Tr_{4 \text{ μm}, CO_2} = 0.996$ (with Table 8.3).

8.3 **[atmosphere transmittance: particle diffusion]** Compute $Tr_{\lambda, d}$ under the following conditions: the distance of interest is 0.5 km, the wavelength is 4 μm, and the visibility is 18 km.

Solution $\gamma_{0.6}$ is first computed with eq. (8.8):

$$\gamma_{0.6} = \frac{3.92}{x} = \frac{3.92}{18} = 0.218$$

Next, γ_λ is obtained with eq. (8.9):

$$\gamma_{4 \text{ μm}} = \gamma_{0.6} \left(\frac{0.6}{\lambda}\right)^{1.3} = 0.218 \left(\frac{0.6}{4}\right)^{1.3} = 0.0185$$

and $Tr_{4 \text{ μm}, d}$ is then obtained from eq. (8.7):

$$Tr_{4 \text{ μm}, d} = \exp(-\gamma_{4 \text{ μm}} x) = \exp(-0.0185 \times 0.5) = 0.991$$

Following eq. (8.9), $Tr_{\lambda, d}$ is inversely proportional to the wavelength.

8.4 **[global atmosphere transmittance]** For Problem 8.3, compute the global transmittance Tr of the atmosphere in the absence of rain and if the rain is light.

Solution   We first compute Tr_{rain}: γ_{rain} for light rain is 0.07 (Table 8.4), with distance $x = 0.5$ km and eq. (8.7):

$$Tr_{rain} = \exp(-\gamma_{rain}x) = \exp(-0.07 \times 0.5) = 0.966$$

In the absence of rain, $Tr_{rain} = 1$ ($\gamma_{rain} = 0$). Finally, $Tr_{4\,\mu m}$ is obtained from eq. (8.6) with values computed in the Problem 8.3:

$$Tr_{4\,\mu m} = Tr_{4\,\mu m, H_2O} Tr_{4\,\mu m, CO_2} Tr_{4\,\mu m, d} Tr_{rain}$$

$$= 0.973 \times 0.996 \times 0.991 \times 0.966 = 0.928 \quad \text{(in the case of rain)}$$

$$= 0.973 \times 0.996 \times 0.991 \times 1 = 0.960 \quad \text{(in the absence of rain)}$$

8.5 **[thermal inertia]** Compute and comment on the thermal inertia of water and brick.

Solution   Thermal inertia P is defined as (in J m^{-2} °C^{-1} s$^{-1/2}$)

$$P = \sqrt{k\rho C_p}$$

where k is the thermal conductivity (W m^{-1} °C^{-1}), ρ the mass density (kg m^{-3}), and C_p the specific heat (J kg^{-1} °C^{-1}). For water $k = 13$ W m^{-1} °C^{-1}, $\rho = 1000$ kg m^{-3}, and $C_p = 4180$ J kg^{-1} °C^{-1}. For concrete $k = 1$ W m^{-1} °C^{-1}, $\rho = 2400$ kg m^{-3}, and $C_p = 800$ J kg^{-1} °C^{-1}. Computations of P gives $P_{water} = 7.4 \times 10^3$ J m^{-2} °C^{-1} s$^{-1/2}$ and $P_{concrete} = 1.4 \times 10^3$ J m^{-2} °C^{-1} s$^{-1/2}$. As mentioned in Section 8.3, since P_{water} is more than five times greater than $P_{concrete}$, water temperatures are more uniform day and night with respect to concrete. Who has never walked by a concrete wall at night and not noticed its higher-than-ambient temperature? Such large daily temperature variations do not occur in, for example, a swimming pool.

PART II

ACTIVE THERMOGRAPHY

CHAPTER NINE

Active Thermography

Contrary to the passive approach, the *active* approach requires an external heat source to stimulate the materials for inspection. Distinctions between the passive and active approaches are not as clear cut as it seems. For example, although maintenance investigation of buildings is generally considered as a passive thermography application, it can be performed by the active approach as well. As it is known, water in a wall activates undesired chemical reactions and thus the degradation process in buildings. In the case of monumental buildings, due to the extremely high cost of restoration, it is first necessary to assess the moisture problem. This can be done by heating the wall uniformly from one side while observing the isotherm pattern on the other side; the thermal map recorded depends in the water content since the heat capacity is greater for water than for dry material (Table 3.7 and Appendix E). It is interesting to note that moisture evaluation of buildings and roofs in particular has been reported by passive methods as well. In this case, thermographic inspection is performed at night or early in the morning and detection is based on the fact that moist areas retained day sunshine heat better than did sound dry areas because of the high thermal capacity of water (Section 11.4.2). In that case, too, the passive label is debatable since solar heating is a form of external thermal stimulation (see below). What is generally admitted is that if one does not need thermal stimulation, the procedure is called passive (and active otherwise). In this chapter we deal with the various active thermal stimulation procedures.

The chapter begins with some theory on thermal waves whose propagation enables subsurface specimen analysis. In the next sections, the common procedures of active thermal stimulation for TNDT are presented. In Section 9.2 we put emphasis on pulsed thermography when a thermal pulse is applied to the material for inspection. Stepped heating in which the specimen is submitted to constant heating during observation is the topic of Section 9.3. Another possible approach makes use of periodic thermal cycling of the specimen inspected. In this case, individual frequencies are tested separately; this is known as lock-in thermography, discussed in Section 9.4. Finally, vibrothermography is the

343

topic of Section 9.5. Tables 1.1 and Table 1.2 list common applications, limitations, and capabilities of these procedures (including passive thermography).

9.1   THERMAL WAVE THEORY

When a surface is heated, highly attenuated and dispersive waves are found inside the material (in a near surface region). These waves, called *thermal waves*, were first investigated by the French mathematician J. Fourier (1768–1830) and the Swedish physicist A. J. Ångström (1814–1874) (Fourier, 1826; Ångström, 1863). Of interest for TNDT is that these waves can be generated and detected remotely. In this section we study briefly the behavior and features of these waves.

In the case of a semi-infinite specimen (planar specimen) onto which a uniform source periodically deposits heat with a modulation (or angular) frequency ω, the mathematical study turns one-dimensional, and the resulting temperature T expressed as a function of depth z and time t due to this stimulation is expressed as (Favro and Han, 1998):

$$T(x,t) = T_0 e^{-z/\mu} \cos\left(\frac{2\pi z}{\lambda} - \omega t\right) = T_0 e^{-z/\mu} \cos\left(\omega t - \frac{2\pi z}{\lambda}\right) \tag{9.1}$$

where μ is the thermal diffusion length, expressed by

$$\mu = \sqrt{\frac{2k}{\omega \rho C}} = \sqrt{\frac{2\alpha}{\omega}} \tag{9.2}$$

with thermal conductivity k, density ρ, specific heat C, modulation frequency ω [$= 2\pi f$ (rad s^{-1}), f the frequency in hertz], thermal diffusivity α, and thermal wavelength λ, defined as

$$\lambda = 2\pi\mu \tag{9.3}$$

The propagation speed of these waves is obtained as

$$v = \lambda \frac{\omega}{2\pi} = \sqrt{2\omega\alpha} \tag{9.4}$$

From the last part of eq. (9.1), we also have the phase ϕ of the thermal wave which is related directly to depth z (phase can also be extracted as a function of the modulation frequency; see below):

$$\phi(z) = \frac{2\pi z}{\lambda} = \frac{z}{\mu} \tag{9.5}$$

Plots of magnitude A and phase ϕ of a thermal wave are shown in Figure 9.1.

Since the thermal diffusion length is inversely proportional to the modulation frequency [eq. (9.2)], this indicates that higher modulation frequencies restrict the analysis in a near-surface region, while low-frequency thermal waves propagate deeper but very slowly [eq. (9.4)]. For example, investigation of paint coatings indicates a depth range up to 40 μm with a modulated heating of 36 Hz, and over 80 μm with 2.25 Hz (Busse, 1994). Interestingly, solar heating also generates thermal waves within soil of Earth. Within a 24-h modulation cycle, the propagation depth reaches about 5 to 15 cm into low-thermal-conductivity materials; over a 365-day modulation cycle, the propagation of these very low frequency waves is greater, about 1 m into the soil (Del Grande and Clark, 1991).

As with other waves, such as ultrasonic waves, thermal waves reflect on subsurface features such as planar gas-filled defects. Reflected thermal waves come back on the specimen surface where the resulting oscillating temperature field can be remotely detected through its thermal infrared emission (by, say, an infrared camera). The reflection coefficient R between two media (subscripts 1 and 2) is expressed by the following equation (in homogeneous materials):

$$R = \frac{b-1}{b+1} \tag{9.6}$$

where b is defined as the ratio of thermal effusivities e between the two media [$b = e_2/e_1$, with $e = \sqrt{kpC}$ (k = thermal conductivity, p = mass density and C = specific heat)].

Example 9.1   As seen in Figure 9.1, thermal waves are highly damped. Compute the decay factor after a propagation of λ under the surface.

SOLUTION   In this case we have $z = \lambda$, and since $\lambda = 2\pi\mu$, we have, from eq. (9.1),

$$T(x,t) = T_0 e^{-2\pi\mu/\mu} \cos\left(\frac{2\pi\lambda}{\lambda} - \omega t\right) = T_0 e^{-2\pi} \cos(2\pi - \omega t) = \frac{T_0}{535} \cos(2\pi - \omega t)$$

As seen, the damping factor is 1/535; thus from the practical point of view, the wave has disappeared.

Example 9.2   Compute the propagation speed of a thermal wave in a CFRP specimen if heating is applied at a frequency of 0.01 Hz.

SOLUTION   CFRP is anisotropical with $\alpha_{\perp} = 0.42$ m^2 s^{-1} and $\alpha_{\parallel} = 3.7$ m^2 s^{-1}. We compute the angular frequency ω with $\omega = 2\pi f = 2\pi(0.01) = 0.0628$ rad/s. Thus we have

$$v_{\perp} = \sqrt{2\omega\alpha} = \sqrt{2 \times 0.0628 \times 0.42} = 0.23 \text{ m s}^{-1}$$

$$v_{\parallel} = \sqrt{2\omega\alpha} = \sqrt{2 \times 0.0628 \times 3.7} = 0.68 \text{ m s}^{-1}$$

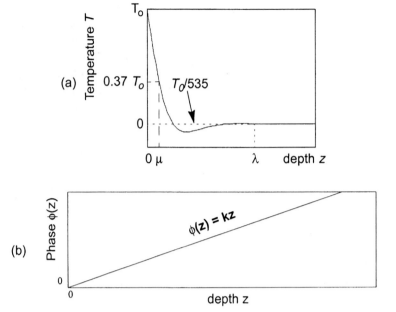

FIGURE 9.1   Plot of a thermal wave showing both (a) the magnitude A and (b) phase for a constant modulation frequency ω (for a semi-infinite plate).

We note that there is almost a threefold propagation factor between speed parallel to the fiber and perpendicular to the fiber. This explains (partly) why it is difficult to probe deep into CFRP laminates.

Example 9.3   Find the reflection coefficient between mild steel and air (considered a defect).

SOLUTION   Let's first compute the thermal effusivity for both media. For steel $k = 46$ W m^{-1} °C^{-1}, $\rho = 7900$ kg m^{-3}, and $C = 440$ J kg^{-1} °C^{-1}, and for air, $k = 0.024$ W m^{-1} °C^{-1}, $\rho = 1.2$ kg m^{-3}, and $C = 700$ J kg^{-1} °C^{-1}:

$$e_{\text{steel}} = \sqrt{k\rho C} = \sqrt{46 \times 7900 \times 440} = 1.26 \times 10^4 \text{ W m}^{-2} \text{ °C}^{-1} \text{ s}^{1/2}$$

$$e_{\text{air}} = \sqrt{k\rho C} = \sqrt{0.024 \times 1.2 \times 700} = 4.49 \text{ W m}^{-2} \text{ °C}^{-1} \text{ s}^{1/2}$$

and then $b = e_{\text{air}}/e_{\text{steel}} = 1/2806$. Finally,

$$R = \frac{b-1}{b+1} = \frac{1/2806 - 1}{1/2806 + 1} \approx -1$$

With such a perfect reflection coefficient, in case of a shallow large planar air defect within a steel slab, the thermal signal strength returning to the surface is expected to be really good. Of course, if the air defect is deep and small, the returning waves will be lost due to their rapid damping (deep defect) and limited returned energy (small size). In such a case, the TNDT application can be in the detection of corrosion, yielding to air presence in a steel piece.

This brief review of thermal wave theory applies to the various deployment procedures described below.

9.2 PULSED THERMOGRAPHY

9.2.1 Pulsed Thermography Concepts

Pulsed thermography (PT) is a popular thermal stimulation method in IR thermography whose protocol consists of pulse heating the specimen and recording the temperature decay. One reason for the popularity of PT is the quickness of inspection, relying on a short thermal stimulation pulse with duration ranging from a few milliseconds for high-conductivity material (such as metal) to a few seconds for low-conductivity specimens (such as plastics, graphite epoxy laminates). Such quick thermal stimulation allows direct deployment on the plant floor with convenient heating sources (reviewed in Section 8.1). Finally, the brief heating prevents damage to the component: Heating is generally limited to a few degrees above the initial component temperature. Obviously, the stimulation source must stay nondestructive and thus must not damage the inspected surface either chemically or physically; this will limit its strength. This can be an issue when high-intensity pulses are deployed, especially when high-intensity pulse heating is directed on delicate surfaces (e.g., frescoes, paintings of Old Masters).

Basically, pulsed thermography consists of heating the specimen briefly and recording the surface temperature decay curve with an infrared camera. Abnormal behavior of this temperature decay curve reveals subsurface defects. Qualitatively, the phenomenon is as follow. The temperature of the material changes rapidly after the initial thermal perturbation because the thermal front propagates, by diffusion, under the surface and also because of radiation and convection losses. The presence of a defect reduces the diffusion rate so that when observing the surface temperature, defects appear as areas of different temperatures with respect to a surrounding sound area once the thermal front has reached them. Consequently, deeper defects will be observed later and with reduced contrast. In fact, the observation time t is a function (in a first approximation) of the square of the depth z (Cielo et al., 1987a) and the loss of contrast c is proportional to the cube of the depth (Allport et al., 1988):

$$t \approx \frac{z^2}{\alpha} \quad \text{and} \quad c \approx \frac{1}{z^3} \tag{9.7}$$

where α is the thermal diffusivity of the material (cf. Table 3.7 and Appendix E for thermal properties of common materials). These two relations show two

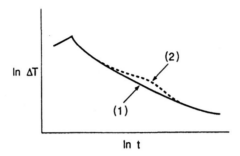

FIGURE 9.2   Temperature evolution curve after absorption of a rectangular heat pulse: Top image: (1) plate made of homogeneous material; (2) same plate containing a sub-surface flaw. Bottom image: difference between curves (1) and (2). (From Maldague, 1993.)

limitations of PT: Observable defects will generally be shallow and the contrasts weak. An empirical rule of thumb says that *the radius of the smallest detectable defect should be at least one to two times larger than its depth under the surface* (Vavilov and Taylor, 1982).

It is now interesting to make the link with thermal wave theory, discussed earlier. Mathematically, a pulse can be decomposed into a multitude of individual sinusoidal components. In that respect, pulse heating of a specimen corresponds to the simultaneous launching into the specimen of thermal waves of various amplitudes and frequencies, in a transient mode. Utilization of a thermal pulse for stimulation of the workpiece is thus practical since thermal waves of many different frequencies are tested simultaneously (the flat frequency spectrum of a Dirac pulse; Figures 10.26 and 10.27). Unscrambling this mix of frequencies to retrieve individual components that are reflected back on subsurface artifacts is also of interest, and this is performed in *pulsed phase thermography* (PPT) processing (discussed in Section 10.4).

The detection phenomenon in pulsed TNDT is illustrated in Figure 9.2. In this figure, the surface temperature evolution following application of the initial thermal perturbation heat pulse is plotted for a thin homogeneous plate (curve 1) and for the same plate when a delamination (air layer) is present under the surface (curve 2). On this logarithmic scale graphic we notice that the central portion of the temperature decay follows a line of slope $(-\frac{1}{2})$. In fact, for a semi-infinite medium, after absorption of a Dirac pulse, this temperature decay conforms to (Carlslaw and Jaeger, 1959)

$$\Delta T = \frac{Q}{e\sqrt{\pi t}} \tag{9.8}$$

where ΔT is the temperature increase of the surface, Q the quantity of energy absorbed, and $e = \sqrt{k\rho C}$ is the thermal effusivity of the material, with k being the thermal conductivity, p the mass density, C the specific heat, and t the time (Table 3.7 and Appendix E list thermal properties of common materials).

It is important to notice that a cold front (heat removal) will propagate in the same fashion as a hot front (heat deposit) inside the material. In some inspection situations, this *cold* approach is more economical: for example, for the inspection of a component already at an elevated temperature (with respect to ambient temperature) due to the manufacturing process, such as for the curing of bonded parts. Moreover, it may be attractive to preheat (if necessary) the surface with a low rate of energy deposition (e.g., by means of electric heating wires). This approach avoids relying on high-intensity local heat sources, which are more costly, susceptible to rapid aging, and potentially dangerous for the surface because of the high thermal stress they induce. Cold thermal sources are also advantageous since they do not generate thermal reflective noise, which can perturb the measurement. Examples of deployment of this cold approach are presented in Section 11.1.

Due to more uniform heating, preheating the specimen in an oven allows us to reveal deeper defects in material of low diffusivity (Wu et al., 1997a). In such experiments, the specimen is introduced in an oven to reach a stable, higher-than-ambient temperature. Next, it is removed from the oven and cooling down of the surface temperature is recorded under ambient conditions. In a reported experiment, a wood sample was inspected both with such a procedure and with lock-in thermography (at $f = 0.0037$ Hz; Section 9.4). While lock-in thermography was able to resolve up to a 4-mm air defect in 5 minutes, the cooling procedure achieved close to 10 mm in 2 minutes. Although this method is less flexible than straight PT, it might be advantageous in specific applications.

The diversity of thermal stimulation sources is very great; authors have reported the use of boiling water, snow, incandescent lamps, lasers, plasma arcs, inductive heating, warm or cool air projection, radiative heating elements (e.g., spiral wires), flash lamps, and aerosol spray (e.g., spray of cold liquid nitrogen vapor). Refer to Section 8.1 for more details on thermal stimulation sources. Progresses could be expected in PT with the availability of uniform high-energy density and small encumbrance thermal stimulation sources, which, moreover, would minimize parasitic (reflected) radiation (see, e.g., Figure 8.11b).

9.2.2 Deployment

Possible deployment for PT (and other approaches discussed below as well) are illustrated in Figure 9.3 (Figure 9.4 shows thermogram aspects in the case of PT). In this figure we present various methods:

(a) *Point heating* (e.g., by means of a laser beam or focused arc lamp). Rather uniform heating is obtained from point to point; the drawback is the necessity to scan the beam if a surface has to be inspected; it is thus a slow method.

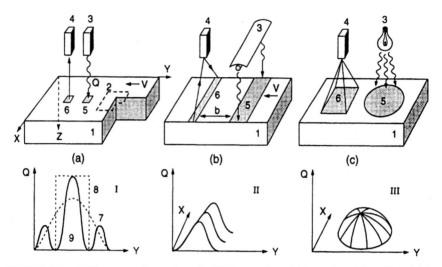

FIGURE 9.3   Different configurations for PT inspection: (a) by point; (b) by line; (c) by surface. I, II, III, energy distribution by the thermal source. 1, sample; 2, defect; 3, thermal source; 4, infrared radiation system; 5, heated area; 6, observation area; 7, 8, 9, heating distributions: Gaussian, uniform, and circular, respectively. (From Maldague, 1993.)

(b) *Line heating* (using a line infrared lamp, heated wire, or line of air jets). Good uniformity and fast inspection rates are obtained because of the lateral scanning.

(c) *Surface heating* (using cinematographic spots, incandescent bulbs, flash lamps, fast laser scanning). It is difficult to obtain uniform heating; the configuration can be either static or mobile.

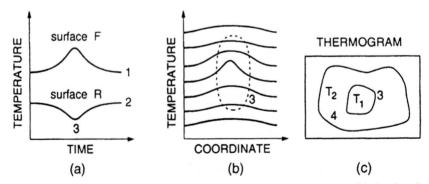

FIGURE 9.4   Shape of the thermograms in PT for (a) point heating; (b) line heating; (c) surface heating. 1, 2, temperature profiles (case of a plate; heating of F, front surface; R, rear surface); 3, zone with defect; 4, heated area. (From Maldague, 1993.)

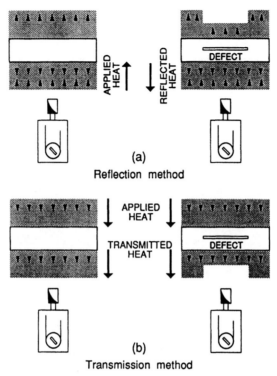

(a)

Reflection method

(b)

Transmission method

FIGURE 9.5 Observation methods for PT: (a) in reflection, (b) in transmission. Note that the thermal front propagation can be either cold or hot. (From Maldague, 1993.)

Position b of the detector (Figure 9.3b) with respect to the heating source is of great importance, since as we have seen before [eq. (9.7)], the optimal time of observation $t_0 = b/v$ is a function of the depth of the defect under observation. Thus point and line heating methods are favorable only when defect detection is limited to cases where the defect depth is known and constant: for example, in bonded assemblies where defects are located between layers of known thickness, presumably at the bonding interface (Section 7.4.1).

Two methods of observation are possible (Figure 9.5): in *reflection*, both the thermal source and the detector unit are located on the same side of the workpiece; in *transmission*, the thermal source and detector unit are located on opposite sides of the workpiece. These two observations methods do not offer the same possibility of detection for defects. In reflection, greater resolution is obtained, but the thickness of the material inspected is small (W. N. Reynolds, 1985). In transmission, a greater thickness of material can be inspected, but the depth information is lost since the thermal front has the same distance to travel whether or not its strength is reduced by the presence of a defect. Moreover, since the resolution is weak, it is necessary to use more sensitive detection equipment. Also, observation in transmission is not always possible, especially for complex structures made of multiple layers (e.g., honeycomb panels). Generally, the approach in reflection is good for the detection of defects located close to the heated surface, while the transmissive approach reveals defects located close to the rear surface (when both sides of a workpiece are available).

Examples of pulsed thermography deployment are presented in Chapter 11. Data analysis in PT is discussed in Chapter 10.

9.3   STEPPED HEATING (LONG PULSE)

In contrast to the thermal stimulation scheme, for which the temperature decay is of interest, here the increase in surface temperature is monitored during application of a stepped heating pulse (the sample is continuously heated, at low power; Figure 9.6). Variations of surface temperature with time are related to specimen features as in pulsed thermography. This technique of stepped heating (SH) is sometimes referred to as *time-resolved infrared radiometry* (TRIR), although strictly speaking, techniques using pulsed heating could also be considered time-resolved. *Time-resolved* means that the temperature is monitored as it evolves during and after the heating process; *stepped-heating* describes the functional dependence of the heating source. SH has numerous applications, such as for coating thickness evaluation (including multilayered coatings, ceramics), integrity of the coating–substrate bond determination or

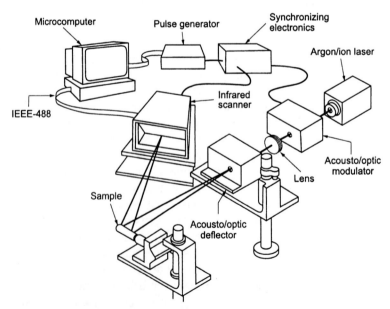

FIGURE 9.6   Experimental setup for time-resolved infrared radiometry. (Adapted from Maclachlan-Spicer et al., 1991.)

evaluation of composite structures, and characterization of airframe hidden corrosion, among others. More details about this technique can be found in Maclachan-Spicer et al. (1991) and Osiander et al. (1998).

Figure 9.6 shows a experimental setup for SH experiments. An argon laser is used to point or line heat the specimen and is time gated using an acousto-optic modulator, allowing a variety of pulse lengths. If the specimen is opaque at the argon wavelength (0.514 μm), surface heating is utilized. Dedicated electronics allow us to synchronize laser heating with respect to the IR camera frame rate (notice that the IR camera operates here in the line mode). Three types of measurement are possible: temperature line scan at a specific time, collection of temperature line scans as a function of time (this is referred to as a TRIR X-time image, showing the temperature development in time over a location of the specimen), and a TRIR X-Y image at a specific time (TRIR X-Y images are constructed on a line-by-line basis with the specimen mounted on a positioning stage moved step by step. Of course, a TRIR X-Y image can be obtained at any time interval desired after starting the heating pulse.)

When SH was developed, only IR scanners were available (Section 4.8). At that time, one of the main advantages of SH was the much faster temporal resolution achievable with respect to full-field imaging at a video rate of about 25 or 30 images/s. This was because in these cases, the vertical mirror of the scanner was stopped so that the device operated only in line scan mode: for example, an 8-kHz line rate mode obtained with an FLIR/Inframetrics 600, with 256 data points (Shepard and Sass, 1990a).

More recently, thanks to the availability of fast FPA IR cameras (Chapter 4), the technique has been simplified. The start of the heating pulse is triggered with respect to acquisition of IR images, which are recorded sequentially for later processing for the entire heating period. In Figure 9.7, the surface is heated with a laser whose beam is enlarged with a beam expander so that a field of view of about 8×8 cm is obtained (Osiander et al., 1998). SH was also dem-

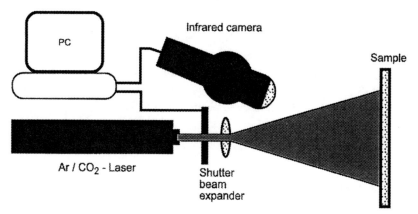

FIGURE 9.7 Experimental setup for time-resolved infrared radiometry with an FPA camera. (Adapted from Osiander et al., 1996.)

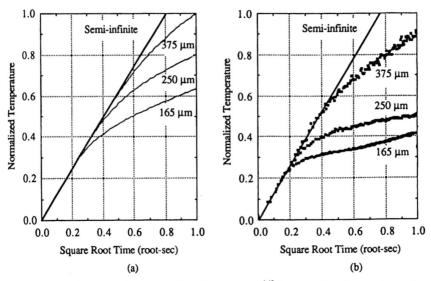

FIGURE 9.8 Curves temperature of surface versus $t^{1/2}$ for a series of zirconia coatings of different thicknesses for a step heating pulse of duration 1 s: (a) theoretical computations; (b) experimental results. (From Maclachlan-Spicer et al., 1991.)

onstrated with induction or microwave heating sources (example: 9 GHz, 20 mW cm^{-2} of power density, Osiander et al., 1995a). In addition to the step and low power involved, experimental deployment of SH is similar to PT (Section 9.2). For example, point or full field heating is possible (Section 9.2.2). Experimental concepts were described in Section 8.1.

Data processing includes evaluation of the thermal transit time t_T, that is, the time the normalized temperature of a given point in the field of view departs from the evolution obtained in the case of a thick, sound specimen. Transit time t_T is related to defect depth.

As an illustration of SH, Figure 9.8 shows the effect of coating thickness on specimen temperature (Maclachan-Spicer et al., 1991). Temperature is plotted versus the square root of time since this linearizes the result for the semi-infinite response. All the curves of Figure 9.8 exhibit the same linear behavior at early times, but at later times each curve begins to drop below the semi-infinite case at a thermal transit time t_T set by the coating thickness L:

$$t_T \sim \frac{0.36L^2}{\alpha} \qquad (9.9)$$

When time is getting larger than t_T, the heating front reaches the substrates (of higher thermal conductivity than the coating for the example shown) and the rate of temperature increase decreases. In the case of a thermally insulated substrate, instead of a drop as shown in Figure 9.8, the rate of temperature increase increases because of heat buildup (the curves would depart above the semi-infinite case). Notice that eq. (9.9) is related to the well-known Parker equation commonly used to establish thermal diffusivity (Section 11.5.1).

Advantages of this approach are that defect depth and thermal properties are accessible from a single measurement without requiring knowledge about a defect-free region (such as to compute the contrast in PT; Sections 5.6 and 10.2). In fact, the early time, before the manifestation of subsurface defects, allows us to calibrate the image in order to take into account spatial variations of emissivity and reflectivity (the constraint of uniform heating is thus relaxed).

Although from the experimental point of view SP is different than PT, from a mathematical point of view, both procedures provide thermal signals that contain exactly the same information (Favro and Han 1998). The choice to rely on PT or SP will thus depend on the applications, and more specifically, on the time scale of the phenomena of interest; slow phenomena are better explored with SH and fast phenomena with PT. Let the period of time required for a given defect to distinguish itself from the background be called Δt_{dist}; then the procedure (SH or PT) that enables to put a maximum energy Q in the specimen during Δt_{dist} will produce the largest signal [eq. (9.8)]. It has been suggested (Favro et al., 1998a) that for a given application, a pulse length slightly smaller than Δt_{dist} be chosen and the pulse strength adjusted so that the maximum surface temperature reached does not damage the specimen. However, other practical criteria, such as limitation of power for continuous sources or limitation in the adjustability of pulse duration for pulsed sources, have to be considered as well. Other examples of SH deployment and data analysis are presented in Section 10.3.

9.4 LOCK-IN THERMOGRAPHY

To introduce the lock-in thermography concept, let's take a linear electric circuit analogy for which in the steady-state regime with a sinusoidal input of angular frequency ω and magnitude E_i, the output E_0 observed is also sinusoidal of the same frequency ω but with specific magnitude A and phase ϕ. The phase corresponds to a shift in the output signal with respect to the input signal. The input parameters are input $= [I, \omega]$ and the output parameters are output $= [A, \phi, \omega]$ (Figure 9.9). Since the parameter ω is left unchanged, it can be put aside. Measurement of parameters $[A, \phi]$ provides various information about the linear circuit involved (for more information on electric circuit analysis, see e.g., D. E. Johnson et al., 1992).

In the case of NDE, this concept is applied as follows, with the specimen acting as the linear circuit. Modulated heat of magnitude Q is deposited on the specimen surface at an angular frequency ω. Such modulated heat launches a thermal wave [eq. (9.1)] which propagates in the specimen and reflects defects back onto the surface, where an IR camera records its manifestation. In this case the input–output parameters are

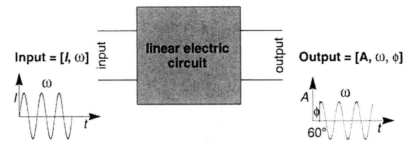

FIGURE 9.9 Concept of lock-in thermography with a linear electric circuit analogy. The input parameters are input = [I, ω], and the output parameters are output = [A, ϕ, ω] (same frequency for the output but different amplitude A and phase). Notice the 60° shift of the output signal on the right.

$$\text{input} = [Q, \omega] \quad \text{and} \quad \text{output} = [A, \phi, \omega] \tag{9.10}$$

Since the parameter ω is left unchanged, it is put aside for now. A is the magnitude of the thermal wave and ϕ the phase.

Lock-in thermography (LT), sometimes called *photothermic radiometry*, is based on the generation inside the specimen under study of these thermal waves, for example, by depositing heat periodically on the specimen surface (Figure 9.10); the resulting oscillating temperature field in the stationary regime can be recorded remotely through its thermal infrared emission by an IR camera (Busse et al., 1992a; Busse, 1994). More specifically, measurement of temperature evolution over the specimen surface permits us to reconstruct the thermal wave and to establish values of A and ϕ from four equidistant temper-

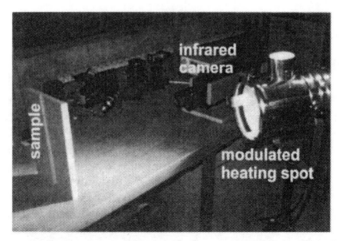

FIGURE 9.10 Lock-in thermography setup. As seen, the positioning of the heating spot is not critical. (Photo courtesy of IKP, Universität Stuttgart, Germany.)

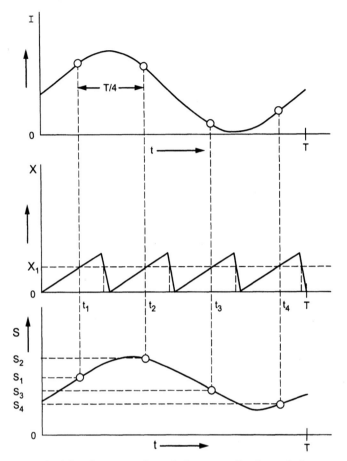

FIGURE 9.11 Principle of computation of phase, amplitude, and thermographic images in lock-in thermography. During each heating cycle of the specimen (light intensity I, top) the infrared camera records four images (pixel, x). Following this process, for any pixel x within the field of view, it becomes possible to reconstruct the local thermal wave. (Redrawn from Busse, 1994, with permission from Elsevier Science.)

ature data points S_1, S_2, S_3, and S_4 recorded over the specimen surface during one cycle of the modulation (Busse et al., 1992a; Busse, 1994; Figure 9.11):

$$\phi = a \tan \frac{S_1 - S_3}{S_2 - S_4} \quad \text{and} \quad A = \sqrt{(S_1 - S_3)^2 + (S_2 - S_4)^2} \qquad (9.11)$$

The thermographic image corresponding to the dc part (or offset) is given by the average of the four points:

$$T = \frac{S_1 + S_2 + S_3 + S_4}{4} \qquad (9.12)$$

Although four points suffice to compute A and ϕ, more points allow us to reduce noise associated with the process. Computer-based commercial systems of LT (such as the lock-in option of FLIR/AGEMA) works on all available points recorded over one or more modulation cycles (1024 points are then involved in the process). Such a system thus provides three images: phase ϕ, amplitude A, and conventional thermography T. The thermographic image is a mapping of the thermal infrared power emitted, the phase image is related to the propagation time, and the modulation image is related to the thermal diffusivity. Obviously, one of the strong points of LT is the phase image, which is relatively independent of local optical and infrared surface features [these features were canceled out in the ratio of eq. (9.11)]. For example, optical features refer to nonuniform heating, while infrared features may concern variability in surface emissivity. Applications of LT are numerous. Just to give an example, the phase was shown to be sensitive to how bolts are (over)tightened (Zweschper et al., 1998, 1999).

The experimental lock-in thermography apparatus shown in Figure 9.10 allows us to observe the effects magnitude and phase of thermal waves on the specimen. The lock-in terminology refers to the necessity to monitor the exact time dependence between the output signal and the reference input signal (i.e., the modulated heating). This can be done with a lock-in amplifier in point-by-point laser heating, or this task can be handled by the controlling computer in full-field deployment (see below). The depth range of the magnitude image is given roughly by the thermal diffusion length μ [eq. (9.2)], while for phase images it is about twice as large. It should be recalled that depth range depends on the material considered [e.g., eq. (9.7)]. For example, in one study in CFRP, phase images were shown to probe 40 to 70% deeper than amplitude images (Vavilov and Marinetti, 1999).

One concern of LT is the acquisition time, which should cover at least one modulation cycle: 1 Hz requires at least 1 s, 0.1 Hz requires at least 10 s, and so on. If the analysis proceeds on point by point through laser heating, for example, these values must be multiplied by the number of pixels in the image, thus still degrading the acquisition time. However, as shown in Figure 9.10, the LT can be applied on a full-field basis by illuminating the entire sample periodically with lamps (to generate the thermal wave) while signal pickup is performed by an IR camera. Reports indicate that such a technique allows us to visualize stringers behind a 2 mm thickness of a CFRP laminate (phase image, 0.03 Hz, time of acquisition: 2 min). On the other hand, since uniformity of illumination is not particularly important for phase images, it is possible to heat a large surface in a given time, up to a few square meters, as shown in Figure 9.12. In such a case, of course, the spatial resolution depends on the IR camera itself: Limited spatial resolution translates into only large-size defects showing up, so that care should be exercised not to miss significant defects (see Section 4.1 on spatial resolution and Section 7.6 on defect size and probability of detection).

Besides the modulation, experimental deployment of LT is similar to PT

heating lamps   infrared camera

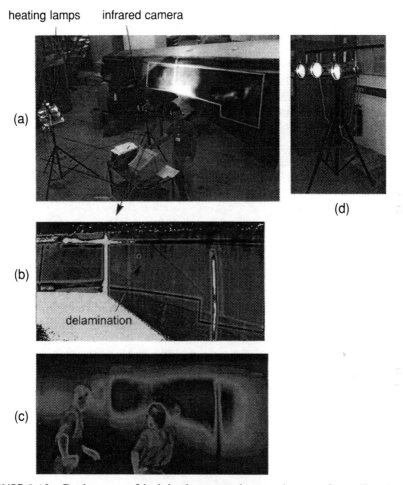

FIGURE 9.12 Deployment of lock-in thermography over large surfaces. Here in the case of a composite helicopter shell: (a) picture of the experimental setup; (b) phase image showing a delaminated area; (c) thermographic image revealing not much of the specimen features; (d) close-up picture of the heating lamps. (Images courtesy of IKP, Universität Stuttgart, Germany.)

(Section 9.3). For example, point or full-field heating is possible (Section 9.2.2). In Section 8.1 we described experimental concepts. Interestingly, in addition to photothermal heating, other heating deployments are possible for various applications. Solar heating cycles were mentioned at the beginning of this chapter. In case of medicine, analysis of blood circulation was demonstrated. Here, a compression cuff located in the upper arm creates the modulation, enabling us to analyze the blood flow in the forearm: repetitive compression is performed through control of the air pressure going into the cuff. Low frequencies (0.03 and 0.015 Hz) reveal blood vessel visualization down to 3 mm under the skin. This has created interest in visualizing the functionality of the circulatory system and especially to discover changes related to a developing pathological process (Wu et al., 1996a).

More recently, it was shown that a suitable thermal stimulation can also be obtained using an ultrasonic transducer (shaker) attached to a specimen (conversely, the specimen can be partly immerged in an ultrasonic bath). With such ultrasonic excitation*, the power is proportional to the stress, and at low stress good power is still available for specimen stimulation so that no damages occur to the specimen. In this deployment, the high-frequency ultrasonic signal (typically, 40 kHz) is modulated with a low-frequency signal. This low-frequency modulation creates a thermal wave of the desired wavelength as in conventional LT (Salerno et al., 1996) while the high-frequency acts as a carrier delivering heating energy right inside the specimen (Rantala et al., 1996a; Dillenz et al., 1999). This technique, referred to as *loss angle lock-in thermography*, is reported to detect deeper and smaller defects, while selective heating allows better discrimination among the defects detected. Typical applications are for detection of corrosion, vertical cracks, and delaminations (Figure 9.13).

A problem sometimes arises with ultrasonic lock-in thermography because the mechanical coupling between the acoustic source and the sample requires that constant pressure be applied by the source on the sample during the entire measuring cycle. To overcome this difficulty, a variant of this technique uses an ultrasonic burst excitation of the sample followed by a cooling-down period. The right spectral components of the resulting response provide information about defects in almost the same way as the lock-in technique but with less sensitivity to coupling problems and with a better signal-to-noise ratio, while

corroded area

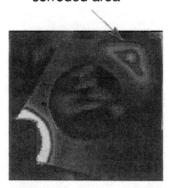

FIGURE 9.13 Ultrasonic lock-in thermography. Image of an aluminum part with corrosion on the back, hidden side. The head of the ultrasonic welding machine is visible (bottom left). Such a part normally bears a bar of the horizontal stabilizer of a small aircraft; it was replaced due to detected corrosion. Arbitrary gray scale units. (Image courtesy of IKP, Universität Stuttgart, Germany.)

* See for instance: www.indigosystems.com.

the measuring duration can be reduced (Dillenz et al., 2000). This is somehow similar to PPT (see Section 10.4).

Another field of application for LT is the characterization of electromagnetic fields, especially their interaction with metallic structures (Nacitas et al., 1994). In this application, the metallic structure of interest is coated with magnetic paint sensitive to the magnetic field or with a relatively thick electrically conductive coating. Heating of these coatings can be related to the electromagnetic field and surface currents. In the steady-state regime the high thermal conductivity of the structure distorts the coating heating, which does not witness the electromagnetic field and surface currents accurately. However, in the modulated regime, if the modulation frequency is sufficiently high to have a diffusion length μ [eq. (9.2)] smaller than the coating thickness, the modulated heating will not be blurred by the thermal diffusion into the metallic substrate. For this application, the magnitude image is of interest since it is proportional to the intensity of the source and thus to the electromagnetic field intensity. As described previously, the phase image which is related to the heat transfer phenomena is not of interest for this application. Finally, it is worthwhile to mention that a pulsed heating approach could have been used instead of the modulation regime. However, even if high-energy pulsed electromagnetic sources are available, a pulsed regime is reported to be difficult to monitor (Balageas et al.,

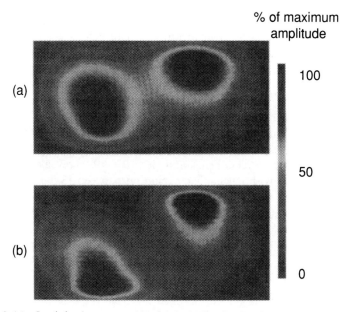

FIGURE 9.14   Lock-in thermography. Analysis distribution in a C-band waveguide, at 8 GHz. The film thickness is 1.5 μm. (a) Constant-amplitude microwave field observed by classical thermography (500 accumulated images); (b) amplitude-modulated microwave field observed by the lock-in thermography system with a modulation frequency of 5.05 Hz. The figure presents the modulus image (amplitude of the modulated component of the temperature) after 5000 accumulated images. (From Balageas et al., 1993.)

1993). Figure 9.14 shows the result of such an analysis performed in a waveguide. In this case, an absorbing screen is inserted into the cavity (Balageas et al., 1993; Matini et al., 1993). For this application, lock-in thermography provides more contrasted values than those of conventional thermography in the stationary regime with a more complex shape distribution available with higher-temperature gradients, while the influence of the cold wall is reduced since a greater part of the film is heated. Other examples of LT deployment and data analysis are presented in Section 10.4.

9.5   VIBROTHERMOGRAPHY

The idea of vibrothermography (VT) originates from German physicist W. E. Weber (1804–1891), who discovered in 1830 that an increase in length of a material results in a decrease in temperature. Such a concept was later formalized by English physicists W. Kelvin (1824–1907) in 1853 and J. Joule (1818–1889) in 1857. Since this effect is reciprocal, a reduction in length causes an increase in temperature. Periodical loading of the material thus generates hot spots at locations of stress concentration.

Vibrothemography (VT) is an active IRT technique based on that principle: Under the effect of mechanical vibrations (0 to 25 kHz) induced externally to a structure, thanks to direct conversion from mechanical to thermal energy, heat is released by friction precisely at locations where defects such as cracks and delaminations are located. Flaws are excited at specific mechanical resonances. Local subplates formed from delaminations resonate independent of the rest of the structure at particular frequencies (Tenek and Henneke, 1991). Consequently, by changing (increasing or decreasing) the mechanical excitation frequency, local thermal gradients may appear or disappear.

Finite element modeling applying 3D equations of linear elasticity permits us to evaluate local energy concentration for components submitted to mechanical loading. For example, studies (Tenek and Henneke, 1991) revealed that thermal patterns appear only with mechanical excitation between 13.5 and 15.0 kHz in the case of a 28×13 cm CFRP (90/0/90) beam attached to an electromagnetic/piezoelectric shaker (coupled to a power amplifier and a frequency generator) from one side. Two simulated delaminations are embedded in the specimen (10×7 mm and 7×7 mm). The result of this analysis is presented on Figure 9.15, where the bigger flaw is excited: At 13.5 and 14.5 kHz, about 20% of the energy released by the beam is concentrated at defect nodes (a temperature increase of about 5°C is noted on the thermogram over the excited delamination).

In some circumstances, mechanical loading provokes breakdown of the specimen. For example, Figure 9.16 illustrates the *ping-pong effect*, where the heat pattern jumps from left to right, up to component failure (Busse, 1994). This is explained by dynamic instability: One side of the hole takes more load than the other and thus becomes hotter, while on the other side, loss of stress

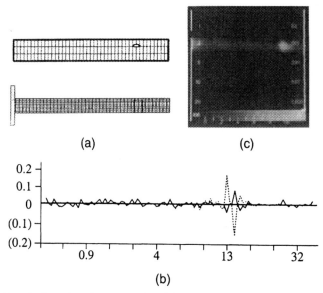

FIGURE 9.15 Active vibrothermography: (a) finite-element discretization of damaged (90/0/90) beam; (b) normalized energy along center of delamination nodes as function of frequency (in kHz); (c) thermogram at 13.5 kHz, showing heat generated by the excited flaw. (Adapted from Tenek and Henneke, 1991.)

stops heat generation (Bauer et al., 1992). Vibrothermography's most significant advantages are the detection of flaws hardly visible by other IR thermography schemes (such as closed cracks), and the ability to inspect large structural areas *in situ* provided that the required mechanical loading can be achieved.

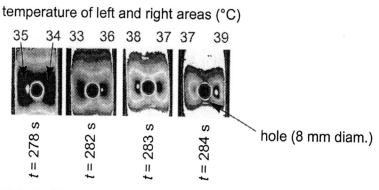

FIGURE 9.16 Thermography under oscillating tensile load: temperature fields of a CFRP 45° tensile laminate of 25 mm width just before failure (time as indicated), stress modulation 10 Hz, 30% of strain. Notice the "ping-pong effect" near the hole. (Adapted from Bauer et al., 1992.)

PROBLEMS

9.1 **[thermal waves: damping factor]** As seen in Figure 9.1, thermal waves are highly damped. Compute the decay factor after a propagation of μ under the surface?

Solution   In this case we have $z = \mu$, and since $\lambda = 2\pi\mu$, we have, from eq. (9.1),

$$T(x,t) = T_0 e^{-x/\mu} \cos\left(\frac{2\pi\lambda}{\lambda} - \omega t\right) = T_0 e^{-1} \cos(2\pi - \omega t)$$

$$= 0.37 T_0 \cos(2\pi - \omega t)$$

The damping factor is 0.37. As mentioned before, in lock-in thermography, phase images are considered to be able to probe up to about 2 µ and amplitude images up to 1 µ (depending on the material inspected).

9.2 **[thermal waves: propagation speed]** Find the propagation speed of a thermal wave in an aluminum specimen.

Solution   The thermal diffusivity of aluminum is $\alpha = 95$ m^2 s^{-1}. We compute the angular frequency ω for, say, three different frequencies: $f = 0.1$, 1, and 10 Hz with $\omega = 2\pi f$. Thus we have

$$v_{0.1} = \sqrt{2(2\pi f)\alpha} = \sqrt{2 \times 6.28 \times 0.1 \times 95} = 11 \text{ m s}^{-1}$$

$$v_1 = \sqrt{2(2\pi f)\alpha} = \sqrt{2 \times 6.28 \times 1 \times 95} = 35 \text{ m s}^{-1}$$

$$v_{10} = \sqrt{2(2\pi f)\alpha} = \sqrt{2 \times 6.28 \times 10 \times 95} = 110 \text{ m s}^{-1}$$

Propagation speed is proportional to the frequency and thus in pulsed heating of an aluminum specimen, thermal waves at different speeds will propagate within the specimen, interact, and reflect on defects. A mix of these frequencies arriving on specimen surface at different times causes blurring at defect edges. Lock-in thermography, stimulating the specimen one frequency at a time, prevents such a blur. In Chapter 10 the pulsed phase thermography processing technique also achieves this through the Fourier frequency transform of images recorded in pulsed thermography.

9.3 **[thermal waves: reflection coefficient]** Find the reflection coefficient between concrete and mild steel.

Solution   Let's first compute the thermal effusivity for both media. For steel $k = 46$ W m^{-1} °C^{-1}, $\rho = 7900$ kg m^{-3}, and $C = 440$ J kg^{-1} °C^{-1}, and for concrete, $k = 1$ W m^{-1} °C^{-1}, $\rho = 2400$ kg m^{-3}, and $C = 800$ J kg^{-1} °C^{-1}:

$$e_{\text{steel}} = \sqrt{k\rho C} = \sqrt{46 \times 7900 \times 440} = 1.26 \times 10^4 \text{ W m}^{-2} \text{ °C}^{-1} \text{ s}^{1/2}$$

$$e_{\text{concrete}} = \sqrt{k\rho C} = \sqrt{1 \times 2400 \times 800} = 1390 \text{ W m}^{-2} \text{ °C}^{-1} \text{ s}^{1/2}$$

and then $b = e_{steel}/e_{concrete} = 9$. Finally,

$$R = \frac{b-1}{b+1} = \frac{9-1}{9+1} \approx +0.8$$

We remember that the reflection coefficient was close to -1 in the case of steel with an air inclusion (Example 9.3). The value obtained now is less favorable, but detection is still possible. Such a situation occurs, for example, in an armed concrete slab in which embedded steel wires increase the mechanical property of the slab (Section 7.3.4). The TNDT application can be aimed at the detection (localization) of the steel wires. The sign is related to the difference in thermal conductivities: here the steel bars (the embedded object) sink the heat energy, while in the case of air (the embedded defect) in Example 9.3, air acts as a thermal barrier. In the case of pulsed thermography deployment on these surfaces, the net result would be a hot spot over the air trapped in the steel slab but a cold spot over the steel wires embedded in the concrete slab.

CHAPTER TEN

Quantitative Data Analysis in Active Thermography

In the preceding chapters, we studied infrared thermography, its theory, and methods and image processing techniques necessary to investigate materials and structures. We also discussed defect detection. Once defects have been located, it is interesting to characterize them quantitatively in order to judge their severity (see Chapter 6 for defect detection procedures). This chapter is dedicated to this study.

In fact, this is known as the *inverse problem*. From the measured thermal temporal and spatial response of a defect detected by TNDT using methods exposed in preceding chapters, we want to evaluate defect size, depth, and thermal resistance. In Section 3.4.2 the *direct problem* was studied by obtaining, through thermal modeling, the thermal response for a defect of known geometry.

In this chapter we first present, in Section 10.1, methodologies to solve the inverse problem, In Section 10.2 we discuss image-processing techniques and experimental procedures needed for the quantitative characterization. Sections 10.2 and 10.3 deal with the pulsed thermography approach, Section 10.3 is concerned with data analysis in step heating, and Section 10.4 presents the pulsed phase thermography together with an inversion procedure and a discussion of lock-in thermography.

10.1 PULSED THERMOGRAPHY: INVERSE PROBLEM

10.1.1 Practical Numerical Approach

Inverse problem solving using thermographic information involves extracting quantitative subsurface defect properties such as depth, thermal resistivity, and size from an experimental TNDT data set. This is an important issue in quantitative evaluation of the severity of a damaged area and it is still on the re-

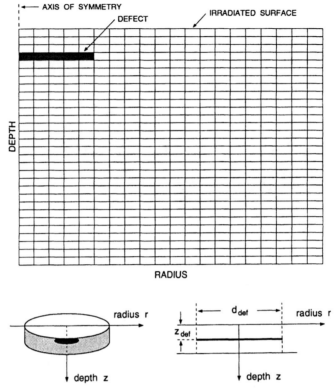

FIGURE 10.1   Modeled geometry for an inclusion-type geometry in cylindrical coordinates, parameters of interest for the defect are also shown. (From Maldague, 1993.)

search agenda (see, e.g., the references at the end of the book). Authors have started to rely on neural networks for inversion of NDE data (this topic is discussed in Section 10.4.5). To study this inverse problem issue we propose here a practical numerical approach that can be programmed in conjunction with the image-processing techniques reviewed in Section 5.6.

As a first approach to this matter, we go back to Section 3.4.2, where a heat transfer model was discussed. We recall that this modeling is used for specimens with subsurface defects. The model for an inclusion-type geometry (for a thermal resistance–type defect) in cylindrical coordinates is shown in Figure 10.1. The subsurface defect is simulated by having a different thermal resistance with respect to the surrounding bulk material. Other geometries are possible, such as the modeling of back-drilled holes considered as subsurface defects when detected from the front surface or Cartesian geometries.

As an application example of the methods discussed, we now study the thermal behavior of graphite epoxy specimens having the same geometry as in the case of the model of Section 3.4.2, but with different thermal resistance R_{def} and depth z_{def}. We recall that for the example reviewed in Section 3.4.2, a graphite epoxy plate was modeled with a defect thickness corresponding to the

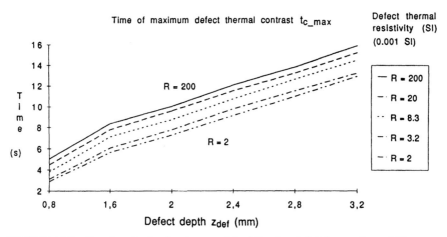

FIGURE 10.2 Results obtained using model of Section 3.4.2 (case of graphite-epoxy specimens), time when thermal contrast reaches a maximum over the defect is plotted for various defect depths and defect thermal resistivity. (From Maldague, 1993.)

thickness of one cell of the mesh. Defect thickness f_{def} can be associated with defect thermal resistance R_{def} by mean of the expression

$$R_{def} = \frac{f_{def}}{k_{def}} \qquad m^2 \, °C \, W^{-1} \quad or \quad SI \qquad (10.1)$$

where k_{def} is the defect conductivity. For example, if an air layer of 200-μm thickness is modeled, since k_{def} is about $0.024 \, W \, m^{-1} \, °C^{-1}$ (see Table 3.1 and Appendix E for thermal values of common materials), the corresponding thermal resistance will be about 8.3×10^{-3} SI.

For the specimens we study, two parameters are varied: defect depth z_{def}, which varies from 0.8 to 3.2 mm (layer 4 to layer 16 in the model of Section 3.5.2), while the thermal resistance of the defects is changed from $R_{def} = 2 \times 10^{-3}$ to 200×10^{-3} SI. Figures 10.2 and 10.3 summarize results obtained for the specimens discussed. On these figures, we plot both the time when thermal contrast reaches a maximum over the defect, parameter t_{c_max} (Figure 10.2) and the value of this contrast (Figure 10.3), parameter $C_{\_max}$. Since only six depths are considered, the curves, particularly on Figure 10.3, are not as smooth as they should be. Considering the geometry of Figure 10.1, temperature evolution is observed (Figure 10.4a) at both the defect center (upper left cell) and far from the defect over a sound area where the effect of the subsurface defect is negligible (upper right cell). This allows us to compute the thermal contrast (Section 5.6) over the defect, which is given by (Figure 10.4b)

$$C(t) = T_{def}(t) - T_{sound}(t) \qquad (10.2)$$

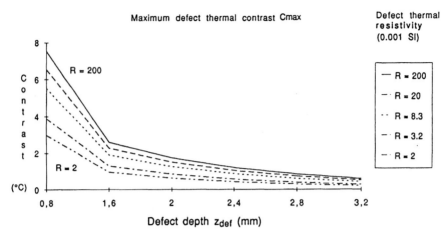

FIGURE 10.3 Results obtained using model of Section 3.4.2 (case of graphite-epoxy specimens), value of the maximum thermal contrast is plotted for various defect depths and defect thermal resistivity. (From Maldague, 1993.)

The variable t is introduced since as shown in these figures, thermal contrast varies with time.

In the example of Figure 10.4b, it is noted that thermal contrast experiences slow variation close to its maximum C_{max}; consequently, uncertainty is associated with the estimation of this parameter and on the associated time of maximum thermal contrast, t_{c_max}. Extraction of the time to reach half the maximum thermal contrast before or after the contrast peak is performed with greater accuracy. These parameters are noted by $t_{c\_1/2max}$ and $t_{c\_max1/2}$.

From data like these used for Figure 10.2 or 10.3, it is possible to fit functions through standard mathematical packages such as Matlab (e.g., function *polyfit* in Matlab) or Mathematica. For example, using such mathematical packages, we computed a function to estimate t_{c_max} with depth z_{def} and thermal resistance R_{def} as input parameters:

$$t_{c\_max} = 1.705 + 1.881 z_{def} + 0.921 z_{def}^2 + 0.000799 z_{def} R_{def}$$
$$+ 0.00138 R_{def} - 8.164 \times 10^{-9} R_{def}^2 \qquad (10.3)$$

Notice that for this equation we selected a quadratic function; this is in agreement with eq. (9.7). In the same fashion, it is also possible to derive functions enabling the inversion process. For example, combining data for both t_{c_max} (Figure 10.2) and C_{max} (Figure 10.3), we can separate the variables R_{def} and z_{def}. The following expression is obtained for z_{def} with C_{max} and t_{c_max} as inputs; this form was originally proposed by Balageas et al. (1987c):

$$z_{def} = 0.6722 \sqrt{t_{c\_max}} (C_{max})^{-0.258} \qquad (10.4)$$

Computed temperature evolution curve

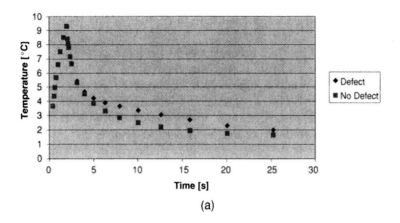

(a)

Contrast evolution over subsurface defect

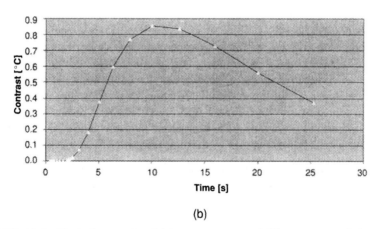

(b)

FIGURE 10.4 Typical example of (a) temperature and (b) contrast evolution curves obtained with model of Section 3.4.2 (case of graphite-epoxy specimen, geometry $z_{def} = 2.4$ mm, $d_{def} = 20$ mm, and $R_{def} = 8.3 \times 10^{-3}$ SI). Plot (b) is obtained using data of (a). (*Note:* Same modeled specimen as for Figure 3.31.)

Refer to the information given at the end of this section to help find the parameters of eq. (10.4).

This kind of analysis, performed first by modeling the analyzed process and than inverting the data computed through mathematical procedures is a simple, though efficient approach to the inverse problem. To be very efficient, however, it is possible to combine more information available on the defect detected, such as $t_{c\_1/2max}$ and $t_{c\_max1/2}$, which are extracted from the temperature evolution curve of the specimen. This extraction process is explained in more detail in

Section 10.2. Of course, this type of analysis yields better results if more cases are analyzed in the direct modeling process. In the present example, only 36 cases were modeled on a restricted span of depths z_{def} and thermal resistances R_{def}.

Equation (10.4) provides accurate figures (to a few percent) on a restricted span of values. Such a figure is generally the best we can expect from such an analysis. This is acceptable, especially if we consider the many other variables present when actual TNDT experiments take place (Sections 4.2, 4.8, and 5.4.1). For example, uneven heating, approximate values for thermal properties of the specimen, variability of the thermal losses (convection, radiation), and difficulty in estimating the emissivity of the surface inspected, to mention only a few. Also, plant floor conditions are more often very different than laboratory conditions, which obviously restricts the limit of the accuracy that such model–experiment comparisons can achieve. This is especially the case for surface-wide inspection. In fact, better agreements are often obtained in point-by-point analysis, due to the better-controlled thermal stimulation conditions. Although convenient, the use of inversion functions such as eq. (10.4) may not be accurate enough in some circumstances. For these situations it is possible to proceed with inverse interpolation using data obtained using the direct model (see, e.g., Krapez et al., 1991).

A good validation procedure for inversion functions involves feeding back data obtained using the direct model. This brings information about the best degree of agreement we should expect and now to determine the effect of possible uncertainties in the input parameters. Such a performance analysis is an important tool. For example, Figure 10.5 presents typical results of such analysis. On the curves of Figure 10.5, it is noted that effects are generally linear and proportional. For evaluation of thermal resistance R_{def}, it is noted that the error tends to amplify; consequently, inversion procedures should yield greater uncertainties regarding thermal resistance than on the depth or size evaluation.

To validate the (direct) model, it is necessary to perform actual experiments in controlled laboratory conditions and compare experimental results with those actually obtained with the model. Ideally, the samples tested have to be destroyed after inspection to determine exactly the nature and geometry of the flaws detected (alternatively, another NDE procedure, such as ultrasonic C-scan, can be used; Section 7.5). If a sufficiently large collections of samples is tested, it is possible to derive a statistical measure of confidence on the inverse functions. This makes it possible to obtain quantitative measurements with a known degree of uncertainty when actual inspection sessions take place [such as the Tanimoto detection criterion, eq. (6.19)].

In this section we insisted on the depth z_{def} of the defect; in the next section we discuss the estimation of both thermal resistance R_{def} and defect size, which can be associated with the corresponding defect diameter d_{def}.

Note on Eq. (10.4). Common mathematical packages cannot fit data based on equations of the eq. (10.4) type. Using logarithms, such formulas simplify to a

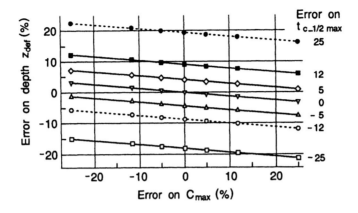

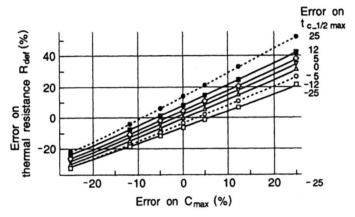

FIGURE 10.5 Effect of the uncertainties introduced on input parameters such as noise on experimental startup data. Cases of defect depth z_{def} and thermal resistance R_{def} are shown (graphite-epoxy specimens). (From Maldague, 1993.)

line whose parameters can be evaluated using just a simple pocket calculator (with the regression function). Let the formula be

$$z_{def} = A\sqrt{t_{c\_max}}(C_{max})^n \qquad (10.5)$$

Taking the logarithm on both sides and rearranging the terms yields

$$\log(z_{def}) - \log(\sqrt{t_{c\_max}}) = \log(A) + n\log(C_{max}) \qquad (10.6)$$

$$\log\left(\frac{z_{def}}{\sqrt{t_{c\_max}}}\right) = \log(A) + n\log(C_{max}) \qquad (10.7)$$

$$\log\left(\frac{z_{def}}{\sqrt{t_{c\_max}}}\right) = B + n\log(C_{max}) \qquad (10.8)$$

If, in the fitting procedure, data are expressed as $\log(z_{def}/\sqrt{t_{c\_max}})$ and $\log(C_{max})$, which at this stage are known, it is only necessary to compute B and n, eq. (10.8) corresponds to the equation of a line and we finally obtain

$$z_{def} = \log(e^B)\sqrt{t_{c\_max}}(C_{max})^n \qquad (10.9)$$

Obviously, $\log(e^B)$ is a constant easily calculated; eq. (10.9) is of the same form as eq. (10.5).

10.1.2 Normalized Variables

Use of normalized variables allows us to generalize the numerical approach presented in the preceding section and to apply these concepts to a wider span of subsurface defect geometries (z_{def}, R_{def}, d_{def}). In fact, investigations and results presented in the preceding section are valid only for a very specific geometry. In some instances it is interesting to rely on broader geometries. Such studies make use of normalized parameters, the Fourier numbers Fo and the Biot numbers Bi.

Characterization of Defect Depth and Thermal Resistance. First, let us recall the effusivity $e(t)$, which describes the temperature evolution of the sample surface. For times t such as $\alpha t/L^2 < 0.1$, it can be expressed as (Balageas et al., 1987, eq. 2; see also Section 9.1)

$$e(t) = \frac{Q}{\Delta T(t)\sqrt{\pi}\sqrt{t}} \qquad (10.10)$$

where Q is the density of the pulsed energy, ΔT the temperature evolution, L the material thickness, and α is the material thermal diffusivity, $\alpha = k/\rho C$, where ρ is the density, C the specific heat, and k the thermal conductivity (Table 3.7 and Appendix E).

The Fourier number Fo_{def} for a defect depth z_{def} is expressed as

$$Fo_{def} = \frac{\alpha t}{z_{def}^2} \qquad (10.11)$$

and the Biot number Bi_{def}:

$$Bi_{def} = \frac{z_{def}}{kR_{def}} \qquad (10.12)$$

In Figure 10.6 we show the evolution of normalized effusivity e_n over the defect for some values of Bi_{def}. The normalized effusivity e_n is given by

$$e_n = \frac{e}{e_{material}} = \frac{e}{\sqrt{k\rho C}} \qquad (10.13)$$

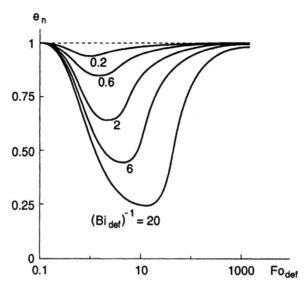

FIGURE 10.6 Evolution of normalized effusivity e_n over the defect (specimen of graphite-epoxy made of two layers with identical thermal properties, the defect, a delamination, is characterized by a specific Biot number Bi_{def}. (From Maldague, 1993.)

It is interesting to note that considering a given combination of defect properties (R_{def}, z_{def}), the effusivity curve is characterized completely, for example by its minimum $(Fo_{def,min}, e_{n,min})$, by its half-decrease point $(Fo_{def,1/2min}, e_{n,1/2min})$, or by its half-rise point $(Fo_{def,min1/2}, e_{n,min1/2})$. Notice that this is similar to the analysis of Section 10.1.1, where we considered the time of maximum or half contrast rise.

From these parameters, interesting inverse functions, such as the subsurface delaminations detected in layered materials, were established (Balageas et al., 1987c):

$$e_{n,min} = [Fo_{min}]^{-0.528} \tag{10.14}$$

$$z_{def} = \sqrt{\alpha}\sqrt{t_{min}}[e_{n,min}]^{0.95} \tag{10.15}$$

where t_{min} is the time when the effusivity curve is minimum. Equation (10.15) is similar to eq. (10.4), studied earlier. Another useful inverse function is

$$z_{def} = 1.61\sqrt{\alpha}\sqrt{t_{1/2min}}[e_{n,1/2min}]^{0.85} \tag{10.16}$$

Where the determination of the minimum of the effusivity curve is not possible with sufficient precision, use of a half-decrease point is preferred.

Thermal resistance characterization is also possible using a similar approach (Delpech and Balageas, 1991). It is, however, not possible to express this parameter R_{def} in a form as compact as that for the defect depth z_{def}. Balageas et al. (1987c) introduced the function ψ relating $e_{n,min}$ and Bi_{def}^{-1} (Figure 10.6):

$$e_{n,min} = \psi(Bi_{def}^{-1}) \tag{10.17}$$

and then they found

$$R_{def} = \frac{1}{e_{material}} \sqrt{t_{min}} [e_{n,min}]^{0.95} \psi^{-1}(e_{n,min}) \tag{10.18}$$

or using the half-decrease point,

$$R_{def} = \frac{1.61}{e_{material}} \sqrt{t_{1/2min}} [e_{n,1/2min}]^{0.95} \psi^{-1}(e_{n,min}) \tag{10.19}$$

Notice the similarity with eqs. (10.15) and (10.16). Using numerical methods similar to those presented in Section 10.1.1, inverse functions can be derived for particular materials and geometries.

10.1.3 Characterization of Defect Size

It is possible to address the characterization of defect size (corresponding to an equivalent defect diameter d_{def}) in a fashion similar to that for defect depth z_{def} and defect thermal resistance R_{def}. Krapez et al. (1991) found that the position of the steeper temperature gradient recorded on the sample surface either at the time of maximum contrast t_{c_max} or at the half-rise time $t_{c\_1/2max}$ is very close to the position of the subsurface defect border. This parameter is called the apparent defect diameter d_{app}. The normalized diameter d_n is then defined as $d_n = (d_{def}/z_{def})(\alpha_z/\alpha_r)^{1/2}$, where α_z and α_r are the thermal diffusivities in the transverse and radial directions of the specimen, respectively.

Heat transfer modeling allows us to construct curves such as those shown in Figure 10.7, relating d_{app} and d_n for various normalized defect resistances R_n ($R_n = R_{def}k/z_{def}$); k is the thermal conductivity and z_{def} is the defect depth. From such curves, once values of z_{def}, R_{def}, and d_{app} (recorded either at t_{c_max} or at $t_{c\_1/2max}$) have been established, recovery of the value of d_{def} is possible using techniques similar to those described in previous sections. Inverse functions $d_{def} = g(z_{def}, R_{def}, d_{app})$ can be derived. Of course, d_{app} depends on both z_{def} and R_{def} since a large deep defect can have the same d_{app} value as a shallow smaller subsurface flaw [this is shown clearly in Figure 6.10j; e.g., for the 10-mm-diameter defects on the image P_D.3, different depths of 1.12 mm (top row) and 2.25 mm (bottom row) lead to different d_{app} values]. If the subsurface defect is not circular in shape, a d_{app}/d_{def} corrective factor may be applied to the apparent defect shape (a steeper temperature gradient on the sample surface) to recover the correct flaw shape.

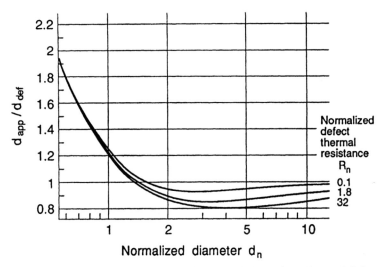

FIGURE 10.7 Computational results showing the relation of the d_{app} and d_n for various normalized defect resistances R_n, graphite-epoxy specimens. (From Maldague, 1993.)

In the next section, we study various procedures and methods necessary for the extraction of experimental parameters.

10.2 PULSED THERMOGRAPHY: EXPERIMENTAL PROCEDURE FOR DATA ANALYSIS

In Section 10.1 we described practical methods useful in solving the inverse problem. We came to the conclusion that knowledge of a few parameters extracted along the temperature evolution curve, a curve recorded at the center of the detected subsurface flaw on the specimen surface, enables the evaluation of defect characteristics, primarily the depth z_{def}, the thermal resistance R_{def}, and the size corresponding to an equivalent defect diameter d_{def} (Figure 10.1). The parameters of interest are the value of the maximum thermal contrast C_{max}, time of maximum contrast t_{c_max}, time of half-rise contrast $t_{c\_1/2max}$, and time of half-decay $t_{c\_max1/2}$. Figure 10.8 shows these parameters over an experimental temperature evolution curve.

A special experimental procedure is required to extract from the recorded thermogram sequence the parameters $t_{c\_1/2max}$, t_{c_max}, $t_{c\_max1/2}$, and C_{max} and to evaluate the subsurface flaw size. Such an experimental procedure is reviewed next.

10.2.1 Thermal Contrast

As related to its timing evolution [such as the contrast $C(t)$], the temperature signal is less affected by noise and is thus a quantity of interest to compute from

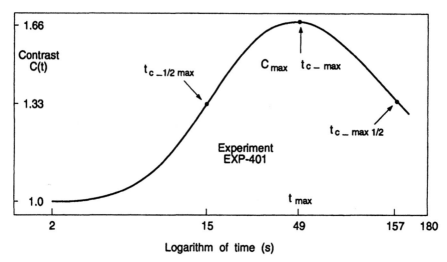

FIGURE 10.8 Experimental thermal contrast evolution curve on which location of interest parameters are indicated (case of a Plexiglas plate with back hole 10 mm in diameter, 1.62 mm below the front surface). (From Maldague, 1993.)

thermographic experiments for both qualitative (defect visualization) and quantitative (defect characterization) purposes. Let us first consider a surface having an emissivity $\varepsilon \sim 1$ initially at temperature T_0. The ambient environment (env) is at temperature T_a. Suppose we heat this surface so that a temperature T_{def} is observed over a defect and a temperature T_{soa} is obtained over a sound area. The radiance signal picked up by the thermal imager and available in an image format is given by, from eq. (4.15),

$$I_{img\_def}(T_{def}) = \varepsilon I_{def}(T_{def}) + (1 - \varepsilon)I_{env}(T_a) \qquad (10.20)$$

$$I_{img\_soa}(T_{soa}) = \varepsilon I_{soa}(T_{soa}) + (1 - \varepsilon)I_{env}(T_a) \qquad (10.21)$$

Contribution   Thermal reflections
from the surface from the surroundings

Before heating (at $t = 0$) the surface was at uniform temperature T_0. The signals were

$$I_{img\_def}(T_0) = \varepsilon I_{def}(T_0) + (1 - \varepsilon)I_{env}(T_a) \qquad (10.22)$$

$$I_{img\_soa}(T_0) = \varepsilon I_{soa}(T_0) + (1 - \varepsilon)I_{env}(T_a) \qquad (10.23)$$

If we subtract eqs. (10.20)–(10.22) and (10.21)–(10.23) to suppress the adverse contributions from the surrounding environment, we have

$$I_{img\_def}(T_{def}) - I_{img\_def}(T_0) = \varepsilon[I_{def}(T_{def}) - I_{def}(T_0)] \qquad (10.24)$$

$$I_{img\_soa}(T_{soa}) - I_{img\_soa}(T_0) = \varepsilon[I_{soa}(T_{soa}) - I_{soa}(T_0)] \qquad (10.25)$$

Then we take the ratio of eq. (10.24) to (10.25):

$$C'' = \frac{I_{\mathrm{def}}(T_{\mathrm{def}}) - I_{\mathrm{def}}(T_0)}{I_{\mathrm{soa}}(T_{\mathrm{soa}}) - I_{\mathrm{soa}}(T_0)} \tag{10.26}$$

The four terms in eq. (10.26) can be converted to temperature using the techniques of Section 4.8 [e.g., eq. (4.21)] and to obtain the contrast C':

$$C' = \frac{T_{\mathrm{def}} - T_{0,\mathrm{def}}}{T_{\mathrm{soa}} - T_{0,\mathrm{soa}}} = \frac{\Delta T_{\mathrm{def}}}{\Delta T_{\mathrm{soa}}} \tag{10.27}$$

We can now add both the spatial variations [since the thermal contrast may change for every pixel (i, j) in the image] and the time variations t (since the thermal contrast evolves during the experiment) to obtain the thermal contrast equation [this is, in fact, the derivation of eq. (5.33)]:

$$C(i, j, t) = \frac{\Delta T_{\mathrm{def}}(i, j, t)}{\Delta T_{\mathrm{soa}}(t)} = \frac{T_{\mathrm{def}}(i, j, t) - T_{\mathrm{def}}(i, j, t = 0)}{T_{\mathrm{soa}}(t) - T_{\mathrm{soa}}(t = 0)} \tag{10.28}$$

We recall that $T_{\mathrm{soa}}(t)$ is the temperature averaged over the entire sound area (soa). Equation (10.28) gives the thermal contrast, which is the temperature differential normalized by the temperature differential over a nondefect area. A unity value will be obtained over an area without defect if the disturbances from emissivity and thermal reflections stay constant during the experiment. Equation (10.28) combines the temporal and spatial reference methods reviewed in Section 5.5 and is a standard definition of thermal contrast. Other contrast definitions are useful as well (Section 5.6).

10.2.2 Logarithmic Time Scale

Equation (10.28) allows us to evaluate for any time t the thermal contrast above the subsurface defect detected. In Section 8.1 we discussed practical limitations of image acquisition systems on both the rate of acquisition and the total amount of memory available to record a complete thermal experiment. Taking these limitations into account, it is often necessary to rely on a *temporal scale* in order to perform inspections, which sometimes might last up to 15 minutes for some low-thermal-conductivity samples.

A logarithmic time scale is selected for two reasons. First, with temporal intervals getting larger with time, we can proceed to inspection sessions which span over long periods of time while having, at the beginning, short time intervals. This allows us to catch fast thermal events which are likely to happen at the beginning of the thermal evolution curve. In other terms, use of a logarithmic time scale permits us to make linear the quadratic law of thermal diffusion [eq. (9.7)].

Second, as time passes, thermal contrasts weaken because of the 3D diffusion of the thermal front, which tends to make uniform the surface temperature distribution, thus increasing the perturbing effect of the noise. Large time intervals containing a large number of images which are averaged together make it possible, in fact, to increase the signal-to-noise ratio without distorting the thermal contrast evolution curve, since for these large time values, thermal events change slowly.

Disjoint Case. The logarithmic time scale is divided in N zones from time t_0 after starting the thermal perturbation (at $t = 0$) to t_f at the end of the experiment:

$$
\begin{array}{ccccccc}
i = 0 & i = 1 & \cdots & i & \cdots & i = N - 1 \\
\end{array}
$$

$$
t_{0-} \quad t_0 \quad t_{1-} \quad t_1 \quad \cdots \quad t_{i-} \quad t_i \quad \cdots \quad t_{f-} \quad t_f
$$

Two other parameters are system dependent: q is the image acquisition rate; this is the time required to frame-grab and save the image temporarily in fast memory (images are later saved on a slower, high-volume memory medium such as a hard disk) and NM is the maximum number of images that the fast temporary memory can hold.

The logarithmic time interval is given by

$$
\Delta \log = \frac{\log t_f - \log t_0}{N - 1} = \frac{1}{N - 1} \log \frac{t_f}{t_0} \tag{10.29}
$$

and for zone t_i corresponding to zone i of the temporal scale, we have

$$
\log t_i = \log t_0 + i\Delta \log \qquad i = 0, 1, \ldots, N - 1
$$

$$
t_i = t_0 \left(\frac{t_f}{t_0} \right)^{i/N-1} = t_0 \mu^i \quad \text{with} \quad \mu = \left(\frac{t_f}{t_0} \right)^{1/N-1} \tag{10.30}
$$

Low borders of time zones are given by $t_{0-}, t_{1-}, \ldots, t_{i-}, \ldots t_{n-}$, so that zone i spans from t_{i-} to t_i. Let's introduce λ, which specifies the zone width:

$$
\log t_{i-} = \log t_i - \log \lambda
$$

$$
t_{i-} = \frac{t_i}{\lambda} \tag{10.31}
$$

Parameter λ is evaluated by an iterating process to ensure that the maximum possible number of images NM will be recorded knowing that we cannot grab a *fraction* of an image. Obviously, there is also at least one image per zone. In the following formula "1" is added since as soon as the image acquisition process is started, it must be ended:

$$NM \approx NM_{estimated} = \frac{t_0 - t_{0-} + 1}{q} + \cdots + \frac{t_i - t_{i-} + 1}{q} + \cdots + \frac{t_f - t_{f-} + 1}{q}$$

which is simplified using eqs. (10.30) and (10.31):

$$NM \approx NM_{estimated} = \frac{t_0}{q} \left(1 - \frac{1}{\lambda_{first\ run}}\right) \left(\frac{1 - \mu^N}{1 - \mu} + N\right) \qquad (10.32)$$

By isolating λ in eq. (10.32), a starting value $\lambda_{first\ run}$ can be computed. Second, the exact value of λ is obtained by adjusting its value and computing the *exact* number of images until exactly NM images are obtained, $NM_{estimated} \rightarrow NM$:

$$NM_{estimated} = \text{integer part} \left[\frac{t_0}{q} \left(1 - \frac{1}{\lambda}\right) + 1\right] + \cdots$$

$$+ \text{integer part} \left[\frac{t_f}{q} \left(1 - \frac{1}{\lambda}\right) + 1\right] \qquad (10.33)$$

There is another constraint on λ: Since time zones cannot overlap others, we must have

$$t_0 < t_{1-} \qquad (10.34)$$

Thus with eqs. (10.30) and (10.31), we have

$$\lambda < \mu \qquad (10.35)$$

In Table 10.1 an example of a logarithmic time scale is computed using eqs. (10.30) to (10.33); time zones are disjoint, as shown.

It is not always possible to respect constraints on all parameters (t_0, t_f, N, NM, and q) simultaneously. This means that acquisitions are done on a non-stop basis, time zones becomes contiguous ($\lambda = \mu$) and the algebraic derivation of the time zones is slightly different, as we see next.

Continuous Case. In the contiguous case the end of a zone equals the beginning of the next. Using eqs. (10.30) and (10.31), we have

$$t_{i-1} = t_{i-} \qquad (10.36)$$

$$t_0\mu^{i-1} = \frac{t_0\mu^i}{t\lambda} \qquad (10.37)$$

TABLE 10.1   Disjoint Time Zone Case[a]

```
Input parameters:
  NM (maximum number of images):284
  q (acquisition + temporary storage, 1 image):0.27s

Input parameters:N (number of zones):10
                  t₀ (starting of acquisition):      1 s
                  t_f (end of experiment):         100 s
```

Computed parameters:

μ:	1.668101
λ(first run):	1.422476
$NM_{estimated}$ (first run):	279
λ(42 nd run):	1.432348
$NM_{estimated}$ (42 nd run):	284

ZONE #	t- (s)	t (s)
0	0.70	1.00
1	1.16	1.67
2	1.94	2.78
3	3.24	4.64
4	5.41	7.74
5	9.02	12.92
6	15.04	21.54
7	25.09	35.94
8	41.85	59.95
9	69.82	100.00

Source: Adapted from Maldague (1993, Table 6.1).

[a] See for instance Matlab script *timezones2.m* available in Appendix F.

$$\frac{t_0 \mu^i}{\mu} = \frac{t_0 \mu^i}{\lambda} \tag{10.38}$$

$$\mu = \lambda \tag{10.39}$$

Before we proceed to the actual time zone computations, it is necessary to check if the parameters supplied by the user (t_0, t_f, N) allow us to obtain at least one image in each zone. For this to occur, following the acquisition of one image we immediately proceed with the next zone. However, the starting time of the next time zone must not be exceeded after the last image acquisition of the preceding time zone:

$$\frac{t_0}{\lambda} \quad t_0 \quad \frac{t_1}{\lambda} \quad t_1 \quad \cdots$$

From this we see immediately that

$$\begin{array}{ccc} \text{start of} & + \text{time to} & < \text{beginning} \\ \text{acquisition} & \text{grab} & \text{of next zone} \end{array}$$

$$\frac{t_0}{\lambda} + q < \frac{t_1}{\lambda} \tag{10.40}$$

with eqs. (10.30), (10.31), and (10.39):

$$t_0 < (t_0 - q)\mu \tag{10.41}$$

and finally,

$$\frac{t_0}{t_0 - q} < \mu \tag{10.42}$$

If eq. (10.42) is respected, this means that there is no overlapping zone and the computations can be done as previously using eqs. (10.30) to (10.33), and eq. (10.35) will be respected. If eq. (10.42) is not respected, the constraints imposed on the parameters are too strong for the given value of q (q is the time to frame-grab an image; it is installation dependent). As mentioned before, this corresponds to the nonstop acquisition case, and t_f and N must be adjusted accordingly. In this case, since we want at least one image per zone, we are at the limit of eq. (10.42), and parameter μ is then computed as follows:

$$\mu = \frac{t_0}{t_0 - q} \tag{10.43}$$

The time for the end of experiment t_f is changed:

$$t_f = \text{minimum}[t_f(\text{specified by user}), t_0 + \text{NM} \cdot q] \tag{10.44}$$

and the number of zones is computed to eq. (10.43) with eq. (10.30):

$$N = \text{rounded}\left[1 + \frac{\log(t_f/t_0)}{\log \mu}\right] \tag{10.45}$$

In this case, all the zones will be contiguous. In Table 10.2 we show a logarithmic time scale computed using eqs. (10.30), (10.31), (10.43), (10.44), and (10.45); the time zones are contiguous. Notice that in Table 10.2, the last zone ends at $t = 81.94$ s instead of 77.68 s because in eq. (10.45), we must have an integer number of zones.

We may now summarize time-zone computations once the user has specified the input parameters t_0, t_f, N, NM, and q. First, eq. (10.42) is checked using eq. (10.30):

TABLE 10.2 Continuous Time Zone Case[a]

```
Installation parameters:
  NM (maximum number of images):284
  q (acquisition + temporary storage, 1 image):0.27s

Desired parameters:
                N (number of zones):           30
                t0 (starting of acquisition):   1 s
                tf (end of experiment):       100 s

Computed parameters:
                μ = λ:                    1.369863
                tf (end of experiment):     77.68 s
                N (number of zones):           15
```

ZONE #	t- (s)	t (s)
0	0.73	1.00
1	1.00	1.37
2	1.37	1.88
3	1.88	2.57
4	2.57	3.52
5	3.52	4.82
6	4.82	6.61
7	6.61	9.05
8	9.05	12.40
9	12.40	16.99
10	16.99	23.27
11	23.27	31.87
12	31.87	43.66
13	43.66	59.81
14	59.81	81.94

Source: Adapted from Maldague (1993, Table 6.2).
[a] See for instance Matlab script *timezones2.m* available in Appendix F.

- For the disjoint case, use eqs. (10.30), (10.31), (10.32), and (10.33).
- For the contiguous case, use eqs. (10.30), (10.31), (10.43), (10.41), and (10.42).

As we saw in Chapter 9, the stimulating heat pulse has, of course, a finite duration. It can be demonstrated that in these conditions, the best origin for the time scale is the one corresponding to the barycenter of the heating pulse. The origin of the time scale is thus the barycenter of the heating pulse, defined as $t = 0$.

Once the thermal inspection is completed, we obtain a certain number of images in each temporal zone. To limit noise, images are averaged in each zone; the idea is to obtain only one *resulting image* associated with each zone *i* corresponding to time t_i. This value t_i depends on the acquisition time of all images averaged together in zone i. Since on the temperature-history curve (Figure 9.2), the temperature T follows in time $1/\sqrt{t}$ after application of the thermal perturbation (heat pulse), we have

$$t_i = \frac{1}{\left(\dfrac{1}{n_i}\displaystyle\sum_{j=1}^{n_i}\dfrac{1}{\sqrt{t_j}}\right)^2} \qquad (10.46)$$

where n_i is the number of images in time zone i and t_j is the time corresponding to image j in zone i.

Example 10.1   Suppose that there are three images recorded at times 8, 9, and 10 s in a given time zone. Find the time associated with this zone.

SOLUTION   Following eq. (10.46), the time associated with this zone is given by

$$t = \frac{1}{\left[\dfrac{1}{3}\left(\dfrac{1}{\sqrt{8}}+\dfrac{1}{\sqrt{9}}+\dfrac{1}{\sqrt{10}}\right)\right]^2} = 8.94 \text{ s}$$

10.2.3   Practical Computation of $C(t)$, C_{max}, $t_{c\_1/2max}$, t_{c_max}, and $t_{c\_max1/2}$

We now have all the elements needed for the experimental computation of thermal contrast. Once the pulsed thermography experiment is performed based on the techniques reviewed in preceding chapters, if a subsurface flaw is present, using an automatic defect detection algorithm (Chapter 6), it is located through the series of thermal images recorded at position (i_d, j_d), corresponding to the center of the defect. Thermal contrast $C(t)$ is computed next using eq. (10.28). Preliminarily, a series of, say, 15 images is recorded and averaged prior to the beginning of heating in order to obtain the *cold image* (Img0). The temperature of this image is also converted, following the techniques of Section 4.8, to obtain $T_0(i, j), \forall i, j$. A region far from the defect called r_far, comprising s elements, is defined in the image field. Also, from cold image Img0 we compute the temperature for all elements l of this zone to obtain row vector r_far$T0(h)$, $\forall l = 0, 1, \ldots s - 1$.

A row vector called Row$T0$ is computed from the cold image Img0. It is the average, pixel by pixel, of three rows converted into temperature and passing through the defect center; these are the rows i_{d-1}, i_d, and i_{d+1}: For $\forall j, j = 0, 1, \ldots, \text{Maxcol} - 1$:

$$\text{Row } T0(j) = \frac{1}{3}\sum_{i=i_d-1}^{i=i_d+1} T0(i, j)$$

For every time zone k, thermal contrast $C_{\text{row}}(k, j)$ along elements j of row i_d passing by the hottest point of the defect (i.e., defect center) is computed using eq. (10.28). In fact, this contrast computation is performed on the three rows, and the three contrast values are averaged together for each position of the row, to obtain the row vector $C_{\text{row}}(k, j)$. For $\forall j, j = 0, 1, \ldots, \text{Maxcol} - 1$, we compute

$$C_{\text{row}}(k, j) = \frac{1}{N_k} \sum_{m=0}^{N_k-1} \left[\frac{\text{Row } T(m, j) - \text{Row } T0(j)}{(1/s) \sum_{l=0}^{l=s-1} [r\_\text{far } T(m, l) - r\_\text{far } T0(l)]} \right] \qquad (10.47)$$

where Row $T(m, j)$ is the jth element $(j = 0, 1, \text{Maxcol} - 1)$ obtained from computation of the average of the rows i_{d-1}, i_d, and i_{d+1} of the mth image belonging to time zone k converted to temperature $T(m, i, j) \forall i, j$. For $\forall j, j = 0, 1, \ldots, \text{Maxcol} - 1$, we compute

$$\text{Row } T(m, j) = \frac{1}{3} \sum_{i=i_d-1}^{i=i_d+1} T(m, i, j)$$

where $r\_\text{far } T(m, l)$ is the lth element $(l = 0, 1, 2, \ldots, s - 1)$ of the sound region defined far from the defect and computed in temperature for the mth image in time zone k and m is the image index for time zone $k, m = 0, 1, 2, N_k - 1$. There are N_k images in time zone k.

The denominator of eq. (10.47) is a scalar number that divides each element of the numerator. All the temperature computations are done following the procedure described in Section 4.8. The row vector $C_{\text{row}}(k, j)$, $\forall j$ corresponds to thermal contrast of the image row passing through the defect center for time zone k; time zone k corresponds to an instant given by eq. (10.46).

Contrast computations thus proceed on an image row basis. From a programming point of view, this format is advantageous since we obtain for all time zones an image of the temporal evolution of the thermal contrast where every row (one row per time zone) is given by eq. (10.47). Since this corresponds to the same format as that of a standard image, [number of rows] × [number of columns], Maxrow × Maxcol (missing rows are set to 0 if the number of time zones is smaller than Maxrow), all display routines, save programs, and so on, are the same as for regular thermograms. Figure 10.9 shows a typical result: This is the temporal thermal contrast evolution recorded over a drilled hole (5 mm diameter, 2 mm beneath the surface in a graphite epoxy plate). There are 40 time zones, from $t_0 = 5$ s to $t_f = 120$ s. In this figure, the rise and decay of the thermal contrast are clearly evident.

From this plot of the temporal evolution of the thermal contrast $(C_{\text{row}}(k, j), \forall j, \forall k)$, which is smoothed using techniques exposed in Section 5.5.5, we extract the value of the maximum contrast above the defect:

$$C_{\text{row}}(k_{\max}, j_{\max}) > C_{\text{row}}(k, j) \forall k, j \qquad (10.48)$$

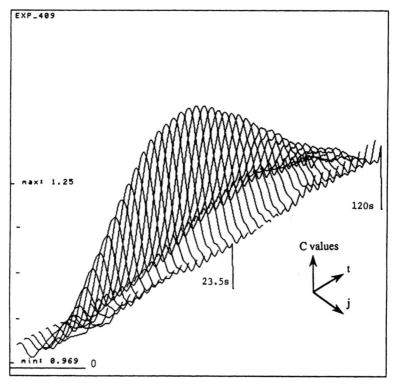

FIGURE 10.9 Temporal contrast evolution curve above graphite-epoxy plate with defect (hole of 5 mm diameter drilled 2 mm beneath the front surface). The 40 time zones are shown from 5 to 120 s. Every line plotted corresponds to the thermal contrast along the image rows (average of three) passing just above the defect for a given time zone. (From Maldague, 1993.)

where k represents the time zone considered and j represents the position along the row. Generally, we have $j_{\max} = j_d$.

The profile of the maximum temporal thermal contrast is then extracted from a plot of the thermal contrast over the defect (Figure 10.10):

$$C_{\text{row}}(k, j_{\max}), \quad \forall \text{ time zones } k = 0, 1, N_k - 1 \qquad (10.49)$$

Figure 10.10 shows the profile $C_{\text{row}}(k, j_{\max})$ obtained in the case shown in Figure 10.9. From the plot of eq. (10.49), $t_{c\_1/2\max}$, $t_{c\_\max}$, $t_{c\_\max 1/2}$, and $C_{\max}$ are extracted. We recall that these parameters are needed for the inversion procedure of Section 10.1.1. For accurate extraction, it is necessary to interpolate (using a line or a parabola) since we only have values at instants corresponding to time zones k:

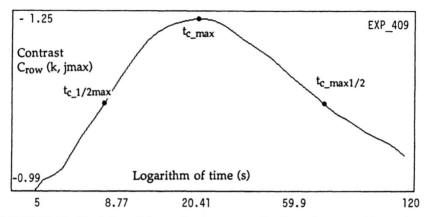

FIGURE 10.10   Evolution of the maximum contrast $C_{row}(k, j_{max})$ extracted from Figure 10.9. (From Maldague, 1993.)

- C_{max}: maximum contrast.
- t_{c_max}: instant of maximum contrast. The two values C_{max} and t_{max} are obtained by computing the parabola passing by the three or five points having the largest value in the contrast vector of eq. (10.49) (refer to Appendix C for details on such computations).
- $t_{c\_1/2max}$: instant at which thermal contrast reaches half its maximum value C_{max}. It is obtained by seeking through the contrast vector of eq. (10.49) for the closest half-rise contrast values and interpolating on two or four neighbors.
- $t_{c\_max1/2}$: instant at which thermal contrast decays to half its maximum value C_{max}. It is obtained by seeking through the contrast vector of eq. (10.49) for the closest half-decay contrast values and interpolating on two or four neighbors.

If we take the example of Figure 10.10, the following values are obtained:

- Maximum contrast C_{max}:

$$[5 \text{ points}] = 1.247 \quad t_{max} = 20.32 \text{ s}$$
$$[3 \text{ points}] = 1.247 \quad t_{max} = 20.42 \text{ s}$$

- Half-rise thermal contrast $t_{c\_1/2max}$:

$$[4 \text{ points}] = 8.744 \text{ s (half contrast} = 1.124)$$
$$[2 \text{ points}] = 8.772 \text{ s}$$

• Half-decay thermal contrast $t_{c\_max1/2}$:

$$[4 \text{ points}] = 59.94 \text{ s}$$
$$[2 \text{ points}] = 59.95 \text{ s}$$

From our definition of thermal contrast, eq. (10.28), a value of 1 corresponds to an absence of defect (i.e., a state of uniform temperature). For example, in the case of Figure 10.10, the thermal contrast above the defect decays to 1.05 (i.e., 5% contrast) 2 minutes after the heating because of the temperature, which tends to become uniform, due to the cooling down of the graphite epoxy plate (i.e., heat-spreading diffusion process within the material and surface losses).

The defect shape (d_{app}) is obtained from the thermal contrast image C_{img} at the time the thermal contrast is maximum. To obtain the contrast image C_{img}, contrast computations are done on the entire image (all the rows). All the images of the time zone k_{max} where thermal contrast is maximum are selected and eq. (10.47) is applied to each pixel $T(m, i, j)$ of the mth image (converted in temperature) of the time zone k_{max} (i = row, j = column in the image): For $\forall j$, $j = 0, 1, (\text{Maxcol} - 1)$ and $\forall i, i = 0, 1, \ldots, \text{Maxrow}$:

$$C_{img}(k_{max}, i, j) = \frac{1}{N_{k_{max}}} \sum_{m=0}^{N_{k_{max}}-1} \frac{T(m, i, j) - T0(i, j)}{\frac{1}{s} \sum_{l=0}^{l=s-1} [r\_far \, T(m, l) - r\_far \, T0(l)]} \tag{10.50}$$

where $N_{k\,max}$ is the number of images in time zone k_{max}. For example, in the case of Figure 10.9, we give in Figure 10.11 three images: (*a*) the raw image (one taken among the images of zone k_{max}), (*b*) the (unsmoothed) contrast image C_{img} obtained using eq. (10.50), and (*c*) the (smoothed) contrast image C_{img} [i.e., image (*b*) smoothed using the sliding Gaussian of Section 5.5.5]. We notice that the smoothed (maximum) contrast image C_{img} is well corrected (uniform background, good defect visibility) and ready for defect shape extraction, as we will see in the next section. The better appearance of the contrast image was foreseeable since this image is obtained after enhancement procedures which reduce radiometric distortions (through temperature computations) and eliminate parasitic effects of thermal reflection and emissivity. Finally, notice that obviously the technique presented in this section will be more accurate if q is high (q is the image acquisition rate: number of thermal images recorded per second).

10.2.4   Defect Shape Extraction

Gradient Computation. We recall that an extraction algorithm that permits us to determine the defect shape was presented in Section 6.3. This algorithm was based on an adjustable threshold gradually lowered around the hottest pixel

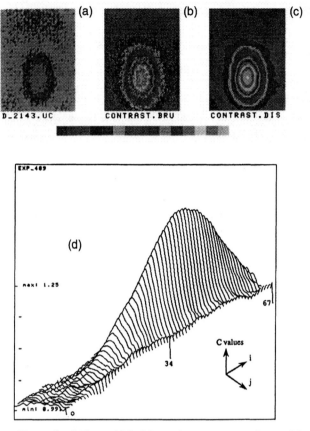

FIGURE 10.11   Example of Figure 10.9: (a) raw images among those of time zone k_{max}; (b) standard contrast image C_{img} computed using eq. (10.50); (c) image (b) smoothed using the technique of Section 5.5; (d) 3D plot of image (c). The gray scale is relative (pixel intensity from 0 to 255). (From Maldague, 1993.)

of the defect (defect center). This method was practical and fast but was only an approximation. As we saw in Section 10.1, the exact defect shape can be obtained from the apparent shape, which corresponds to the steepest temperature gradient computed on the sample surface over the defect detected. In this section we discuss gradient computation and defect shape extraction. The gradient will be computed from the smoothed maximum contrast image C_{img} (as that shown in Figure 10.11c and d).

It is necessary to differentiate the maximum contrast image [eq. (10.50)], once smoothed (Section 5.5.5), in order to obtain the gradient image. One of the simplest differentiation methods consists of computing the Roberts gradient (see also Section 5.5), which is defined with respect to the cross difference of Figure 10.12a and which is defined as follows (Gonzalez and Woods, 1992,

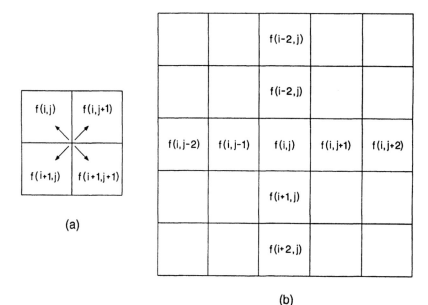

FIGURE 10.12 Pixel relations for (a) Roberts gradient; (b) gradient computed using eq. (10.52). (From Maldague, 1993.)

p. 199, Sec. 5.5):

$$|G| = G[f(i, j)] = \{[f(i, j) - f(i + 1, j + 1)]^2 + [f(i + 1, j) - f(i, j + 1)]^2\}^{1/2} \tag{10.51}$$

In this formula f is the image on which the gradient $|G|$ is computed and i and j are image indexes. As we will see below, this first-order approximation produces noisy gradient images from which defect extraction is more delicate. Consequently, it is better to rely on a higher degree of approximation. In Appendix D a higher-order gradient approximation is derived. Using this method, the gradient image $|G|$ is computed by means of the following formula applied on both image f rows and columns as shown on Figure 10.12b [eq. (D.4), Appendix D): For $\forall i, i = 2, 3, \ldots, \text{Maxcol} - 3$ and $\forall j, j = 2, 3, \ldots, \text{Maxrow} - 3$, compute

$$|G| = \left\{ \begin{array}{l} [f(i, j - 2) - 8f(i, j - 1) + 8f(i, j + 1) - f(i, j + 2)]^2 \\ + [f(i - 2, j) - 8f(i - 1, j) + 8f(i + 1, j) - f(i + 2, j)]^2 \end{array} \right\}^{1/2} \tag{10.52}$$

and the edge pixels are obtained through duplication. Such a method is more accurate than the Roberts gradient [eq. (10.51)] without requiring much more time for computation (see below). Sensitivity to residual noise presents in the smoothed contrast image is one of the main disadvantage of this method.

If a smoother gradient image is wanted, another method can be used which is based on a second-order fit over the few pixels taken around the pixel of interest at location indexes (i, j) in image f (Figure 10.12b; refer to Appendix C for fitting computations). For each direction (row i, column j), the second-order fit function can be expressed by

$$f_r(i) = a_r + b_r i + c_r i^2$$
$$f_c(j) = a_c + b_c j + c_c j^2$$

where a_r, b_r, c_r and a_c, b_c, c_c are the parameters found by the fitting procedure (along the row and column, respectively). Notice that to avoid numerical errors in the fitting procedure for pixels of high indexes, it is necessary to replace the exact index location by $-3, -2, -1, 0, 1, 2, 3$, where 0 is for the pixel considered (i, j). The first derivative is then given by

$$f_r'(i) = b_r + 2c_r i \sim b_r$$
$$f_c'(j) = b_c + 2c_c j \sim b_c$$

and then $|G|$ is given by

$$|G(i, j)| = [f_r'(i)^2 + f_c'(j)^2]^{1/2} \tag{10.53}$$

Figure 10.13 presents typical results of gradient computations. As expected, Roberts gradient [eq. (10.51)] has a strong noise level. A higher-degree approximation [eq. (10.52)] shows improvement in this respect (labeled *ROBERTS_2nd* in Figure 10.13). Better results are obtained if the method of eq. (10.53) is applied (labeled *FIT_2* in Figure 10.13), particularly if the fitting procedure spans more pixels, such as using a 11 by 11 kernel. The method of Equation (10.53) is, however, costly in terms of processing time (Table 10.3). See also Section 5.5.1.

Defect Shape Extraction. From this gradient image $|G|$, the next step is to extract the defect shape, which corresponds, to a given factor (Section 10.1.2), to the locus of maximum derivative around the defect center located in i_d, j_d (the center of the crater; Figure 10.13). This operation can be performed by a simple technique. From the defect center, a line segment D of orientation θ is rotated $0 < \theta < 360°$ and the values (pos x, pos y) corresponding to the maximum gradient along D are found (Figure 10.14):

$$\text{pos } y = j_d - (\text{pos } x - i_d) \tan \theta \tag{10.54}$$

For a given θ, pos x is incremented to 1 at each iteration (pos $x = i_d, i_d + 1$, $i_d + 2, \ldots$) until the maximum $|G|(\text{pos } x, \text{pos } y)$ is found through a search done within all the values obtained along D. The algorithm presented in Section 6.3 can be adapted to perform this maximum search successfully. (Remember that

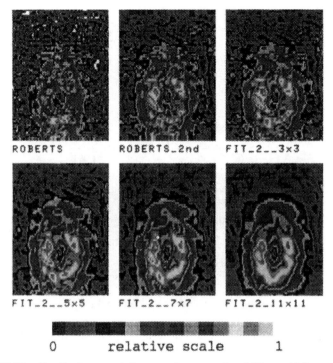

FIGURE 10.13 Gradient computations on the example of Figure 10.9c and d. (From Maldague, 1993.)

since the differentiation process increases noise content, gradient images are more delicate to work with; a simple comparison between Figures 10.11 and 10.13 shows this clearly.) We proceed in this fashion with a *rotating line* all around the defect center. At the end of this process, we obtain the list of coordinates for the vertices of defect shape:

$$\text{pos } x[i] \quad \text{and} \quad \text{pos } y[j] \tag{10.55}$$

TABLE 10.3 Relative Computation Times for the Gradient Image

Method of Obtaining the Gradient Image $|G|$	Time
Method of eq. (10.51), Roberts gradient	1
Method of eq. (10.52)	1.5
Method of eq. (10.53), 3 × 3 cross	8
Method of eq. (10.53), 5 × 5 cross	11
Method of eq. (10.53), 7 × 7 cross	12
Method of eq. (10.53), 11 × 11 cross	16

Source: Adapted from Maldague (1993, Table 6.3).

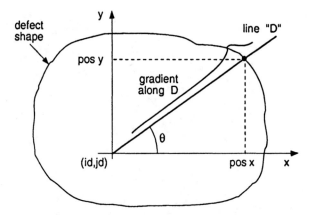

FIGURE 10.14   Method used for contour extraction of the defect. (From Maldague, 1993.)

where $i = 0, 1, 2, P - 1$,

$$P = \text{number of angular position around } (i_d, j_d)$$

To avoid problems related with infinite values of tan 90° in eq. (10.54), we use eq. (10.54) only for $0° < \theta < 45°$, $135° < \theta < 225°$, and $315° < \theta < 360°$. For orientations $45° < \theta < 135°$ and $225 < \theta < 315°$, we work with θ shifted of 90° with respect to θ (Figure 10.15):

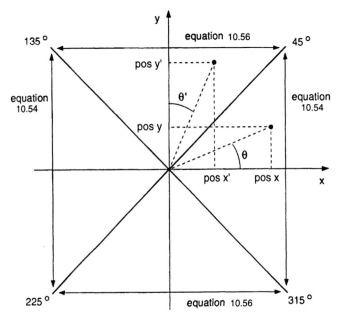

FIGURE 10.15   Illustration of eqs. (10.54) and (10.56). (From Maldague, 1993.)

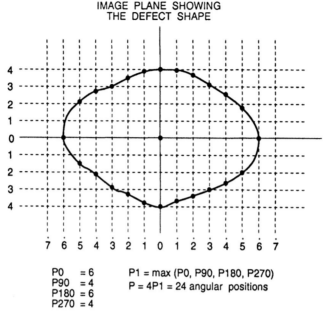

FIGURE 10.16 Computations needed to establish the number of steps that define the defect contour. (From Maldague, 1993.)

$$\text{pos } x' = i_d - (j_d - \text{pos } y) \tan \theta' \tag{10.56}$$

The number of angular steps P is specified by an initial maximum gradient search at $\theta_1 = 0°$, $\theta_2 = 90°$, $\theta_3 = 180°$, and $\theta_4 = 270°$ in gradient image $|G|$ in order to find the number of columns (or rows) available for the increment of pos x (or pos y') so that no point lays on the same column or row as its neighbor (Figure 10.16):

$$G_{\text{max}, \theta=0°} \Leftrightarrow (i_d, j_{\text{max}\_\theta=0°})$$
$$P_0 = (j_{\text{max}\_\theta=0°} - j_d)$$

and P_{90}, P_{180}, and P_{270} are computed in the same way. The maximum value is selected next, along rows and columns: $P_1 = \max(P_0, P_{90}, P_{180}, P_{270})$ and

$$P = 4P_1 \tag{10.57}$$

This method to extract the apparent defect contour by finding the list of vertices [eq. (10.55)] is advantageous since it makes it easy to correct the distortion effects of image fields not having the same number of rows and columns, as we will see now. A practical *trick* to measure the field of view is to place, at the correct focus length, a warm bright metallic ruler with engraved black painted lettering in the infrared camera field of view. Due to difference in emissivities, the lettering is seen with a good contrast with respect to the bright metal.

Without considering the LSF or SRF (Section 4.1.7), it is possible to compute a geometric factor that relates pixels with apparent size in the field of view and compensate for the unequal number of rows and columns. For example, if

$$105 \text{ columns} = \text{Maxcol, horizontal field size} = 32.9 \text{ mm}$$

$$68 \text{ rows} = \text{Maxrow, vertical field size} = 35.9 \text{ mm}$$

then

$$\frac{32.9 \text{ mm}}{105 \text{ pixels}} = 0.313 \text{ mm/pixel along columns } (x)$$

$$\frac{35.9 \text{ mm}}{68 \text{ pixels}} = 0.528 \text{ mm/pixel along rows } (y)$$

In this example, there is a distortion in the field of view since the *apparent* size of a pixel is different along rows and columns. For instance, if a square object placed in the field of view fills a 20×12 pixel surface on the IR monitor screen, then

$$\text{along } x \rightarrow 20 \text{ pixels represents } 6.26 \text{ mm}$$

$$\text{along } y \rightarrow 12 \text{ pixels represents } 6.34 \text{ mm}$$

If we multiply dimensions along y by the factor $f_y = 0.528/0.313 = 1.687$ to display the object without apparent spatial distortion, then

$$\text{along } x \rightarrow 20 \text{ pixels represents } 6.26 \text{ mm}$$

$$\Rightarrow 6.26/20 = 0.313 \text{ mm/pixel}$$

$$\text{along } y \rightarrow 12 \times 1.687 = 20.24 \text{ pixels represents } 6.34 \text{ mm}$$

$$\Rightarrow 6.34/20.24 = 0.313 \text{ mm/pixel}$$

and the object is displayed correctly as a square on the screen (the image displayed is of equal size in x and y), f_y can be expressed by

$$f_y = \frac{\text{Maxcol} \times \text{vertical field}}{\text{Maxrow} \times \text{horizontal field}} \tag{10.58}$$

The method based on a rotating line segment to extract defect shape allows easy correction of the unequal vertical and horizontal image fields since it is only necessary to multiply all values of pos y [i] of the vertices list [eq. (10.55)] by f_y [eq. (10.58)].

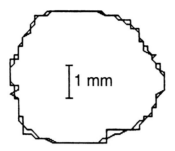

FIGURE 10.17   Example of Figure 10.9, extraction of apparent defect contour d_{app}. The shape is shown raw and smoothed. In this case eq. (10.52) was employed to compute the gradient, thus explaining the noisy shape obtained prior to smoothing. Graphite-epoxy anisotropy is visible. (From Maldague, 1993.)

For example, Figure 10.17 shows the result of such defect shape extraction performed with methods reviewed in this section (this is the example of Figure 10.10). Superimposed on the defect contour is a smoothed plot obtained using closed spline functions [complete derivation of this smoothing algorithm can be found in Laurendeau (1982, p. 62)]. Median filtering or sliding Gaussian noise smoothing techniques (discussed in Section 5.5.5) can also be used with radii computed between points of interest (pos x [i], pos y [i]) and the defect center (i_d, j_d) to correct for uneven shapes. The difficulty is to obtain a closed shape without a steep transition at the extremity (two smoothing passes with different starting points around the defect center can solve the difficulty).

What we saw in this section concerning defect sizing is, in fact, often referred to as the *traditional approach for defect sizing* based on the gradient image, for which the gradient of the contrast image C_{img} is of interest, with C_{img} computed at its maximum value. The basis of this approach is the following argument: Defect visibility is maximum at peak contrast when defect detection is more likely to occur (Figure 10.11). This is particularly true for weak defects. It is reported that such a gradient method is accurate to about 5% for polymer and CFRP components (next section, Table 10.4). Other approaches are, however, possible. For example, *early detection* has been reported, which is based on the following argument: If we wait for maximum contrast to develop, the visibility of a defect will be enhanced, but waiting for this to happen also blurs the edges of defects due to the 3D spreading of the thermal front (Section 9.2). In this respect, Krapez and Balageas (1994) suggested relying on the contrast image as soon as it emerges from the noise or when the slope of the contrast curve is at its peak (Favro et al., 1995b) (Figure 10.10). Yet another refinement of the early detection procedure was proposed by means of an iterative technique to correct the shrinkage due to thermal front spreading during the elapse time of defect detection: A series of thermograms are recorded following pulse heating, and for each thermogram the defect size is taken as half the maximum amplitude. A plot of defect size as a function of the square root of the time is then

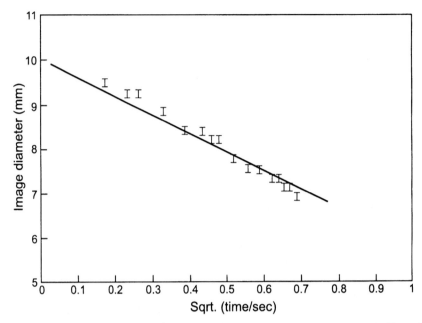

FIGURE 10.18   Experimental transient thermography image diameters at half amplitude *I* of a 10-mm-diameter defect in mild steel compared with prediction (solid line).
(Adapted from Almond and Lau, 1994.)

established and the fit line at time zero yields to the defect size with a small
overestimation (4% was noted in the case of mild steel inspection). It is important to notice that this iterative approach fails for very small defects or at very
long times where edge effects at opposite sides of the defect overlap (Almond
et al., 1994a,b). This procedure is shown in Figure 10.18. Interestingly, whatever
method for defect sizing is selected, the detailed procedure described here can be
adapted (e.g., taking the peak slope contrast instead of the maximum contrast).

In this section we studied the various operations required to obtain the information needed to characterize quantitatively subsurface defects detected (the
detection steps proceed as explained in Chapter 6). These operations concern
the acquisition of the thermogram sequence (particularly the time scale) and all
subsequent computations. Once this experimental information is obtained
about the sample inspected, the *inverse problem* can be attacked as explained in
Section 10.1. In the next section we discuss briefly the expected accuracy of
procedures reviewed in Sections 10.1 and 10.2.

10.2.5   Quantitative Characterization Procedure

In this section we discuss some factors that are of interest when a quantitative
inspection analysis is performed. As mentioned before, one potential difficulty

TABLE 10.4 Expected Accuracy of Quantitative Characterization for Selected Parameters

Parameter Evaluated	Accuracy (%)	
	Krapez and Cielo (1991)	Balageas et al. (1987c)
Depth	8.1	13.8
Thermal resistance	24	126
Size	4.6	[a]

Source: Adapted from Maldague (1993, Table 6.4).
[a] Not available since the analysis is unidimensional.

is to select a sound area within an image [vector $r\_$far, eq. (10.47)]. This has implications due to the 3D spreading of the heat front. In particular, contrast computation can be affected, thus corrupting the quantitative evaluation. In fact, since contrast computation is performed between the zone of interest (above any defects detected) and a sound area, so that unity contrast is obtained in the absence of a defect [eq. (10.28)], thermal contrast values will be corrupted if the sound area is selected too close to the defect and is thus submitted to its influence (the thermal contrast computed will be of smaller amplitude). This effect will be more evident if thermal images are small.

To provide some idea of the expected accuracy of quantitative inversion procedures discussed in for previous sections, Table 10.4 presents some typical results graphite epoxy samples with inclusion-type defects [inclusions made by inserting Teflon film (100 μm thick) between two plies of prepreg] using a pulsed thermography approach in reflection. The numbers indicated correspond to the percentage difference between defect nominal values and defect measured values using the inverse method. A major cause of discrepancy in these values is the account of the thermal losses that takes place during the experiments between the sample and the surrounding. These losses cannot be ignored if the experiment lasts longer than a few seconds. They are, however, difficult to handle in the inverse problem, especially if the erratic behavior of air convection phenomena that take place during thermal inspection experiments have to be taken into account.

As seen from the numbers above, the expected accuracy of inverse analysis performed with experimental data is a few percent on depth and size evaluation; it is worse in the case of thermal resistance (as expected from Figure 10.5). One reason for this is that during the preparation of samples, air can be trapped with the artificial defect inclusion. This can be the case if Teflon implants are inserted between plies of graphite epoxy prepreg in order to simulate an inclusion. The presence of air layers not accounted for perturb the inversion procedure. Many authors have noticed the presence of air layers in samples after destructive testing of such CFRP specimens.

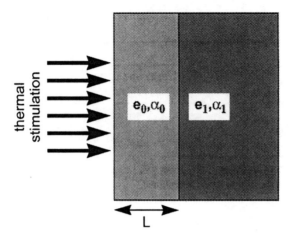

FIGURE 10.19   Diagram of a two-layer specimen.

As a last consideration, it is worthwhile to mention the necessity to perform digitalization of thermal images on a sufficient number of bits. In the past, systems were offered on 6, 7, or 8 bits. Current figures are 12 bits, but 14-bit digitalization is emerging seriously. Such a high level of digitalization brings a reduced noise level at the acquisition stage, thus permitting us to capture faint thermal events and to obtain better agreement of the inversion procedure with real data.

10.3   STEP HEATING: DATA ANALYSIS

As discussed in Section 9.4, in step heating the increase insurface temperature is monitored during the application of a stepped heating pulse (the sample is heated continuously, at low power; Figure 9.5) (J. W. M. Spicer et al., 1996). For the two-layer specimen shown in Figure 10.19, the temperature evolution on the surface due to such heating is given by (J. C. Murphy et al., 1992)

$$T(t) = C_c\sqrt{t}\left\{1 + \sum_{n=1}^{\infty} 2(-\Gamma)^n\left[\exp\left(-\frac{n^2 L^2}{\alpha_o t}\right) - \frac{nL\sqrt{\pi}}{\sqrt{\alpha_o t}}\,\text{erfc}\left(\frac{nL}{\sqrt{\alpha_o t}}\right)\right]\right\} \quad (10.59)$$

where α is the thermal diffusivity ($\alpha = k/\rho C$, where ρ is the density, C the specific heat, and k the thermal conductivity; Table 3.7 gives thermal parameters for common materials), e the thermal effusivity [eq. (10.10)], the subscripts o and 1 represent the first and second layer respectively, and C_c is the constant term related to the energy absorption. For laser heating (Figure 9.6), C_c is expressed as (Osiander et al., 1996)

$$C_c = \frac{\varepsilon(1 - R)I}{4\pi^{3/2}e_o} \quad (10.60)$$

where ε is the surface emissivity at the wavelength of the camera, R the reflectivity for the laser wavelength, and I the laser intensity.

Finally, Γ is the thermal mismatch factor between the two layers. It is defined as (Osiander et al., 1996)

$$\Gamma = \frac{e_1 - e_o}{e_1 + e_o} \tag{10.61}$$

In the case of a second layer with a much lower thermal effusivity e_1 than that of the surface layer, Γ is -1. This is the case for a disbond (air presence) behind a metallic layer. Γ is 0 for two layers of the same e value, and Γ is $+1$ if the second layer has a much higher thermal effusivity e_1 (ceramic coating on metal).

Example 10.2   Compute the thermal mismatch factor in the following situations: aluminum–air, Plexiglas–aluminum.

SOLUTION   The thermal effusivity is computed with eq. (10.10): $e = \sqrt{k\rho C}$, with k the thermal conductivity, ρ the mass density, and C the specific heat. These values are available in Table 3.7 and Appendix E for common materials. Let's first compute e_{Al}, e_{air}, and e_{Plex}:

$$e_{Al} = \sqrt{k\rho C} = \sqrt{230 \times 2700 \times 880}$$

$$= 24{,}000 \text{ J m}^{-1}\,{}^{\circ}\text{C}^{-1}\,\text{s}^{-1/2} = 2.4 \text{ J cm}^{-1}\,{}^{\circ}\text{C}^{-1}\,\text{s}^{-1/2}$$

$$e_{air} = \sqrt{k\rho C} = \sqrt{0.024 \times 1.2 \times 700}$$

$$= 4.5 \text{ J m}^{-1}\,{}^{\circ}\text{C}^{-1}\,\text{s}^{-1/2} = 4.5 \times 10^{-4} \text{ J cm}^{-1}\,{}^{\circ}\text{C}^{-1}\,\text{s}^{-1/2}$$

$$e_{Plex} = \sqrt{k\rho C} = \sqrt{0.2 \times 1200 \times 667}$$

$$= 400 \text{ J m}^{-1}\,{}^{\circ}\text{C}^{-1}\,\text{s}^{-1/2} = 0.04 \text{ J cm}^{-1}\,{}^{\circ}\text{C}^{-1}\,\text{s}^{-1/2}$$

and then the thermal mismatch factor:

$$\Gamma_{Al-air} = \frac{e_{air} - e_{Al}}{e_{air} + e_{Al}} = \frac{4.5 - 24{,}000}{4.5 + 24{,}000} = -1$$

$$\Gamma_{Plex-Al} = \frac{e_{Al} - e_{Plex}}{e_{Al} + e_{Plex}} = \frac{24{,}000 - 400}{24{,}000 + 400} = 0.97 \sim +1$$

These two cases represent two extremes cases for which Γ -1 or $+1$. As a backing layer, air with such a small thermal effusivity e_{air} always yields, $\Gamma = -1$, as shown above.

Equation (10.59) is a one-dimensional approximation that is valid in the case of a layer with a large lateral extension (semi-infinite case) and also early, when

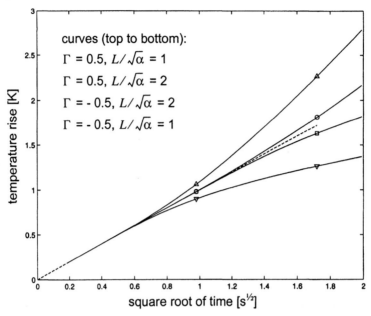

FIGURE 10.20   Temperature rise of specimen surface as a function of square root of time for different layer thickness. Drawn with Matlab m-script Plotsh3. (Data from Osiander et al., 1996.)

thermal diffusion occurs only in the top layer of the specimen, for which the summation term in eq. (10.59) is small (Osiander et al., 1996).

In Figure 10.20 we plot the temperature rise as a function of the square root of the time for different values of Γ and $L/\sqrt{\alpha}$. Interesting features are observed. The semi-infinite case for which $\Gamma = 0$ turns eq. (10.59) to $C_c\sqrt{t}$, which corresponds to the straight dashed line in Figure 10.20 (notice that the horizontal scale is expressed as a square root of the time, thus making this curve linear). Note also that departing from the semi-infinite case at a time called the transit time t_T are curves corresponding to layer thicknesses (parameter $L/\sqrt{\alpha}$). This transit time corresponds to the time the interface starts to influence the surface temperature; in fact, this also corresponds to the time the thermal reflection on the interface extends up to the surface. As shown in the figure, t_T is not affected by the thermal mismatch factor, although logically, it influences the extent of the deflection ($\Gamma = 0$, no deflection, etc.).

As stated earlier, at early time t ($t < t_T$), all the curves follow the semi-infinite case ($C_c\sqrt{t}$), and this allows us to calibrate the data recorded. Dividing eq. (10.59) by $C_c\sqrt{t}$ and subtracting 1 yields to the normalized temperature rise T_N, which is only a function of the thermal mismatch factor Γ and of the normalized time t_N ($t_N = \sqrt{\alpha t}/L$). Figure 10.21 plots T_N versus t_N for which the semi-infinite case is now 0. In this figure, the departing point is seen to be

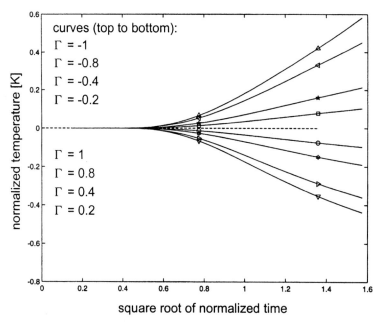

FIGURE 10.21   Normalized temperature rise of specimen surface as a function of square root of normalized time for different values of the thermal mismatch factor. Drawn with Matlab m-script Plotsh4. (Data from Osiander et al., 1996.)

about 0.6:

$$\sqrt{t_N} = 0.6 = \frac{\sqrt{\alpha t}}{L}$$

$$L = \frac{\sqrt{\alpha t}}{0.6} \rightarrow t = \frac{0.36 L^2}{\alpha}$$

which corresponds to eq. (9.9). It is possible to redraw Figure 10.21 in terms of the thermal mismatch factor Γ. This allows for more direct comparisons between the experimental data and the computations. For example, in that figure, it is noted that assuming an air backing for which $\Gamma = -1$, and for a normalized time $t_N = 1 = \sqrt{\alpha t}/L$, the normalized temperature rise T_N is 0.1816*. From this, $L/\sqrt{\alpha}$ can be determined easily by finding the time t when the normalized temperature has this value (i.e., 0.1816; see the example below).

The curves plotted in Figure 10.22 allow us to establish the thermal transit time t_T and thus the thickness layer L if Γ is known. Alternatively, if L and t_T are known values, find the thermal properties of the interface (parameter Γ). Such a method can be applied on a pixel-by-pixel basis. However, such a pro-

*This can be computed with available Matlab function *sh4.m* with parameters: sh4(1,−1).

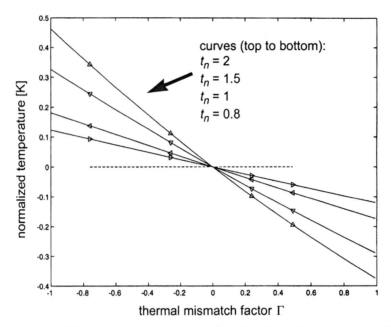

FIGURE 10.22   Normalized temperature as a function of the thermal mismatch factor for different values of the normalized time. Drawn with Matlab m-script Plotsh5. (Data from Osiander et al., 1996.)

cedure is valid only when the one-dimensional assumptions hold. To illustrate the procedure, Figure 10.23 shows an experimental investigation of SH using an expanded CO_2 laser beam that heats CFRP specimens with flat bottom-hole defects of various depths (0.8, 1.4, and 2.8 mm); the hole diameters are 12.4 mm (this example is taken from Osiander et al., 1996). A raw image at $t = 15.6$ s is

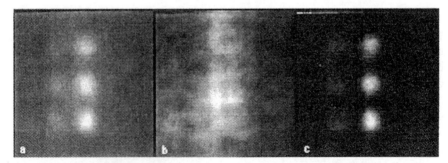

FIGURE 10.23   Infrared images of flat-bottomed holes in graphite-epoxy composite panel after 15.6 s of heating with a CO_2 laser source: (a) raw temperature data; (b) slope image (intensity distribution); (c) normalized temperature image. (From Osiander et al., 1996.)

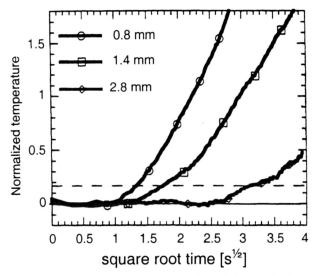

FIGURE 10.24   Normalized temperature as a function of time on flat-bottomed holes of different depths in CFRP (case of Figure 10.23). The horizontal line at ordinate 0.18 is used in determining the thermal transit time. (From Osiander et al., 1996.)

shown in Figure 10.23*a*. The calibration constant is computed at an early time, as mentioned before. The slope image shown in Figure 10.23*b* corresponds to the intensity distribution of the heating source. The normalized temperature image T_N appears in Figure 10.23*c*. Computing different normalized temperature images T_N at different times allows us to plot the graph of Figure 10.24, which shows the normalized temperature as a function of the square root of time (values taken at the centers of defects). In Figure 10.24 the horizontal scale is the square root of the time, not the square root of the normalized time as for the general case of Figure 10.21. Flat-bottomed holes correspond to a case of $\Gamma = -1$, and for a normalized time $t_N = 1$, the normalized temperature rise T_N is 0.18. The line $T_N = 0.18$ is plotted in Figure 10.24, for which the intersection points for the deflection at the three depths can be evaluated, enabling us to retrieve the defect depth if the thermal diffusivity of the material is known.

Example 10.3   Following the described procedure, compute the defect depth for the first deflected curve in Figure 10.24.

SOLUTION   At the intersection $T_N = 0.18$ we noticed that $\sqrt{t} = 1.3$ s$^{1/2}$. This intersection corresponds to the case for which $t_N = 1 = \sqrt{\alpha t}/L$ and thus (with $\alpha = 0.42 \times 10^{-6}$ m^2 s^{-1})

$$L = \sqrt{\alpha t} = \sqrt{\alpha}\sqrt{t} = \sqrt{0.42 \times 10^{-6}} \times 1.3 = 0.84 \text{ mm}$$

which is close to the reported value of 0.8 mm.

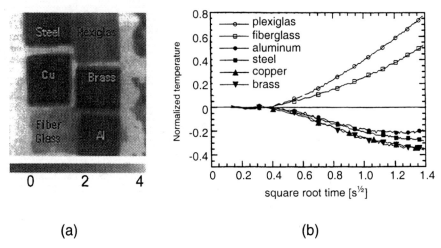

(a) **(b)**

FIGURE 10.25 (a) Normalized temperature image for different materials beneath a 1-mm-thick fiberglass-epoxy coating after a 2-s heating; (b) plots of normalized temperature as a function of square root of time for locations on the different materials shown in (a). (From Osiander et al., 1996.)

The SH procedure can also be used to establish thermal parameter values. An example is shown in Figure 10.25. In that case, different materials are beneath a 1-mm-thick fiberglass–epoxy layer. The plot of the normalized temperature versus square root of the time shows the same transit time t_T due to the same depth of 1 mm in all cases but with different deflections following the different thermal mismatch factors Γ, as explained for Figure 10.20. As expected, on this plot, the slope sign changes for thermally conductive material (aluminum, brass, copper, steel) and for the thermally insulating materials (fiberglass, Plexiglas). In these cases, the metal thickness is too thin for the two-layer model of Figure 10.19 to be valid. A more complex model is needed (J. W. M. Spicer et al., 1996).

10.4 PULSED PHASE THERMOGRAPHY: DATA ANALYSIS

10.4.1 Principles and Features

In Sections 9.4 and 9.2, the operating principles of both LT and PT were reviewed, including thermal wave theory in Section 9.1. As we have noted, reasoning in terms of frequencies, a link can be made between PT and LT. In fact, thanks to the time–frequency duality, the frequency content of an ideal temporal pulse of null duration has a frequency spectrum with uniform energy distribution between all frequencies from 0 to ∞ (Figure 10.26). Of course, a real thermal pulse is different from that of an ideal Dirac pulse since its duration and amplitude have finite values. For example, Figure 10.27 shows a

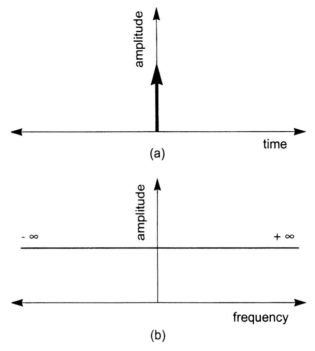

FIGURE 10.26 Time-frequency duality: (a) ideal temporal pulse of null duration with infinite amplitude; (b) corresponding frequency spectrum with uniform energy distribution between all frequencies from $-\infty$ to $+\infty$.

square thermal stimulation pulse centered on $t = 0$ s. With amplitude A_p and duration Δt_p, the shape observed in the frequency domain is defined by the equation

$$F(f) = \frac{A_p \Delta t_p \, \sin(\pi f \Delta t_p)}{\pi f \Delta t_p} = A_p \Delta t_p \, \text{sinc}(\pi f \Delta t_p) \qquad (10.62)$$

where f is the frequency variable $[\omega \; (\text{rad s}^{-1}) = 2\pi f \; (f \text{ in (hertz)})]$. Going back and forth between the temporal and frequency domains is possible with mathematical tool such as the well-known Fourier transform, for which a fast Fourier transform (FFT) algorithm is available (Gonzalez and Wintz, 1977; Gonzalez and Woods, 1992). In Figure 10.27, it is seen that in real situations, not all frequencies are present, while they exhibit different amplitudes as well. It is also noted that longer pulses in the time domain have a narrower spectrum in the frequency domain, the available energy becoming concentrated in lower frequencies. As discussed in Section 9.1, lower frequencies propagate deeper under the surface. The uneven energy distribution with frequency is not a major concern in PT since it is generally of interest to have more energy at low frequencies to enhance the visibility of deeper structures under the surface. Such a

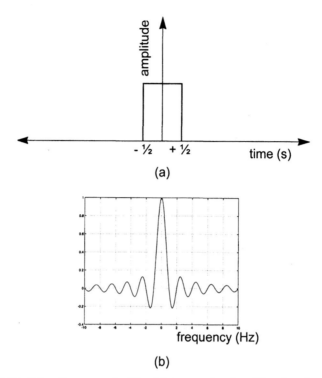

FIGURE 10.27 Time-frequency duality: (a) temporal pulse of 1-s duration centered at time = 0; (b) corresponding frequency spectrum (drawn with Matlab m-script plot_sinc (1,1,10,'k')). In the case of a 5-ms duration pulse, the horizontal scale on plot (b) spans from −1000 to +1000 Hz [then drawn with plot_sinc(200,0.005,1000,'k')].

discussion shows that, in fact, in PT many thermal waves of different frequencies are launched in the specimen simultaneously, whereas in LT, one thermal wave frequency is tested at a time. Of course, PT operates in the transient regime, whereas LT is deployed in stationary regime. Nevertheless, the possibility of linking both techniques was found interesting, and this was called pulsed phase thermography (PPT) processing (Maldague and Marinetti, 1996).

The fundamental idea of PPT processing is thus to extract and analyze the response of the specimen on the frequency domain, based on the frequency spectrum available in the thermal stimulation pulse (Figure 10.27). Extraction of the various frequencies is performed with a discrete one-dimensional Fourier transform on each pixel of the thermogram sequence (Figure 10.28):

$$F_n = \sum_{k=0}^{N-1} T(k)e^{2\pi ikn/N} = \text{Re}_n + i\,\text{Im}_n \qquad (10.63)$$

where i is the imaginary number ($i = \sqrt{-1}$), Re and Im are the real and imaginary parts of the transform, respectively, and the subscript n designates the

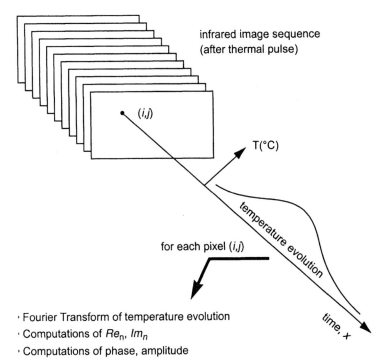

FIGURE 10.28   Principle of PPT computations.

frequency increment. First, the time evolution of each pixel (x, y) in the field of view is extracted as a vector $T(k)$ of N values [$T(k)$ is the temperature at location (x, y), for the kth thermogram of the sequence]. Next, eq. (10.63) is applied to compute the real Re_n and imaginary Im_n parts. Amplitude A_n and phase ϕ_n are then computed with

$$A_n = \sqrt{\text{Re}_n^2 + \text{Im}_n^2} \quad \text{and} \quad \phi_n = a \tan \frac{\text{Im}_n}{\text{Re}_n} \qquad (10.64)$$

Amplitude and phase images are formed by repeating this process for all pixels (x, y) of the field of view. With N time increments available (i.e., N thermograms in the sequence), $N/2$ frequency values are available (due to the symmetry of the Fourier transform). The discrete frequencies available f_n are given by

$$f_n = \frac{n}{N\Delta} \qquad (10.65)$$

where N is the number of thermal images in the sequence, Δ is the time interval between thermal images, and $n = 0, 1, \ldots, N/2$. The maximum frequency available is thus limited by the acquisition rate, while the minimum frequency is limited by the duration of the experiment. In practice, only the first few frequencies are of interest since most of the energy is concentrated in the low frequencies, higher frequencies exhibiting a higher noise level, precluding their use (Figure 10.27).

Direct use of eq. (10.63) is not advisable due to lengthy computations. It is better to rely on the FFT algorithm, which is available in common mathematical packages such as Mathlab. For example, processing time for a sequence of 50 images of format 160×120 is about 19 s [down to 7 minutes for direct use of eq. (10.63) on a SUN Sparc 4].

Example 10.4   Compute the maximum theoretical frequency available assuming a time interval between thermal images of 17.5 ms and that thermal sequences are 50 thermograms long.

SOLUTION   Here we have $N = 50$ and $\Delta = 17.5$ ms. Following eq. (10.65), we have

$$f_n = \frac{n}{N\Delta} = \frac{n}{50 \times (17.5 \times 10^{-3})} = 1.14n$$

Thus the available frequencies will span from $f_0 = 0$ to $f_{50} = 57$ Hz. As mentioned above, the maximum frequency is related directly to the acquisition frequency of the IR camera (57 Hz = 1/17.5 ms). This is a theoretical maximum. Also to be considered is the thermal pulse length. For example, if a 5-ms thermal pulse (using photographic flashes) is used, the first lobe of the sinc function crosses the zero axis at 200 Hz with the maximum amplitude concentrated in the domain 0 to 180 Hz. A 57-Hz camera will not be able to take into consideration these high-frequency thermal waves, thus causing aliasing. The sampling theorem stipulates that to avoid distortion, a signal has to be sampled at at least twice the maximum frequency it contains. This is clearly not the case here ($57 \times 2 < 200$). The Fourier transform of a 5-ms duration pulse can be obtained with the m-file Matlab *plot_sinc.m* with parameters plot_sinc(200,0.005,1000,'k') (Figure 10.27).

It is worthwhile to mention than although the same information is available in PT and PPT, results, especially phase images in PPT as seen below, provide interesting capabilities with respect to the more traditional contrast approach used in PT (reviewed in Section 10.2). The reason is that in PT the frequency mix is kept, while contrast computation is not as powerful as the frequency analysis of PPT, which sorts the available information coherently in term of frequencies. Moreover, with respect to LT, PPT makes available several frequencies following a single experiment, while in LT one experiment is needed for each frequency inspected.

Figure 10.29 shows a comparison between PPT and PT in case of a plastic specimen with a tilted slot of 6 mm width and depth varying from 1 to 3 mm

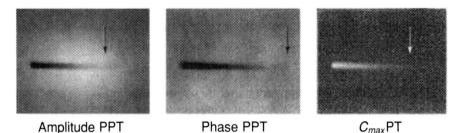

Amplitude PPT   Phase PPT C_{max}PT

FIGURE 10.29 Amplitude, phase, and maximum contrast image; case of a plastic specimen containing a subsurface slot 6 mm wide of variable depth from 1 to 3 mm ($f = 0.063$ Hz). Strong heating nonuniformities are visible in amplitude image. Arrows indicate maximum depth resolved. (From Maldague and Couturier, 1997.)

beneath the front surface. On the amplitude image, nonuniform heating is easily seen, whereas the phase image is flat except for the defect. The contrast image [C_{max}, eq. (10.28)] reveals the slot but with a higher noise content.

10.4.2 Particular Features of PPT

Experiments reveal particular features of PPT (Maldague and Couturier, 1997).

Energy and Frequency. More energy in the low frequencies may be needed to reveal deeper defects, as discussed in Section 9.1. A possibility is thus to lengthen the duration of the thermal pulse (Figure 10.30) to concentrate energy in the low frequencies. Of course, this reduces the energy content in the high frequencies, with as a consequence, the possible requirement to repeat the experiment with another pulse duration to cover the complete frequency span of interest. However, for a given image sequence, phase images become noisier as the frequency increases. This is explained by the inverse relationship between

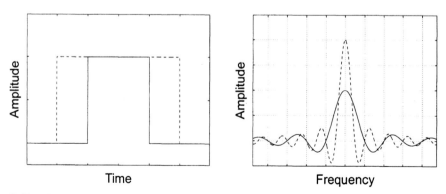

Time   Frequency

FIGURE 10.30 Rectangular thermal pulse of different lengths and corresponding frequency spectra (arbitrary units). (From Maldague and Couturier, 1997.)

available energy and frequency (more energy at lower frequencies), while the division involved in eq. (10.64) contributes further to noise degradation of phase images.

Depth Selection. Phase and amplitude images in PPT have a selective, frequency-dependent probing capability. A difference is, however, observed between phase and amplitude. Amplitude images resolve up to a certain depth for a given frequency (low-pass filter behavior), while phase images select a maximum depth with a reduced visibility for intermediate depths (bandpass filter behavior). These phenomena are illustrated on Figure 10.31: a graphite epoxy plate 4 mm thick with an artificial insert of Teflon [20 mm diameter, 2 (defect 1) and 3 (defect 2) mm under the surface]. Amplitude images are less noisy but are sensitive to nonuniform heating (see below). For comparisons, LT images are shown as well: phase results are similar in PPT and LT, especially the *bandpass* filter effect; experimental frequencies are also similar (PPT phase images in PT were obtained with a single experiment, whereas LT required four). As seen, the frequency per frequency observation of PPT bypasses the problem of the single propagation length associated with an observation performed at a fixed frequency (LT case). The maximum contrast image C_{max} of the PT approach reveals only the shallower defect, due to the blurring effect of the mixed frequencies (Section 9.1).

Depth Probing Capabilities. As discussed in Section 9.1, with respect to amplitude or thermographic images, phase images (either in LT or in PPT) allow us to probe more deeply under the specimen surface, about twice. An example of this is seen in Figure 10.29. Figure 10.31, which compares PPT with LT, shows that similar results are obtained.

Heating Nonuniformity. Since PPT phase images are less affected by heating nonuniformities (see e.g., Figure 10.29), experimental constraints about positions of thermal sources are relaxed. A greater field of view is thus achievable.

Surface Features. As discussed in Section 9.4, phase images are less sensitive to degradation from optical and infrared surface features. This can also be understood with respect to the bandpass filter behavior observed for phase images. Surface artifacts can be considered as defects at zero depth; thus selection of a lower frequency (deeper probing) reduces disturbing surface effects. This is shown in Figure 10.32 for a plastic specimen with a 3-mm-deep 10-mm-wide slot. The specimen was coated uniformly with a high-emissivity (~ 0.9) black paint, and then a 12-mm-wide line of white paint was applied on the plate. The white line reduces the deposit of optical energy over the surface due to its low absorptivity. The amplitude image shows both the line artifact and the slot, while the phase image is sensitive almost only to the subsurface defect (the slot) and not to the uneven thermal stimulation. In additional to relative insensitivity to nonuniform heating, uniform phase images are also obtained in the case of nonflat specimens, at least at frequencies not causing reflections deep into the

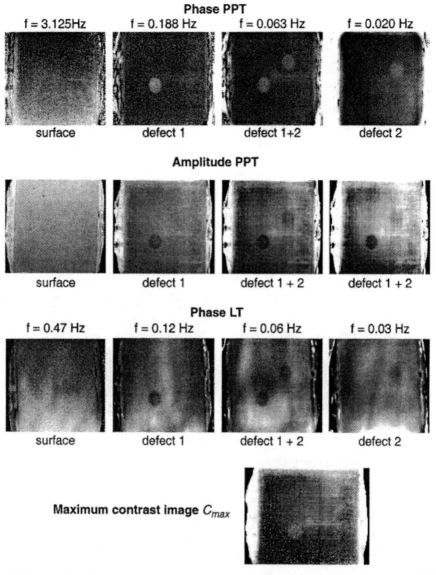

FIGURE 10.31 Phase and amplitude images obtained for different frequencies in PPT and LT, maximum contrast image C_{max} is shown as well (graphite-epoxy plate; see the text). (From Maldague and Couturier, 1997.)

specimen (Figure 10.33 for phase images at $f = 0.45$ Hz with respect to $f = 0.0013$ Hz).

Zero Reference. The Fourier transform [eq. (10.63)] involved in PPT refers not to time increments but to image number in a thermal sequence. Contrary to LT, where the zero-time value is well defined with respect to the time origin of the

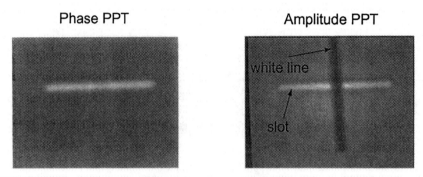

FIGURE 10.32 Phase and amplitude image, specimen with surface artifact (see the text) ($f = 0.049$ Hz). (From Maldague and Couturier, 1997.)

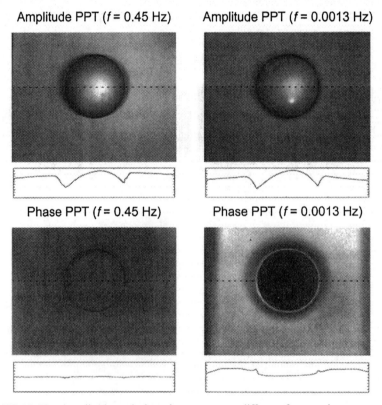

FIGURE 10.33 Amplitude and phase images at two different frequencies over a plastic specimen made of a plate with a bonded half-sphere. Profiles across the half-sphere are shown as well (same scale for both phase and amplitude). Notice the dead pixel top left in the images (black spot) and the subsurface defect (middle right). (From Maldague and Couturier, 1997.)

sinusoidal thermal stimulation, in PPT the definition of zero often causes problems. In PPT, zero corresponds to the acquisition of the first thermogram, but in some instances, the first thermograms are saturated and must be discarded in the Fourier transform, with, as a consequence, a change in the zero value, which degrades significatively the performance of quantitative inversion methods (see below). On the other hand, a good point is that no sound area reference is needed to compute a contrast, as in straight PT analysis (Section 10.1.3).

10.4.3 Data Inversion

Once phase (or amplitude) images are available [eq. (10.64)], the next step consists generally of determining the presence of defects, and characterize them. Various methods have been proposed for data inversion in PPT: neural networks (Maldague et al., 1998), statistical methods (Vallerand et al., 1999; Vallerand and Maldague, 2000), dedicated calibration (Vavilov and Marinetti, 1999; Busse, 1994; Galmiche and Maldague, 1999). Before reviewing these methods, it is interesting to analyze thermal material behavior as a function of parameters of interest (phase, specimen thickness, frequency).

Figure 10.34 shows a graph with phase evolution as a function of frequency and specimen thickness (case of aluminum) in the case of a 14-Hz IR image acquisition rate f_s. Cuts in the three directions enable us to study the quantitative behavior of the material (data from Largouët, 1997; thermal model similar to that of Section 3.4.2). Figure 10.35 shows the spectral distribution of the phase as a function of the frequency for thicknesses of 1, 3, 5, 7, and 10 mm

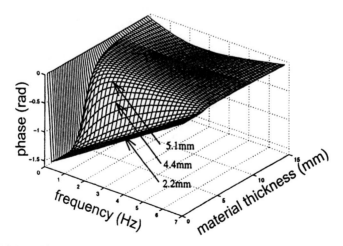

FIGURE 10.34 Phase spectra as a function of depth ($f_s = 14$ Hz, aluminum). (From Maldague et al., 1998, reprinted with permission from Editions Elsevier.)

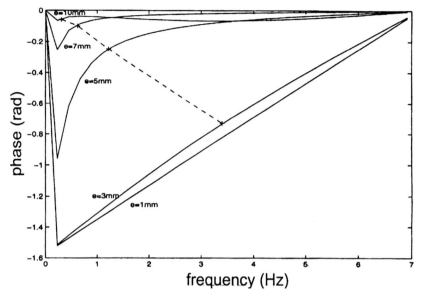

FIGURE 10.35   Phase evolution as a function of frequency for different material thicknesses e ($f_s = 14$ Hz, aluminum). (From Maldague et al., 1998, reprinted with permission from Editions Elsevier.)

(drawn from cuts in Figure 10.34). The dashed line corresponds to frequencies for which the thermal diffusion length [eq. (9.2)] is equal to the material thickness d_{th} (i.e., $\mu = d_{th}$). For small thicknesses d_{th} (<3 mm) spectra are linear. Also, phase values get smaller as frequency increases, behavior related to the thermal diffusion length μ, which is inversely proportional to the square root of the frequency. At high frequencies (as mentioned in Section 9.1), inspection is limited to close to the surface material layers with small phase value. As material thickness increases, maximum phase values become smaller, with hyperbolic curve behavior starting to mix (thus making the inversion difficult; see Section 10.4.7). Interestingly, at low frequencies, the phase get close to $-\pi/2(\sim -1.5)$; this is a classically known result in the case of a finite plate.

Finally, Figure 10.36 shows phase profiles as a function of material thickness for various frequencies. Curve shapes are similar. At each frequency, the phase is constant up to about 2.6 mm and then decreases with thickness. An important point is shown here since the phase spectra do not allow us to differentiate among material thicknesses smaller than 2.6 mm with a sampling frequency f_s of 14 Hz. A plot of Figure 10.34 with a faster acquisition rate reduces this flat area. For example, with a sampling frequency 10 times faster ($f_s = 140$ Hz), the flat unusable area reduces to thicknesses below 1 mm (Figure 10.37). The issue of experimental sampling frequency is thus of paramount importance if quantitative analysis is sought in PPT, especially in the case of high-thermal-conductivity specimens such as aluminum.

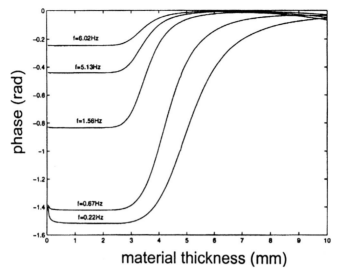

FIGURE 10.36   Evolution of phase as a function of thickness e ($f_s = 14$ Hz, aluminum). (From Maldague et al., 1998, reprinted with permission from Editions Elsevier.)

10.4.4   Statistical Method

Although we discussed PPT primarily here, the method we are going to review is general and can be applied on thermographic (temperature), amplitude, phase, or in fact a wide variety of experimental data and properties (not only depth retrieval). This method takes into account the variability of the mea-

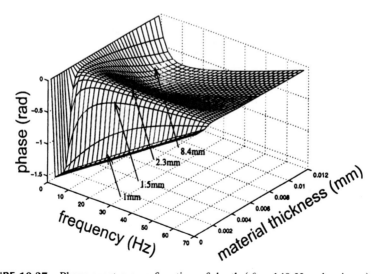

FIGURE 10.37   Phase spectra as a function of depth ($f_s = 140$ Hz, aluminum). (From Largouët, 1997.)

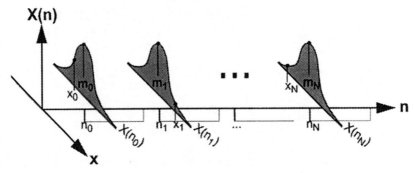

FIGURE 10.38   Gaussian random process $X(n)$. (From Vallerand and Maldague, 2000, reprinted with permission from Elsevier Science.)

surement, expressed by the mean and variance, in order to establish the value of a given property, such as defect depth. Like many signal and noise processes, temperature, phase, and amplitude data can be modeled as Gaussian random processes (Grinzato et al., 1995b). A Gaussian random process is specified completely by a vector of means and a covariance matrix. The probability of having a certain vector $x = [x_0, x_1, \ldots, x_N]^{\mathrm{T}}$ (T is the transposition operator), where x_0 is a realization of $X_0 = X(n_0)$, x_1 is a realization of $X_1 = X(n_1)$, $\ldots, x_N$ is a realization of $X_N = X(n_N)$, and where X_i is a Gaussian random process of mean m_i and covariance $C_x(n_i, n_i)$ is given by (Figure 10.38)

$$f_{X_0, X_1, \ldots, X_N}(x_0, x_1, \ldots, x_N) = \frac{e^{-(1/2)(x-m)^{\mathrm{T}} K^{\mathrm{T}}(x-m)}}{(2\pi)^{N/2}|K|^{1/2}} \qquad (10.66)$$

The vector of value x, the vector of means m, and the covariance matrix K are defined as follows (Leon-Garcia, 1994):

$$x = \begin{bmatrix} x_0 \\ x_1 \\ \cdots \\ x_N \end{bmatrix} \qquad m = \begin{bmatrix} m_0 \\ m_1 \\ \cdots \\ m_N \end{bmatrix}$$

$$K = \begin{bmatrix} C_X(n_0, n_0) & C_X(n_0, n_1) & \cdots & C_X(n_0, n_N) \\ C_X(n_1, n_0) & C_X(n_1, n_1) & \cdots & C_X(n_1, n_N) \\ \cdots & \cdots & \cdots & \cdots \\ C_X(n_N, n_0) & C_X(n_N, n_1) & \cdots & C_X(n_N, n_N) \end{bmatrix}$$

(10.67)

Assuming that all Gaussian random variables $X_0, X_1, \ldots, X_N$ are independent, the problem reduces significantly to (global probability equates the product of individual probabilities; Leon-Garcia, 1994):

$$f_{X_0, X_1, \ldots, X_N}(x_0, x_1, \ldots, x_N) = f_{X_0}(x_0) f_{X_1}(x_1) \cdots f_{X_N}(x_N) \qquad (10.68)$$

with the known Gaussian distribution expressed as

$$f_{X_i}(x_i) = \frac{e^{-(x-m_i)^2/2s_i^2}}{\sqrt{2\pi} s_i} \qquad (10.69)$$

The probability that X_i lies in a small interval h in the vicinity of x_i is approximately equal to $f_{X_i}(x_i)h$ (Leon-Garcia, 1994):

$$P[x_i < X_i \le x_i + h] \approx f_{X_i}(x_i)h \qquad (10.70)$$

From this it is established that the probability that a vector x is the realization of the Gaussian random variable X is calculated by

$$P(X_0 = x_0, X_1 = x_1, \ldots, X_N = x_N)$$
$$\approx P[x_0 < X_0 \le x_0 + h, x_1 < X_1 \le x_1 + h, \ldots, x_N < X_N \le x_N + h]$$
$$P[x_0 < X_0 \le x_0 + h, x_1 < X_1 \le x_1 + h, \ldots, x_N < X_N \le x_N + h]$$
$$\approx h^N f_{X_0}(x_0) f_{X_1}(x_1) \cdots f_{X_N} \qquad (10.71)$$

Practically speaking, to implement this statistical approach, in the learning step, temperature images of known flaws (of particular depths) are obtained, and for each known flaw, including the background (sound area) as well, mean m and variance s are computed at each discrete time n for all classes (Figure 10.39a). For all pixels x within a defined zone, m and s are computed with the usual formulas (W. J. Dixon and Massey, 1983):

$$m = \frac{\sum_{i=1}^{N} x_i}{N} \qquad s = \sqrt{\frac{\sum_{i=1}^{N} (x_i - m)^2}{N - 1}} \qquad (10.72)$$

Next, in the analysis step, all pixels in the image sequence are analyzed one by one (Figure 10.39b). For each, at each time step n, the probability of being any of the known flaws or background is computed with the Gaussian probability equation (10.69) and previously established means and variance. All probabilities of being a given known defect or background at each time step are multiplied together [eq. (10.71)], so that we obtained for a given pixel its global probabilities. The winning category (i.e., the category with which it is associated) corresponds to the largest probability value. This identifies the unknown pixel location as the more probable known flaw (or nondefect). This process is repeated for all pixels in the field of view.

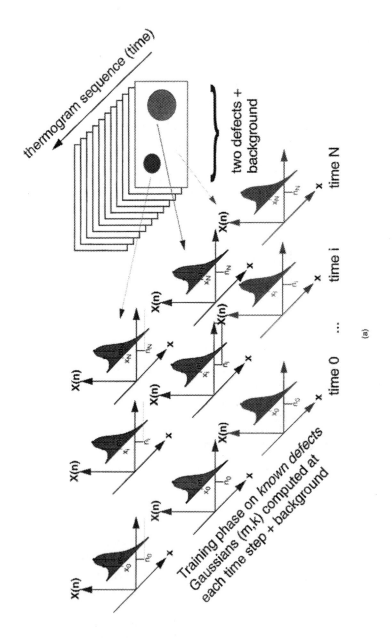

thermogram sequence (time)

two defects + background

X(n) time N

time i

...

time 0

(a)

Training phase on *known defects*
Gaussians (m,k) computed at
each time step + background

420

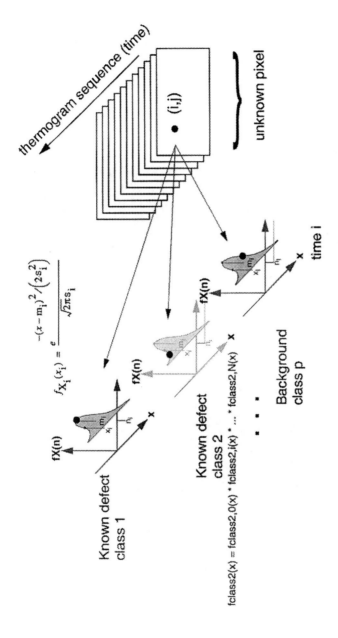

$$f_{X_i}(x_i) = \frac{e^{-(x - m_i)^2/(2s_i^2)}}{\sqrt{2\pi}s_i}$$

fclass2(x) = fclass2,0(x) * fclass2,i(x) * ... * fclass2,N(x)

The pixel is assigned to the class with largest probability

FIGURE 10.39 Principle of the statistical method: (a) calibration; (b) analysis.

(b)

421

Such a procedure can be applied directly to phase data at each discrete frequency f to estimate defect depth (it can also be applied on amplitude or directly on the temperature sequence). Alternatively, it is recommended (Bison et al., 1998) to detect defects first and than estimate their depth. Best reported results use phase data for detection stage and amplitude for characterization (Figure 10.40). In a study (Vallerand, and Maldague, 2000) performed on aluminum specimens with embedded simulated corrosion from 0 to 93% (flat-bottomed holes 25 mm in diameter), the percent of correct characterization

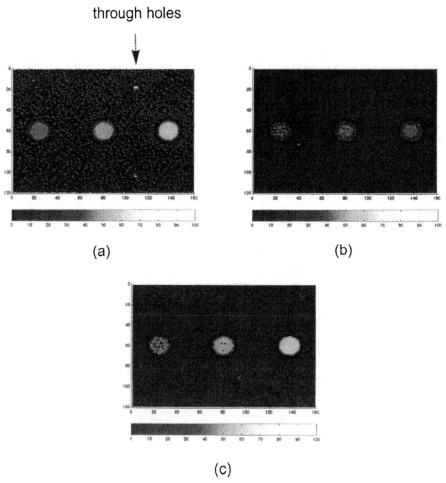

FIGURE 10.40   Results obtained with the statistical method. Computed corrosion images in aluminum (corrosion defects in aluminum simulated by flat-bottomed holes, from left to right, corrosion: 43%, 52%, and 62%, equivalent respectively to remaining thicknesses of metal: 0.061 cm, 0.051 cm, and 0.041 cm for a total plate thickness of 0.106 cm): (a) with amplitude data only; (b) with phase data only; (c) detection with amplitude data and characterization with phase data. Bright points are through holes in the plate. (From Vallerand, 2000.)

TABLE 10.5   Percent of Correct Characterization[a]

Data Set	Statistical Method (%)	Perceptron (%)
Temperature	28:28:35	84:87:69
Phase	94:99:62	77:80:57
Amplitude	64:64:65	82:83:74
Detection with *phase* followed by characterization with *amplitude*	96:99:76	Not tested

Source: Adapted from Vallerand and Maldague (2000, Table 2).
[a]Categories are over whole image:over sound area regions:over flaw regions.

(PCC) was measured over whole images (defect/nondefect), over sound area regions, and over flaw regions (Table 10.5). This percentage is established by computing the ratio of all correctly categorized pixels over all pixels in the region of interest (here over whole images, over sound area regions, over flaw regions). In Table 10.5, the statistical method is applied on temperature, phase, or amplitude data for complete characterization or, alternatively, detection on phase followed by characterization with amplitude data (last column of the table) if a flaw is detected. The procedure made it possible to differentiate among flaws with thickness differences down to 100 µm. Tests performed by moving the specimen (± 10 cm) from a reference position in front of the IR camera indicated that PCC results are affected by about 4% (Vallerand and Maldague, 2000, Table 3, p. 312).

For temperature, thermograms at any time t in the sequence are first normalized by the average temperature of the first image $T_{t=0}$:

$$T_{n,t} = \frac{T_t}{\text{average of } (T_{t=0})} \tag{10.73}$$

This allows us to compute the normalized temperature $T_{n,t}$ image at time t. Of course, to apply eq. (10.73) successfully, the effects of subsurface defects should not yet be present in the first image. Such normalization allows us to compensate for variations in heating from one experiment to another.

10.4.5 Neural Network

In Chapter 6, the theory of neural networks (NNs) was presented. As mentioned, NNs are particularly attractive for data inversion of ill-posed problems, such as in the case of phase analysis in PPT. In fact, many authors have reported the use of NNs in TNDT; see for example, the work of Prabhu et al., 1992; Prabhu and Winfree, 1993; Bison et al., 1994a; Grinzato et al., 1995a; Tretout et al., 1995, 1997; Santey and Almond, 1997; Bison et al., 1998; Maldague et al., 1998). To demonstrate the usefulness of NNs for such purposes, the problem of corrosion evaluation in aluminum is now addressed with a three-layer perceptron NN (see Figure 6.17).

The number of layers used was set experimentally. For temperature data, the structure was 50–10–7 (each number represents the number of neurons for each respective layer: first, second, third). In the case of amplitude and phase data, the structure was 25–10–7, since, due to the symmetrical property of the Fourier transform, N time-domain values provide $N/2$ complex frequency-domain values.

Before using an NN, weighting factors need to be adjusted with the back-propagation learning algorithm (Section 6.5). Each input vector in the NN is associated with a particular value of aluminum corrosion and is thus associated to the class with the most similar corrosion characteristics. In this example, eight corrosion classes are defined, and thus the NNs have seven outputs: 0% (no defect), 17%, 26%, 42%, 52%, 62%, 83%. The NN training is supervised since training samples are provided. Once trained, the NNs are ready to assign unknown input data to a specific corrosion value (i.e., to a specific output class). Table 10.5 summarizes the performance; Figure 10.41 shows a typical result. The statistical method with phase data or phase data for detection followed by amplitude data for characterization were shown to be superior to perceptron NN (Figure 10.41c).

10.4.6 Wavelets

The first mention of *wavelets* dates to 1909 in a thesis by A. Haar. The present wavelet data analysis was proposed by J. Morlet, the main algorithms to perform discrete wavelet transform (WT) efficiently are from S. Mallat in 1988. Since then, wavelets have been disseminated in many fields of data analysis. Basically, as Fourier analysis decomposes a signal into sine waves of various frequencies, wavelet analysis breaks up a signal into shifted and scaled versions of the original (or mother) wavelet (Figure 10.42). One of the main advantage of wavelet analysis lies in its ability to perform local analysis, to reveal trends and breakpoints (Misiti et al., 1997). In Section 10.4.3 we mentioned that one difficulty related to data inversion in PPT is, in fact, related to the loss of time information that takes place with the Fourier transform. To preserve this information, one can substitute the Fourier transform for the wavelet transform which maintains this parameter.

The WT of function $f(t)$ based on function $h_{S,T}(t)$ is defined by (Sheng, 1996; Misiti et al., 1997)

$$W_f(S, T) = \int_{-\infty}^{+\infty} f(t) h_{S,T}^*(t)\, dt = \mathrm{Re} + i\,\mathrm{Im} \qquad (10.74)$$

where * represents the complex conjugate, i is an imaginary number ($i = \sqrt{-1}$), and Re and Im are the real and imaginary parts of the transform, respectively.

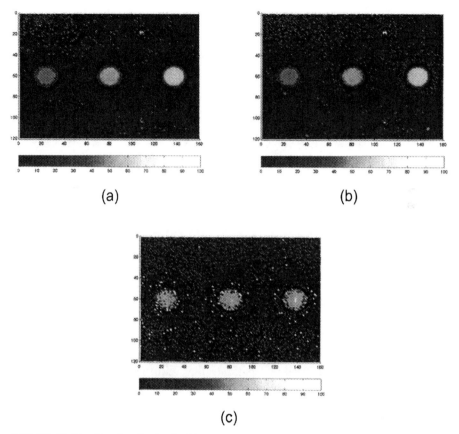

FIGURE 10.41 Results obtained with a perceptron neural network. Computed corrosion images in aluminum (same specimen as Figure 10.40): (a) with temperature data; (b) with amplitude data; (c) detection phase data. (From Vallerand, 2000.)

The function $h_{S,T}(t)$ is generated by translating and scaling the mother wavelet $h(t)$:

$$h_{S,T}(t) = \frac{1}{\sqrt{S}} h\left(\frac{t-T}{S}\right) \tag{10.75}$$

where S is the scaling factor that is related to the frequency and T the translation factor associated to the time. Generally, a positive scaling factor is used. If S is larger than 1, the wavelet is dilated; if it lies between 0 and 1, it is contracted. In a reported application, the mother wavelet was selected as the Morlet wavelet, expressed as (Galmiche et al., 1999):

$$h(t) = \exp(-j\omega_0 t) \exp\left(-\frac{t^2}{2}\right) \tag{10.76}$$

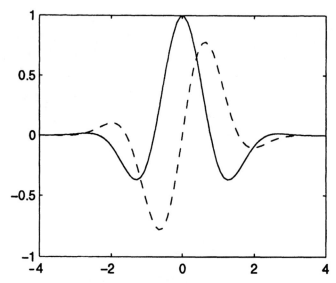

FIGURE 10.42   Plot of the Morlet wavelet showing both real (continuous line) and imaginary parts (dashed line).

Basically, the principle is as follows. First, a value of ω_0 for the mother wavelet is selected. This value represents the frequency at which the signal is analyzed. As discussed previously, low frequencies concentrate most of the energy available in the stimulation pulse (Figure 10.27). It is therefore preferable to choose a small value of ω_0. In the case of the simulation shown in Figure 10.43, a value of $\omega_0 = 2$ rad s^{-1}, corresponding to a frequency of 0.32 Hz, was selected (we recall that $\omega = 2\pi f$). The WT is computed with eq. (10.74); the phase is computed with eq. (10.64). Next, the value of the translation factor yielding to a maximum contrast between defect and no-defect areas is found. In Figure 10.43 the phase value of the WT for a 3-mm depth defect (flat-bottomed hole in aluminum) is shown as a function of the translation factor for a line passing over the defect center (scaling factor set to 1); the defect is clearly visible. The white line in that figure indicates the translation factor yielding to the maximum phase value. Figure 10.44 gives the evolution of the translation factor as a function of the defect depth for the maximum phase value. Since a quasilinear relationship is observed between the two parameters, data inversion becomes simple. In preserving the time information, the WT thus also maintains depth information.

10.4.7   Calibration for Lock-in Thermography

In LT the analysis is performed in steady state, whereas in PPT it is transient. Nevertheless, methods reviewed in previous sections can be applied as well to LT data, due to the similarity with PPT data (Maldague and Couturier, 1997).

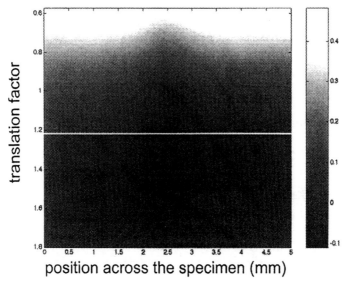

FIGURE 10.43 Wavelet transform. Simulation of phase in an aluminum specimen with a 5-mm-diameter flat-bottomed hole located 3 mm below the surface. In this case $\omega_0 = 2$ and $S = 1$. The image shows the phase profile across the defect as a function of the translation factor. The line indicates the maximum phase value. (Adapted from Balmiche et al., 2000a.)

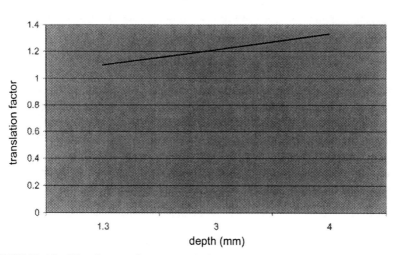

FIGURE 10.44 Wavelet transform. Translation factor evolution as a function of depth, for maximum phase contrast.

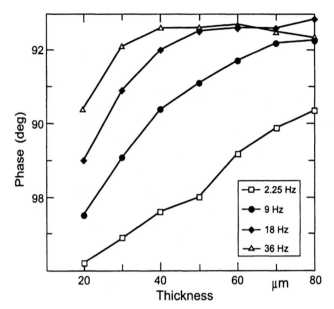

FIGURE 10.45   Relationships between phase value and thickness of paint on a polymer substrate in LT, for various modulation frequencies. (Redrawn from Busse, 1994, with permission from Elsevier Science.)

Interestingly, generally experimenters often wait only one cycle of the heating modulation to get the steady-state regime in LT.

As a final topic in data inversion, authors have also reported direct experimental calibration of the phase data with depth. For instance, in Figure 10.45, phase variation related to paint thickness applied on a polymer substrate is shown for lock-in thermography (Busse, 1994). Following such calibration, the polynomial fit function $g_{calib}(\cdot)$ on these data allow us to retrieve the parameter of interest—here the paint thickness TH_{paint}:

$$TH_{paint} = g_{calib}(phase, frequency) \qquad (10.77)$$

Notice, however, that the nonuniqueness of phase as a function of paint thickness as seen in Figure 10.45 at $f = 36$ Hz restricts the validity of such an analysis to a limited span of paint thickness values. This is a general problem when using only the phase for data inversion (as also discussed in Section 10.4.3). The two-step approach discussed in Section 10.4.4, with detection on phase data followed by characterization with amplitude data if a flaw is detected avoids this problem, as confirmed by better-obtained inversion figures, as shown in Table 10.5; the same is true for the wavelet approach in Section 10.4.6.

10.5   THERMAL TOMOGRAPHY

The idea of *thermal tomography*, introduced by Vavilov in 1986 (Vavilov et al., 1990, 1992a), is to slice the specimen along depth layers corresponding to the

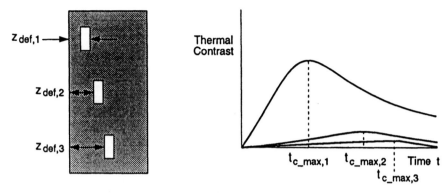

FIGURE 10.46 Occurrence of time of maximum contrast t_{c_max} as a function of subsurface flaw depth. (From Maldague, 1993.)

distribution of thermal properties at specific depths. By analogy with computed x-ray tomography (CT), such slices are called *thermal tomograms.* However, tomography principles cannot be applied directly to the heat transfer process, which occurs not through a straight direction but according to a diffusion propagation scheme. Instead of being based on angular projections as for CT, thermal tomography (TT) is based on the surface temperature evolution of the component inspected following the initial thermal perturbation [pulsed thermography (PT) stimulation; Section 9.3]. As an analogy with CT, time increments are associated with angular projections. In fact, TT is essentially a different method of processing the thermogram sequence and presenting the PT data.

TT principles can be understood with respect to Figure 10.46. In this figure, thermal contrasts beneath a sample surface are drawn (observation is in reflection following the initial pulse thermal perturbation). Recalling eq. (9.7), it is noted that the occurrence of the time of maximum thermal contrast t_{c_max} is proportional to the square of the depth (at least in homogeneous materials); consequently, deeper defects experience longer t_{c_max}. If from the thermogram sequence, we extract for every pixel (i, j) in the field of view, the time t_{c_max} when the thermal contrast is maximum at this location (i, j), we obtain the distribution of all the t_{c_max} values for all the pixels in the image. This distribution is called the *timegram* TGM_{c_max}. Figure 10.47 illustrates a method of computing the timegram image TGM_{c_max}. Interestingly, timegrams can be obtained with respect to different parameters, such as time for half-contrast, time for maximum contrast slope, and so on. Assuming uniform heating of the surface inspected, areas of the specimen having uniform thermal properties (such as thermal conductivity) will have the parameter of interest show up in the same time window in the timegram. On the contrary, subsurface flaws having different thermal properties will experience different values of the parameter of interest and, consequently, will exhibit different time values in the timegram. This yields to possible defect detection if the timegram is sliced. Such

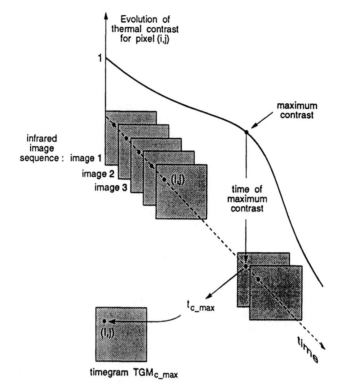

FIGURE 10.47 Thermal tomography principle: computation of timegram TGMc_max. (From Maldague, 1993.)

a slice, called a *thermal tomogram*, is obtained by thresholding timegram TGM_x in a fixed gate time $[t_1, t_2]$, x being the parameter of interest (maximum contrast, half-contrast, etc.). An illustration of this analysis is shown in Figure 10.48.

The TT technique is another way to look at thermal information in terms of time rather than amplitude as in *standard* PT processing. It has some advantages in the interpretation of results since structures detected appear directly in terms of time (or depth if converted from the time domain), with sharp detected edges.

10.6   PROCESSING AND EXPERIMENT FOR NONPLANAR SURFACE INSPECTION

Most of the time, application of TNDT to nonplanar surfaces results in distortions of the temperature images collected. These distortions come from two effects. First, if the surface is heated, areas parallel to the heating device will receive more energy with respect to areas perpendicular to it. Second, similarly,

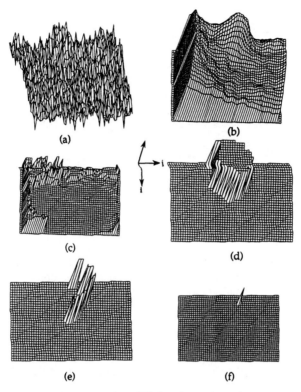

FIGURE 10.48 Thermal tomography of Teflon insert in carbon-epoxy specimen using the tc_max parameter: (a) raw image; (b) smoothed image; (c) timegram TGMc_max; (d) tomogram of the layer at 0.8 to 1.5 mm; (e) tomogram of the layer at 1.4 to 1.8 mm; (f) tomogram of the layer at 1.8 to 2 mm. Specimen: 28 4.25-mm-thick layers black-painted CFRP panel with a 10-mm-diameter Teflon implant inserted at the eighth layer (1.2 mm under the front surface). Arbitrary amplitude units. (From Maldague, 1993.)

areas parallel to the IR camera will emit more energy in its direction. Although some algorithms (such as the one reviewed in Section 6.3 and also LT and PPT) and PPT phase computations at certain frequencies (Figure 10.33) can compensate partially for nonuniform temperature, most of the techniques reviewed in this book generally assume a flat surface and thus uniform collected temperature images. The question of nonplanar surface inspection is thus interesting, although little work has been done in this area. Figure 10.49 illustrates the problem. On the left, a visible image shows part of the tail section of an aircraft, which is inspected by TNDT. On the right, the corresponding thermogram is confusing. First, the thickness variation of the internal structure of the airplane is visible (e.g., notice the two plies of aluminum, the left pile being cooler). Moreover, due to the aircraft geometry, a discontinuous heating pattern is observed and not having Figure 10.49a at hand, a misinterpretation of the various zones in the IR image of Figure 10.49b could have resulted.

18.1 °C   28.1 °C

(a)   (b)

FIGURE 10.49   The problem of shape curvature in TNDT: (a) visible image of rear part of tail section of a DASH-8 aircraft; (b) corresponding thermogram recorded 150 ms after flash and showing all subsurface structure (flash duration is 15 ms). (Adapted from Maldague et al., 1994a, with permission from Kluwer Academic Publishers.)

The idea is to make use of the TNDT apparatus to get information about specimen geometry. Once specimen geometry is known, it becomes possible to correct thermograms for nonplanar effects, such as the ones shown in Figure 10.49b. This section is thus concerned with the question of sensing of nonplanar 3D surfaces by the TNDT apparatus itself. We call such a system an *intrinsic 3D sensing system* (3DSS) as opposed to active extrinsic 3DSSs, which rely on energetic illumination devices such as laser beams. In active 3DSSs, the great number of photons collected by the sensor from the illumination device allows us to achieve great performances (Pitre et al., 1993). Since intrinsic 3D sensing relies mainly on the thermographic apparatus and on the poorly energetic infrared radiation to extract object shape [e.g., eq. (2.4) with long involved wavelengths in infrared thermography with respect to the visible spectrum], it is not surprising that the accuracy and resolution reported are limited. However, appropriate reconstruction processing partially alleviates this problem. Generally, these techniques require processing of an early recorded thermogram (ERT) in which potential defect contrasts have not yet developed. Following the analysis of the ERT, the entire thermogram sequence can be corrected for surface curvature.

In the next sections, four different techniques are described for shape reconstruction and thermogram correction in TNDT applications. Table 10.6 summarizes the characteristics of the four methods discussed. From this we see that shape from heating is probably the most attractive method, although it requires more complex processing than the other methods, especially for the segmentation task. Attractive fields of applications of theses methods are in the aero-

TABLE 10.6 Characteristics of Methods for Shape Correction in TNDT

Method	θ	R	Comments
Point-source heating	$\cos\theta$ assumed to be close to 1	Computed	• Works on limited distances in front of IR camera • Absolute calibration • Point-source heating required
Video thermal stereo vision	Computed	Assumed without much influence on restricted distances	• Video camera and illumination device required • Two-parameter calibration • Orthogonal heating
Direct thermogram correction	Computed	Assumed without much influence on restricted distances	• No extra hardware • Two-parameter calibration • Orthogonal heating
Shape from heating	Computed	ΔR is computed	• No extra hardware • Relative calibration (not mandatory) • Orthogonal heating • More complex processing

Source: Adapted from Maldague et al. (1994a, Table 1).

space industry and also for pipe inspection (Pelletier et al., 1996; Pelletier and Maldague, 1996).

10.6.1   Thermal Stimulation Background (Curved Objects)

As reviewed in Section 8.1, thermographers have experimented many ways to stimulate samples: from snow, water bags, and liquid nitrogen, to more conventional heating devices such as lamps, the simplest probably being the common household light bulb, whose heating distribution can be expressed by the well-known $1/R^2$ relationship (derived from Lambert's law of cosines):

$$\Delta T_p \sim \varepsilon \frac{P \cos \theta \, \Delta t}{4\pi R^2 \rho C \, dz} \quad {}^\circ C \tag{10.78}$$

where R is the distance between the point heating source and the sample surface (meters), θ the angle between the normal to the sample surface and incident stimulating rays (radians), P the heating power of the point source (watts), ρ the specimen density (kg m^{-3}), Δt the thermal pulse length (seconds), dz the depth of penetration of the heating front during Δt (meters), C the specific heat (J kg^{-1} $^\circ$C^{-1}), α the thermal diffusivity (m^2 s^{-1}), and ε the spectral emissivity, and where dz is given approximately by

$$dz \sim \sqrt{\alpha \, \Delta t}. \tag{10.79}$$

As seen in eq. (10.78), two parameters (R and $\cos \theta$) in particular are involved in the inspection of curved objects and contribute to the distortion of thermograms. The correction mechanism thus implies that we must find a way to derive these parameters before correcting thermograms in order to suppress inhomogeneities due to shape curvature, thus improving the reliability of subsequent interpretation. Depending on the radiative heating device, different behaviors are obtained. For example, if a parabolic reflector is placed behind a point source, thus concentrating the heating rays, dependence on the distance R may either disappear or become linear (on restricted distances), thus simplifying the correction process (orthographic projection; Section 5.2). An example of such a source was shown in Figure 8.7.

10.6.2   Point-Source Heating Correction

This method is derived directly from eq. (10.78) and on dependence of the rising temperature on parameters R and θ. In this case a high-powered bulb (e.g., 1000 W) is used as a thermal stimulation device. It is mounted on top of the IR camera (Figure 10.50). In a first step, a calibration phase is necessary. The calibration procedure consists of obtaining the *early recorded thermogram* (ERT) in the 3D inspection volume located in front of the IR camera. The ERT is the earliest unsaturated thermogram recorded just after the thermal stimulation is

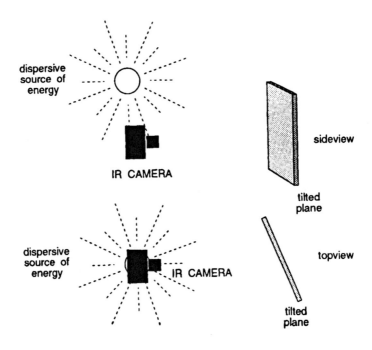

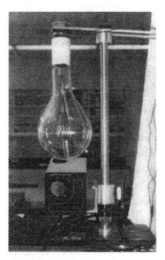

FIGURE 10.50 Schematic diagram of the experimental setup needed for the point-source heating correction method (the tilted plane is used for calibration) and picture of the high-powered bulb. (From Maldague, 1993.)

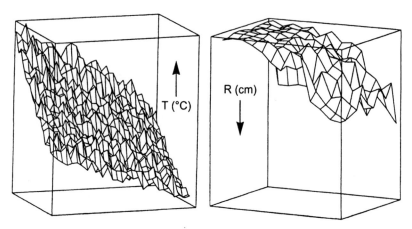

FIGURE 10.51 Thermal NDE for nonplanar surfaces: experimental results obtained on a tilted Plexiglas plate: left image is the temperature image, right image is the computed 3D range image of this scene (arbitrary units). (From Maldague, 1993.)

completed (IR image recorded before shallow defects start to perturb the surface temperature distribution). Typically, in the calibration phase, the ERTs are recorded on a plane covered with high-emissivity paint (the same paint as that used on the specimens) at various distances in front of the IR camera, enabling us to constitute a complete data bank of the expected temperature distribution within the inspection volume (Maldague et al., 1991c; Kiiskinen et al., 1992).

Once the database is completed, the actual inspection proceeds and the shape correction consists of matching specimen ERT with the ERT of the database on a pixel-per-pixel basis. This matching produces a range image of the scene inspected, which can be used further to correct subsequent thermograms from the specimen. Figure 10.51 shows the result of such (ERT → range) image conversion.

Drawbacks of this technique are twofold. A high-powered source is required, and only a fraction of the available power is used, due to the inherent spreading of the radiation in the entire space; consequently, the inspection volume is restricted. Experiments described by Maldague et al. (1991c) reported a maximum workable distance of about 20 cm in front of the IR camera and a resolution of about 1 cm. Moreover, the calibration procedure is slow if the calibration plane is to be tested at many different distances in front of the camera. A possibility discussed by Kiiskinen et al. (1992) is to use a tilted plane covering the entire distance span in front of the IR camera. In such a case, one experiment may be sufficient (however, this is not valid if a point-source-like heating device is used, due to the nonuniform heating pattern of such a device). Moreover, the database has to be recalibrated periodically as any calibration-based technique (e.g., due to lamp aging). Finally, since the method does not include the parameter $\cos \theta$ of eq. (10.78), it is limited to objects of restricted curvature ($\cos \theta \sim 1$ for $\theta = 90 \pm 20°$).

On the other hand, the technique is interesting since the thermograms allow us to extract information that would otherwise not be available, and in addition to the database being calibrated, it does not require extra hardware with respect to the standard TNDT procedure. Moreover, if the geometry of an object is known (a reasonable hypothesis), the parameter $\cos\theta$ becomes available for a more accurate correction.

10.6.3 Video Thermal Stereo Vision

One of the principal disadvantages of the previous correction technique is related to the point-source-like heating device, which provides nonuniform heating patterns and for which most of the heating power is lost, thus reducing the amplitude of the thermal contrasts available. In most cases it is better to rely on a uniform heating device, and then a possible choice for thermogram correction is to rely on some additional hardware. An interesting approach is based on a video thermal stereo vision method (Nouah et al., 1992; Wiecek and Zwolenik, 1998). To apply this technique, in addition to the standard TNDT apparatus (heating device and IR camera), a video camera and an illumination device are required, along with the usual computer control and frame-grabbing facilities (Figure 10.52).

The reflectance information of the scene present in a visible image makes it possible to determine θ. This method is based on the work of Nandhakumar and Aggarwal (1988) and assumes that observed surfaces are Lambertian and objects opaque (these hypotheses are valid in most TNDT applications). In

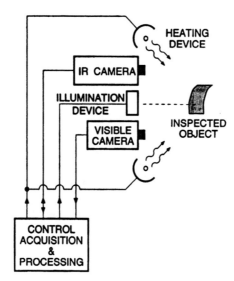

FIGURE 10.52 Schematic diagram of the experimental setup needed for the video thermal stereo correction method. (From Maldague et al., 1994a, with permission from Kluwer Academic Publishers.)

these conditions,

$$L = K_d \cos \theta + C_d \tag{10.80}$$

where L is the digitized intensity value of the visible image and K_d and C_d are the overall calibration constants of the visible imaging system. K_d and C_d are specific to a given experimental apparatus; they are found experimentally using the following procedure. A visible image of a scene with two surfaces of known orientation is digitized. Substitution of known orientations and digitized values for the two surfaces in eq. (10.80) allows us to write a two-equation system that solves for K_d and C_d. Once K_d and C_d have been calibrated for a given apparatus, the orientation image $\cos \theta$ can be computed with eq. (10.80) from the visible image of the scene inspected.

Before proceeding to curvature correction of the thermograms, it is necessary to align visible and IR image formats (this is sometimes called *image registration*). This is essential to have a direct pixel-to-pixel correspondence between the two images. This registration step is based on the following transformation between the coordinates (x, y) and (u, v) for infrared and visible images, respectively. For the u coordinate,

$$u = \sum_{i=0}^{N} \sum_{j=0}^{N-i} a_{ij} x^i y^j \tag{10.81}$$

with a_{ij} polynomial coefficients and N the degree of the polynomial (usually, 2). The v coordinate is computed in a similar fashion (see also Wiecek et al., 1998a).

A calibration step is required to compute the polynomial coefficients. A set of common points between two images of the same scene, visible and infrared, are selected and eq. (10.81) is solved. Finally, curvature correction of thermograms is done by dividing successive thermograms by $\cos \theta$ following eq. (10.78).

This double image correction raises several problems. First, it is necessary to cover a specimen surface with a paint that has both good pseudo-Lambertian properties in the visible spectrum and acceptable emissivity values in the sensibility band of the IR camera. Second, the illumination device must have limited dependence with parameter R for eq. (10.80) to be valid. Besides these difficulties, this method is an effective correction tool, as illustrated by Figure 10.53 (correction process performed in the case of a plastic tilted plane).

10.6.4 Direct Thermogram Correction

The idea of the direct thermogram correction is to suppress the extra hardware (illumination device and video camera) required in the video thermal stereo vision correction method while keeping a similar correction scheme. The idea is as follows. For opaque bodies at close to room temperature, one of the main

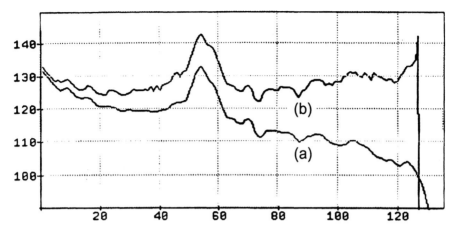

FIGURE 10.53   Results of the video thermal stereo correction method profile with defect extracted from a thermogram of a plastic tilted plane (a) before and (b) after correction. Arbitrary vertical units; horizontal scales: pixel coordinates. (From Maldague et al., 1994a, with permission from Kluwer Academic Publishers.)

differences between visible and IR images lies in their formation process. Visible images are formed due to a reflection phenomenon while IR images are formed by an emission mechanism. Nevertheless, if the heating device provides little dependence on distance R, eq. (10.80) can be adapted to TNDT applications, with L the digitized value of the thermogram, K_d and C_d the overall calibration constants of the IR imaging system, and θ the angle between the normal to the surface patch (for a given pixel set x, y) and the direction of observation with the IR camera. Equation (10.80) thus provides a reasonable estimation of local surface orientation θ once parameters K_d and C_d are calibrated using a technique similar to the one described in the preceding section (with tilted planes of various orientation).

The orientation correction process is completed next using an ERT image after a thermal pulse. With calibration parameters K_d and C_d and eq. (10.80), the orientation image is formed and all subsequent thermograms are divided by this orientation image. Figure 10.54 shows a result of this method. An important point is that thermal pulse specifications (duration, power) must, of course, be the same for both calibration and correction steps.

The disadvantage of this method is that heating distance (parameter R) is assumed sufficiently large so that local depth variations on the specimen do not affect the IR emission process. This hypothesis is not always valid. Nevertheless, there is an obvious interest in such a simple correction process, at least for (qualitative) defect detection purposes.

10.6.5 Shape from Heating

This method is based on *shape from heating* theory, originally developed in one dimension and in which thermograms are analyzed row per row (E. Barker

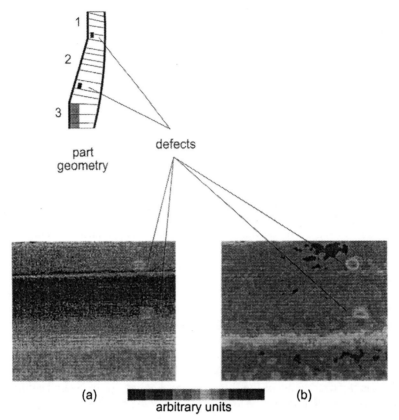

FIGURE 10.54 Results of the direct thermogram shape correction: (a) raw thermogram of an aircraft composite door (Kevlar–foam) with two subsurface defects visible; (b) corresponding thermogram after correction. Duration of heating 2 s, observation time 10 s, depth of defect is about 2 mm under the surface (heating device of Figure 8.7). The part geometry profile is shown. Notice how the correction is effective in zones 1 and 2 made of same material. Zone 3 is made of a different material assembly thus explaining the different response obtained both in (a) and (b).

et al., 1995). Interestingly, shape from heating is somehow inspired by *shape from shading* theory, well known in digital vision due, for example, to its ability to restore shape but not absolute distance measurements (Horn, 1982; Bruckstein, 1988). To reconstruct object shape, shape from shading extracts shape gradients $p = dz/dx$ and $q = dy/dz$ from a single reflectivity image (z being the depth coordinate).

As in the other correction methods reviewed for nonplanar surfaces, this method is based on two-dimensional analysis of the early recorded thermogram in the thermogram sequence. This method relies only on the thermographic apparatus to extract local surface orientation. The following hypotheses apply. The assumed projection is orthogonal (i.e., perpendicular to axis or projection

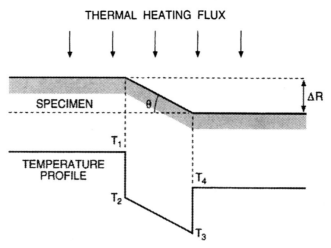

FIGURE 10.55 Heating analysis of a flat surface specimen: experimental configuration and corresponding thermal profile. (From Maldague et al., 1994a, with permission from Kluwer Academic Publishers.)

plane; Section 5.2), the heating flux is assumed orthogonal to the specimen (i.e., heating perpendicular to specimen axis), the heating has a little dependence on distance (the dependence being more on the local orientation φ of the surface), and the surface emissivity is high (e.g., by using blackpainting; Section 8.2).

Linear Segments. The early thermogram of a specimen is analyzed row by row (or divided into linear patches; see below). Each row is divided in segments which are classified as to whether or not they are linear. If a segment is linear, this implies that variation in the distance R (heating device to specimen distance) is greater than the orientation angle θ. This is the case for objects with flat surfaces. Figure 10.55 depicts this case. From Figure 10.55 and with eq. (10.78), assuming that before heating the specimen surface temperature is T_i, we can write

$$T_2 - T_i = (T_1 - T_i) \cos \theta$$
$$T_3 - T_i = (T_4 - T_i) \cos \theta$$
(10.82)

from which we obtain

$$\cos \theta = \frac{T_2 - T_3}{T_1 - T_4}$$
(10.83)

Notice that T_1 and T_4 values are not always available; think of a tilted plate, for example. In such a case, a variant procedure is to be used, as shown below.

Provided that the experimental curve temperature rise versus distance R is known (through appropriate calibration), slope S becomes a known parameter with the following definition (on TNDT workable distances):

$$S = \frac{\Delta T}{\Delta R} \qquad (10.84)$$

With respect to eqs. (10.83) and (10.84),

$$\Delta R = \frac{T_1 - T_4}{S} \qquad (10.85)$$

and finally, with eqs. (10.82) and (10.85),

$$\Delta R = \frac{T_2 - T_3}{S \cos \theta} = \frac{\Delta T}{S \cos \theta} \qquad (10.86)$$

Using eq. (10.83), it is possible to compute surface orientation θ, while eq. (10.86) relates both orientation to height variation together (in an orthogonal projection). Figure 10.56 depicts the case of a single plane segment. From Figure 10.56, the following relationship is obtained:

$$\Delta R = d \tan \theta \qquad (10.87)$$

where d is the length of the projection and is measured by the number of pixels N associated with the segment, $d = N$ *step*, with *step* being the horizontal distance per pixel in the field of view, at focus. This relationship is valid only for

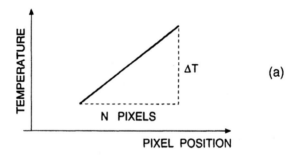

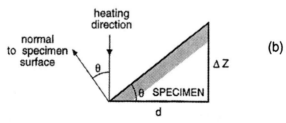

FIGURE 10.56 Heating analysis for a tilted plane: (a) experimental configuration; (b) corresponding thermal profile. (From Maldague et al., 1994a, with permission from Kluwer Academic Publishers.)

orthogonal projections. Combining eqs. (10.86) and (10.87), we obtain

$$\sin \theta = \frac{\Delta T}{dS} \tag{10.88}$$

which allows us to determine orientation θ for cases where T_1 and T_4 are not available.

To summarize, for linear row segments, orientation θ is first computed using eq. (10.83). Next, height variations ΔR can be found with either eq. (10.85) or (10.86) if the rising temperature versus distance curve is available. Otherwise, an approximated solution of ΔR is given by eq. (10.87) and θ is found using eq. (10.88).

Nonlinear Segments. In these cases, variations of distance R are much less than variations due to orientation θ. It is then assumed that for small segments, temperature variations are proportional only to orientation θ:

$$T - T_i = (T_{max} - T_i) \cos \theta \tag{10.89}$$

where T_{max} is the local maximum temperature close to the pixel of interest, which is at temperature T. Knowing T_i (the before-heating temperature) it becomes possible to compute the local surface orientation. Figure 10.57 illustrates the geometry studied. Following this scheme, nonlinear segments are divided

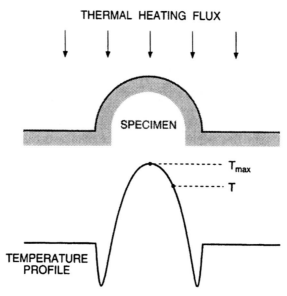

FIGURE 10.57 Heating analysis of a curved specimen: experimental configuration and corresponding thermal profile. (From Maldague, 1994a, with permission from Kluwer Academic Publishers.)

into elementary segments for which local orientation is found with eq. (10.89). The next segment size is derived using eq. (10.87), with $d = step$ for one-pixel-segment size. Complete shape reconstruction proceeds by merging all segments one after the other. Image correction can proceed as described in Section 10.6.4 once the orientation image is generated. Some results are presented in Figure 10.58. Figure 10.58a depicts the experimental setup: a plastic part constituted of a plane and a half-cylinder, tilted by 30° and heated orthogonally. Figure 10.58b shows the corresponding raw temperature image of this part geometry. Figure 10.58c is a profile of the middle row of the thermogram of Figure 10.58b. Due to the high level of noise, the thermogram must be smoothed before processing (using a 5×5 averaging filter; Section 5.5), Figure 10.58d shows the corresponding smoothed profile. Figure 10.58e presents the result of the shape reconstruction; line 1 is the reconstructed profile superimposed on the exact profile (line 2). The maximum reported error is 2° for global orientation, and the two shapes (exact and reconstructed) match quite well.

Linear Patches. When objects being inspected can be divided into linear patches, it becomes preferable to proceed with a *shape from heating* analysis taking advantage of that geometry. The processing of the ERT is as follows: first, noise smoothing using convolution operators (of size set to 1.25 times the standard deviation of noise; Section 5.4.2) and then image splitting in 5×5 patches on which planes are fitted using the equation

$$T(x, y) = Ax + By + C \qquad (10.90)$$

where A and B are the thermal gradients along the x and y directions and T is the temperature variable. Patches with abnormal standard deviation are discarded [e.g., as in the case in nonheated (shadow) areas]. The remaining valid patches are sorted into linear and nonlinear categories since these two cases correspond to different processing. For linear patches, depth is considered more important than surface orientation. The *linearity criterion* is based on the thermal slope $S = dT/dz$ (S is obtained by experimental calibration). If for a given patch the variation dT/d is greater than S, the nonlinear case is assumed (d is the distance in the thermogram; refer to our earlier discussion for the nonlinear segment case).

With all linear patches identified, the next step consists of agglomerated patches having similar orientation to obtain more global regions. A region-growing algorithm is used for such purpose. First, a seed is established to start the agglomeration. An error function E is computed for all unagglomerated patches and the seed is set in the patch having the smallest E value. The region growing proceeds by agglomerating adjacent patches having similar A, B, C coefficients. The geometry of agglomerated regions is next extracted with $p = dz/dx = \tan\theta_x$ and $q = dz/dy = \tan\theta_y$ knowing A and B, with $\tan\theta_x = dz/dx = \tan(\sin^{-1} A/S)$ and $\tan\theta_y = dz/dy = \tan(\sin^{-1} B/S)$. An angle ϕ between the surface normal and the direction of the stimulation becomes

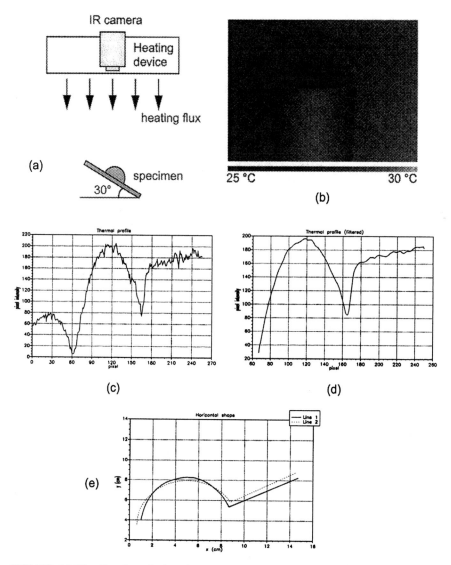

FIGURE 10.58 Results of the shape from heating reconstruction method (line): (a) experimental configuration; (b) early recorded thermogram, heating pulse (heating device of Figure 8.7) for $t = 0$ to 125 ms, observation at $t = 167$ ms; (c) raw extracted profile: middle row of image (b); (d) profile (c) after smoothing; (e) result of the shape reconstruction: line 1 is reconstructed profile and line 2 is exact specimen shape. (From Maldague, 1994a, with permission from Kluwer Academic Publishers.)

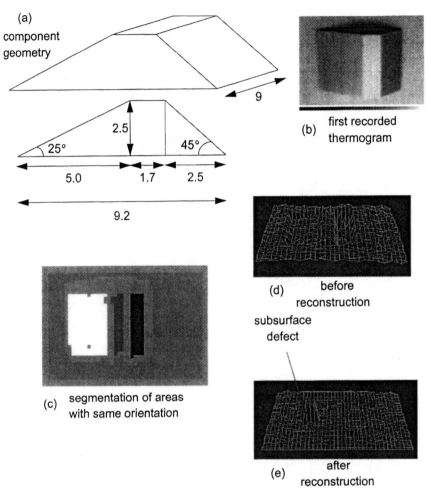

FIGURE 10.59 Results of the shape from heating reconstruction method (linear patches): (a) component geometry; (b) first recorded thermogram (ERT); (c) result of segmentation showing areas with same orientation; (d) 3D profile of thermogram (b); (e) 3D profile of a later thermogram with subsurface defect identified. (From Pelletier and Maldague, 1996.)

available:

$$\phi = \sqrt{p^2 + q^2} \qquad (10.91)$$

Correction of orientation proceeds with division of thermogram intensities I by $\cos \phi$ [eq. (10.78)], while correction for depth is based on singular points (vertices) I_{sing}, $I_{\text{corr}} = I + Sz$ with $z = (I_{\text{sing}} - I)/S$. Thermograms afflicted by shape distortions and corrected in this fashion experience a flattened perspective, thus making the defect detection step and subsequent quantitative characterization more reliable. An illustration of the correction process is shown in Figure 10.59. The subsurface defect appears clearly in the corrected image.

TABLE 10.7   Matlab M-Scripts Available in Appendix F

Name of Matlab M-Script	Short Description, Purpose

timezones2.m
out = timezones2(NM, q, N, to, tf, p);
	Compute a logarithmic time scale. NM is the maximum of images. q is the image acquisition + temporaray storage rate, N is the number of zones wanted, to is the starting of acquisition, and tf is the end of the experiment. Parameter p specifies for which case the time zones are computed:
	    1: disjoint case
	    All other values: continuous case
	Example of use: **out = timezones2(284, 0.27, 10, 1, 100, 1);**

plotsh3.m
out = plotsh3(g,la,symbol);
	Plot values for Figure 10.20 (temperature rise versus square root of time) in step heating.
	g is mismatch factor gamma.
	la is L/sqrt(thermal diffusivity).
	symbol is to overprint on the solid line a symbol. Example: ko is black (k) with circles (o). Type 'help plot' in the Matlab command window for a complete list.
	Example of use:
	    Plotsh3(0.5, 1, 'ko');
	    hold (maintain the plot)
	    Plotsh3(0.5, 1, 'ko'); (replot)
	    Plotsh3(−0.5, 1, 'ks'); (plot next curve)
	This function uses the sh3(..) function, which performs the computations.

plotsh4.m
out = plotsh4(g,symbol);
	Plot values for Figure 10.21 (4) (normalized temperature rise versus square root of normalized time) in step heating.
	g is mismatch factor gamma.
	symbol is to overprint on the solid line a symbol. Example: ko is black (k) with circles (o). Type 'help plot' in the Matlab command window for a complete list.
	Example of use:
	    Plotsh4(1, 'ko');
	    hold (maintain the plot)
	    Plotsh4(1, 'ko'); (replot)
	    Plotsh4(−1, 'ks'); (plot next curve)
	This function uses the sh4(..) function, which performs the computations.

plotsh5.m out = plotsh(t,symbol);	Plot values for Figure 10.22 (normalized temperature rise versus thermal mismatch factor) in step heating. t is the normalized time. g is mismatch factor gamma.

(Continued)

TABLE 10.7 *(Continued)*

Name of Matlab M-Script	Short Description, Purpose
	symbol is to overprint on the solid line a symbol. Example: ko is black (k) with circles (o). Type 'help plot' in Matlab command window for a complete list. Example of use: Plotsh5(1, 'ko'); hold (maintain the plot) Plotsh5(1, 'ko'); (replot) Plotsh5(2, 'ks'); (plot next curve) This function uses the *sh4(..)* function, which performs the computations.
plot_sinc.m; out = Plot_sinc(A, DT, length, symbol)	Display sinc *x* (Fourier transform of a square pulse). Parameters: *A* is pulse amplitude *DT* is pulse duration *length* is length of the plot from negative length to positive length. *symbol* is to plot the curve with, for instance -k plot a continuous line in black. Example of use: Plot_sinc(1, 1, 20, '-k'); This function uses the *sinc_x(..)* function, which performs the computations.

In the shape from shading theory, determining whether a surface is convex or concave is a problem. The same problem arises here, although the known direction of heating helps; moreover, it is possible to have a priori information on object shape to solve this ambiguity. The advantage of this reconstruction process is that it requires only a single early recorded thermogram and is more robust to thermal drifts than an absolute temperature method, since it is based on temperature differences within the thermograms (this is not the case with the point-source correction method described above and based on absolute values). On the other hand, only relative heights are computed, as in the shape from shading technique; hence ΔR becomes available, but not R. Finally, calibration is not mandatory.

Table 10.7 lists the Matlab m-scripts discussed in this chapter.

PROBLEMS

10.1 **[time zones]** Suppose that there are four images, recorded at 18, 19, 20, and 21 s, in a given time zone (a logarithmic time scale as described in Section 10.2.2). Calculate the time associated with this zone.

Solution   Following eq. (10.46), the time associated with this zone is given by

$$t = \frac{1}{\left(\frac{1}{4}\left[\frac{1}{\sqrt{18}} + \frac{1}{\sqrt{19}} + \frac{1}{\sqrt{20}} + \frac{1}{\sqrt{21}}\right]\right)^2} = 19.5 \text{ s}$$

10.2 **[step heating]** Compute the thermal mismatch factor in the following situations: aluminum–GRP (glass fiber–reinforced plastic) and epoxy–GRP.

Solution   The thermal effusivity is computed with eq. (10.13): $e = \sqrt{k\rho C}$ (with k the thermal conductivity, ρ the mass density, and C the specific heat). These values are available in Table 3.7 and Appendix E for common materials. Let's first compute e_{Al}, e_{GRP}, and e_{epox}:

$$e_{Al} = \sqrt{k\rho C} = \sqrt{230 \times 2700 \times 880}$$

$$= 24,000 \text{ J m}^{-1} \, ^\circ\text{C}^{-1} \, \text{s}^{-1/2} = 2.4 \text{ J cm}^{-1} \, ^\circ\text{C}^{-1} \, \text{s}^{-1/2}$$

$$e_{GRP} = \sqrt{k\rho C} = \sqrt{0.3 \times 1900 \times 1200}$$

$$= 830 \text{ J m}^{-1} \, ^\circ\text{C}^{-1} \, \text{s}^{-1/2} = 0.08 \text{ J cm}^{-1} \, ^\circ\text{C}^{-1} \, \text{s}^{-1/2}$$

$$e_{epox} = \sqrt{k\rho C} = \sqrt{0.2 \times 1300 \times 1700}$$

$$= 670 \text{ J m}^{-1} \, ^\circ\text{C}^{-1} \, \text{s}^{-1/2} = 0.07 \text{ J cm}^{-1} \, ^\circ\text{C}^{-1} \, \text{s}^{-1/2}$$

and then the thermal mismatch factor:

$$\Gamma_{GRP-Al} = \frac{e_{Al} - e_{GRP}}{e_{Al} + e_{GRP}} = \frac{24,000 - 830}{24,000 + 830} = 0.9$$

$$\Gamma_{GRP-epox} = \frac{e_{epox} - e_{GRP}}{e_{epox} + e_{GRP}} = \frac{670 - 830}{670 + 830} = -0.1$$

These two cases illustrate somehow what is shown in Figure 10.20, with curves departing up ($\Gamma < 0$) and down ($\Gamma > 0$) from the semi-infinite case.

10.3 **[step heating]** Following the procedure described, compute the defect depth for the third deflected curve in Figure 10.24.

Solution   At the intersection $T_N = 0.18$ we noticed that $\sqrt{t} = 3.3 \text{ s}^{1/2}$. This intersection corresponds to the case for which $t_N = 1 = \sqrt{\alpha t}/L$ and thus (with $\alpha = 0.42 \times 10^{-6} \text{ m}^2 \text{ s}^{-1}$):

$$L = \sqrt{\alpha t} = \sqrt{\alpha}\sqrt{t} = \sqrt{0.42 \times 10^{-6}} \times 3.3 = 2.1 \text{ mm}$$

The value reported is 2.8 mm. The same computation for the second curve of that figure ($\sqrt{t} = 1.75$ s$^{1/2}$) gives a depth of 1.1 mm. As seen (also with Example 10.3), the error increases due to degradation of the one-dimensional assumption with depth.

10.4 **[pulsed phase thermography]** Compute the maximum theoretical frequency available assuming a time interval between thermal images of 175 ms and that the thermal sequences are 100 thermograms long.

Solution   Here we have $N = 100$ and $\Delta = 175$ ms. Following eq. (10.65), we have

$$f_n = \frac{n}{N\Delta} = \frac{n}{100 \times (175 \times 10^{-3})} = 0.057n$$

Thus the frequencies available will span from $f_0 = 0$ to $f_{100} = 5.7$ Hz. As mentioned above, the maximum frequency is related directly to the acquisition frequency of the IR camera (5.7 Hz = 1/175 ms). If a 1-s thermal pulse is used, the first lobe of the sinc function crosses the zero axis at 1 Hz with maximum amplitude in the domain 0 to 0.9 Hz. The 5.7-Hz camera should not cause aliasing. The Fourier transform of a 1-s duration pulse can be obtained with available m-file Matlab *plot_sinc.m* with parameters: *plot_sinc*(1,1,5,'k') (Figure 10.27).

PART III

ACTIVE AND PASSIVE THERMOGRAPHY: CASE STUDIES

CHAPTER ELEVEN

Applications

In this final chapter of the book, various applications are presented, including both active and passive infrared TNDT in a variety of fields, such as maintenance, electric utilities, and quality control. This material illustrates how to make use of the concepts described in preceding chapters.

11.1 ACTIVE PULSED THERMOGRAPHY WITH EXTERNAL THERMAL STIMULATION

As discussed previously, in pulsed thermography, the workpiece being inspected is submitted to a thermal pulse and the subsequent temporal temperature response of the surface analyzed to discover eventual anomalies linked to subsurface flaws. As noted in (Section 9.2), the optimum time of observation is proportional to the square of the depth z_{def} for the defect considered and the TNDT method is especially sensitive to relatively large size defects (with respect to the defect depth). In this section we focus our attention on the classical techniques needed for defect detection and localization based on this principle. More specifically, we make use of thermal sources external to the material; in the next section we study applications where thermal sources are internal (i.e., located inside the workpiece).

As we mentioned often in earlier chapters, recorded TNDT images are often corrupted by various sources of noise. Moreover, in the active approach, signals are of small amplitude and further degraded by temperature spreading due to the 3D diffusion of the thermal front under the surface. Considering this, special processing is needed to enhance TNDT images either for quantitative characterization, for a traditional operator-assisted procedure (e.g., to present images in a more comprehensive format), or for automated inspection.

First, pulsed thermography with external stimulation is used for graphite-epoxy workpieces and then aluminum bonded laminates. The particular choice of these two materials is based on the fact they have very opposite thermal behaviors and properties, especially with regard to thermal conductivity and their isotropic (aluminum) or anisotropic nature (graphite epoxy) (Table 3.7). Consequently, this study permits us to cover a broad spectrum of materials, especially since it is not possible to review all possible materials. The cases illustrated are representative of major active TNDT possibilities, and utilization of these techniques on other materials should not pose a particular difficulty. Finally, it is important to note that even if illustration of one technique is studied for, say, a graphite-epoxy composite, it could be applied as well on an aluminum laminate, and vice versa (of course, with particular thermal behaviors). This is true for most of the techniques presented in this chapter.

11.1.1 Graphite–Epoxy Composites

As shown in Sections 7.1 and 7.4, graphite–epoxy structures are used more and more because of their numerous qualities. In this section we see some TNDT techniques that can be applied to the inspection of graphite–epoxy structures.

Impact Damages. In the aerospace industry, the service life of a graphite-epoxy component can be reduced drastically as a result of impact damage such as a bird strike. In fact, an impact of sufficient energy on such a composite can induce both fiber breakage and delaminations (Vavilov et al., 1998a). Such damage can result from various accidents, from a tool hitting a surface by accident to damage caused under combat conditions (such as by artillery). Generally, on the side where the impact takes place there is only a barely visible scratch, while the opposite surface can be broken totally. This is the reason that such damage is often called *blind-side impact damage.* An important point to note is that the opposite side of the component is often not accessible once the complete structure is assembled. TNDT is thus well suited for *in-service* damage detection in CFRP structures since it is well suited to detect shallow delaminations and can be deployed directly *on site*, requiring only external access.

To illustrate TNDT deployment in such a context, tests of impact damage have been executed on Narmco 5217 graphite epoxy plates whose plies were stacked in various layups. Impact damage was produced with a Dyna Tup drop weight impact tester. Figure 11.1 shows a typical result: Large delaminated zones have been produced even at low-energy impact (6 J) on unsupported thin samples: eight plies 1.14 mm thick. In part (a) the fiber orientation layup was $(0 \pm 45, 90)_s$ and $(0, 90)_{2s}$ in part (b). Samples were supported only by a foam pad. The area imaged is about 5×5 cm and the heating stimulation is about $15°C$ above room temperature. This thermal stimulation took place with a 1-s pulse from one 1000-W projector (Figure 8.14). On the thermograms, serious damage is visible on both sides of the impact area, with a typical *butterfly* defect shape and preferred orientation along the fiber of the first ply (from top to bottom in Figure 11.1). In some instances, damage can extend deeply. For example, in Figure 11.1b, a wide horizontal delamination is visible. In the case of thicker, supported graphite–epoxy sheets, damage is less important.

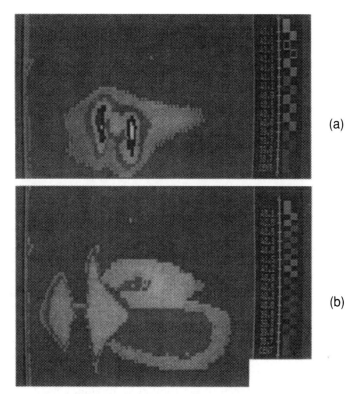

FIGURE 11.1   Impact damage on eight-plie graphite-epoxy plates. The field of view is 5.5 × 5.0 cm, impact energy was about 6 J: (a) layup: (0, ±45,90) s; (b) layup (0,90) 2 s. (From Maldague, 1993.)

Evaluation of Fiber Content. Thermal methods are useful both for defect detection and to characterize industrial materials. In this section we present an example of such an investigation for the evaluation of fiber content in graphite-epoxy composites. The problem consists of evaluating the fiber/resin ratio in a graphite-epoxy piece once it has been cured. Inappropriate temperature, pressure, or assembly conditions can cause evacuation of a significant amount of epoxy resin out of the composite. Areas having a poor fiber/resin ratio present reduced mechanical properties.

A typical configuration is presented in Figure 11.2. The workpiece is point heated with a laser beam while the thermal pattern is recorded with the infrared camera. If the material has an oriented structure with anisotropic properties as for a unidirectional graphite-epoxy sheet, an elliptical pattern will be observed. The ratio b/a between the two ellipse axes is related to the square of the diffusivities ratio along longitudinal and transverse directions (Cielo et al., 1987b) and thus to the fiber orientation: A test of an isotropic material would give a circle instead of an ellipse. This property can be used to evaluate fiber orienta-

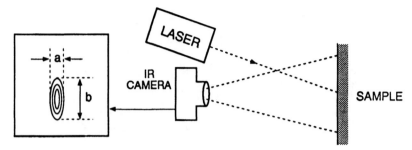

FIGURE 11.2 Thermal analysis of fiber orientation in composite materials. (From Maldague, 1993.)

tion in extruded or molded parts and to evaluate thermal conductivities of composite materials. The latter approach is illustrated in Figure 11.3. Thermograms (a) and (b) show the thermal patterns obtained after 20 s of heating using an 0.5-W argon laser on an eight-ply unidirectional layup Narmco 5217 sheet cured (a) under high (690 kPa) pressure and (b) under reduced (69 kPa) pressure. The high pressure induces low-resin-content areas, which have high thermal conductivity. For the rich epoxy content sample (Figure 11.3b), the pattern is smaller size and the central temperature is higher than for the low-resin-content pattern. Consequently, the image difference between the two patterns (Figure 11.3c) is positive on the periphery and negative in the center. Such temperature inversion is typical of a variation of thermal conductivity and is independent of both the heating source and the absorptivity or emissivity of the surface.

Delaminations. Appropriate modeling (Section 3.4.2) allows to predict, up to a certain extent, the thermal behavior of a particular workpiece under TNDT inspection, particularly when the analytical study becomes too complex. Modeling of graphite-epoxy delaminations is a typical example of this situation (Figure 11.4). This configuration is modeled on Figure 11.5. In this figure we note that the presence of delamination does not affect the surface temperature distribution at $t = 1.6$ s. After 6.3 s, a thermal contrast appears, revealing the presence of the defect, and after some 25 s, the radial thermal flow (bottom graphics on Figure 11.5) is inverted: positive above the defect and negative below. At first it is located in a restricted volume around defect borders, expanding in a larger volume with time. As a consequence, the thermal distribution just above the defect will be smoother on the surface ($z = 0$) than at the defect interface (top graphic in Figure 11.5). This smoothing effect reduces defect visibility and increases as the radius/depth and axial/radial thermal conductivity ratios are reduced. These results can be confirmed experimentally.

   To check for the limit of detectability of defects and illustrate the applicability of TNDT to CFRP components, tests were performed on graphite-epoxy plates in which artificial known defects were inserted at the production stage.

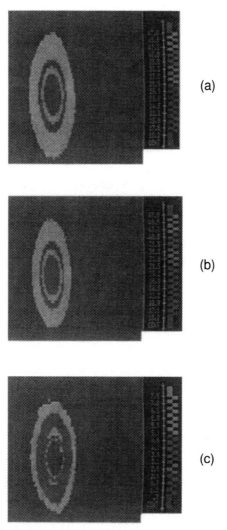

FIGURE 11.3 Elliptical thermal patterns obtained with point laser heating on a graphite-epoxy plate cured with pressure (a) normal and (b) low. Image (c) shows the difference between the two patterns. The field of view is 25 cm^2. (From Maldague, 1993.)

Such defects are made by inserting at a known depth, before curing, two pieces of Teflon films 50 μm thick between two plies of prepreg. The two films allow us to simulate a delamination rather than an inclusion.

Figure 11.6 (top row) shows thermograms obtained on samples in which implants of size (a) 20×20 mm, (b) 10×10 mm, and (c) 3×3 mm were inserted 0.3 mm beneath the surface. Heating by reflection (Figure 9.5a) is performed using the rosette of six projectors described in Section 8.1 (Figure 8.14). Thermograms show the central portion of the heated area. The thermal pulse

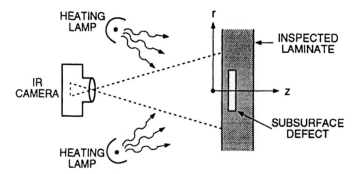

FIGURE 11.4 Schematic diagram of the thermographic inspection configuration used for thermal modeling. (From Maldague, 1993.)

duration is fixed at 200 ms. The visibility of defects is adequate even if heating is restricted to a few degrees above room temperature. We must point out that for the 3×3 mm defect, the heating pulse was a little longer, to improve its visibility. This can be explained by the 3D spreading of the thermal front (as noted above; Figure 11.5), by the limited spatial resolution of the infrared camera available, and by the possible infiltration of epoxy along the edges of this small Teflon implant.

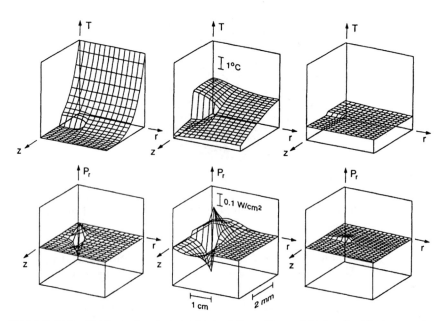

FIGURE 11.5 Tridimensional plots obtained from the model, showing the temperature distribution (upper row) and radial flux (bottom row) computed for the plane r–z indicated on Figure 11.4 at 1.6 s (on left), 6.3 s (center), and 25 s (right) after beginning of heating, respectively. (From Maldague, 1993.)

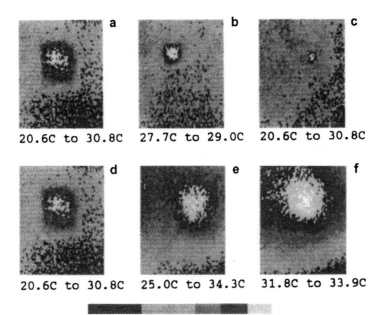

FIGURE 11.6 Raw thermal images recorded on a graphite-epoxy plate containing Teflon implants. Top row, implant size: (a) 20 × 20 mm; (b) 10 × 10 mm; (c) 3 × 3 mm inserted 3 mm under the front surface. Bottom row, implant size 20 × 20 mm inserted; (d) 0.3 mm; (e) 1.12 mm; (f) 2.25 mm under the front surface. (From Maldague, 1993.)

These 3D propagation effects are analyzed experimentally with respect to Figure 11.6 (bottom row). In this case, Teflon implants of 20 × 20 mm are inserted at depths of 0.3, 1.12, and 2.25 mm beneath the surface. The edges become more and more blurred as depth increases. The same is true for the thermal contrast above the defect: It tends to vanish. This is a direct consequence of thermal front propagation through diffusion. In this sense, defect visibility weakens when the thermal propagation defect-surface distance becomes similar to the propagation distance corresponding to the defect radius. This is why in an isotropic material, defects whose radius/depth ratio is less or close to unity are difficult to detect. Graphite epoxy is an anisotropic material for which thermal diffusivity parallel to the fibers is typically 10 times greater than diffusivity perpendicular to the fibers (Table 3.7). This thermal anisotropy increases the minimum ratio needed for detection by a given factor, as we can also deduce from eq. (9.7). Image processing techniques reviewed in Chapter 5 are useful to enhance results (see, e.g., Figures 5.18 and 5.19).

11.1.2 Aluminum Laminates

Pulsed thermography, which is characterized by an excellent inspection rate, is an interesting alternative NDE technique in the aluminum industry, where high production volumes are the rule. In this study we analyze the case of bonded

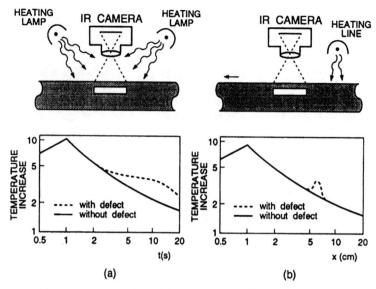

FIGURE 11.7   Thermal inspection with configuration (a) static and (b) mobile. Shape of the signals obtained is also shown: time pulse (a) and spatial pulse (b). (From Maldague, 1993.)

laminates: aluminum–aluminum and aluminum–foam. They are used increasingly in many industries (e.g., transport). As described in Section 7.4, these laminates are made of thin aluminum sheets (a few millimeters thick) epoxy bonded on either a foam core or to another aluminum sheet.

Unpainted aluminum is characterized by low emissivity ($\varepsilon \sim 0.05$). Consequently, thermal inspection will be possible only if high-emissivity paint is first applied on the surface. The paint can serve as a primer or can be eliminated after inspection, if required, by high-pressure water jets. In the case of planar surfaces, we reviewed in Section 8.2 a method based on transfer imaging which precludes such painting operations. Note that to solve this low-emissivity problem, all the aluminum samples tested in the book were covered with high-emissivity black paint ($\varepsilon \sim 0.9$). In this section we illustrate two typical configurations (Figure 11.7): the static configuration (time-domain thermal pulse) in reflection or in transmission, and the mobile configuration (space-domain thermal pulse) in reflection or in transmission.

Static Configuration. As shown in Section 3.4.2, modeling is a useful tool to predict defect visibility and select optimal inspection parameters. Figure 11.8 shows such analysis of heat-front thermal propagation in aluminum samples. The inspection geometry is shown in part (a): the aluminum laminate is heated on one side using a rosette of projectors (such projectors are pictured in Figure 8.14). Surface temperature can be recorded either on the same side that thermal

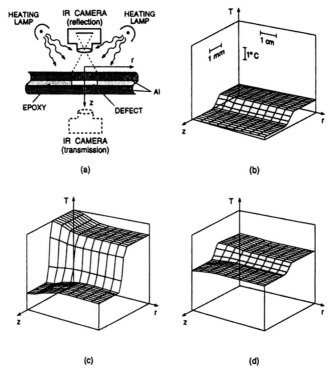

FIGURE 11.8 Modeling of the inspection of an Al–epoxy–Al sandwich panel: (a) geometry; (b–d) thermal distribution computed 0.1, 1.6, and 4 s after beginning heating. (From Maldague, 1993.)

perturbation is applied (i.e., in reflection) or on the opposite side (in transmission). A bonding defect consists of a lack of adhesive, which can be modeled by a thin air layer of high thermal resistivity. The parameters for the modeling of Figure 11.8 are as follows:

- *Thickness of aluminum sheets:* 1 mm, thickness of the epoxy layer: 0.2 mm, radius of the defect: 1 cm.
- *Heat deposit:* 2 W cm^{-2} for 1 s.
- *Thermal conductivities:* 2 and 0.002 W cm^{-1} K^{-1}, specific heat: 1 and 1.5 J kg^{-1} K^{-1}, density: 2.7 and 1 g cm^{-3} for aluminum sheets and epoxy layer, respectively. Thermal resistivity of the interface: 83 cm$^2 \cdot$ K W^{-1}; this corresponds to thermal conduction through an air layer of 200 μm.
- *Losses through radiation:* $5.67 \times 10^{-12}(T^4 - T_r^4)$ W cm^{-2} where T and T_r are the surface and room temperatures, respectively (expressed in Kelvin); convective surface losses: $10^3(T - T_r)$ W cm^{-2}.

Figure 11.8b to d displays temperature distribution in the r–z plan pictured in Figure 11.8a, (a) 0.1, (b) 1.6, and (c) 4 s, respectively, after starting heating. As

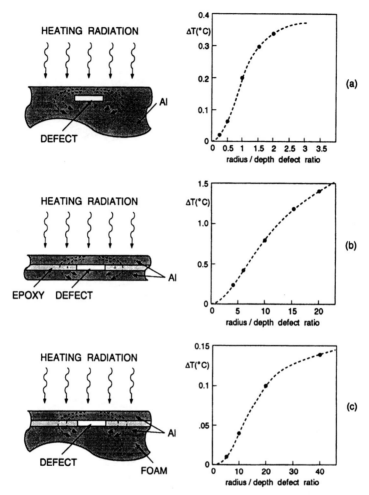

FIGURE 11.9   Modeling of defect visibility in the case of (a) an homogeneous Al spec-imen; (b) an Al–Al laminate; (c) an Al–foam laminate. Thermal pulse lengths are 0.1, 1.0, and 1.5 s and observation times are 0.16, 1.6, 2.7 s, respectively, for (a) to (c). Heat deposit: 2 W cm^{-2} for 1 s. (From Maldague, 1993.)

we can see, initially the temperature distribution is uniform over both surfaces, thus preventing defect detection (Figure 11.8b). The defect visibility increases up to a maximum (Figure 11.8c) before vanishing through lateral diffusion (Figure 11.8d).

The effect on defect visibility of 3D heat flow spreading is particularly evi-dent in Figure 11.9. For the three cases illustrated, the temperature difference between the center of the delaminated area and the uniformly heated sound surrounding area is plotted for different specimen geometries. As already men-tioned, as a general rule, for a defect to be visible, the radius/depth ratio must

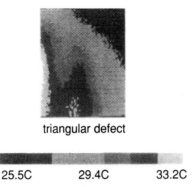

triangular defect

25.5C   29.4C   33.2C

FIGURE 11.10   Detection of lack of adhesive (zone of triangular shape) in an Al–foam panel. The effect of the poor heating uniformity is apparent on this thermogram. (From Maldague, 1993.)

be such that the radial flow around the defect is substantially smaller than the longitudinal flow through the bonded interface. For homogeneous materials, this condition limits the radius/depth ratio of a defect to values greater than or close to unity. This is the case in Figure 11.9a. For bonded laminates (Figure 11.9b), this ratio is close to 10 because of the low thermal conductivity of the bonding layer compared with the high thermal conductivity of the aluminum skin. Greater radius/depth ratios are required in the case of aluminum and foam structures (Figure 11.9c) because of the refractory properties of foam (thermal conductivity, ~ 0.0004 W cm^{-1} K^{-1}, specific heat, 2 J g^{-1} K^{-1}; and density, 0.1 g cm^{-3}). This study shows, more formally than in the case of bonded aluminum laminates, that TNDT is limited to the detection of large shallow defects.

A configuration similar to that of Figure 11.8 has been used to inspect a panel constituted of two sheets of aluminum (of 1 mm thickness) epoxy-bonded on a polymethacrylimid foam core of 5-cm thickness. Artificial defects have been made by removing the epoxy adhesive film at specific locations before curing. Figure 11.10 shows a thermogram of the panel 2.5 s after starting heating (fixed configuration in reflection, incident heating: 1 W cm^{-2} for 1 s, field of view: 19×19 cm, triangular defect). The spatial subtraction technique (Section 5.5) is used here to suppress the nonuniform heating pattern (Figure 11.11). It is important to note that the spatial subtraction technique (for which a defect-free image recorded under the same conditions as described for the specimen inspected is subtracted systematically from the images, to analyze) does not always bring such a dramatic improvement in defect visibility as the one presented in Figure 11.11.

Figures 11.12 and 11.13 illustrate other investigations performed on aluminum laminates. Figure 11.12 shows bonding defect detection using the transmission configuration. In this case a flat bonded joint between two aluminum sheets of 1-mm thickness is inspected; a 1.5×1.5 cm area without adhesive is

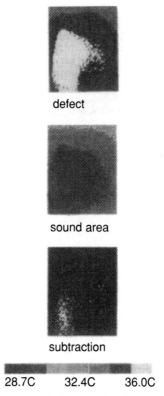

defect

sound area

subtraction

28.7C   32.4C   36.0C

FIGURE 11.11 Spatial reference technique used to improve defect visibility. Top, raw image recorded over the defect; center, image recorded over a sound area; bottom, the subtracted image delineates clearly the position of the triangularly shaped zone with missing bonding. (From Maldague, 1993.)

present. On the thermogram, the hot area on the right corresponds to the single aluminum sheet (one-sheet thickness), which heats more rapidly than the bonded area (two-sheet thickness). The heating must be rapid to avoid lateral spreading of the thermal front from the hot plates to the bonded area. Special masks blocking the heating radiation from the lamp allow us to limit this disturbing phenomenon by stimulating only the joint area; this is the case for the left side of the bonding line visible in Figure 11.12. This is especially useful for the repetitive inspection of identical samples.

In the case of components that are already at an higher temperature than room temperature because of various fabrication steps, for example, a cold thermal perturbing source can be used advantageously, as discussed in Section 8.1. An example of such an analysis is presented in Figure 11.13. The structure is initially at a temperature some 10°C above room temperature, and a line of air jets is used to quickly cool the inspected area. The image shows the hotter

FIGURE 11.12 Thermographic inspection in transmission for an Al laminate (two sheets bonded). Top, experimental apparatus; bottom, thermal image showing a zone without adhesive of 1.5×1.5 cm. The field of view is 4×5 cm. (From Maldague, 1993.)

central area, which corresponds to the bonded area (oriented vertically on the thermogram). The cooler region at the center of the bonding line reveals a bonding defect (lack of adhesive).

Mobile Configuration. The mobile configuration (Figure 11.7b) allows to reach high inspection rates. It is employed in reflection or transmission. An interesting application concerns the inspection of flat joints in Al–Al laminates. The configuration in transmission is shown in Figure 11.14a. The bonded joints are inspected by horizontal scanning of the panel, the joint being placed perpendicularly to the scanning direction. A 18-cm-long 530-W line heating lamp serves for the stimulation. The vertical band in the central part of the image (Figure 11.14b) corresponds to the bonding line. It is cooler because of the

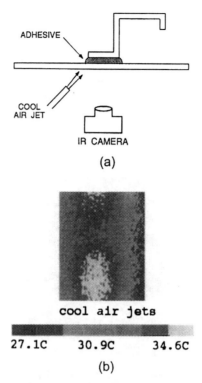

FIGURE 11.13 Thermographic inspection by propagation of a cold front for the inspection of bonded Al structures: (a) experimental apparatus; (b) thermal image showing a zone of 1.5 × 1.5 cm without adhesive. (From Maldague, 1993.)

greater material thickness of the workpiece in this area. On this image, a much cooler region corresponds to a bonding defect: The back surface thermal front takes more time to reach the front surface, due to the increased thermal resistance of the subsurface defect that it encounters on its way. The two aluminum plates have a thickness of 1 mm, while the overlapping zone is 2.7 cm in width.

An example of stimulation by propagation of a cool front in reflection is shown in Figure 11.15. In this case it is necessary to detect an unbonded area in a aluminum–foam laminate. The panel is first heated uniformly at a temperature about 10°C above room temperature. The surface is then cooled by a line of cool air jets during lateral movement of the panel (at a constant speed of 2.4 cm s^{-1}). In the figure, two thermograms are shown, one for a sound area and one for an area where circular disbonding (4 cm in diameter) is present. The defect is clearly visible at the image center.

The same panel was tested in reflection using a line heating thermal source (Figure 11.16). The panel is scanned from right to left in front of the static line heating (here a 30-cm-long 2500-W lamp), which is oriented perpendicular to the figure plane. A polished back reflector is used to concentrate the thermal

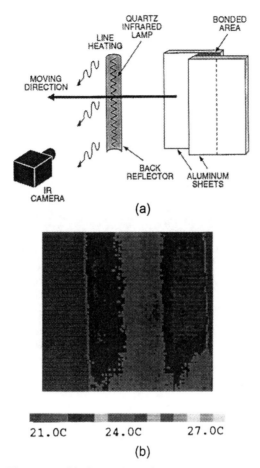

FIGURE 11.14 Thermographic inspection of a bonded line between two aluminum sheets using the mobile configuration: (a) experimental apparatus; (b) the thermogram shows a bonding defect at the image center. The lack of uniformity at the bottom of the thermogram is due to an insufficient heating in this area (the lamp used—530 W, 18 cm long—was a little short in regard to specimen size). (From Maldague, 1993.)

stimulation on the workpiece surface. This configuration makes possible a 5°C increase on the stimulated surface while the panel is moving at a constant speed of 11.5 cm s^{-1}.

By comparison with Figure 11.14 (or Figure 6.14), the fivefold-less-powerful 530-W lamp produces a mere 1°C contrast. This is a good illustration of the direct relationship between the temperature differential obtained and the amount of power deposited on the stimulated surface [e.g., eq. (10.10)]. Powerful stimulation devices should be use whenever it is possible with however concerns not to damage inspected surface through overheating.

It is also interesting to compare the radiative-heat injection (Figure 11.16)

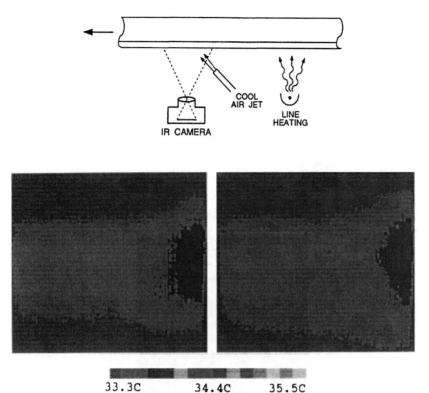

FIGURE 11.15 Thermographic inspection using a line of air jets. Top, experimental apparatus; bottom, the thermogram shows on the left an Al–foam panel without defect, and on the right a similar panel with a circular 4-cm-diameter lack-of-adhesive defect visible on the image center. (From Maldague, 1993.)

and convection-heat removal approach (Figure 11.15). A reduced thermal contrast is obtained with the cool air stimulation, due to both the reduced heat capacity of air (Table 3.7) and the smaller temperature differential between the thermal perturbing source and the surface inspected (case of air versus high-temperature radiative source). Recall of eqs. (3.11) and (3.15) is also of interest to understand such differences: radiation stimulation is a function of the temperature differential expressed at the fourth power, while convection depends on a temperature differential expressed at the first power.

11.1.3 Pulsed Thermography of Space Launch Vehicles

Quality control consists of measuring the conformity of a product to a standard. Often, the component is at uniform temperature, and thus the active techniques described previously are well adapted for such a task. In this section we describe an interesting application of pulsed thermography for inspection of

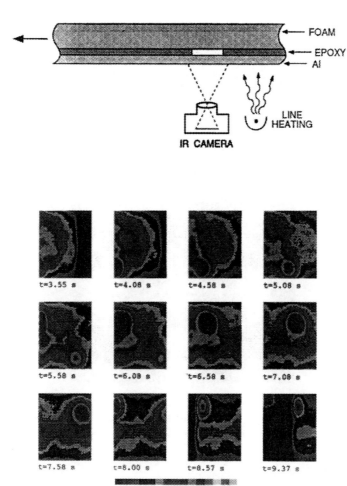

FIGURE 11.16 Mobile configuration, Al–foam panel, line heating 30 cm long, 2500 W. Top, experimental apparatus; bottom, a few thermograms recorded at a specific time are displayed while the specimen was moving across the field of view. Careful examination of the sequence reveals three defects. Same specimen as on Figure 6.11b, where three unbonded areas are visible. (From Maldague, 1993.)

space launch vehicles, based on work by Burleigh et al. (1994). For safety reasons, components of space vehicles and aircraft are generally inspected 100%. Common NDT techniques in that field are ultrasonics (such as with squirter water jets), x-ray, and "coin tapping" (a defect alters the frequency of noise, as when knocking on a cracked cup). Due to its attractive advantages, as discussed previously (fast, remote, noncontact, portable, real time with recording capability, etc.), thermography is thus of increasing use in this field.

The Atlas rocket, the original ICBM (intercontinental ballistic missile), has been in production since 1955 (Figure 11.17). In 1957 the weight of the rocket

FIGURE 11.17   Picture of the Atlas rocket which was the original ICBM (interconti-nental ballistic missile). It has been in production since 1955. The composite thrust structure at the bottom of the rocket is inspected by TNDE. (Courtesy of D. D. Burleigh.)

was reduced by replacing the thrust structure with one made of lightweight composite parts. This design is still in use with only minor changes. The parts to inspect are as large as 2.5 × 3 m and are made of complex assemblies of alu-minum honeycomb covered with fiberglass–phenolic facesheets with various skin thicknesses and densities of honeycomb. As mentioned above, the original NDT method for these parts was coin tapping on a 1.25-cm grid. Ultrasonic was never considered, since the parts could not be in contact with water (moreover, ultrasonic does not work well with fiberglass–phenolic material, which is highly attenuating to mechanical vibrations).

Since 1990, thermography has been a production tool at General Dynamics for use on the composite parts of the Atlas thrust structure. It is able to detect the following: facesheet disbonds, facesheet disbonds under the cork insulation on doors, cork disbonds, honeycomb core splices, differences between the two densities of aluminum core found in the nacelle doors, epoxy injected for repair, metal hardware below the skin, and phenolic blocks below the skin (Figure 11.18). Inspection proceeds on bare fiberglass (emissivity > 0.95) or painted

FIGURE 11.18 Cross section of an aluminum honeycomb panel with core defects detected. (Courtesy of D. D. Burleigh.)

parts (emissivity > 0.9). Sometimes, spurious reflections degrade thermograms, although these reflections are constant during a thermographic test while the thermal contrasts change. Discrimination is thus possible, such as by subtraction of subsequent thermograms (known as the *temporal reference technique*; Section 5.5). The inspection of these large structures is challenging since access is sometimes remote. The best technique developed is to have operators working in pairs, one climbing on the launcher to heat the surface 10 to 15°C with a 1750-W heat gun equipped with a triangular fishtail diffuser, while the second operator monitors the thermal contrast evolution with an infrared camera (often equipped with a telephoto lens for remote inspection—up to 5 m is reported), Figure 11.19 shows the inspection procedure. Such a heating scheme has shown to be the best choice for this application since it is safe and portable. The scanning rate is rapid, using such an approach: Up to 45 m^2 h^{-1} is reported in a mobile configuration of bare production parts moving in front of static inspection apparatus and 5 to 10 m^2 h^{-1} when inspecting directly on the launch pad at Cape Canaveral (defects detected slow the inspection rate, since it is then necessary to document and mark them). Sometimes, the inspection team also relies on a cold pulse by spraying liquid nitrogen on the part (Figure 8.9).

Due to their high cost, defective parts are not necessary discarded; instead, they are fixed, for example by injection of epoxies (some containing aluminum powder). In some instances, thermographic inspection with imaging capability enables us to understand why defects happen and thus to correct the fabrication techniques. This was, for example, the case for some flexed honeycombs in which disbonds and empty core splices were found (Figure 11.18). A secondary NDT technique was required in some instances. The use of a self-illuminated, small-diameter (2.5 to 5 mm) rigid borescope with a 70 or 90° mirror at the tip is reported. To observe the core, the borescope is inserted in the suspected part through a hole drilled in the skin, the hole being used later to inject epoxy for repairs. Thermography was found to provide a greatly improved inspection method at a lower cost per vehicle than that of the previous coin tap inspection method. It should be mentioned that the inspection task described here is suitable for aircraft components as well.

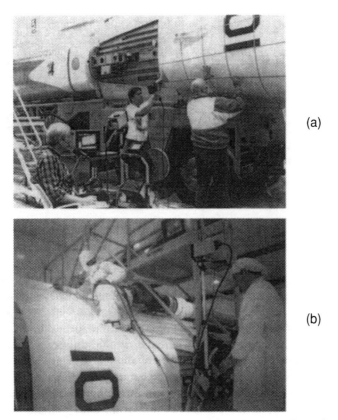

(a)

(b)

FIGURE 11.19 TNDE inspection procedure. In (a) an inspector is performing coin tapping concurrently. (From SPIE 2245.)

11.1.4 Detection of Rolled-in Scale on Steel Sheets

In this section we present an application of active thermography to detect rolled-in scale on steel sheets (see Section 11.4.3 for a passive application in the steel industry). Scale patch formation on steel sheets occurs during the reheating step prior to hot rolling (at about 1300°C in the reheating furnace). These scale patches can reach up to 1 mm in thickness and are essentially made of FeO, FeO_4, and Fe_2O_3. Chemical, mechanical, and hydraulic techniques such as acid pickling, steel surface sweeping with chains, and high-pressure water spraying are used in the steel industry to prevent oxide formation and remove scale patches. However, if adherence of patches on steel is good, they can stay on the surface and thus be rolled in the steel sheet during hot rolling. Figure 11.20 depicts the problem. Such steel sheets have a degraded aesthetic appeal which reduces their commercial value. Even painted, such rolled-in scale patches remain visible on steel sheets. Figure 11.21 shows pictures taken in the visible spectrum of (Figure 11.21a) rolled in scale (5 × 30 mm): The elongated

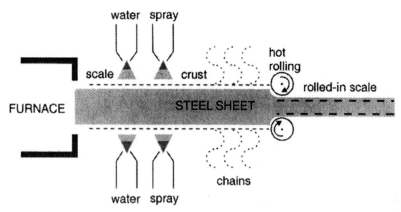

FIGURE 11.20 Formation process of scale crusts on steel sheets during hot-rolling operations. (From Maldague, 1993.)

shape is due to the rolling process, which tends to spread the scale material along the rolling direction, and (Figure 11.21b) scale crusts form on the steel sheet surface after furnace heating.

An inspection system that can recognize the presence of scale before and after rolling would be very interesting for process and quality control. More often, such inspection is performed visually by an operator after the fact, causing expensive production of wasted material. Obviously, considering the hostile environment of the hot-rolling process (high temperature, heavy water vapors, large vertical excursions of the sheets), a noncontact technique is required. The thermographic approach is a good candidate for this application, as it takes advantage of the naturally high temperature of the steel sheets during hot rolling.

If the scale patches do not adhere perfectly on the steel surface, temperature distribution recorded over the sheet allows us to detect residual scale patches which are at a lower temperature than the metal because of the larger thermal resistance of the scale–steel interface. However, high-scale emissivity with respect to steel can compensate for this temperature difference. It is then necessary to rely on active thermal stimulation by surface cooling using gas jets, taking advantage of the low thermal conductivity of the scale, whose temperature will be reduced significantly while bulk steel will stay at the higher temperature (this is the *cool* approach described in Section 8.1). As a general rule, thermal resistance effects are more important at the descaling station, where scale patches are not yet rolled-in within the metal. On the contrary, after rolling, thermal resistance is reduced, and consequently, emissivity effects are prominent. These principles can be analyzed experimentally.

To demonstrate the inspection method, steel sheets are heated uniformly by immersion in a warm water tank (water is about 5°C above room temperature). Figure 11.22 shows the comparison between Figure 11.22a, a video image of rolled-in scale crust (visible image, diffuse illumination), and Figure 11.22b, an

(a)

(b)

FIGURE 11.21 Pictures taken in the visible spectrum: (a) rolled in scale (5 × 30 mm): the elongated shape is due to the rolling process, which tends to spread the scale material along the rolling direction; (b) scale crusts present on the surface prior to rolling. (From Maldague, 1993.)

infrared image. The infrared image shows more clearly the high-emissivity defect and is less sensitive to surface discoloration patterns, which are clearly noticeable in Figure 11.22a (they seem to be due to FeOOH deposits formed by the repetitive contact of the steel with warm water during experiments). At room temperature, scale defect visibility is much lower because emissivity differences are compensated by variations in the surface reflectance opposite to those from the ambient temperature background.

When warm steel sheets are inspected with a emissivity variation criterion related to the presence of rolled-in scale defects, the observation angle is an important factor to take into consideration. Figure 8.31 displays polar emis-

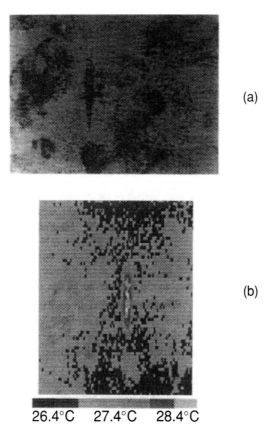

26.4°C   27.4°C   28.4°C

FIGURE 11.22   Comparison between (a) a video image of rolled-in scale crust (visible image, diffuse illumination) and (b) an emittance image recorded with an infrared camera. (From Maldague, 1993.)

sivity diagrams for common industrial materials. In this figure, we notice that emissivity is rather constant for nonmetallic material for a wide variation of the observation angle from normal incidence. For smooth metallic surfaces, emissivity tends to be lower at normal incidence than at grazing incidence. This suggests that rolled-in scale defect visibility will be higher at normal incidence when emissivity of the smooth steel surface is low with respect to scale emissivity, which is higher and uniform. This result is confirmed in Figure 11.23, which compares the emissivity of steel sheet and rolled-in scale defects for different orientations with respect to the infrared camera.

Figure 11.24 shows a comparative temperature evolution curve for two adjacent zones over a steel sheet: one above a rolled-in scale defect and the other over the smooth steel surface. These temperatures were recorded while an air jet was cooling the surface. It is noted that apparent temperature of scale is steadily superior to apparent steel temperature; however, the temperature dif-

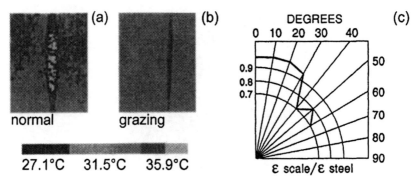

FIGURE 11.23 IR image of the scale defect at (a) normal incidence and (b) near-grazing incidence. The polar diagram (c) shows the variation of the normalized emissivity ratio as a function of observation angle with respect to the normal. (From Maldague, 1993.)

ferential between the two curves tends to decline as the sheet temperature falls. These temperature differences are due partially to emissivity differences as indicated above and confirm the low thermal resistance at the scale deposit–steel interface. Greater temperature differences are present if scale patches are partly loose and the surface is observed under forced cooling conditions; Figure 11.25

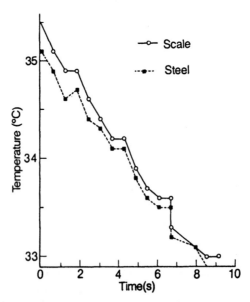

FIGURE 11.24 Comparative temperature evolution of two points, one over a rolled-in scale patch and one over bare steel without scale while inspected steel sheet is subject to a cool airstream. (From Maldague, 1993.)

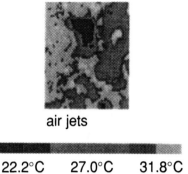

air jets

22.2°C   27.0°C   31.8°C

FIGURE 11.25 Cool-thermal wave approach: an artificially oxidized, partially descaled plate is uniformly heated and subsequently surface cooled by an air jet. Colder regions correspond to the partially loose scale patches. Notice that the temperature scale is only indicative, because of emissivity variations across the field of view. (From Maldague, 1993.)

illustrates this. A steel sheet covered with scale patches formed during passage in the furnace is observed with the infrared camera while air jets cool the surface. The temperature of the steel sheet is slightly above room temperature, and the steel surface was partly descaled, so that free scale steel areas are present, as shown in Figure 11.21b. On Figure 11.25 areas of low temperature correspond to loose scale patches, which have greater thermal resistance at the scale–steel interface. Areas of higher temperature correspond to bare steel, whose thermal inertia prevents fast cooling.

This cool thermal stimulation approach has many advantages, as mentioned previously: low cost for the thermal stimulation apparatus, improved security, and no spurious thermal reflection from the thermal source. Note that for this application, air is not a good stimulation medium since it activates scale formation at such high rolling temperature; consequently, an inert gas is preferred. This application shows that infrared thermography is an attractive technique for scale detection either before or after rolling. Rugged infrared imaging equipment is now commercially available and is capable of surviving in a harsh environment such as the operating conditions found in the steel industry.

In addition to the scale detection method presented here, it should be pointed that temperature measurement of the hot strip during cooling is important to maintain a certain grain size before and after transition from one metallurgical structure (e.g., thick slab) to another (e.g., thin sheet) (Section 11.4.3). Holmsten and Houis, (1990) specifie the various locations where temperature measurement is critical in the steel industry: at the exit of reheating furnace, at the rougher and descalers (as specified above), after the last finishing stand, and before coiling (see Section 11.4.3).

11.2   ACTIVE PULSED THERMOGRAPHY WITH INTERNAL THERMAL STIMULATION

In previous chapters we explained how pulsed thermography can be deployed for the inspection of materials and structures by mean of a thermal transient perturbation. In Section 11.1 the thermal stimulation was accomplished by means of lamps, thermal radiators, lasers, air/water jets, and other apparatus; in all cases, an external thermal stimulation was performed (with respect to the workpiece). We saw that thermal contrasts obtained over defective areas were generally small, on the order of a few degrees only, since the thermal stimulation heated (cooled) the surface inspected by at most 10°C (e.g., Figure 11.10). As discussed in Section 1.3, one of the limitation of pulsed thermography is precisely the difficulty to obtain a uniform, short, high-intensity energy deposit over a wide surface with the added constraint not to damage it.

There exist situations where access is available from inside the structure itself. In these situations, the structure can be inspected in transmission (Figure 9.5b) rather than in reflection. Moreover, instead of using a thermal perturbation source of a radiative type such as a lamp or a radiator, thermal perturbation can be accomplished, if possible, by changing the temperature of a fluid circulating inside the structure. Depending on the heat capacity of the fluid, large temperature differences can be deployed (within the acceptable limit to avoid component damage) to produce high thermal contrast, leading to reliable thermal inspection.

In this section we illustrate the internal stimulation technique by reviewing two typical applications that use common fluids: liquid (water) and gas (air). More specifically, these examples concern evaluation of corrosion damage in pipes due to the flow of corrosive liquids and analysis of internal structure of jet turbine blades. This will permit us to introduce various experimental methods as well as image-processing techniques.

11.2.1   Cavitation Damage of Pipes

It is estimated that in industrialized countries, losses caused by corrosion and wear account for around 3% of GNP. In plants where industrial processes require circulation of corrosive fluids, the corrosion problem is thus an important issue. In fact, circulation of corrosive fluids can damage walls considerably because of *local erosion*. This phenomenon is more prominent in bent portions of pipes, where turbulences and vortexes often take place, increasing wear surface significantly. The phenomenon takes place in three stages (Kvernes et al., 1988): a long period of time during which the surface is attacked, an active period during which damage is no longer superficial (the core starts to be damaged and cracks can appear due to fatigue), and *catastrophic failure* when the thinned wall cannot support the internal pressure any longer and actually breaks down. Such wall thinning must be taken into consideration to avoid accidents which may happen if corroded walls explode under internal pressure, especially in nuclear or petrochemical industries.

Traditionally, the industry relies on the statistical replacement of pipe sections judged potentially dangerous after a certain period of time: This method may reveal costly if intact sections are replaced. Other methods are used, such as visual inspection performed with boroscopes introduced directly inside pipe sections. Such a method is not always practical because of the dismantling involved.

Other NDE techniques (especially ultrasonic and x-rays) are good candidates for this task. Some problems of accessibility and of surface curvature slow point-by-point operation of ultrasonic, and security aspects of x-rays may restrict its deployment. For this application, TNDT offers many advantages (Section 1.3): no contact, surface scanning (rapidity), no harmful radiation involved, and ease of deployment.

It can be demonstrated that, in stationary conditions, for a pipe in which a fluid is circulating, temperature differences between zones of different thicknesses are very small. A much more important contrast can be obtained in transient conditions. If the surface of a thick piece is heated starting at $t = 0$ with a source delivering W watts per surface unity, the temperature difference at depth z beneath the surface follows the expression [Carlslaw and Jaeger, 1959, Chap. 10, eq. (9)]

$$T - T_a = 2W\left[\sqrt{\frac{t}{\pi k\rho C}}\exp\left(\frac{-z^2}{4\alpha t}\right) - \frac{z}{2k}\operatorname{erfc}\left(\frac{z}{2\sqrt{\alpha t}}\right)\right] \qquad (11.1)$$

where k is the thermal conductivity, ρ the density, C the specific heat, $\alpha = k/\rho C$ the thermal diffusivity of the material, and T_a the temperature before the perturbation. Because of the abrupt spatial distribution of the erfc($\cdot$) function, the argument of the function becomes prominent in thermal transient conditions so that after a time t (starting from the beginning of heating), the thermal front will reach a depth z_{th}:

$$z_{th} \approx \sqrt{\alpha t} \qquad (11.2)$$

This relation was seen in eq. (9.7). It indicates that for a pipe section whose wall thickness has been reduced by a factor of 2 due to the corrosion, the thermal perturbation will reach the exterior surface after a period of time fourfold shorter than for an intact uncorroded wall section. After a time (counted from the beginning of the thermal perturbation) on the order of z_c^2/α [eq. (9.7)], where z_c is the thickness of the corroded wall, this wall section will have reached a different temperature with respect to nearby sound sections submitted to the same perturbation and still at the temperature T_a (temperature before the perturbation). Consequently, if measurements of the propagation time are performed, the wall thickness can be evaluated.

To demonstrate these concepts, a simple method is possible. A transient thermal perturbation is obtained inside a pipe section by changing the temperature of the circulating fluid. The temperature distribution on the external surface is observed with an infrared camera. The temperature of pipe sections having a thinner wall will be affected first.

We now study an application of this method in the case of a corroded 90° bent section of stainless steel pipe. The experimental setup is shown in Figure 11.26. The original wall thickness was 5 mm. The pipe section is connected to two water supplies—cold (6°C) and warm (40°C)—with a flow of nearly

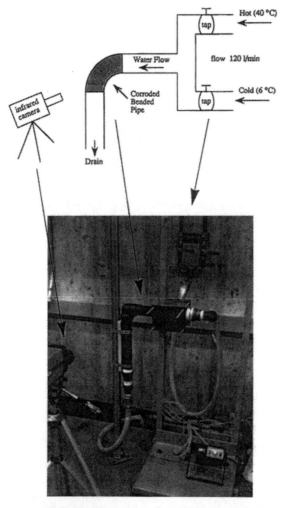

FIGURE 11.26 Experimental apparatus with infrared camera visible on left as well as the bent pipe section connected to the cold and front water supply taps (top right). (From Maldague, 1999. Copyright © 1999 The American Society for Nondestructive Testing, Inc.; reprinted with permission.)

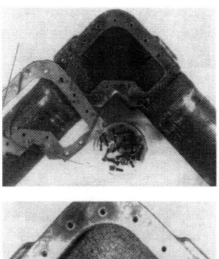

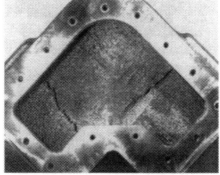

FIGURE 11.27 Rear-side pictures of the corroded pipe of Figure 11.26. The plastic window is clearly visible. Wall-thickness differences are not evident, although up to 75% of the total thickness is missing at some locations (original thickness was 5 mm).

120 L min^{-1}. Thermal inspection proceeds as follows: Flow is circulated in the pipe until the thermal permanent regime is reached on the external surface; next, the flow temperature is changed and a sequence of infrared images is recorded during the thermal transition. High flow and large temperature differences yield an excellent thermal contrast.

Different configurations are tested: cool or warm transition, direct or reverse flow direction. The same thermal signature is obtained in all cases. All these experimental conditions revealed the same abnormal temperature pattern, which corresponds to corroded areas. This last fact was checked after a visual examination was performed on the corroded pipe section cut into two pieces over its length (Figure 11.27). Figure 11.28 shows a typical thermal transition from hot to cold temperature. The corroded areas are visible from $t = 1.09$ s to $t = 3.53$ s. The still-hot areas present at $t = 22.57$ s correspond to a sound area four times thicker than the damaged area, thus explaining the long delay required for the cold front to reach the outside surface at these locations. As noted before, in the permanent regime (Figure 11.28, $t = 0$ s and 50 s), no temperature difference is observed and thus no damage can be detected. Notice

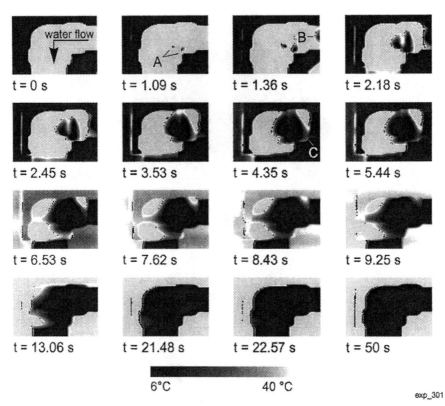

FIGURE 11.28   Infrared image sequence recorded during the active inspection of the corroded bent pipe section of Figure 11.26 using the transmissive approach. Temperature transition from uniformly hot top left image, 40°C to uniformly cold bottom right image, 6°C. Direction of water flow as indicated. Times of observation of structures pointed by highlighted letters shown on Figure 11.30. (From Maldague, 1999. Copyright © 1999 The American Society for Nondestructive Testing, Inc.; reprinted with permission.)

that vortexes responsible for corrosion in specific spots because of the increased water flow at these locations amplify the thermal transient effect observed by the infrared camera on the outer surface. Figure 11.29 shows the image recorded at $t = 1.14$ s during cold-to-warm transition and reverse flow direction, the same pattern that is observed during warm-to-cold transition (Figure 11.28, $t = 1.09$ s). As stated before, based on eq. (11.2), an approximate quantitative wall thickness evaluation is possible for an isotropic material such as steel. On Figure 11.30 the wall thickness (depth) versus time of observation is plotted with steel thermal diffusivity set as $\alpha_{asteel} = 10 \times 10^{-6}$ m^2 s^{-1}. Also indicated on that figure are the times of observation of a few relevant structures highlighted in Figures 11.28 and 11.29, with corresponding measured exact wall thicknesses obtained from visual inspection after the corroded pipe was cut into pieces (Figure 11.27).

This thermal method, in transmission, is fast and easy to deploy if the

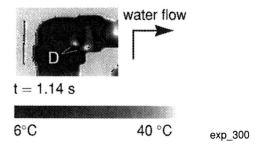

FIGURE 11.29 Method in transmission. Cold-to-hot sudden thermal transition starting at $t = 0$ s, apparatus of Figure 11.26, direction of water flow as indicated (to be compared with Figure 11.28, thermograms at $t = 1.09$ s). Time of observation of structures pointed by highlighted letters shown on Figure 11.30. (From Maldague, 1999. Copyright © 1999 The American Society for Nondestructive Testing, Inc.; reprinted with permission.)

changing temperature of the circulating fluid is possible and does not require extended dismantling of piping. Moreover, since the thermal perturbation source is located inside the pipe, it does not generate any thermal noise. This is an advantage with respect to a radiative source of thermal stimulation (e.g., Figure 11.7). Finally, such a thermal perturbation method allows us to obtain large thermal contrasts, due to the large temperature differentials of the stimulation source. Alternatively, if a change in water flow is not possible, a reflective

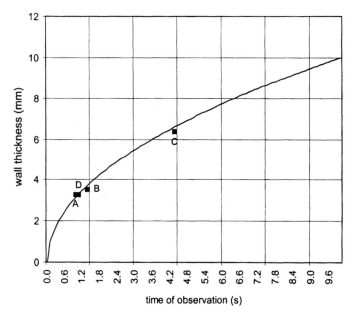

FIGURE 11.30 Wall thickness versus time of observation following eq. (11.2). Letters point to structures identified in Figures 11.28 and 11.29.) (From Maldague, 1999. Copyright © 1999 The American Society for Nondestructive Testing, Inc.; reprinted with permission.)

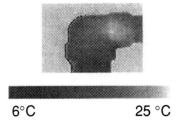

6°C   25 °C

FIGURE 11.31   Method in reflection (heat gun thermal pulse). Cold-to-warm sudden thermal transition starting at $t = 0$ s; same pipe as in Figure 11.26. (From Maldague, 1999. Copyright © 1999 The American Society for Nondestructive Testing, Inc.; reprinted with permission.)

method can be implemented. For example, the same pipe specimen was tested on reflection in Figure 11.31; a Master 1500-W heat gun, model HG 501, induced thermal contrasts externally. Although the damaged area is visible, the contrast is not as good as with the method in transmission, due to the much smaller temperature differences involved. This TNDT method can also be applied to other hollow structures, as we will see in the next section.

11.2.2   Inspection of Jet Turbine Blades

In this section we illustrate another possible deployment of the internal perturbation technique for the inspection of the inner structure of jet engine turbine blades. In this study we use two different thermal stimulation media: a gas rig (air) and a liquid rig (water). Following thermodynamic laws, engine efficiency depends on the temperature differential of the thermodynamic cycle. Jet engines are thus designed to operate at temperatures in the range 550 to 1000°C, and metal protection is ensured through the use of thermal barriers and appropriate cooling methods. This is the case of turbine blades, which must be cooled continuously through internal circulation of a fluid in order to sustain the extreme operating temperatures present in the engine. This is especially valid for blades exposed to combustion gases.

Considering the high rotation speed of blades during operation, the failure of one blade can have dramatic consequences. To prevent these types of accidents, blade temperature can be monitored continuously, in service, through a fiber optic pyrometer. This allows early detection of sudden abnormal blade temperature rise before blade failure. It is also important to check blade integrity at the engine manufacturing stage. In particular, it is important to ensure that blade cooling ducts are unblocked.

Several nondestructive evaluation methods have been proposed and used for blade inspection tasks, including techniques using ultrasonics, neutron radiography, x-ray radiography, computerized tomography, or manual investigation. Manual testing is performed by inserting thin wires through blade channels to check open passages. This method is labor intensive and cannot be automated.

The internal perturbation technique described in Section 11.2.1 can be

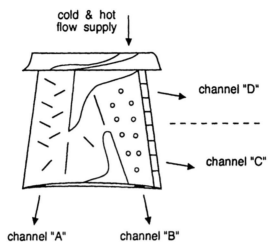

cold & hot
flow supply

channel "D"

channel "C"

channel "A"   channel "B"

FIGURE 11.32 Simplified schematic diagram of the turbine blades tested showing various ejection channels on the blade edges (pressure side). (From Maldague, 1993.)

applied for evaluation of internal structures of jet engine blades. Figure 11.32 shows a simplified schematic of the internal structure of turbine blades tested; note the various ejection channels on the blade edges. Defective blades can have these ejection channels blocked by tiny solidified pieces of metal debris following injection molding during the blade manufacturing process.

Flow circulation of a liquid (water) or a gas (air) at various temperatures in the transient thermal regime makes it possible to detect blocked passages by means of delayed arrival of the thermal front. If a portion of the blade wall thickness is increased by a factor of 2 though machining errors or other causes, the thermal disturbance will reach the outer surface of this portion after a time period four times longer than that required for it to reach the surrounding unblocked portions. This is the basic principle of the method. Areas of the blade where the cooling flow is more efficient (unblocked) will show greater outer surface temperature fluctuations after a time period t on the order of $t \sim z_{\text{th}}^2/\alpha$, following eq. (9.7), where z_{th} is the blade wall thickness.

Water is a good medium for thermographic inspection because of its good cooling properties, and consequently, it is often used to investigate blades or other hollow structures thermally. Conversely, air offers poor thermal removal efficiency. For a simplified one-dimensional stationary analysis, we can write for thermal exchanges between the flow medium and the blade structure, considering only the convection [eq. (3.15)],

$$q_c = \bar{h} A_S (T_S - T_F) \qquad (11.3)$$

where q_c is the rate of heat transferred from the surface at uniform temperature T_S to the fluid of temperature T_F, A_S the surface area, and $\bar{h}$ the coefficient of heat transfer. With the reduced thermal capacity of air, in order to obtain a similar thermal contrast with respect to water stimulation, the temperature differential $(T_S - T_F)$ for an air rig should be greater than for a water rig and the flow rate through the blade must be more important. If we were to compute (Thomas, 1980, p. 374) the coefficient $\bar{h}$ for a circular structure 1 cm in diameter and 2 cm in length, we would obtain $\bar{h} = 96$ W m^{-2} °C^{-1} for air at a rate of 10 m s^{-1} and $\bar{h} = 350$ W m^{-2} °C^{-1} for water at a rate of only 1 cm s^{-1} (a flow 1000-fold smaller). As a result of this simplified analysis, we expect fewer visible time-delayed effects with air stimulation than in the case of water stimulation.

Experimental Analysis. These concepts can be applied as follows. A thermal transient is generated inside a blade under inspection by changing the temperature of the circulating fluid (water or air) and then observing the transient temperature distribution on the outside blade surface with an infrared camera (pressure or suction side). Areas with thinner walls will have their temperature affected first, blocked areas will be affected later: The thermal signature of the blade will depend on the flow circulation inside the blade. Any blocked passage will thus produce a different thermal response, leading to possible detection and recognition of which channel is blocked. The basic arrangement is shown in Figure 11.33. Cool and hot flow lines serve for stimulation of the blades; two computer-controlled electric valves allow either the cool or hot flow to circulate through the blade. For an air rig, a room-temperature air feed line can be used for the cool line, while the hot line can be obtained by diverting the flow of the room temperature air tap through a serpentine held in an oven. For a water rig, two temperature-controlled recirculating baths can supply cool and warm water. Due to the repetitive nature of computer-controlled thermal stimulation, several thermal transitions [or "runs," hot to cold (or the reverse)] can be repeated successively with the images of each run averaged to increase the signal-to-noise ratio (SNR).

For thermal transitions, two images are gathered for each run. This allows us to implement the time subtraction technique of Section 5.5: Two images obtained at different times are subtracted to remove unwanted reflections and background (considered constant at the two acquisition times). The first image is recorded during the warm permanent regime, while the second image is recorded during the thermal transient (from warm to cool). The opposite scheme—the first image recorded during the cool permanent regime and the second image recorded during the cold-to-warm thermal transient—yields the same kind of thermal signature (Section 5.5). The image subtracted shows more clearly the thermal signature of the blade. This subtraction scheme is particularly helpful if unpainted blades are inspected (see Figures 11.37 and 11.38). In this case the low emissivity ε of about 0.3 for such a nickel alloy material makes the blade surface act as a mirror (reflection coefficient: $1 - \varepsilon$) and unwanted thermal reflections caused by warm surrounding bodies be superimposed on the thermal emission of the blade itself, making the analysis harder if not impossible. To increase surface emissivity, blades can be covered with a paint of high emissivity in the spectral region of interest before the thermographic in-

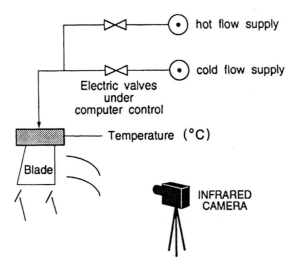

FIGURE 11.33 Schematic diagram and illustration of the experimental apparatus for thermographic inspection of turbine blade using internal stimulation (transmission approach). Hot and cold water supplies are visible; they are controlled directly by the computer using electric valves. This allows us to obtain good reproducibility of the measurements. (From Maldague, 1993.)

spection is undertaken (Section 8.2). Although convenient at a research stage, this painting step is totally inappropriate on the plant floor, where no foreign material is allowed in engine components. The subtraction technique is thus essential in these conditions.

Typical results obtained for the five categories of blades are shown in Figure 11.34 (air stimulation) and Figure 11.35 (water stimulation): Very characteristic thermal signatures, depending on which of the cooling passage is blocked, are

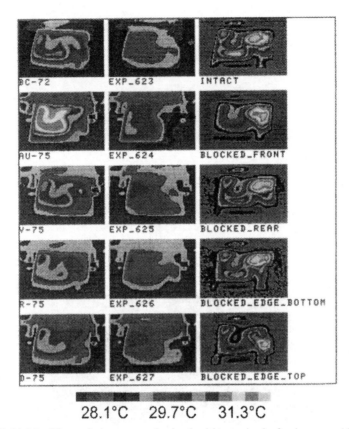

28.1°C   29.7°C   31.3°C

FIGURE 11.34 Thermal signatures obtained with an air rig for intact turbine blades and for some blades with blocked channels. Acquisition of two images; pressure side: one at $t = 1.8$ s after starting heating (left images) and one during cooling of the blade (middle images, $t = 3.95$ s after starting cooling). The time subtraction image is shown for all cases (right images). (From Maldague, 1993.)

obtained. In the case of air stimulation in Figure 11.34, differences are particularly clear if profiles along a bottom row and an edge column are drawn (Figure 11.36).

Figure 11.37 (air stimulation) and Figure 11.38 (water stimulation) show the result of inspection for painted and unpainted blades. The images subtracted reveal a thermal signature similar to that for painted blades in both cases (the uneven emissivity distribution over the unpainted surface in Figure 11.38 is due to the black lettering used for blade marking at production stages). When inspecting unpainted surfaces, it is important to note that the time subtraction approach will work only if the unwanted thermal reflections are the same for the two acquisition times.

Figure 11.38 (water stimulation) and Figure 11.39 (air stimulation) show how averaging many consecutive *runs* together helps to increase the definition

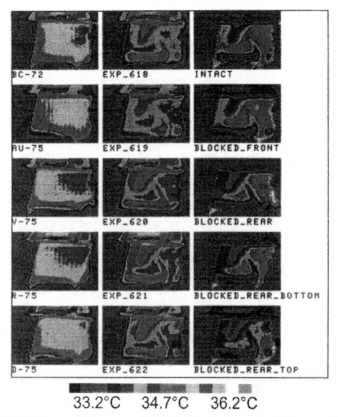

$$33.2°C \quad 34.7°C \quad 36.2°C$$

FIGURE 11.35 Thermal signatures obtained with the water rig for intact turbine blades and for some blades with blocked channels. Acquisition of two images; pressure side: one in the warm permanent regime (left images) and one during the cooling of the blade (middle images, $t = 1.3$ s after starting cooling). The time subtraction image is shown for all cases (right images). (From Maldague, 1993.)

of the thermal signature by limiting the noise content. We recall that SNR is increased by the square root of the number of summed runs. However, even taken individually (no averaging), the thermograms offer the same basic geometric organization. Note that the thermograms shown in Figures 11.34, 11.35, and 11.37 are the result of averaging many consecutive runs together.

Thermal Signature Analysis. The detection algorithm presented in Section 6.3 can be used to discriminate automatically between the different blade signatures obtained in the preceding section: The idea is to be able to recognize automatically which channel is blocked, if any, in a given thermogram, for example, front channel A blocked, rear channel B blocked, and so on. We recall that this algorithm detects blobs in an image; examination of Figures 11.34 and 11.35

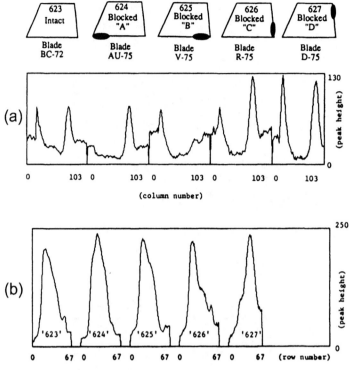

FIGURE 11.36 (a) Profiles of one row at the bottom of the subtracted images of Figure 11.34 (row 53). The peaks are related to the presence of open cooling channels at the bottom of the blades. (b) Profiles of one column at the blade edge of the subtracted image of Figure 11.34 (column 77). The position and width are related to the presence of open cooling channels at the edge of the blades. (From Maldague, 1993.)

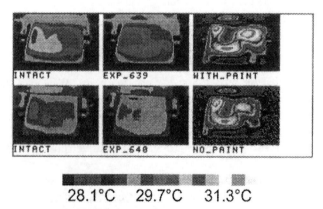

FIGURE 11.37 Test on intact blades with the air rig. Upper row blades are black-painted to increase surface emissivity; bottom row blades are without paint on the surface. The subtracted images on the right reveal the same thermal signatures, although attenuated in the unpainted case. Disposition of images and parameters are as in the case of Figure 11.34. (From Maldague, 1993.)

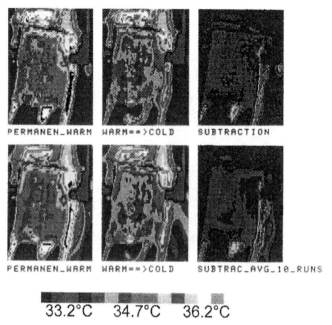

FIGURE 11.38   Test on an intact unpainted blade with a water rig; acquisition of two images, pressure side. Top row, result of one run; bottom row, average of 10 consecutive runs. Disposition of images and parameters as in the case of Figure 11.35. (From Maldague, 1993.)

reveals that location and amplitude of blobs in the subtracted image (Figure 11.35) or on profiles extracted from the subtracted image along the bottom row and edge column (Figure 11.34) depend on the type of blade that is tested. Determination of blob location and amplitude leads to possible detection and recognition.

Water Rig. In the case of a water rig, many different tests (78 in total) were carried out with the five categories of blades (one intact and four with artificial defects of types A to D; Figure 11.32). For all the runs of a given category, the

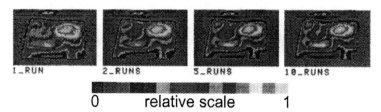

FIGURE 11.39   Intact blade, air rig; same parameters as for Figure 11.34. Increased signal-to-noise ratio is obtained by averaging many consecutive runs together. (From Maldague, 1993.)

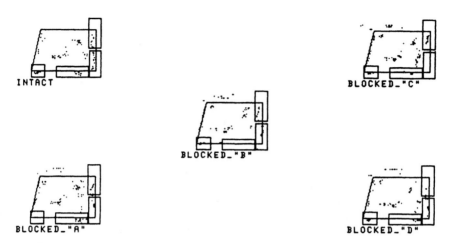

FIGURE 11.40 Water rig: automatic interpretation of the inspection procedure (see the text). (From Maldague, 1993.)

computer program plots a drawing showing the sketch of the blade as well as the positions where peaks are detected within the image (Figure 11.40). We recall that one *run* consists of one test of a blade: acquisition of two images during warm-to-cold thermal transition subtraction of these two images, and detection of the high-intensity peaks by the algorithm in the image subtracted. Due to the close proximity of features of interest present within the images, the MND parameter is set to 5 in all cases [eq. (6.13)].

In Figure 11.40, in addition to the sketch of the blade, four boxes are drawn which correspond to the various possible blocked channels. It can be noted that depending on which of the blade channels is blocked, more peaks will be located in the other boxes (with the exception of the type D defect in Figure 11.40), making the interpretation very simple: If no peak is detected within a box, this means that the corresponding blade opening is blocked (defect). The reason behind this can be understood as follows: If a channel is blocked, the water flow will divert toward the other open channels. Figure 11.40 is plotted in black and white so that the color-coded intensities of the peaks detected (intensity of the brightest pixel within the peak, i.e., seed intensity) are missing, thus preventing visualization of peaks detected on box borders. This is, for example, the case of the few peaks present in the type D defect (Figure 11.40), which can be rejected with a simple validation criterion based on a minimum intensity of the peaks detected in a given box. Using the algorithm of Section 6.3, with its *box intensity criterion*, a reliability factor of 95% is obtained following the Tanimoto criterion, eq. (6.19).

Air Rig. For an air rig, it was found more convenient to operate on profiles (bottom row and edge column) extracted from the subtracted image of a run. A typical investigation using the algorithm of Section 6.3 is shown in Figure

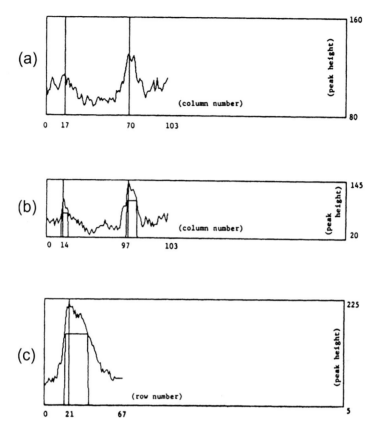

FIGURE 11.41   Investigation by the 1D-peak finder algorithm: (a) detection along row 53 (run 5 of experiment 626, Figure 11.34); (b) detection and peak width (at 3 dB) along row 53 (run 11 of experiment 626, Figure 11.34); (c) detection and peak width (at 3 dB) along column 77 (run 5 of experiment 623, Figure 11.34). (From Maldague, 1993.)

11.41. The correct behavior of the algorithm, independent of the absolute values, is particularly clear in Figure 11.41a, where the two main modes of the profile are located correctly even under noisy conditions. In all cases, the MND parameter [eq. (6.13)] is set to 20. A large value is retained this time since it is necessary to extract low spatial frequency peaks from the profiles.

Following peak detection and location, the peak width (at 3 dB of the maximum) is computed (Figure 11.41b,c). In Figure 11.42, sketches of blades for categories A to D are plotted and analyzed; row and column are also indicated. Also shown on these sketches are the position and the 3-dB width of the peaks detected by the algorithm for all 78 blades tested. The peak widths are indicated by the length of the line segments, and the position of peaks detected are indicated by the location of these line segments. The line segments marked with an "s" are separator marks used to differentiate the five blade categories. The

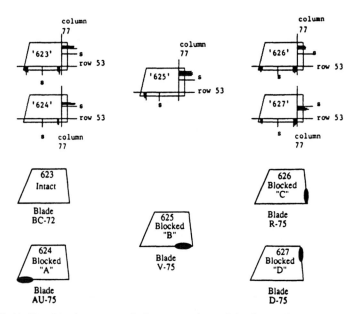

FIGURE 11.42   Air rig: automatic interpretation of the inspection procedure (see the text). (From Maldague, 1993.)

positions of the separators are computed by statistical analysis (average and standard deviation) for the cases considered. In Figure 11.42 it is clear that the distinction between intact (experiment 623), blocked-A (experiment 624), blocked-B (experiment 625), and blocked-D (experiment 627) categories is easily made because of the presence or absence of peak(s) represented as line segment(s) at a particular location on one side or the other of a given separator. Distinction of the blocked-C (experiment 626) category is also possible by considering the 3-dB peak width along the column of interest (Table 11.1). Using this classification scheme, all 78 blades were classified correctly.

The choice of the row (here number 53) and column (here number 77) analyzed is not found to be critical. An improved SNR value can even be obtained by averaging a few (e.g., 3) rows (or columns) together prior to the analysis. If

TABLE 11.1   Parameter Selection[a] for Automatic Defect Detection in Turbine Blades

	Average	Standard Deviation	Acceptable Separator
Blocked-C category	12.9	1.0	15
All other categories	22.2	4.1	

Source: Adapted from Maldague (1993, Table 5.1).

[a] 3-dB peak width along column 77.

blades of identical shapes are inspected and clamped in the same position with the blade holder, the row and column analyzed stay the same.

Discussion. Studies reported in this section show how internal thermal thermography can be implemented for the detection of blocked channels in turbine blades. The same kind of analysis can be implemented for other hollow structures. It was shown how a computerized water or air rig associated with dedicated image processing could sort defective blades automatically. Investigations done on the suction side of the blades revealed thermal signatures similar to those depicted in the various figures in this section, which correspond to experiments recorded on the pressure side.

Although both stimulation media (air and water) permit us to achieve similar results, air is probably a more advantageous choice since it avoids contaminating the inspected part with a foreign substance (water). On the other hand, as stated previously, due to its good thermal capacity, water analysis yields to improved thermal contrast. If air analysis is selected, greater flow and a higher temperature differential (for cool-to-warm thermal stimulation) are needed to obtain acceptable thermal contrasts with respect to the same procedure as that undergone with an water rig.

In this section we studied the *internal pulsed thermal stimulation* approach for the inspection of hollow workpieces. This approach provides enhanced thermal contrasts due to the possible high-temperature differential between hot and cold stimulating flow (liquid or gas) inside the structure inspected. Inspection proceeds during the thermal transient (warm to cold or the reverse). Since the thermal perturbing device is inside the component inspected, it does not generate any radiative noise, such as in the case of a hot external thermal perturbing device (e.g., with lamps); this is another advantage of this approach. We also reviewed how the image-processing techniques of Chapter 6 could be applied for the enhancement and interpretation of recorded thermograms.

11.3 LOCK-IN THERMOGRAPHY AND PULSED PHASE THERMOGRAPHY

In this section we present an application of lock-in thermography (LT) and pulsed phase thermography (PPT) for wood inspection. The content of this section is based on work of Wu (1994) and Wu et al. (1997a).

Wood has a wide span of applications, from the building industry to furniture. Due to its high cost, natural wood is often restricted to external layers with cheaper material making up the inside (e.g., mix of waste wood chips and glue). This is the case of veneer. However, during the manufacturing process, it is important to assert proper bonding between the core and the wood skin for both esthetical and structural reasons. Wood is a composite layered material for which ultrasonics and x-ray do not proceed well. On the contrary, lock-in thermography offers interesting inspection capabilities in this case. As discussed previously (Sections 9.4 and 10.4), phase images have two noticeable advantages: limited sensitivity to surface disturbances (such as wood grain) and deeper thickness of material that can be probed (with respect to conventional pulsed thermography).

For this application, the lock-in principles were deployed with a thermal wave source based on modulated warm air, thus making unnecessary the blackpainting step sometimes used to increase the emissivity (Section 8.2), which would not be possible here, for obvious reasons (however, wood has an emissivity of about 0.9). Tests demonstrated the ability to detect subsurface structures up to 2 mm in depth and as small as 4 mm in diameter. Measurements of the following is reported: knots, disbonds, substrate of different woods, embedded material, and thickness of paint (at a 50-μm resolution). The right-hand sides of the three parts of Figure 11.43 show some results. For now the technique is restricted as an off-line quality control tool, due to the length of time required for phase image generation (3 to 4 minutes at 0.03 Hz). Interestingly, PPT (Section 10.4) is helpful in this respect. For example, Figure 11.43-left, a comparison of results is shown between LT and PPT. In the case of PPT, a long thermal pulse of 20 to 30 s was used. Such a long thermal pulse has a richer low-frequency content, suitable for the detection of deeper artifacts [eq. (9.2)]. Moreover, the long pulse compensates for the low wood absorptivity of the optical energy.

11.4 PASSIVE THERMOGRAPHY

11.4.1 Preventive Maintenance for Electrical Utilities

It is a common to say that prevention is more effective than fixing. This is particularly true for case of electrical utilities. Although electrical utilities manage regular preventive maintenance programs, not all complement these with continuous IR surveys, although the payback can be very significant. For example, in one survey reported, the programmed shutdown of a nuclear power plant saved $2 million for a utility with respect to a nonplanned shutdown that would have occurred if the IR survey that discovered a hot (100°C) 500-kV switch had not been performed (Hurley, 1994). Although such a spectacular situation rarely occurs, undetected hot spots consume valuable energy worth detecting, while, of course, no detection can result in (costly) damage. Several problems detailed below can be detected in electrical utilities through the use of passive thermography. This section was inspired by many sources. Kaplan (1993) and Hurley (1994) were found particularly helpful, and we invite interested readers to refer to these documents, which contain more information than it is possible to cover in a single section.

Considerations in IR Surveys. Good practice leads to meaningful surveys. Below we discuss important survey parameters. During IR surveys, particular

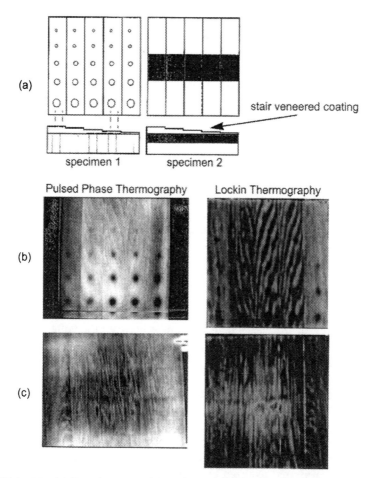

FIGURE 11.43   (a) Sample geometries: stair veneered coating of variable thickness: 0.5, 1, 1.5, 2, and 2.5 mm bonded over the wood substrate of thickness: 15 mm. Specimen 1: five rows of holes of diameter 2, 4, 6, 8, and 10 mm (top to bottom as shown). Specimen 2: knot in wood substrate. Comparison between pulsed phase thermography and lock-in thermography. (b) Analysis of specimen 1: PPT ($f = 0.004$ Hz) left and LT ($f = 0.015$ Hz) right. (c) Analysis of specimen 2: PPT ($f = 0.017$ Hz) left and LT ($f = 0.03$ Hz) right. Similar results are obtained in both cases. (From Couturier et al., 1998. Copyright © 1998 The American Society for Nondestructive Testing, Inc.; reprinted with permission.)

attention must be paid to spatial resolution, especially in aerial surveys (Section 8.4.3). If a thermogram is recorded far from the problem discovered, it becomes difficult to pinpoint its source. The principles of spatial resolution discussed in Section 4.1 apply here.

A good practice recommended in order to document the report later is to record a corresponding visible picture of the IR scene of interest simultaneously. By seeing directly what is inspected, the chances of misinterpretation are reduced. Some IR camera systems offer simultaneous recording of IR and visible properly registered images with overlaying capabilities (e.g., 50%–50%; Figure 11.49). In Section 10.6.3 we saw how such registration can be computed.

Knowledge of whether or not electrical equipment is energized and data regarding the loading state are important in assessing the findings exactly. For example, a nonenergized component will be at close to ambient temperature (unless submitted to solar loading, below), with no information about its operating state even if the component is damaged. Moreover, at peak consumption, momentary high loading is likely to cause the components inspected to be warmer with respect to normal consumption periods, and larger observed temperatures are then not necessarily regarded as significant (unless really abnormal; see below). Exact information on the loading state is not easy to obtain, although voltage and current readings can be obtained with appropriate (high voltage) measurement apparatus (voltage/current meter). One has to remember that in an ac system the average power (i.e., the power that dissipates watts) is (D. E. Johnson et al., 1992)

$$P_{avr} = V_{AC}I_{AC} \cos \theta \qquad (11.4)$$

where V_{AC} and I_{AC} are voltage and current phasors, respectively and $\cos \theta$ is called the power factor, with θ being the angle between the voltage and current phasors (e.g., θ can be measured with an oscilloscope; alternatively, P_{avr} can be measured by a wattmeter). The product $V_{AC} I_{AC}$, known as the complex power, has units of VA (volt-ampere).

As discussed in Section 2.9, emissivities affect apparent temperature so that observation of electrical components having different emissivities can lead to erroneous readings. Fortunately, such effects are often relatively obvious, since one can easily notice visually changes of observed surfaces leading to such differences. This is the case, for example, of a newly replaced shining bolt with respect to a weatherized oxidized bolt (in Table 8.1 we see that emissivity for a shiny steel is 0.07 compared to 0.8 for oxidized steel (for copper these figures are 0.03 and 0.7, respectively), the same as for a busbar made of copper or aluminum (emissivity values of about 0.05 for polished aluminum surfaces and 0.25 for oxidized surfaces). As discussed above, simultaneous recording of a corresponding visible picture of the IR scene might help to remove such a misinterpretation.

Finally, as shown in Section 8.4.3, in the case of outdoor surveys, the long-wavelength band (8 to 12 μm) located beyond the peak emission of solar illumination (0. 5 μm) is preferred to limit the disturbing effect of the sun (Figure 2.6). Indoor surveys can proceed either in the short- or long-wavelength bands.

Decision Criteria. The temperature differential (often called ΔT) is the criterion taken into consideration in order to assess the condition of an piece of electrical equipment being inspected. ΔT is the temperature difference between the temperature of the piece of equipment suspected and a reference temperature, which is the temperature of similar nearby equipment connected in the

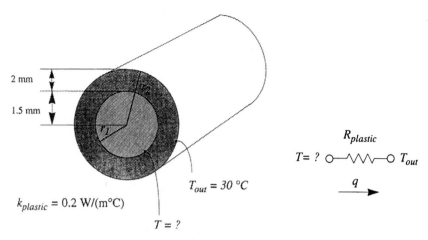

FIGURE 11.44   Geometry for Example 11.1.

same phase as the piece of equipment. Such a comparison helps in removing the effects of both solar loading (see below) and the environment (such as wind effects). For oil-immerged transformers, ambient temperature is used as the reference temperature. Since ΔT is affected by wind (losses by convection) and rain (losses by conduction), it is not recommended IR surveys be performed under heavy wind and rain conditions. Practical studies indicate that a ΔT value of 20°C can generally be considered as the manifestation of a problem (with the exception of momentary conditions of peak loading; see below). Obviously, a 20°C ΔT value recorded on the casing of a piece of equipment corresponds to an even greater temperature differential inside the casing. Nevertheless, the hottest point detected corresponds generally to where the problem is or close to it. Finally, when measuring these temperatures, attention should be paid to surface emissivities, as noted above.

Example 11.1   A piece of plastic 2 mm thick covers a 5-cm-long cylindrical connector for electrical insulation purposes. Temperature readings indicate 30°C in steady-state conditions on the plastic. A 10-A current is circulating through the copper connector (with $k = 386$ W m^{-1} °C^{-1}), electrical resistivity $\rho = 30$ μΩ · cm, and diameter 3 mm). Figure 11.44 shows the geometry involved. Find the temperature at the copper–plastic interface.

SOLUTION   Practical heat transfer calculation principles (Chapter 3) are useful here. Since the plastic is also a thermal insulator, a higher temperature is expected. All the power circulating through the connector is dissipated in the air by convection and radiation:

$$P = i^2 R = q$$

The electrical resistance of the connector is calculated with

$$R = \rho \frac{L}{A} = \frac{(30 \times 10^{-6}) \times 5}{\pi(0.15)^2} = 0.002122 \ \Omega$$

The dissipated power (average power) is thus

$$P = i^2 R = (10)^2 \times 0.002122 = 0.2122 \ \text{W}$$

Let's now calculate the radial conduction resistances using eq. (3.24), for $L = 0.05$ m and with (from Figure 11.44) $r_1 = 0.0015$ m and $r_2 = 0.0035$ m:

$$R_{x_{\text{rad}} \cdot \text{plastic}} = \frac{\ln(r_2/r_1)}{2\pi L k_{\text{plastic}}} = \frac{\ln(0.0035/0.0015)}{2\pi(0.05 \times 0.2)} = 13.5°\text{C W}^{-1}$$

Now we can compute the heat transfer rate:

$$q_r = \frac{\Delta T}{R} = \frac{T - 30}{13.5} = 0.2122 \ \text{W}$$

and thus solving for T yields

$$T = 30 + (0.2122 \times 13.5) = 32.9°\text{C}$$

As a rule of thumb, a value of ΔT greater than 75°C on exposed metal or 30 to 50°C on an isolated oil-submerged piece of equipment requires immediate attention, with appropriate action to fix the problem detected. Table 11.2 gives more details; these values are for 50% of the plated load and thus must be corrected accordingly with the following factor (Kaplan, 1993, p. 90):

$$\text{load factor} = \left(\frac{50}{\text{actual load value in \%}}\right)^2 \tag{11.5}$$

$$\Delta T_{\text{table\_11.2}} = \Delta T \times \text{load factor} \tag{11.6}$$

TABLE 11.2 Classes of Reported Faults

Class of Faults	ΔT (°C)	Comment
I	>0.5	First stage of overheating, to be kept under control and repaired at next scheduled maintenance
II	5–30	Developed overheating, to be fixed as soon as possible
III	>30	Strong overheating, to be fixed rapidly

Source: Adapted from Kaplan (1993, Table 7.1, p. 90).

TABLE 11.3 Compensation for Wind

Wind Speed		Correction Factor[a]
m s^{-1}	km h^{-1}	
1 or less	4	1.00
1.4	5	1.15
2	7	1.36
2.8	10	1.57
3	11	1.64
4	14	1.86
4.2	15	1.90
5	18	2.06
5.6	20	2.16
6	22	2.23
7	25	2.40
8	28.8	2.50
8.3	30	2.53
9 or more	32	Measurements not recommended

Source: Adapted from Kaplan (1993, Table 7.2, p. 91).
[a]The following third-order relationship applies: correction factor ≈ 0.0014 ws^3 − 0.037 ws^2 + 0.4425 ws + 0.6, where ws = wind speed m s^{-1} in range 1 to 9 m s^{-1}.

It is also possible to take the wind factor into account (see the following examples), although IR surveys are not recommended if the wind speed is faster than about 32 km h^{-1} (Table 11.3).

Example 11.2   State what you conclude from a ΔT value of 10°C if the equipment on which this reading is obtained is 100% loaded (no wind).

SOLUTION   Following eq. (11.5), we obtain a correction factor of $(50/100)^2 = 0.25$, and thus $\Delta T_{\text{table\_11.2}}$ is calculated as $0.25 \times 10 = 2.5$°C, corresponding to the first stage of overheating (Table 11.2).

Example 11.3   An IR survey reports a ΔT value of 4°C on a busbar (50% loaded). The wind is 30 km h^{-1}. State your conclusion here.

SOLUTION   From Table 11.3 we obtain an empirical correction factor of 2.53; the ΔT value is thus corrected: $4 \times 2.53 = 10$°C. Without taking into account the effect of the wind, this finding was a bare overheating, whereas after correction it appears more severe (Table 11.2).

High Resistance. The high electrical resistance of a connection will indeed provoke local overheating, a hot spot, that could be detected by IR thermography. This is easily explained with the known relationship that links the dissipated power P with the electrical resistance R (in ohms) crossed by an electrical current I (in amperes) [with respect to eq. (11.4), $\cos \theta = 1$ since both phasors I and V are in phase in the case of resistances, $\theta = 0$]:

$$P = VI = RI^2 = \frac{V}{R^2} \qquad \text{watts} \qquad (11.7)$$

where V (in volts) is the voltage drop across R. Thus in the case of a connection point with electrical resistance R crossed by a current I, P watts will be dissipated as heat, causing local overheating of the connection itself. This energy propagates locally in the air (by convection and radiation; Chapter 3) but also through metal (by conduction; Chapter 3); this eases detection in the case of a hidden fault (e.g., a bad connection behind a casing). A side effect of high resistance for an electric utility is, of course, the resulting voltage drop $V = RI$ (due to Ohm's law; D. E. Johnson et al., 1992).

Examples of such high electrical resistances are, for example, loose connections (e.g., an untighted bolt; Figure 11.45) and broken strand(s) of wire. With time, vibrations of electrical connections can sometimes loosen bolts not well tightened in the first place. In the case of broken stranded wires, the high electrical resistance between strands due to oxidation of conductors sometimes makes the current noncirculating through the remaining strands. The heating pattern observed thus follows the spiral pattern of the remaining energized strands, while the amplitude of overheating depends on the cross-sectional wire diameter that remains. This effect is often called *spiral heating* (Figure 11.46). Finally, arcing is another case of high resistance and thus also causes hot spots to develop, resulting in rapid deterioration of components affected. Arcing sometimes results from bad design.

Two particular cases of abnormal resistances result in problems: $R = 0$ (short circuits) and $R \to \infty$ (open circuits). Short circuits often involve large currents, leading to high dissipated power and corresponding ΔT. They generally do not last long: Either a circuit breaker opens, removing the power, or rapid deterioration takes place, sometimes up to total breakdown. Short circuits in winded equipment (transformer, motor, generators) can last and TNDT can detect them (Figure 11.47). Another example of short circuits is the case of energized grounds after insulation failure. Although their exact origin might be elusive, they are easily detected by TNDT, due to the large values of ΔT involved.

Open circuits are also problematic since an electrical component not working can cause disruptions elsewhere in the electrical system. This condition can also be detected and reported (if unenergized conditions are not wanted).

Excessive Loading. As indicated by eq. (11.7), since P is proportional to I^2, a small change in current I results in a significant effect in the dissipated power P. In the case of a conductor of constant cross-sectional area connected to a load, overloading will thus provoke a generalized increase in conductor tem-

0   relative   1
scale

FIGURE 11.45 Preventive maintenance for electrical utilities. A 12-kV primary elbow indicates excessive resistance heating in the screw type, 90° angle connection. Both infrared and visible images are shown. (From Hurley, 1994, with permission from Gordon and Breach Publishers.)

perature connected to that load. Just as an example, this effect is sometimes noticed at home if an electrical equipment is powered with an extension cord of insufficient rating (e.g., 14 AWG instead of 12 AWG energizing a small heater). In that case, the cord can be felt at the touch, as it becomes warmer.

In the case of three-phase systems commonly used in industrial environments, charge imbalances are detected by TNDT. For example, in the case of a three-phase system connected to a three-phase motor, charge imbalance occurs if the windings associated with one phase become disconnected (e.g., broken wire or fuse, worn brushes; Figure 11.48). However, unbalanced heating can also result when one phase is more loaded with respect to others while under normal conditions. As mentioned earlier, knowledge of the electrical load status is thus helpful in such a case to assess whether or not a problem is present.

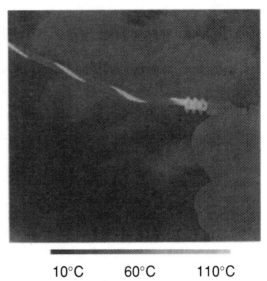

10°C   60°C   110°C

FIGURE 11.46 Preventive maintenance for electrical utilities. Example of spiral heating, a single strand of a multistranded wire carries the electrical load. Both infrared and visible images are shown. (Courtesy of Flir Systems.)

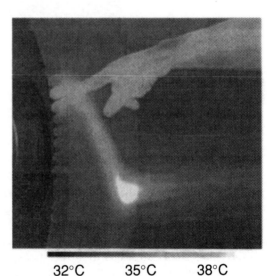

32°C   35°C   38°C

FIGURE 11.47 Preventive maintenance for electrical utilities. Example of a shorted winding isolated on a generator stator. Both infrared and visible images are shown. (Courtesy of Flir Systems.)

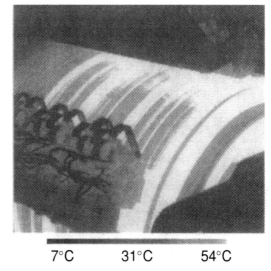

7°C   31°C   54°C

FIGURE 11.48   Preventive maintenance for electrical utilities. Example of worn brushes in a dc generator as revealed by warm streaks. Both infrared and visible images are shown. (Courtesy of Flir Systems.)

Solar Loading. Some equipment is heated by the sun. This can add to heat readings, which should then be regarded with care. One inspection possibility is to wait for the sun to move out so that the equipment is in the shadows. Moreover, in some instances, a screen can be positioned so that equipment is protected from the sun. Obviously, it is important to wait long enough for equipment to cool back down to its operating temperature. Performing IR surveys under overcast conditions (no shadow), early in the morning (before the sun rises), or after the sun sets (after the effects of solar loading has vanished) is probably the best approach. It is not recommended to conduct IR surveys in the presence of solar loading since such a perturbing effect can mask problems of small ΔT values (such as leaking current; see below). If inspection has to proceed under direct sunlight illumination anyway, it is then suggested that, if possible, the readings obtained be corrected using those from similar equipment (emissivity, surface material, size), energized or not, and to consider only significantly higher ΔT values while taking note of possible undetected problems (e.g., low ΔT) due to unfavorable inspection conditions.

Example 11.4   Infrared inspection of an energized transformer casing under direct sunshine illumination gives a 40°C reading; the ambient temperature is 30°C. A similar nonenergized transformer nearby, also under direct sunshine illumination, gives a reading of 36°C. State your conclusion.

FIGURE 11.49   Preventive maintenance for electrical utilities. A visible and IR mix reveals the voltage tracking on a terminator. (From Hurley, 1994, with permission from Gordon and Breach Publishers.)

SOLUTION   In this case, a $40 - 36 = 4°C$ differential (ΔT) is noted. This does not look abnormal (see the severity criteria below and Table 11.2).

Leaking Current. This phenomenon is often referred to as *tracking*. Leak currents sometimes occur in damaged pieces of equipment in which currents find a path to ground. They cause a limited temperature differential (around 2°C and sometimes less) and are thus difficult to detect by TNDT, although they can eventually lead to failure and thus represent a danger. Examples of leak current are found, for example, on isolators (through cracks; Figure 11.49) and wooden poles. Leak current through a defective bushing of transformer support can char a wooden pole. It can also be due to environmental conditions. For example, along coasts, salty mists are deposited on equipment, necessitating special design to avoid leaking (e.g., insulators are taller). This confirms what was previously mentioned concerning the suggestion that the operator search for abnormal hot spots, even on nonenergized components.

Eddy-Current Inductive Heating. Energized electrical components produce electrical and magnetic fields. This is particularly the case for winded devices, in which such a field is proportional to the number of spires. Although most of this field is generally collected in the device itself (through the armature), some might escape in the neighborhood. If the distance to ground is not appropriate, these fields might generate inductive currents in nearby ferrous components such as ungrounded steel casings and stands (Figure 11.50). In turn, electrical resistance of the material in which the induced current circulates dissipates

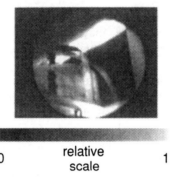

0   relative   1
          scale

FIGURE 11.50 Preventive maintenance for electrical utilities. Eddy-current inductive heating found on the adjacent metal cabinet enclosing an harmonic filter. Both infrared and visible images are shown. (From Hurley, 1994, with permission from Gordon and Breach Publishers.)

power [eq. (11.7)] and releases heat that can be detected by TNDT. This phenomenon is not necessarily dangerous but has to be documented. Inductive heating is more important in high-powered systems, where ΔT can be quite high. Even if IR surveys are directed primarily toward electrical components (for both production and distribution), it is recommended that the inspector making the survey check what is going on in nearby equipments as well. Discovery of a hot spot in nonenergized equipment can, in fact, reveal a problem of inductive heating. In some instances, local heating due to inductive heating is so severe that paint degrades. An example of inductive heating is observed, for example, when a nonferrous bolt is replaced by a steel bolt.

Solar Magnetic Disturbances. Solar flares produce solar winds, causing magnetic disturbances to charged particles located over Earth's poles in the high atmosphere (100 to 300 km). Motion of these particles generates high currents (we recall that an electrical current I is defined as a motion of charged particles dQ in time dt ($I = dQ/dt$; the differential d refers to the quantities involved; (D. E. Johnson et al., 1992). In turn, these large currents (up to 10^6 A has been observed in the ionosphere) lead to fluctuations of the magnetic field, causing dc voltage gradients called *earth surface potential* (ESP) in Earth's crust. ESP can last several minutes, with an amplitude of several volts per kilometer. Following Ohm's law [see the discussion of eq. (11.7)], such an ESP results in solar-induced currents (SICs), whose strength depends on Earth's electrical resistance. SIC depends on both latitude and geology. For example, igneous rock is characterized by high electrical resistivity, and electrical currents prefer low-resistance paths such as those found in grounded electrical systems. SIC are found, for example, in grounded transformers, generating corruptive harmonics and saturating cores (thus causing loud noises). Since these phenomena are transient, they are hard to observe directly, although damaged (burnted) paint on the ground-side attachment of a transformer can demonstrate an SIC effect (Figure 11.51). Solar activity runs on a 11-year cycle with the last peaks observed in 1989 and 2000. In 1989, magnetic disturbances caused by the sun provoked a power blackout in a large region of Canada and the United States.

Miscellaneous Applications. For oil-submerged equipment, IR imaging can detect the oil level in the tank due to temperature transition observed between the top (air-filled) and the bottom (oil-filled). This complements visible inspection in case oil leakage is observed (Figure 11.52). Under energized conditions, the air-filled space will be hotter than the oil-filled bottom, which cools the equipment.

Transmission lines are good candidates for IR surveys since they are essential to distribution. Both aerial (Section 8.4.3 and below) and ground inspection are helpful. Such surveys permit us to detect problems such as poor electrical connections, leaking currents across insulators, induced voltages to ground, wire overloading, unbalanced loads (three-phase), and spiral heating (Figure 11.53). The ideal period to perform such surveys is after the hunting season, as this allows to locate broken insulators that served as targets to hunters.

Fiber optic sensors have begun begun to be embedded with wires (or within concrete slabs), enabling continuous, long-term, distributed monitoring of strain and temperature, made possible by the Brillouin scattering effect (Thévenaz, 2000). Finally, it is worthwhile to mention that the phenomena of electromagnetic emission and arcing are best detected by radio receivers.

11.4.2 Passive Thermography for the Construction Industry

Another important field of inspection by TNDT is concerned with construction. This section was inspired by many sources, but Kaplan (1993) and Ljungberg

FIGURE 11.51 Preventive maintenance for electrical utilities. Hot ground streak caused by solar flares found on a transformer core. (From Hurley, 1994, with permission from Gordon and Breach Publishers.)

(1994a) were found to be particularly helpful, and we invite interested readers to refer to these documents, which bring more additional information than it is possible to cover in a single section.

When the first IR cameras became available in 1965 (scanning radiometer by AGA, now Flir), several researchers had already become interested such applications. Activity increased dromatically following the energy crisis of 1973, with, for example, in the development of standards (Table 11.4) and of large survey programs aimed at the detection of thermal losses and subsequent improvement in thermal building insulation to save energy. Subsequently, activities have expanded, with the detection of trapped moisture in walls, roofs, and bridge decks; historical building refurbishment (including fresco evaluation

0   relative   1
           scale

FIGURE 11.52   Preventive maintenance for electrical utilities. Unusual heating pattern correlated as faulty contact points of the internal tap changer. The midway heat line on the thermogram can be construed as a low oil level (notice the oil-stained shell of the unit). Both infrared and visible images are shown. (From Hurley, 1994, with permission from Gordon and Breach Publishers.)

(Bison et al., 1996); and so on. Detecting such problems by TNDT allows for early repair, making possible savings on energy and maintenance costs.

Although most of these applications involve passive thermography, some make use of active thermography. In the latter case, thermal transients are generated within a building, for example, by blowing hot air on inside walls and monitoring temperature evaluation from the outside with an IR camera (Büscher et al., 1998; Rosina et al., 1998). Active thermography was also demonstrated for bridge deck inspection. In this case, hot concrete bricks (60°C) are applied on a bridge deck for a period of time, after which they are removed, enabling pulsed IR inspection to proceed in order to reveal subsurface defects following the principle shown in Figure 9.2. An example of such an investigation

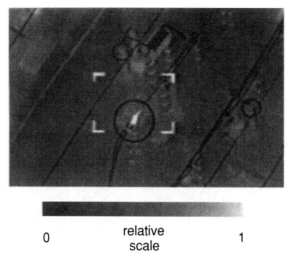

0   relative   1
             scale

FIGURE 11.53 Preventive maintenance for electrical utilities. Aerial inspection and evaluation of a transmission system with anomaly pinpointed. (From Hurley, 1994, with permission from Gordon and Breach Publishers.)

TABLE 11.4 ASTM Standards for Infrared Thermography and Imaging

Standard Number	Standard Title
E1934-99	Standard Guide for Examining Electrical and Mechanical Equipment with Infrared Thermography
D4788-88(1997)	Standard Test Method for Detecting Delaminations in Bridge Decks Using Infrared Thermography
E1933-99a	Standard Test Methods for Measuring and Compensating for Emissivity Using Infrared Imaging Radiometers
E1897-97	Standard Test Methods for Measuring and Compensating for Transmittance of an Attenuating Medium Using Infrared Imaging Radiometers
E1862-97	Standard Test Methods for Measuring and Compensating for Reflected Temperature Using Infrared Imaging Radiometers
E1543-00	Standard Test Method for Noise Equivalent Temperature Difference of Thermal Imaging Systems
E1213-97	Standard Test Method for Minimum Resolvable Temperature Difference for Thermal Imaging Systems
C1060-90	Standard Practice for Thermographic Inspection of Insulation Installations in Envelope Cavities of Frame Buildings
C1046-95	Standard Practice for In-Situ Measurement of Heat Flux and Temperature on Building Envelope Components
C1153-97	Standard Practice for Location of Wet Insulation in Roofing Systems Using Infrared Imaging
E1311-89(1999)	Standard Test Method for Minimum Detectable Temperature Difference for Thermal Imaging Systems

Source: American Society for Testing and Materials: www.astm.org.

(a)

(b)

(c)

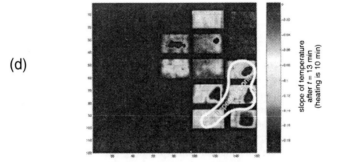

(d)

FIGURE 11.54   Active thermography bridge deck inspection: (a) IR camera in position over an area heated by contact with warm concrete bricks (60°C); (b) position of the bricks on the bridge deck with chalk mark references; (c) bricks being heated on gas burners; (d) composite processed thermogram indicating a suspected area (corresponding thermal modeling indicates a 0.5- to 1.5-cm-deep delamination for a slope −0.11 to −0.15 as circled). See also Figure 8.12 on this method.

is shown in Figure 11.54. Such an approach is particularly helpful when weather conditions are not appropriate for passive thermography to be deployed.

When possible, IR surveys can be complemented by other methods to identify the source of problems: for example, introduction of a visible borescope through a wall to pinpoint an insulation problem (Figure 11.55). In addition to thermal insulation assessment (below) and roof moisture detection (below),

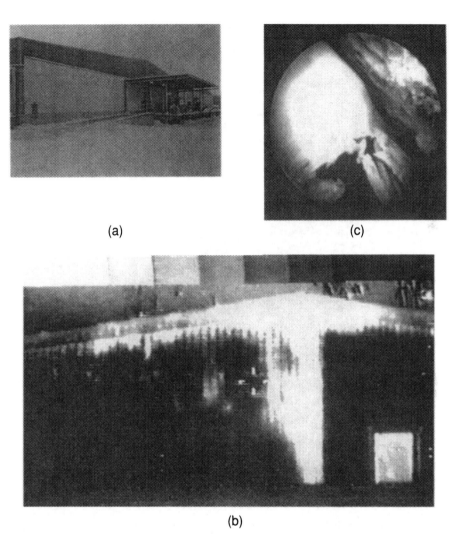

(a)

(c)

(b)

FIGURE 11.55 Passive thermography for the construction industry. (a) Visible picture of an all-concrete construction assembled from structural elements and lightweight walls connected to a slab on the ground. (b) The thermogram shown indicates high-temperature anomalies and heavy energy leakage for a large area on the external wall of the building shown in (a). (c) Photographic inspection of the interior building envelope using a boroscope revealing damages to the vapor barrier. (From Ljungberg, 1994a, with permission from Gordon and Breach Publishers.)

TNDT allows detection of failures in district heating, sewer systems, wastewater pipes, and in canals and aqueducts through detection of leaks of different temperatures with respect to ambient. Such detection can proceed either from outside (aerial surveys) or from inside with an IR camera mounted in a sled pulled downstream within the pipe being inspected. For example, during the

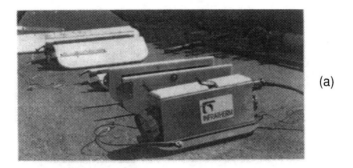

(a)

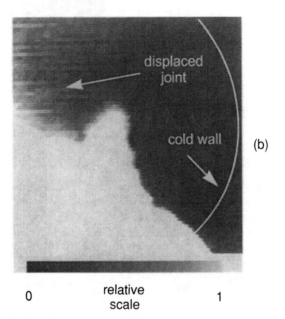

displaced
joint

cold wall   (b)

0   relative   1
          scale

FIGURE 11.56   Passive thermography for the construction industry. (a) An encased IR camera mounted on a sled and pulled downstream in a wastewater pipe to locate surface water and groundwater leaking in the pipe; notice the dismantled casing in the background. (b) Thermogram indicating surface water leakage into a wastewater pipe; the cold wall is due to surface water around the sewer and a displaced joint. (From Ljungberg, 1994a, with permission from Gordon and Breach Publishers.)

cold seasons, sewer water is warmer than its surroundings (Figure 11.56). Unless involved temperature differentials are very large, underground detection is always more challenging (Svedamar, 1985; Okamoto et al., 1994c, 1995, 1997). Nevertheless, some successful underground applications have been reported (Figure 11.57; Galmiche and Maldague, 1999).

Other applications reported for TNDT are for road and bridges (Figure 11.58). In these cases detection of delaminations, breakdown of cement matrix,

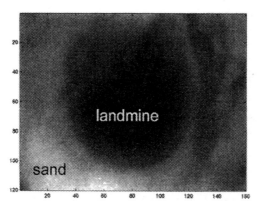

FIGURE 11.57 Buried landmine detection by infrared thermography.

and disbondings are possible, due to differences in thermal conductivities. Moisture trapped at an asphalt–concrete interface is difficult, although sometimes possible, to detect (Weil, 1985). For these applications, visible images usefully complement an IR survey to, for example, pinpoint specific areas, such as those with oil spills or cracks. For example, ASTM standard D4788 (1997) defines a test method for detecting delaminations in bridge decks using infrared thermography.

Good compaction of asphalt requires a temperature of about 140°C. A cooler temperature results in reduced quality. Asphalt paving can also be monitored by TNDT: for instance, with an IR camera mounted on a (4-m-tall) mast, for example (see Figure 8.47). It should be remembered that for outdoor measurements, long wavelengths are preferred unless surveys proceed at night or before sunrise, when short wavelengths can also be used (Section 4.7).

Thermal Insulation Assessment. Missing or weak thermal insulation is a problem of bad design and/or construction (Figure 11.59). In the case of new construction it is a good practice, if possible, to conduct the survey during construction (e.g., before outside facing is installed), so that problems detected early can be fixed at lower cost. In passive thermography, thermal gradients are naturally present on the surface observed, due to the temperature gradients ΔT_{i-o} between the inside and outside temperatures. Following the heat transfer principles discussed in Chapter 3, in the steady-state regime, if the ΔT_{i-o}, thickness of material, radiation, and convection values are known, the thermal properties (and then the insulation factor R) of a wall or roof can be determined. TNDT can proceed from the outside or from the inside, with, of course, opposite temperature differential findings (Figure 11.60). Moisture due to water infiltration from the outside or due to piping leakage can also cause thermal insulation to deteriorate. Techniques similar to those used to detect roof moisture can reveal the problem (see below).

(a)

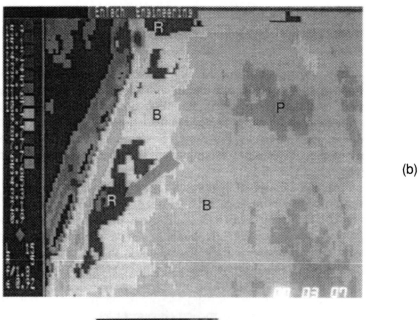

(b)

(c)

FIGURE 11.58 Passive thermographic bridge deck inspection: (a) areas inspected; (b) thermogram highlighting: good areas in blue (B), indeterminate areas in purple (P), deteriorated areas in three shades of red (R); (c) verification pavement penetration test performed where indicated by the arrow. (From Weil, 1985.)

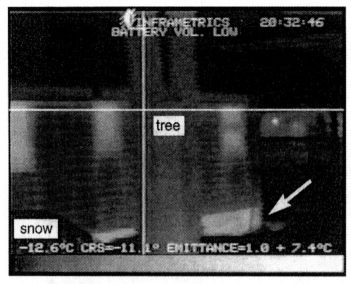

FIGURE 11.59 Passive thermography for the construction industry. Thermogram recorded at night in wintertime on a bungalow indicating missing insulation (arrow) in the basement, especially considering the lower concrete emissivity with respect to top wall in wood painted planks with higher emissivity (exterior temperature −17°C; a tree is seen in front of the house, snow is blocking the view on the left).

Example 11.5   Recalling Example 3.11, compute the R-value of the concrete–steel assembly of Figure 3.16.

SOLUTION   In the building industry, the R-value, defined as

$$R\text{-value} = \frac{\Delta T}{q/A}$$

is used to classify insulation performance. It differs from the thermal resistance $R(R = \Delta T/q)$ introduced in Chapter 3 by the heat flow, which is per unit area. With this we recall that R_{total} for the assembly was found as

$$R_{\text{total}} = 0.51 \;^{\circ}\text{C W}^{-1}$$

For a cross-sectional area of $A_{\text{concrete}} = 0.5 \text{ m}^2 + A_{\text{steel}} = 0.1 \text{ m}^2 = 0.6 \text{ m}^2$ with $\Delta T = 25°C$ and $q = 49$ W (Figure 3.16), the R-value is computed as (the conversion factor is $1 \text{ ft} \cdot \text{h} \cdot {}^{\circ}\text{F/Btu} = 0.1762 \; ({}^{\circ}\text{C W}^{-1}) \cdot \text{m}^2$)

$$R\text{-value} = \frac{\Delta T}{q/A} = \frac{25}{49/0.6} = R \times A = 0.51 \times 0.6$$

$$= 0.31 \;^{\circ}\text{C W}^{-1} \cdot \text{m}^2 = 1.76 \text{ ft} \cdot \text{h} \cdot {}^{\circ}\text{F/Btu}$$

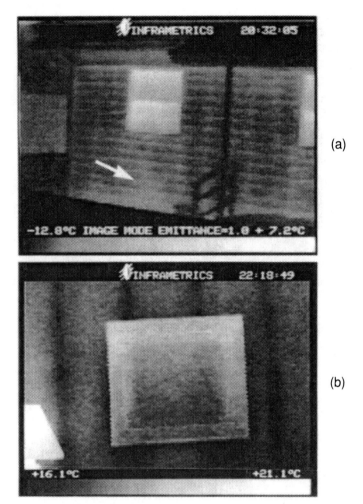

FIGURE 11.60 Passive thermography for the construction industry. (a) Thermogram recorded at night in wintertime on a bungalow, revealing energy leakage through wood studs of the 4-in (10-cm) wood frame (studs are located every 16 in, 40 cm). (b) Inside inspection thermogram reveals the same phenomenon with a temperature of 17°C between studs, 16.7°C over studs, and room temperature of 19°C (exterior temperature −17°C, objects blockage is visible). Notice in (a), indicated by an arrow, the heat losses through the wall in front of the warm air outlet from the central heating system located inside, under the window.

which is quite small indeed. Minimum R-values (in Canada) are 20 ft · h · °F/ Btu for walls and 40 π · h · °F/Btu for attics, and less for windows (Table 11.5). Higher R-values can be achieved with the addition of insulating material (see, e.g., Example 3.9). However, the addition of thermal insulation is sometimes useless if other negative effects, such as those of chimneys (drain of warm air to

TABLE 11.5   R-Value for Windows[a]

Type of Glass	R-Value (ft·h·°F/Btu)[b]
Single	0.9
Double [12.7 mm ($\frac{1}{2}$ in.) air space]	2.1
Double [12.7 mm ($\frac{1}{2}$ in.) air space] and low-emissivity ε coating, $\varepsilon = 0.4$	2.5
Double [12.7 mm ($\frac{1}{2}$ in.) air space] and low-emissivity ε coating, $\varepsilon = 0.1$	3.2
Triple [12.7 mm ($\frac{1}{2}$ in.) air space]	3.2
Caloriglass double [12.7 mm ($\frac{1}{2}$ in.) air space]	4.5
Caloriglass double [9.5 mm ($\frac{3}{8}$ in.) air space]	4
Caloriglass double [6.47 mm ($\frac{1}{4}$ in.) air space]	3.2

Source: Hyalin Inc., St. Hyacinthe, Quebec, Canada.
[a] R-value at the center of the window.
[b] The conversion factor is 1 ft·h·°F/Btu = 0.1762 (°C W^{-1})·m^2.

the outside), air infiltration and exfiltration (air leaks), and thermal bridges (i.e., parts with no thermal insulation connecting two areas of different temperature), are not taken into consideration and fixed properly. It is thus important while making a thermal survey to consider a building globally. In the case of air leakage, tracer gas can be used to measure the changing rate.

Conditions for IR Surveys. It is recommended that the $\Delta T_{i\text{-}o}$ value be at least 10°C for at least 3 h before IR measurement (to be in the steady-state thermal regime and generate sufficiently high gradients). Inspection can proceed either from the inside (e.g., for small buildings) or from the outside (e.g., for large buildings). Inside building inspection is less dependent on the weather, although it is more time consuming, due to difficult access, including the necessity to move cumbersome objects (Figure 11.60). Outside TNDT surveys are best conducted at night or before sunrise with no direct sun on the surface, limited wind (Table 11.3), and no rain for at least one day before the survey. In the case of inside or outside inspection conducted at night or early in the morning, as discussed in Section 8.4.2, short-wavelength deployment is possible. Long wavelengths are preferred otherwise.

Roof Moisture Detection. It is possible to detect trapped moisture in a roof using TNDT. Moisture generally comes from rain infiltrating deteriorated roofing or from leaking pipes. While adding considerable weight to a building (1 L of water weights 1 kg) such moisture also reduces the thermal insulation (due to the greater heat capacity of water; Table 3.7). By making moisture detection possible, IR surveys allow us to pinpoint and localize a problem, sometimes allowing us to fix only those zones degraded rather than proceeding to a full repair at greater cost. Obviously, fixing a moisture problem secures a building. Two detection methods can be adopted.

The first method is based on solar heating. During the day, due to solar heating, roof structures store heat, which is released at night. Dry areas store less energy, and their temperature drops earlier than those in zones with a higher moisture content (due to the higher thermal capacity of water; Table 3.7). The good thing about this approach is that recorded IR images are generally not affected by heating ducts, fans, and other machinery located beneath a roof structure (Figure 11.61)

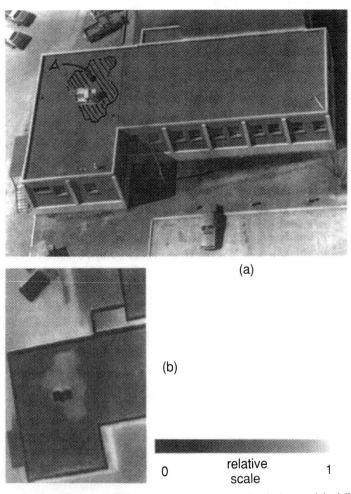

(a)

(b)

0   relative   1
          scale

FIGURE 11.61 Passive thermography for the construction industry: (a) visible picture with an overlay showing potentially wet area identified in the thermogram; (b) aerial thermogram of a roof indicating moisture damage. (From Ljungberg, 1994a, with permission from Gordon and Breach Publishers.)

The second method of detection is based on the inside–outside temperature gradient ΔT_{i-o}. Because moist areas conduct heat better than dry areas, warmer temperatures are observed over moist areas and cooler temperatures in dry areas. Since ΔT_{i-o} is relevant here, recorded thermal images tend to be affected by heating ducts, fans, and other machinery located beneath roof structures. Particular attention has to be taken regarding activities that are going on inside a building since this may affect the specific thermal signatures recorded on the roof in the presence of cold storage, production areas, or office space, for example.

Roof inspection by TNDT is now well established, with several companies offering such expertise. For example, ASTM standard C1153 (1997) defines procedures for location of wet insulation in roofing systems using infrared imaging (such standards require drilling cores to confirm IR findings).

Conditions for IR Surveys. In the method based on solar heating, it is recommended that IR surveys be performed when the ΔT_{i-o} value is about 0°C for at least 24 h before the measurement (to be in the steady-state thermal regime). The best time to perform the survey is late at night or early in the morning, when moist areas have not yet released their energy. The method based on ΔT_{i-o} requires a ΔT_{i-o} value of about 10°C for at least 24 h prior to the survey. The best time to perform the survey is late at night or early in the morning, and overcast conditions the day prior to measurement are preferred, to avoid the perturbing effects of solar heating.

Example 11.6 Compute the temperature evolution for a 4-cm thickness of both water and concrete, following pulse heating of 1 W cm^{-2} for 1 s.

SOLUTION   The thermal model of Section 3.4.2 can be used here, replacing all parameters (defect, sound area) by the thermal properties of water and concrete. This is shown in Figure 11.62. Although truly academic, such an example illustrates the fact that when water temperatures remain higher longer, detection of moisture trapped in a roof structure is simplified.

Air Surveys. Aerial surveys by airplanes or helicopters enable a group of buildings or a large building to be inspected rapidly, provided that the spatial resolution is sufficient to locate problems. Infrared line scanners are an interesting alternative to 2D imagers when the survey is to cover a large area (the flight path furnishing the second dimension). If surveys are done at night, it is recommended that the building also be flown over during the day, to record corresponding visible images, to simplify the location of findings on IR imagery later. As mentioned previously, surfaces to be surveyed should be free of moisture. Typically, an IR survey costs about $200 per km. Finally, an alternative to aerial survey is to place an IR camera on a mast maintained over the surface to be inspected. Section 8.4.3 provides additional information about aerial surveys.

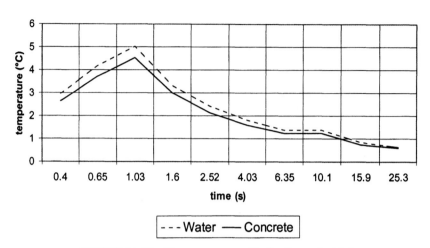

Pulsed heating profiles

FIGURE 11.62   Temperature profile for Example 11.6.

11.4.3   Temperature Measurement of Hot Steel Strip Mills

In this section we describe the use of passive thermography in hot steel strip mills. In writing this section, the work of Holmsten and Houis (1990) was found to be particularly helpful, and we invite interested readers to refer to this document, which includes more information than it was possible to cover in a single section.

For such an application, the temperature of interest varies from 1300 to 600°C (before coiling), and of most interest is measurement of the edge temperature. Figure 11.63 shows locations of interest for temperature measurement. It is relevant to mention that the purpose of hot rolling is much more than transforming a thick slab into a long thin strip (up to 1 km). Certain mechanical

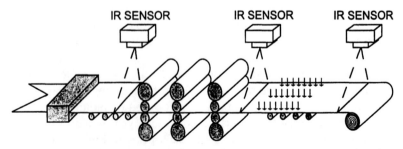

FIGURE 11.63   Typical locations for surface temperature measurements in a hot-strip mill (left to right): after the rougher and descalers, after the last finishing stand, and after the last row of coiling sprays before coiling. (Data from Holmsten and Houis, 1990.)

properties have to be maintained, and usually improve during the process, such as tensile strength, yield strength, ductility, toughness, creep resistance, and fatigue life. To achieve this, thermomechanical treatments combine a controlled amount of plastic deformation with a controlled heating–cooling cycle. Due to the high speed of the strip (up to 1000 m per minute), an infrared line scanner is preferred over a 2D imager (moving the strip produces the second dimension). Moreover, a fast line scanner allows us to locate the points on the strip where measurements are made (this permits us, for example, to better adjust cooling sprays). Uncooled silicon CCD detectors (Figure 4.26) are candidates for such an application, although they have a limited dynamic response (a few hundred levels of information from 800 to 1300°C, especially if 8-bit digitalization is used. The response from a silicon detector drops sharply at about 1.1 μm, while blackbodies at 500 and 1000°C have 70% of the total emitted radiation in the bands at 3 to 7 μm and 2 to 9 μm, respectively (Figure 2.6), making the band 2 to 5 μm an ideal candidates for such applications (the band at 8 to 12 μm works with a small portion of the total energy available at these temperatures). Important features of the imager include measurable temperature difference in a wide temperature span without saturation (this makes a 12- or 14-bit system preferable), spatial resolution, speed, temperature measurement accuracy, width measurement capability, flexibility (e.g., possibility of implementing various spectral filters), and emissivity correction. Once the steel is solidified, one of the most serious difficulties for temperature measurement is the presence of scale (crust oxide) as a layer of variable thickness (Section 11.1.5). Software techniques such as peak detection and interpolation enable us to suppress scale pixels. Another important issue is the width measurement capability of the system. This refers to the slit response function introduced in Section 4.1.7. It is important to recall that if a manufacturer specifies a number of pixels per line (say, 2100), this generally refers to a modulation of 50%, and if 90% is required for accurate measurement with, say, 7 mrad at 90%, this will reduce the spot size measurement to [object distance × tg (7 mrad) ~ object distance × 7/1000], and consequently, the effective number of measurement points per line will be reduced. Finally, we should mention that recent technological advances make scanning technology reliable and durable (see, e.g., products from Servo-Robot, Flir, HGH), and calibration of line-scanning devices is easy (for single detector devices).

As an illustration of such an analysis, the ATL020 infrared line scanner from HGH is used to perform temperature measurements in such an environment. Specifications for the device are an SRF of 0.75 mrad (at $m = 50\%$) and 6.7 mrad (at $m = 90\%$), 2100 points (at $m = 50\%$), a 90° field of view, 6°C accuracy at 800°C, a scanning rate of 20 to 80 Hz, and thermoelectric cooling. Figure 11.64 shows the temperature profile of a steel strip with an M shape, which is caused by the presence of coarser grain in the middle and extreme edges of the strip. Surface temperature monitoring of the strip enables us to correct and evaluate coarse-grain presence and depth. Such a profile allows width measurements also. Similar temperature measurements can be made on float glass

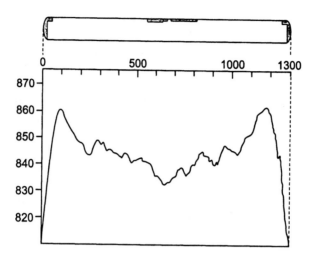

FIGURE 11.64 Micrographic versus thermal analysis: extreme edges and middle section of strip coarser-grain areas (shaded bar at top) correspond in the thermal profile as rapidly dropping temperature levels. Note also the good agreement between width measurement thermally and physically (ruler: width in millimeters). (Adapted from Holmsten and Houis, 1990.)

for thermal stress monitoring or for paper drying, plastic-film extrusion, paint drying, and so on.

11.4.4   Temperature Measurement of Glass (Bottles and Lamps)

In this section we described the use of passive thermography in glass inspection. In writing this section, the work of Wilson (1991) and Wallin (1994) was found to be particularly helpful, and we invite interested readers to refer to these documents, which include more information than it was possible to cover in a single section.

The purpose of this study is to determine the parameters necessary to improve glass distribution in a bottle and thus achieve weight reduction. A standard 12-ounce bottle weighs about 185 g, and the company wants to reduce the weight to 175 g and eventually to 170 g. The bottle-forming process is as follows. Glass is melted in a furnace (temperature about 1500°C) and channeled through a refiner to stabilize its temperature. From there it goes to a spout and orifice ring containing three holes. A predetermined amount of glass is fed through each hole and cut to form gobs, which are fed to a section machine, then transferred to the mold side, where they are blown into bottles (a *gob* is a certain quantity of hot glass ready for forming at about 1000°C). Contact measurement is not possible for this application, so thermography is used to analyze the cooling process and its associated variables. A better set of variables, such as mold changes, allows a more consistent temperature, yielding more even glass distribution, characterized by improved bottle strength, enabling a reduction in the total amount of glass necessary to form a bottle and making cost savings possible.

The problem with glass is that it is partially transparent to infrared radiation. However, it is known that most glasses do not transmit infrared radiation above 4.8 μm. On the other hand, glass reflectance is high for wavelengths over 8 μm, thus preventing the use of long waves (8 to 12 μm) for glass temperature measurement [a short-wavelength (2 to 5 μm) imager is required]. According to Kirchhoff's laws (Section 2.9.5), both transmittance and reflectance must be low to allow high emissivity, thus permitting temperature measurement. Accordingly, a glass filter with a middle wavelength set at 5 μm and a half-width of 145 nm is required for temperature measurement of glass whose emissivity (in this wavelength band) is 0.97.

In the lamp industry, it is of interest to measure both the temperature distribution of the envelope and the temperature of the structures located inside the lamp. The good transmission of glass (under 4.8 μm) allows such investigation. In this case, use of a 2.35-μm filter (half width of 100 nm) is reported. Knowledge of emissivities of the inner (metal) structures is also necessary to obtain accurate temperature figures. For example, the emissivity of tungsten is between 0.03 and 0.2, depending on the temperature. Another important factor to consider is the attenuation through the lamp shell, which can be measured experimentally by looking at a calibrated blackbody through the object with a reference image taken from the source directly. Comparing the two images, it is possible to calculate the transmission loss, the only difference with the real measurement being that the radiation is attenuated by two walls. The correct transmission value for measurement through one wall is the square root of the two-wall measured transmission value. Finally, the extreme temperature range to be covered (from room temperature to over 2000°C for a common household bulb) is challenging. Especially if the turn-on/turn-off cycle is to be measured, 12-bit direct digital output systems (72 dB dynamic range) are preferred.

11.4.5 Thermal Inspection of High-Temperature Industrial Structures

In this section we describe the use of passive thermography for thermal inspection of high-temperature industrial structures. For these applications, the structure inspected is already at a given temperature, and the temperature distribution recorded over its surface contains useful information regarding its operating regime or physical integrity.

More specifically, in this section we see how infrared thermography can be useful to help solve design problems for heat exchangers used to warm warehouses. Figure 11.65 shows such a unit, made of aluminum–steel elements internally heated by gas burners. The heat is transmitted to the ambient environment by forcing air to circulate between the elements by means of a fan located at the back face of the unit.

A thermal image of the unit yields a map of the heat distribution within the elements. An example of such a thermogram is shown in Figure 11.66. Hotter

FIGURE 11.65 Passive infrared thermography deployed to help designing heat exchanger units used to heat warehouses. (From Maldague, 1993.)

areas are visible on the center at a position determined by the geometry of the convection coefficient, which depends on the hot gas circulation inside the elements. A slight displacement of the maximum temperature distribution toward the right part of the unit is also visible in this thermogram. This is due to the counterclockwise rotation of the propeller fan.

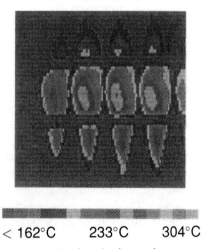

$< 162°C \qquad 233°C \qquad 304°C$

FIGURE 11.66 Thermal image showing the internal structure of the heat exchanger unit of Figure 11.65 during operation. (From Maldague, 1993.)

Many tests can be performed with the heat exchanger operating under various operation regimes and with different orientations of the vent panels (visible as cold structures in Figure 11.66). The heating rate during startup or shutdown can be analyzed as well, in order to specify optimum operating conditions, such as burner/fan turn-on cycles for both fast response and reduced thermal-shock fatigue of specific elements. In general, such an analysis is a very powerful tool to assist the designer in the conception of new units and to locate areas of extreme thermal sensitivity which require reinforcement or geometrical modification.

11.5 EVALUATION OF MATERIAL THERMAL DIFFUSIVITY

In this section we present methods to evaluate the thermal diffusivity of materials. Evaluation of material properties is a field of great interest for TNDT. This is an important subject for at least two reasons. First, in order to predict the behavior of a given component or solve the "inverse problem," it is necessary to model the actual experiments. Such modeling requires knowledge of the thermophysical properties of the materials being inspected.

On-site measurement of thermophysical material properties is thus advantageous, since it allows us to better simulate the thermal behavior of components. Values given in tables and handbooks [e.g., such as *Thermophysical Properties of Matter* by Touloukian and DeWitt (1970)] give only an estimation of such values, since material properties may change due to the variability of the fabrication processes. A more accurate figure is obtained by direct evaluation.

This brings us to another motivation for the determination of material properties. In fact, it can be shown that in some instances, measurement of one thermophysical property is closely related to the value of another material property. For example, Peralta et al. (1991) indicated there are consistent differences in the thermal diffusivity of high-purity aluminum specimens, depending on the degree of specimen recrystallization.

In the same way, Heath and Winfree (1989) have shown that in the through-ply direction, a roughly linear functional dependence exists between porosity and thermal diffusivity in carbon–carbon composites. This is an interesting result since the porosity of such components is reduced through repetitive processing cycles. (A processing cycle consists of immersion in phenolic resin followed by pyrolyzation: The resin penetrates the pores and upon pyrolysis is reduced to carbon residues, which fill the pores, thus reducing porosity.)

Thermal diffusivity δ (m^2 s^{-1}) is a good candidate for porosity assessment since it is related to mass density:

$$\alpha = \frac{k}{\rho c} \tag{11.8}$$

where k is the thermal conductivity (W m^{-1} °C^{-1}), ρ the density (kg m^{-3}), and c the specific heat (J m^{-3} °C^{-1}). Since diffusivity can be measured easily in a noncontact fashion through infrared thermography, it is thus an attractive technique to discuss and is the subject of this section. Finally, note that TNDT is not the only method for diffusivity measurement; others methods, such as photoacoustic microscopy, have been developed as well (see, e.g., Balageas et al., 1991; Zhang et al., 1995; Ahuja et al., 1996; Ouyang et al., 1998; Bison et al., 2000; Muscio et al., 2000).

11.5.1 Classical Thermal Diffusivity Measurement Method

Several methods have been developed to compute thermal diffusivity from standard infrared thermography experimentation (see below). Probably the best known technique was derived by W. J. Parker et al. (1961). The method consists of heating a sample and observing the temperature evolution of either the front or back face. If the back face is monitored, W. J. Parker et al. showed that for a sheet of thickness L, the time $t_{1/2}$ (this is the time for the back surface to rise to half its maximum value) can be expressed by (assuming a unidimensional heat flow within a semi-infinite medium, an acceptable hypothesis if experimental conditions reproduce such a scheme)

$$t_{1/2} = \frac{1.38L^2}{\pi^2 \alpha} \tag{11.9}$$

W. N. Reynolds and Wells (1984) and Hobbs (1991) computed $t_{1/2}$ for a wide range of materials. They concluded that, for example, in the case of a copper plate of 1 cm thickness, $t_{1/2}$ is only 0.13 s, whereas it is 33 s for a 1-cm-thick CFRP plate (measured perpendicular to the fibers). For high-diffusivity materials such as copper, this imposes a certain requirement on the acquisition apparatus, especially if conventional equipment is to be used.

The same analysis can be performed if the front surface is being observed. In this case, assuming a unidimensional heat flow within a semi-infinite isotropic medium and after absorption of a Dirac pulse [eq. (9.8)],

$$\Delta T = \frac{Q}{e(\pi t)^{1/2}} \tag{11.10}$$

where Q is the absorbed energy and e is the thermal effusivity [eq. (10.10)]. A plot of this curve will show a line of $-\frac{1}{2}$ slope on a log-log scale (Figure 9.2). For a finite-thickness sample the linear cooling decay is followed by a horizontal plateau (Figure 11.67). From this curve, time t^* can be derived. Time t^* corresponds to the point of intersection of the two asymptote lines in the temperature decay plot. It is linked to the diffusivity by (Delpech et al., 1990)

$$\alpha = \frac{L^2}{\pi t^*} \tag{11.11}$$

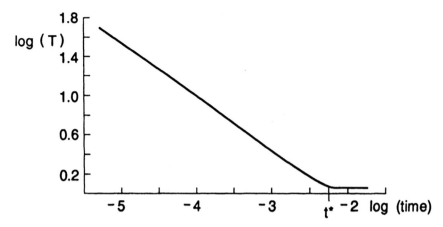

FIGURE 11.67   Temperature evolution curve for a finite thickness sample after pulse heating of the surface. (From Maldague, 1993.)

The main problem with diffusivity measurements using front [eq. (11.11)] and back [eq. (11.9)] surface approaches is to estimate time $t_{1/2}$ or t^*, especially if thermograms are noisy.

11.5.2   Diffusivity Measurement Method Based on the Laplace Transform

The Laplace transform (D. E. Johnson et al., 1992), well known in electric circuit analysis, has found new fields of applications in TNDT (see, e.g., Houlbert, 1991). Delpech et al. (1990) improved classical thermal diffusivity measurement significantly through introduction of the Laplace transform (after P. S. Laplace, French physicist, astronomer, and mathematician, 1749–1827). In fact, the Laplace transform, through the integration it implies, substantially reduces noise and authorizes finer diffusivity measurement.

In Laplace space, eq. (11.10) becomes

$$\bar{T} = \frac{Q\,\text{ch}(qL)}{kq\,\text{sh}(qL)} \tag{11.12}$$

with the same definition as in Section 11.5.1, with $q = \sqrt{s\alpha}$, where s is the Laplace variable. A plot of the normalized curve $s\bar{T}/s_0\bar{T}_0$ as a function of $L\sqrt{s}$ reveals two asymptotes (as in case of Figure 11.67 in the temporal domain), from which the intersection point can be found in s^* (s_0 is the limit value when s tends to zero as for T_0). Diffusivity is then given by

$$\alpha = (L\sqrt{s^*})^2 \tag{11.13}$$

Figure 11.68 illustrates the method. Since Laplace transform is obtained

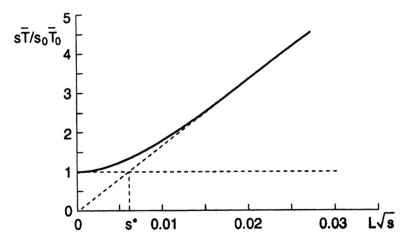

FIGURE 11.68 Temperature evolution curve in the Laplace transform domain (see the text). (From Maldague, 1993.)

through integration, a substantial reduction in the noise present in the thermograms is obtained. The reliability of the diffusivity evaluation is thus improved.

The Laplace transform of the thermogram $T(t)$ is obtained by numerical computation of the integral:

$$T(s) = \int_0^\infty T(t)e^{-st}\,dt \tag{11.14}$$

As specified in Delpech et al.'s paper, there is a problem in computing eq. (11.14) since the integral must be evaluated from time zero. This is a problem since the first thermogram acquisition starts at time Δt, which corresponds to one acquisition period. They solve the problem by separating eq. (11.14) into three terms (with t_{max} being the time for end of acquisition):

$$T(s) = \int_0^{t_i} T(t)e^{-st}\,dt + \int_{t_i}^{t_{max}} T(t)e^{-st}\,dt + \int_{t_{max}}^\infty T(t)e^{-st}\,dt \tag{11.15}$$

The second and third terms can be computed without difficulty. The first term can computed by adjusting the constant p ($t_i = p\Delta t$) so that the asymptote crosses the origin (Figure 11.69). In fact, the great sensibility to p of the transformed thermogram is observed in this figure.

To determine s^* with reliability (i.e., to determine the intersection point of the two asymptotes), it is necessary to evaluate $s\bar{T}_0$ close to the origin, in order to normalize the curve as shown on Figure 11.68. This can be done by performing a second-order fit for values close to the origin (Appendix C) and extrapolating the value $s\bar{T}_0$. Results reported for silicon carbide and magnesium–

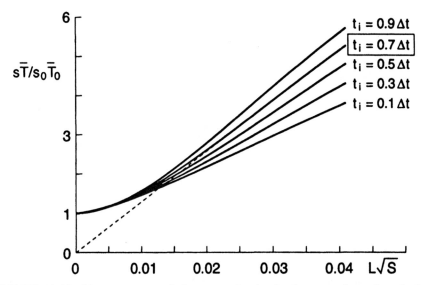

FIGURE 11.69 Temperature evolution curve in the Laplace transform domain for different values of the initial time t_i. (From Maldague, 1993.)

carbon composites reveal agreement with the classical method of Section 11.5.1 of about 6%. It is noted, however, that the Laplace transform method should provide more accurate findings since it is less sensitive to noise.

11.5.3 Diffusivity Measurement Method Based on Phase Measurement

Heath et al. (1989) developed an alternative noncontact method for thermal diffusivity measurements based on the phase delay between a thermal wave source (e.g., pulsed ND-YAG laser) and a spatially offset detection region. For in-plane measurements, temperature evolution is recorded on the sample surface in reflection (same side as the perturbation) using an infrared camera. The measurement setup is shown in Figure 11.70 Through-ply measurement is also possible by monitoring the temperature evolution on the back face. Part of the beam needed as a timing reference is directed on the back face with a beam splitter (Figure 11.71). A typical spatial profile of the point-source heating obtained by taking a cross section of a thermogram is plotted in Figure 11.72 (carbon–carbon composite material). Figure 11.73 shows a typical temperature evolution of a point spatially offset from the source.

Notice that stimulation performed with a Nd:YAG laser operating at 1.06 μm is advantageous since it is invisible for either a short-wave (3 to 5 μm) or a long-wave (8 to 12 μm) IR camera. Reported frequencies of heating pulse are 1 Hz for front-face measurement and 0.08 Hz for back-face measurement. Low diffusivity in the through-ply measurement is responsible for this low-frequency heating. The standard video rate of 60 or 50 Hz limits application

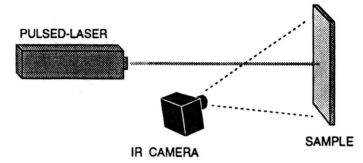

FIGURE 11.70 Experimental apparatus for in-plane diffusivity measurement. (From Maldague, 1993.)

of this method to low- to medium-diffusivity materials; superior frame rates enable higher-thermal-diffusivity materials to be tested.

If a sample is stimulated with a periodic heat source, Carlslaw and Jaeger (1959) demonstrate that observation of the temperature at a position sufficiently distant from the source approaches a linear function. The slope m of this line is then proportional to the thermal diffusivity of the material (Heath et al., 1989, eq. 1):

$$\alpha = \frac{\pi f}{m^2} \qquad (11.16)$$

where f is the frequency of the periodic heat source. For example, a typical plot of the phase delay as a function of the heating source distance is drawn in Figure 11.74. In this figure it is seen that the phase function becomes roughly linear for a certain distance to the source. At greater distances, the signal-to-

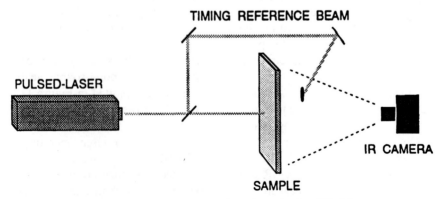

FIGURE 11.71 Experimental apparatus for through-ply diffusivity measurement. (From Maldague, 1993.)

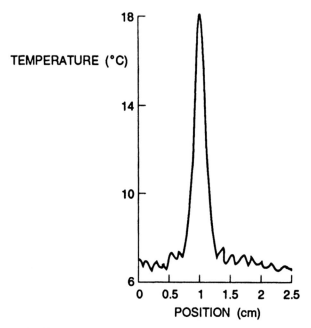

FIGURE 11.72 Spatial temperature distribution of the point source (in-plane measurement). (From Maldague, 1993.)

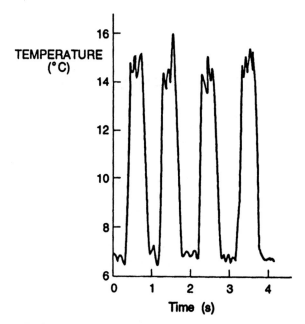

FIGURE 11.73 Typical temperature evolution of a point spatially offset from the source (in-plane measurement). (From Maldague, 1993.)

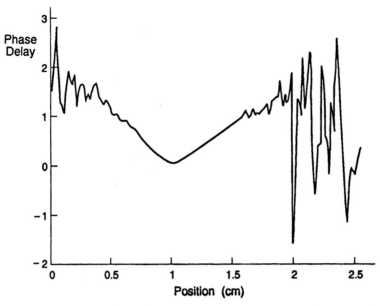

FIGURE 11.74   Typical phase delay plotted as a function of the heating source distance (in-plane measurement). (From Maldague, 1993.)

noise ratio (SNR) gets smaller and no measurement is possible. From this figure, the slope m of the line can be computed through standard fitting procedures and thermal diffusivity is obtained using eq. (11.16). Diffusivity values obtained using this phase-delay method are reported to be in good agreement with values obtained using other methods (such as the those of earlier sections).

Other interesting references on thermal diffusivity measurements are as follow: Bison et al., 2000; Muscio et al., 2000; Ouyang et al., 1998a; Poncet et al., 1998; Winfree et al., 1998.

PROBLEMS

11.1 **[heat transfer]** A piece of plastic 20 mm thick covers a 5-cm-long cylindrical connector for electrical insulation purposes. A temperature reading on the plastic indicates 30°C under steady-state conditions. A 200-A current is circulating through the copper connector (with $k = 386$ W m^{-1} °C^{-1}, electrical resistivity $\rho = 30$ μΩ · cm, and radius 3 cm). Figure 11.75 shows the geometry involved. Find the temperature at the copper–plastic interface.

Solution Practical heat transfer calculation principles described in Chapter 3 are useful here. Since the plastic is also a thermal insulator, a higher temperature was expected, as calculated below. All the power circulating through the connector is dissipated in the air by convection

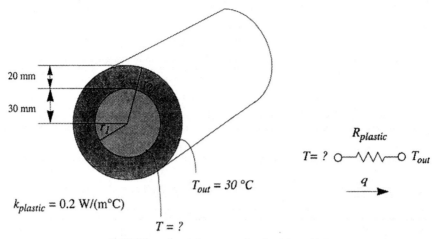

FIGURE 11.75   Geometry for Problem 11.1.

and radiation:

$$P = i^2 R = q$$

The electrical resistance of the connector is calculated as

$$R = \rho \frac{L}{A} = \frac{(30 \times 10^{-6}) \times 5}{\pi (3)^2} = 5.305 \times 10^{-6} \ \Omega$$

The power dissipated is thus

$$P = i^2 R = (200)^2 \times 5.305 \times 10^{-6} = 0.2122 \ \text{W}$$

Now let's calculate the radial conduction resistances, using eq. (3.24), for $L = 0.05$ m and with (from Figure 11.75) $r_1 = 0.03$ m and $r_2 = 0.05$ m:

$$R_{x_{\text{rad}}, \text{plastic}} = \frac{\ln(r_2/r_1)}{2\pi L k_{\text{plastic}}} = \frac{\ln(0.05/0.03)}{2\pi(0.05 \times 0.2)} = 8.13 °\text{C W}^{-1}$$

Now we can compute the heat transfer rate:

$$q_r = \frac{\Delta T}{R} = \frac{T - 30}{8.13} = 0.2122 \ \text{W}$$

and solving for T yields

$$T = 30 + (0.2122 \times 8.13) = 31.7°\text{C}$$

11.2 **[solar loading]** Two identical energized transformers (connected on the same phase) are inspected by TNDT. Transformer A is under direct solar illumination (for a long time) and gives a 75°C reading. Transformer B is in the shadows (for a long time) and gives a 35°C reading. The ambient temperature is 30°C. State your conclusion.

Solution   In this the temperature differential (ΔT) is $75 - 35 = 40$°C. However, under direct solar heating, such a reading is obviously distorted. Nevertheless, a 40°C differential is quite high, and it is thus recommended that it be reported as a potential problem.

11.3 **[wind effect]** An IR survey reports a ΔT value of 10°C on a transformer (50% loaded). The wind is 5 km h^{-1}. State your conclusion.

Solution   From Table 11.3 we obtain an empirical correction factor of 1.15. The ΔT value is thus corrected as $10 \times 1.15 = 11.5$°C. According to Table 11.2, this corresponds to a developed overheating that should be documented and fixed at the next scheduled maintenance.

11.4 **[load and wind effects]** A ΔT value of 10°C is obtained for equipment that is 10% loaded (with a wind speed of 7 m s^{-1}).

Solution   From Table 11.3 we obtain an empirical correction factor of 2.40. The ΔT value is thus corrected as $\Delta T_{\text{table}} = 10 \times 2.4 = 24$°C. Next, following eq. (11.5), we obtain a correction factor of $(50/10)^2 = 25$, and thus ΔT_{table} is calculated as $25 \times 2.4 = 60$°C, corresponding to a severe stage of overheating that should be fixed as soon as possible (Table 11.2).

11.5 **[R-value]** Compute the R-value for a window of $A = 1$ m^2, $\Delta T = 3.5$°C, and $q = 10$ W.

Solution   The R-value is expressed as

$$R\text{-value} = \frac{\Delta T}{q/A} = \frac{3.5}{10/1} = 0.35 \ (°\text{C W}^{-1}) \cdot \text{m}^2 = 1.98 \text{ ft} \cdot \text{h} \cdot °\text{F/Btu}$$

Such a value corresponds to a double-pane glass window, for which $R \sim 2$. Note that the conversion factor is $1 \text{ ft} \cdot \text{h} \cdot °\text{F/Btu} = 0.1762$ $(°\text{C W}^{-1}) \cdot \text{m}^2$.

References and Bibliography

This lists includes all references referred to in the book as well as many others, to cover as completely as possible what has been published on TNDT in recent years. This list is offered with search capabilities on the Web.*

Note: SPIE, Society of Photo-Interpretive Engineers, Bellingham, Washington; QIRT, quantitative infrared thermography.

Abel IR, "Radiometric accuracy in a forward looking infrared system," *Optical Engineering*, **16**[3]:241–248, 1977.

Abramova EV, Budadin ON, Panin VF, "Automated thermal NDT system applied to internal defects inspection of sheet rolled metal in manufactures," *Proc. QIRT-96, Eurotherm Seminar*, **50**:304–308, 1996.

Abutaleb AS, "Automatic thresholding of gray-level pictures using two-dimensional entropy," *Computer Vision, Graphics, and Image Processing*, **47**[1]:22–32, 1989.

Agam U, Gal E, Markevich N, Grimberg E, "Hot spot detection probability dependence on thermal imager parameters," in Spiro IJ, ed., *Infrared Technology XIV, SPIE Proc.*, **972**:188–194, 1988.

Agassi E, Yosef NB, "Effect of the thermal infrared data on principal component analysis of multi-spectral remotely-sensed data," *International Journal of Remote Sensing*, **19**[9]:1683–1694, 1998.

Aglan H, Shroff S, Abdo Z, Ahmed T, Wang L, Favro LD, Thomas RL, "Cumulative fatigue disbond of adhesive joints and its detection using thermal wave imaging," in Thompson DO, Chimenti DE, eds., *Review of Progress in Quantitative Nondestructive Evaluation*, Vol. 14, **14**:431–438, 1995.

Aguilera R, "256 × 256 hybrid Schottky focal plane arrays," in Buser RG, Warren FB, eds., *SPIE Proc. Infrared Sensors and Sensor Fusion*, **782**:108–113, 1986.

Ahmed T, Feng ZJ, Kuo PK, Hartikainen J, Jaarinen J, "Characterization of plasma sprayed coatings using thermal wave infrared video imaging," *Journal of Nondestructive Evaluation*, **6**[4]:169–175, 1987.

Ahmed T, Jin HJ, Chen P, Kuo PK, Favro LD, Thomas RL, "Real-time thermal-wave imaging of plasma-sprayed coatings and adhesive bonds using a box-car video tech-

---

* http://irndt.gel.ulaval.ca/

537

nique," in Murphy JC, Maclachlan-Spicer JW, Aamodt LC, Royce BSH, eds., *Photoacoustic and Photothermal Phenomena II*, **62**:30–32, 1989a.

Ahmed T, Kuo PK, Favro LD, Jin HJ, Thomas RL, Dickie RA, "Parallel thermal wave IR video imaging of polymer coatings and adhesive bonds," in Thompson DO, Chimenti DE, eds., *Review of Progress in Quantitative Nondestructive Evaluation*, **8A**: 1385–1392, 1989b.

Ahuja S, Ellingson WA, Stuckey J, Koehl ER, "Determining thermal diffusivity and defect attributes in ceramic matrix composites by infrared imaging," in Burleigh DD, Spicer JWM, eds., *Thermosense XVIII, SPIE Proc.*, **2766**:249–257, 1996.

Aksenov VP, Isaev YN, "Reconstruction of laser radiation intensity distribution from temperature along target surface," in Allen LR, ed., *Thermosense XV, SPIE Proc.*, **1933**:298–308, 1993.

Aksenov VP, Isaev YN, Zakharova EV, "Spatial–temporal reconstruction of laser beam intensity distribution from the temperature along surface of the heated target," in Burleigh DD, Spicer JWM, eds., *Thermosense XVIII, SPIE Proc.*, **2766**:346–356, 1996.

Albright GC, Stump JA, McDonald JD, Kaplan H, "True temperature measurements on microscopic semiconductor targets," in Lemieux DH, Snell JR, eds., *Thermosense XXI, SPIE Proc.*, **3700**:245–251, 1999.

Allport J, McHugh J, "Quantitative evaluation of transient video thermography," in Thompson DO, Chimenti DE, eds., *Review of Progress in Quantitative Nondestructive Evaluation*, **7A**:253–262, 1988.

Allred LG, "Identification of age degradation in EPROM chips using infrared thermography," in Snell JR, Wurzbach RN, eds., *Thermosense XX, SPIE Proc.*, **3361**: 35–39, 1998.

Allred LG, "Minimal entropy reconstructions of thermal images for emissivity correction," in Allred LG, Lemieux DH, Snell JR, eds., *Thermosense XXI, SPIE Proc.*, **3700**:417–424, 1999.

Allred LC, Howard TR, "Application of thermal imaging to electronic fault diagnosis," in Snell JR, Jr., ed., *Thermosense XVI, SPIE Proc.*, **2245**:224–230, 1994.

Allred LC, Jones MH, "Need for image processing in infrared camera design," in Dinwiddie RB, Lemieux DH, eds., *Thermosense XXII, SPIE Proc.*, **4020**:289–292, 2000.

Allred LG, Howard TR, Serpen G, "On-the-fly neural network construction for repairing F-16 flight control panel using thermal imaging," in Burleigh DD, Spicer JWM, eds., *Thermosense XVIII, SPIE Proc.*, **2766**:284–294, 1996.

Almond DP, Lau SK, "A quantitative analysis of pulsed video thermography," *Proc. QIRT-92, Eurotherm Seminar*, **27**:207–211, 1992.

Almond DP, Lau SK, "Edge effects and a method of defect sizing for transient thermography," *Applied Physics Letters*, **62**[25]:3369–3371, 1993.

Almond DP, Lau SK, "Defect sizing by transient thermography. I: An analytical treatment," *Journal of Physics D: Applied Physics*, **27**[5]:1063–1069, 1994.

Almond DP, Patel PM, "A quantitative thermal wave assessment of the characteristics of subsurface defects," *Proc. QIRT-92, Eurotherm Seminar*, **27**:367–370, 1992.

Almond DP, Patel PM, Reiter H, "The testing of plasma-sprayed coatings by thermal-wave interferometry," *Materials Evaluation*, **45**[4]:471–475, 1987.

Almond DP, Patel PM, Lau SK, "Edge-effects and a method of defect sizing for transient thermography," in Thompson DO, Chimenti DE, eds., *Review of Progress in Quantitative Nondestructive Evaluation*, **13A**:435–439, 1994a.

Almond DP, Saintey MB, Lau SK, "Edge-effects and defect sizing by transient thermography," *Proc. QIRT-94, Eurotherm Seminar* **42**, 247–252, 1994b.

Almond DP, Hamzah R, Delpech P, Peng W, Behesty MH, Saintey MB, "Experimental investigations of defect sizing by transient thermography," *Proc. QIRT-96, Eurotherm Seminar* **50**:233–238, 1996.

Alzofon FE, "An infrared nondestructive testing system for rocket motors," *Materials Evaluation*, **23**[11]:537–539, 1965.

Alzofon FE, McDonald AD, "Infrared evaluation of microweld quality," *Materials Evaluation*, **25**[8]:183–184, 1967.

Ammer K, "Reliability of thermal imaging in rheumatology and neuromuscular disorders," *Proc. QIRT-98, Eurotherm Seminar*, **60**:4–8, 1998.

Anbar M, "Dynamic area telethermometry and its clinical applications," in Semanovich SA, ed., *Thermosense XVII, SPIE Proc.*, **2473**:312–322, 1995a.

Anbar M, "Multiple wavelength infrared cameras and their biomedical applications," in Semanovich SA, ed., *Thermosense XVII, SPIE Proc.*, **2473**:323–331, 1995b.

Anderson CJ, "Automated infrared scanning in Cray Y-MP production," in Semanovich SA, ed., *Thermosense XII, SPIE Proc.*, **1313**:207–216, 1990.

Ångström, MAJ, "New method of determining the thermal conductibility of bodies," *Philosophical Magazine*, **25**:130–142, 1863.

Anthony S, Ramamoorthy PA, Grogan TA, "Median-filters-optical implementation using symbolic substitution," in Tescher AG, ed., *SPIE Proc. Applications of Digital Image Processing*, **829**:140–143, 1987.

Anttonen H, Kauppinen T, Lehmuskallio E, Rintamäki H, "Use of thermography in testing skin creams," in Snell JR, Jr., ed., *Thermosense XVI, SPIE Proc.*, **2245**:252–261, 1994.

Appleby R, Lettington AH, "Infrared and millimetre wave imaging," *IEE Colloquium on MM-Wave and IR Applications Digest*, **65**[11]:1–4, 1987.

Arconada A, Argiriou A, Papini F, Pasquetti R, "La mesure en thermographie infrarouge: calibration et traitement du signal," *Journal of Modern Optics*, **34**[10]:1327–1335, 1987.

Arpaci VS, *Conduction Heat Transfer*, Addison-Wesley, Reading, MA, 1966.

Ashauer M, Ende J, Glosch H, Haffner H, Hiltmann K, "Thermal characterization of microsystems by means of high-resolution thermography," *Microelectronics Journal*, **28**[3]:327–335, 1997.

Aspnes DE, Theeten JB, Hottier F, "Investigation of effective-medium models of microscopic surface roughness by spectroscopic ellipsometry," *Physical Review*, **B20**:3292–3302, 1979.

Astarita T, Cardone G, Carlomagno GM, "IR heat transfer measurements in a rotating channel," *Proc. QIRT-96, Eurotherm Seminar*, **50**:147–152, 1996.

Attorre DR, Parker JD, Williams NT, "Heat development in resistance seam welding of mild steel using continuous current," *Ironmaking and Steelmaking*, **23**[2]:136–149, 1996.

Auric D, Hanonge E, Kerrand E, de Miscault JC, "Thermal imaging system for material processing," in Kreutz E, Quenzer A, Schuocker D, eds., *SPIE Proc. High Power Lasers,* **801**:354–358, 1987.

Ayers WA, "Back to basics: nonintrusive infrared testing of high-voltage switchgear," *Materials Evaluation,* **49**[5]:561–565, 1991.

Badenas C, "Review and improvement of an algorithm for determining emissivity of a heterogeneous cavity in thermal infrared remote sensing," *International Journal of Remote Sensing,* **19**[4]:731–741, 1998.

Bahraman A, Chen CH, Genecko JM, Shelstad MH, Ting RN, Vodicka JG, "Current state of the art in InSb infrared staring imaging devices," in Caswell RS, ed., *SPIE Proc. Infrared Systems and Components,* **750**:27–31, 1987.

Bailey SJ, "Instruments connected to plant networks control process temperatures," *Control Engineering,* Jan., pp. 57–61, 1988.

Baird GS, "ASNT thermographer certification update," in Allen LR, ed., *Thermosense XV, SPIE Proc.,* **1933**:238–239, 1993.

Baird GS, Mack RT, "ASNT thermographic certification update," in Eklund JK, ed., *Thermosense XIV, SPIE Proc.,* **1682**:98–100, 1992.

Baker DC, Aggarwal JK, Hwang SS, "Geometry guided segmentation of outdoor scenes," in Trivedi MM, ed., *Applications of Artificial Intelligence VI, SPIE Proc.,* **937**:576–583, 1988.

Baker IM, Crimes G, Ard C, Jenner MD, Pearsons JE, Ballingall RA, Elliott CT, "Photovoltaic CdHgTe–silicon focal planes," in *Proc. 4th International Conference on Advanced Infrared Detectors and Systems,* London, June 1990.

Balageas DL, Levesque P, "EMIR: a photothermal tool for electromagnetic phenomena characterization," *Revue Générale de Thermique,* **37**[8]:725–739, 1998.

Balageas DL, Luc AM, "Transient thermal behavior of directional reinforced composites: applicability limits of homogeneous property model," *AIAA J.,* **24**[1]:109–114, 1986.

Balageas DL, Krapez JC, Cielo P, "Pulsed photothermal modeling of layered materials," *Journal of Applied Physics;* **59**[2]:348–357, 1986.

Balageas DL, Déom AA, Boscher DM, "Contrôle nondestructif des composites carbone-époxy par méthode photothermique impulsionnelle," *Revue Générale de Thermique Fr.* **301**[Jan.]:37–41, 1987a.

Balageas DL, Bosher DM, Déom AA, "Temporal moment in pulsed photothermal radiometry: application to carbon epoxy NDT," 5th International Topical Meeting on Photoacoustic and Photothermal Phenomena, Heidelberg, July 27–30, *Springer Series in Optical Sciences,* **58**:500–502, 1987b.

Balageas DL, Déom AA, Boscher DM, "Characterization and nondestructive testing of carbon-epoxy composites by a pulsed photothermal method," *Materials Evaluation,* **45**[4]:466–465, 1987c.

Balageas DL, Bosher DM, Déom AA, Fournier J, Henry R, "La thermographie infrarouge: un outil quantitatif à la disposition du thermicien," *Revue Genérale de Thermique Fr.,* **322**[Oct.]:501–510, 1988.

Balageas DL, Bosher DM, Déom AA, "Measurement of convective heat-transfer coefficients on a wind tunnel model by passive and stimulated infrared thermography," *Infrared Technology XVI, SPIE Proc.,* **1341**:19, 1990.

Balageas DL, Bosher DM, Déom AA, "Photoacoustic microscopy by photodeformation applied to the determination of thermal diffusivity," in Baird GS, ed., *Thermosense XIII, SPIE Proc.*, **1467**:278–289, 1991.

Balageas DL, Levesque P, Déom AA, "Characterization of electromagnetic fields using a lock-in infrared thermographic system," in Allen LR, ed., *Thermosense XV, SPIE Proc.*, **1933**:274–285, 1993.

Balageas DL, Levesque P, Nacitas M, Krapez JC, Gardette G, "Microwaves holography revealed by photothermal films and lock-in IR thermography: application to NDE of dielectric and radar absorbing materials," *Proc. QIRT-96, Eurotherm Seminar*, **50**:73–78, 1996.

Bales MJ, "Dynamic testing of rotary structures using phase-locked infrared imaging," in Snell, JR, Jr., ed., *Thermosense XVI, SPIE Proc.*, **2245**:131–134, 1994.

Bales MJ, Bishop CC, "Pulsed infrared imaging: a new NDT methodology for above-ground storage tanks," *Materials Evaluation*, **52**[7]:814–815, 1994.

Bales MJ, Bishop CC, "Corrosion/erosion detection in steel storage vessels using pulsed infrared imaging," in Semanovich, SA, ed., *Thermosense XVII, SPIE Proc.*, **2473**:295–297, 1995.

Balfour LS, "Infrared polarization thermometry using an imaging radiometer," *Proc. QIRT-94, Eurotherm Seminar*, **42**, 103–107, 1994.

Ball RJ, Almond DP, "The detection and measurement of impact damage in thick carbon fibre reinforced laminates by transient thermography," *NDT&E International*, **31**[3]:165–173, 1998.

Ballard DH, Brown CM, *Computer Vision*, Prentice Hall, Upper Saddle River, NJ, 523 p., 1982.

Ballingall RA, Review of infrared focal plane arrays, in Lettington AH, ed., *SPIE Proc. Infrared Technology and Applications*, **1320**:70–87, 1990.

Banerjee P, Chin BA, "Infrared sensor-based on-line weld penetration control," in Allen LR, ed., *Thermosense XV, SPIE Proc.*, **1933**:83–92, 1993.

Bantel T, Bowman D, Halase J, Kenve S, Krisher R, Sippel T, "Automated infrared inspection of jet engine turbine blades," in Kaplan H, ed., *Thermosense VIII, SPIE Proc.*, **581**:18–23, 1986.

Baranowski P, Mazurek W, Walczak RT, "The investigation of actual evapotranspiration with the use of thermography," *Proc. QIRT-98, Eurotherm Seminar*, **60**:272–277, 1998.

Barker E, Maldague X, Laurendeau D, "Object shape determination from a single infrared thermal image," in Snell JR, Jr., ed., *Thermosense XVI, SPIE Proc.*, **2245**:95–105, 1994.

Barker E, Maldague X, Laurendeau D, "Shape reconstruction from a single thermal image," *Optical Engineering*, **34**[Jan.]:154–159, 1995.

Barker KM, Knettel MG, "Moisture entrapment in buildings with EIFS and stucco facades in the state of Florida," in Lemieux DH, Snell JR, eds., *Thermosense XXI, SPIE Proc.*, **3700**:463–470, 1999.

Barnard KJ, Boreman GD, "Synthesis of infrared spectral signatures," *Optical Engineering*, **29**[3]:233–239, 1990.

Barnsley MF, Sloan AD, "A better way to compress images," *Byte*, Jan., pp. 215–223, 1988.

Barre JC, "Thermographie infrarouge et maintenance," *Achats et Entretien*, **405**[Mar.]: 42–48, 1988.

Barth K, Spring RW, "Certification of UAW/Ford thermographers," in Dinwiddie RB, Lemieux DH, eds., *Thermosense XXII, SPIE Proc.*, **4020**:91–93, 2000.

Batsale JC, Mourand D, Gobbé C, "Estimation of thermophysical properties of thin plates with averaging techniques and two temperature model," *Proc. QIRT-96, Eurotherm Seminar*, **50**:46–51, 1996.

Bauer M, Guntrum Ch, Ota M, Rippel W, Busse G, "Thermographic characterisation of defects and failure in polymer composites," *Proc. QIRT-92, Eurotherm Seminar*, **27**:141–144, 1992.

Baughman SR, "Applications for thermal NDT on advanced composites in aerospace structures," in Snell JR, Wurzbach RN, eds., *Thermosense XX, SPIE Proc.*, **3361**:311–319, 1998.

Baughn TV, Johnson DB, "A method for quantitative characterization of flaws in sheets by use of thermal-response data," *Materials Evaluation*, **44**[June]:850–858, 1986.

Beaudoin JL, Bissieux C, "Theoretical aspects of the infrared radiation," in Maldague XPV, ed., *Infrared Methodology and Technology*, International Advances in Nondestructive Testing Monograph Series, Chap. 2, pp. 9–51, Gordon and Breach, New York, 1994.

Beaudoin JL, Merienne E, Dartois R, Danjoux R, "Quantitative photothermal imaging: a new method for the non-invasive characterization of thermally thin and layered materials," *Proc. 5th Infrared Information Exchange*, New Orleans, LA, pp. 29–31, Oct. 11–18, 1985.

Beaudoin JL, Henry JF, Mérienne E, Danjoux R, Egée P, Coste P, "Energy deposit by a xenon flash lamp illuminator dedicated to photo thermal IR thermography: calculations and measurements," *Proc. QIRT-92, Eurotherm Seminar*, **27**:243–250, 1992.

Bednarczyk S, Hervé P, Adam P, "Analysis of atmospheric pollution by quantitative infrared thermography," *Proc. QIRT-92, Eurotherm Seminar*, **27**:319–325, 1992.

Bedrossian J, Slazas P, "Practical applications of infrared thermometry," *Sensors*, Jan., pp. 24–31, 1987.

Beghdadi A, Le Negrate A, "Contrast enhancement techniques based on local detection of edges," *Computer Vision, Graphics, and Image Processing*, **46**[2]:162–174, 1989.

Bein BK, Gu JH, Mensing A, Sommer T, Wunderlich B, Pelzl J, Seidel U, "Modulated infrared radiometry of rough surfaces at high temperatures," *Proc. QIRT-92, Eurotherm Seminar*, **27**:393–399, 1992.

Bell IG, "A high performance infrared thermography system based on class II thermal imaging common modules," in Baird GS, ed., *Thermosense XIII, SPIE Proc.*, **1467**:438–447, 1991.

Bell JF, Hamilton-Brown R, Hawke BR, "Composition and size of Apollo asteroid 1984 KB," *Icarus*, **73**[3]:482–486, 1988.

Bell ZW, "An image segmentation algorithm for non film radiography," in Thompson DO, Chimenti DE, eds., *Review of Progress in Quantitative Nondestructive Evaluation*, **6A**:773–778, 1987.

Bell ZW, "Evaluation of thresholding heuristics useful for automated filmless radiography," in Thompson DO, Chimenti DE, eds., *Review of Progress in Quantitative Nondestructive Evaluation*, **7A**:739–746, 1988.

Ben J, Kröse A, van der Smagt P, *An Introduction to Neural-Network*, 7th ed., Chaps. 3–5, University of Amsterdam, Faculty of Mathematics and Computer Science, 1995.

Bendada A, Maillet D, Degiovanni A, "Nondestructive transient thermal evaluation of laminated composites: discrimination between delaminations, thickness variations and multidelaminations," *Proc. QIRT-92, Eurotherm Seminar*, **27**:218–223, 1992.

Beniger JR, Robyn DL, "Quantitative graphics in statistics: a brief history," *American Statistician*, **32**[1]:1–11, 1978.

Benkö I, "Applications of infrared thermogrammetry in thermal engineering," *Proc. QIRT-92, Eurotherm Seminar*, **27**:343–352, 1992.

Benkö I, Köteles GJ, Németh G, "Thermal imaging of the effects of beta-irradiation on human body surface," *Proc. QIRT-96, Eurotherm Seminar*, **50**:354–362, 1996.

Benkö I, Köteles GJ, Németh G, "New infrared histographic investigation of the effects of beta-irradiation in medical field," *Proc. QIRT-98, Eurotherm Seminar*, **60**:40–45, 1998.

Bentz DP, Martin JW, "Thermographic imaging of surface finish defects in coatings on metal substrates," *Materials Evaluation*, **50**[2]:242–246, 252, 1992.

Berardi PG, Cuccurullo G, "A procedure to measure thermal conductivities of anisotropic laminates by infrared thermography," *Proc. QIRT-94, Eurotherm Seminar*, **42**:81–85, 1994.

Berardi PG, Cuccurullo G, Di Maio L, "Thermography for polymers film blowing," *Proc. QIRT-98, Eurotherm Seminar*, **60**:265–271, 1998.

Berger H, "Characteristics of a thermal neutron television imaging system," *Materials Evaluation*, **24**[9]:475–481, 1966.

Berger H, "100 years of x-rays: industrial use of NDT and the role of ASNT," *Materials Evaluation*, **53**[11]:1253–1260, 1995.

Berger H, Kraska IR, "Characteristics of an infrared vidicon television system," *Materials Evaluation*, **24**[4]:197–200, 1966.

Bergstrom L, Baeu D, "Nondestructive testing by high-speed thermography," *Materials Evaluation*, **27**[5]:25A–28A, 1969.

Bernascolle P, Pelous G, Grenier L, "Remote detection of natural gas clouds in open fields by IR imagery," in Lemieux DH, Snell JR, eds., *Thermosense XXI, SPIE Proc.*, **3700**:417–424, 1999.

Berry SA, Bouslog SA, Brauckmann GJ, Caram JM, "Shuttle Orbiter experimental boundary-layer transition results with isolated roughness," *Journal of Spacecraft and Rockets*, **35**[3]:241–248, 1998.

Berz R, "Regulation thermography: a survey," *Proc. QIRT-98, Eurotherm Seminar*, **60**:18–25, 1998.

Bevington PR, *Data Reduction and Error Analysis for the Physical Sciences*, McGraw-Hill, New York, 336 pp., 1969.

Beynon TGR, "Radiation thermometry applied to the development and control of gas turbine engines," in Schooley JF, ed., *Temperature: Its Measurement and Control in Science and Industry*, Vol. 5, part 1, pp. 471–477, American Institute of Physics, New York, 1982.

Biagioni PA, Lamey PJ, "Electronic infrared thermography as a method of assessing herpes labialis infection," *Acta Dermato-Venereologica*, **75**[4]:264–268, 1995.

Bielecki Z, Chrzanowski K, Matyszkiel R, Piatkowski T, Szulim M, "Infrared pyrometer for temperature measurement of objects, emissivity of which depends on wavelength and time," *Proc. QIRT-98, Eurotherm Seminar,* **60**:316–321, 1998.

Bieman LH, "Three-dimensional machine vision," *Photonics Spectra,* May, pp. 81–85, 1988.

Biermann R, "Thermocouples," *Standardization News,* **16**[5]:40–42, 1988.

Bingley B, "Ergonomics: rules of work," *Photonics Spectra,* Aug. 1988.

Bison PG, Bressan C, Grinzato EG, Marinetti S, Vavilov VP, "Active thermal testing of moisture in bricks," in Allen LR, ed., *Thermosense XV, SPIE Proc.,* **1933**:207–214, 1993.

Bison PG, Bressan C, Di Sarno R, Grinzato E, Marinetti S, Manduchi G, "Thermal NDE of delaminations in plastic materials by neural network processing," *Proc. QIRT-94, Eurotherm Seminar,* **42**:214–219, 1994a.

Bison PG, Grinzato EC, Marinetti S, "Moisture evaluation by dynamic thermography data modeling," in Snell JR, Jr., ed., *Thermosense XVI, SPIE Proc.,* **2245**:176–182, 1994b.

Bison PG, Braggiotti A, Bressan C, Grinzato EG, Marinetti S, Mazzoldi A, Vavilov V, "Crawling spot thermal nondestructive testing (NDT) for plaster inspection and comparison with dynamic thermography with extended heating," in Semanovich SA, ed., *Thermosense XVII, SPIE Proc.,* **2473**:53–66, 1995.

Bison PG, Dezzi Bardeschi M, Grinzato EC, Kauppinen T, Rosina E, Tucci G, "Survey of facades based on thermal scanning: Palazzo della Ragione, Milan, Italy," in Burleigh DD, Spicer JWM, eds., *Thermosense XVIII, SPIE Proc.,* **2766**:55–64, 1996.

Bison PG, Bressan C, Cavaccini G, Ciliberto A, Crinzato EC, "NDE of composite materials by thermal method and shearography," in Wurzbach RN, Burleigh DD, eds., *Thermosense XIX, SPIE Proc.,* **3056**:220–229, 1997.

Bison PG, Marinetti S, Manduchi G, Grinzato E, "Improvement of neural networks performances in thermal NDE," in Maldague X, ed., *Advances in Signal Processing for Nondestructive Evaluations of Materials,* 3rd Quebec Workshop (1997), Quebec City, American Society of Nondestructive Testing Press, *TONES,* **3**:221–227, 1998.

Bison PG, Muscio A, Grinzato EG, "Thermal parameter estimation by heating and cooling and by thermographic measurement," in Lemieux DH, Snell JR, eds., *Thermosense XXI, SPIE Proc.,* **3700**:402–408, 1999.

Bison PG, Grinzato EG, Marinetti S, Muscio A, "Diffusivity measurement of thick samples by thermography and heating–cooling technique," in Dinwiddie RB, Lemieux DH, eds., *Thermosense XXII, SPIE Proc.,* **4020**:137–142, 2000.

Bjornsson S, Arnason K, "State of the art of aerial thermography in Iceland: thermography of geothermal areas during the past 20 years," in Semanovich SA, ed., *Thermosense XVII, SPIE Proc.,* **2473**:2–14, 1995.

Blackwell J, Botts S, Laband A, Arnold H, "An affordable 128 × 128 InSb hybrid focal plane array," in Spiro IJ, ed., *Infrared Technology XV, SPIE Proc.,* **1157**:243–249, 1989.

Blais F, Rioux M, Beraldin JA, "Practical considerations for a design of a high precision 3-D laser scanner system," *SPIE Proc. Optomechanical and Electro-optical Design of Industrial Systems,* **959**:225–246, 1988.

Blatt FJ, "Hall effect," in Parker SP, ed., *Concise Encyclopedia of Science and Technology,* pp. 829–830, McGraw-Hill, New York, 1984.

Blazquez CH, "Detection of problems in high-power voltage transmission and distribution lines with an infrared scanner/video system," in Snell JR, Jr., ed., *Thermosense XVI, SPIE Proc.*, **2245**:27–33, 1994.

Bobo S, "Recent developments in use of infrared for nondestructive tire testing," *Materials Evaluation*, **32**[7]:147–152, 1974.

Boccara AC, Fournier D, Guitonny J, Le Liboux M, Mansanares AM, "Differential stimulated infrared radiometry: application to remote detection of cracks," *Proc. QIRT-92, Eurotherm Seminar*, **27**:382–387, 1992.

Bodnar JL, Egée M, Menu C, Pigeon M, Le Blanc A, "Détection de fissures par radiométrie photothermique sour excitation sinusoïdale," *Revue Pratique de Contrôle Industriel*, **179**:50–54, 1993.

Bogomol'nyi VM, "Dynamic theory of electrothermal degradation and NDT of defects in metal–dielectric–metal (MDM) structures," in Lemieux DH, Snell JR, eds., *Thermosense XXI, SPIE Proc.*, **3700**:436–443, 1999.

Boivin D, Laurendeau D, Comeau F, Richards C, "A computer-vision based apparatus for the measurement of planar movement: an application in physiotherapy," *Proc. Vision Interface '89*, pp. 37–44, 1989.

Boizumault F, Harmand S, Desmet B, "Experimental determination of the local heat transfer coefficient on a thermally thick wall downstream of a backward-facing step," *Proc. QIRT-96, Eurotherm Seminar*, **50**:141–146, 1996.

Bolte J, Bein BK, Pelzl J, "IR detection of thermal waves: effect of imaging conditions on the background fluctuation limit," *Proc. QIRT-96, Eurotherm Seminar*, **50**:9–14, 1996a.

Bolte J, Simon K, Pelzl J, "Separation of thermal and optical properties and error limits in thermal depth profiling of fibre-reinforced materials," *Proc. QIRT-96, Eurotherm Seminar*, **50**:245–250, 1996b.

Bonnet M, Maigre H, Manaa M, "Numerical reconstruction of interfacial defects and interface thermal resistances using thermal measurements," *Proc. QIRT-92, Eurotherm Seminar*, **27**:266–271, 1992.

Boogaard J, "Need and necessity of NDT," in Maldague XPV, ed., *Infrared Methodology and Technology*, International Advances in Nondestructive Testing Monograph Series, Chap. 1, pp. 3–7, Gordon and Breach, New York, 1994.

Boras I, Svaic S, Galovic A, "Mathematical model for simulation of defects under a material surface applied to thermographic measurements," *Proc. QIRT-98, Eurotherm Seminar*, **60**:53–58, 1998.

Bosher DM, Balageas DL, Déom AA, Gardette G, "Nondestructive evaluation of carbon epoxy laminates using transient infrared thermography," *Proc. 19th Symposium on Nondestructive Evaluation*, NTIAC, San Antonio, TX, 1988.

Bosher DM, Baudouy B, Déom AA, Coet MC, Delery J, Balageas D, "Infrared thermography characterization of Gortler vortex type patterns in hypersonic flows," *Proc. QIRT-92, Eurotherm Seminar*, **27**:92–97, 1992.

Bouchardy AM, Durand G, Gauffre G, "Processing of infrared thermal images for aerodynamic research," *Report ONÉRA*, tp 1983-32, SPIE, Geneva, Apr. 18–22 6 pp., 1983.

Bougeard D, Vermeulen JP, Baudoin B, "Spatial resolution enhancement of an IR system by image restoration techniques," *Proc. QIRT-94, Eurotherm Seminar*, **42**:3–8, 1994.

Bouvier CG, "Investigating variables in thermographic composite inspections," *Materials Evaluation*, **53**[5]:544–551, 1995.

Bow ST, *Pattern Recognition*, Marcel Dekker, New York, 1984.

Bowen MW, Osiander R, Spicer JWM, Murphy JC, "Thermographic detection of conducting contaminants in composite materials using mirowave excitation," in Thompson DO, Chimenti DE, eds., *Review of Progress in Quantitative Nondestructive Evaluation*, **14**:453–460, 1995.

Bracewell RN, "The fast Hartley transform," *Proc. IEEE*, **72**[8]:1010–1018, 1984.

Brady GS, Clauser HR, *Materials Handbook*, 12th ed., McGraw-Hill, New York, 1038 pp., 1986.

Brady JD Jr, "Mobile infrared gas analyzer for gas transmission line leakage surveys," *NDT&E International*, **20**[5]:332–334, 1962.

Brady RP, Kulkarni MR, "Determination of thermal diffusivity distribution for three types of materials by transient thermography," *NDT&E International*, **29**[4]:205–211, 1996.

Braim SP, Cuthbertson GM, "A dual waveband imaging radiometer," *Infrared Technology XII, SPIE Proc.*, **685**:129–137, 1986.

Breitenstein O, Iwig K, Konovalov I, Wu D, "Lock-in contact thermography on solar cells: comparison with IR measurements," *Proc. QIRT-96, Eurotherm Seminar*, **50**:383–388, 1996.

Breitenstein O, Iwig K, Konovalov I, "Evaluation of local electrical parameters of solar cells by dynamic (lock-in) thermography," *Physica Status Solidi (A) Applied Research*, **160**[1]:271–282, 1997.

Breiter R, Cabanski WA, Ziegler I, Walther M, Schneider H, "High-performance focal plane array modules for research and development," in Dinwiddie RB, Lemieux DH, eds., *Thermosense XXII, SPIE Proc.*, **4020**:257–266, 2000.

Bremond P, "La thermographie infrarouge pour voir les contraintes," *Mesures*, **673**:53–56, 1995.

Brownson J, Gronokowski K, Meade E, "Two-color imaging radiometry for pyrotechnic diagnostics," in Madding RP, ed., *Thermosense IX, SPIE Proc.*, **780**:194–201, 1987.

Bruckstein AM, "On shape from shading," *Computer Vision, Graphics, and Image Processing*, **44**[2]:139–154, 1988.

Bruening RJ, Mordfin L, "Infrared thermography standards for nondestructive testing," in Snell JR, Jr., ed., *Thermosense XVI, SPIE Proc.*, **2245**:220–223, 1994.

Brüggemann G, Demus J, Benziger Th, "Simultaneous in-process control of weld pool geometry and heat affected zone based on thermal-optic imaging for welding by concentrated energy fluxes," *Proc. QIRT-96, Eurotherm Seminar*, **50**:309–314, 1996.

Buchanan RA, Condon P, Klynn L, "Recent advances in digital thermography for nondestructive evaluation," in Semanovich SA, ed., *Thermosense XII, SPIE Proc.*, **1313**:134–142, 1990.

Buchlin JM, "Natural and forced convective heat transfer on slender cylinders," *Revue Générale de Thermique*, **37**[8]:653–660, 1998.

Buchlin JM, Dubois M, "Heat transfer of impinging multijet system: an appilcation of the quantitative thermography," *Proc. QIRT-92, Eurotherm Seminar*, **27**:117–120, 1992.

Buchlin JM, Laperches M, "Detailed investigation of aerothermal behaviour of confined impinging jet," *Proc. QIRT-98, Eurotherm Seminar,* **60**:258–264, 1998.

Buchlin JM, Meyers M, "Infrared thermography study of a confined impinging circular jet," *Proc. QIRT-96, Eurotherm Seminar,* **50**:159–164, 1996.

Buchlin JM, Tasse R, "Convective heat transfer along slender cylinders," *Proc. QIRT-96, Eurotherm Seminar,* **50**:129–134, 1996.

Buchlin JM, Prétrel H, Planquart H, Langer H, Thiry F, "Infrared thermography study of a thermal anti-icing system," *Proc. QIRT-94, Eurotherm Seminar,* **42**:133–138, 1994.

Buczylko K, Wiecek B, Chwala C, "Application of thermography for evaluation of the allergen provocation," *Proc. QIRT-98, Eurotherm Seminar,* **60**:9–14, 1998.

Bumbaca F, Smith KC, "A practical approach to image registration for computer vision," *Computer Vision, Graphics, and Image Processing,* **42**:20–23, 1988.

Burch SF, "Digital enhancement of video images for NDT," *Nondestructive Testing International,* **20**[1]:51–56, 1987.

Burch SF, Burton JT, Cocking SJ, "Detection of defects by transient thermography: a comparison of predictions from two computer codes with experimental results," *British Journal of Non-Destructive Testing,* Jan., pp. 36–39, 1984.

Burger C, Babak R, "Nondestructive evaluation through transient thermographic imaging," *Proc. 15th Symposium on Nondestructive Evaluation,* San Antonio, TX, Apr., **23–25**:56–67, 1985.

Burgess DE, Manning PA, Watton R, "The theoretical and experimental performance of a pyroelectric array imager," *Infrared Technology XI, SPIE Proc.,* **572**:2–6, 1985.

Burgess D, Nixon R, Ritchie J, "The pyroelectric vidicon: ten years on," *Infrared Technology XII, SPIE Proc.,* **685**:100–107, 1986.

Burleigh D, "A bibliography of NDT of composite materials performed with infrared thermography and liquid crystals," *Thermosense IX, SPIE Proc.,* **780**:250–255, 1987.

Burleigh D, "Bibliography of the application of infrared thermography to electronic and microelectronic circuits," *Thermosense X, SPIE Proc.,* **934**:99–100, 1988a.

Burleigh D, "Bibliography of the application of infrared thermography to welding," *Thermosense X, SPIE Proc.,* **934**:190–193, 1988b.

Burleigh DD, "Practical aspects of thermal nondestructive testing," in Burleigh DD, Spicer JWM, eds., *Thermosense XVIII, SPIE Proc.,* **2766**:158–163, 1996a.

Burleigh DD, "Tech tip: practical (nontechnical) aspects of NDT, using thermographic NDT as an example," *Materials Evaluation,* **54**[11]:1266–1269, 1996b.

Burleigh DD, "Thermographic testing used on the X-33 space launch vehicle program by BF Goodrich Aerospace," in Lemieux DH, Snell JR, eds., *Thermosense XXI, SPIE Proc.,* **3700**:84–92, 1999.

Burleigh D, De La Torre W, "Thermographic analysis of anisotropy in the thermal conductivity of composite materials," in Baird GS, ed., *Thermosense XIII, SPIE Proc.,* **1467**:303–310, 1991.

Burleigh DD, Kuhns DR, Cowell SD, Engel JE, "Thermographic nondestructive testing (TNDT) of honeycomb composite structural parts of Atlas space launch vehicles," in Snell JR, Jr., ed., *Thermosense XVI, SPIE Proc.,* **2245**:152–163, 1994.

Burton M, Benning C, "Comparison of imaging infrared detection algorithms," in Narendra PM, ed., *SPIE Proc. Infrared Technology for Target Detection and Classification,* **302**:26–32, 1981.

Büscher KA, Wiggenhauser H, Wild W, "Amplitude sensitive modulation-thermography the new way of moisture measurement in building materials," *Proc. QIRT-98, Eurotherm Seminar*, **60**:184–190, 1998.

Busse G, "Nondestructive evaluation of polymer materials," *NDT&E International*, **27**[5]:253–262, 1994.

Busse G, Eyerer P, "Thermal wave remote and nondestructive inspection of polymers," *Applied Physics Letters*, **43**[4]:355–357, 1983.

Busse G, Bauer M, Rippel W, Wu D, "Lockin vibro thermal inspection of polymer composites," *Proc. QIRT-92, Eurotherm Seminar*, **27**:154–159, 1992a.

Busse G, Wu D, Karpen W, "Thermal wave imaging with phase sensitive modulated thermography," *Journal of Applied Physics*, **71**[8]:3962–3965, 1992b.

Büyüköztürk O, "Imaging of concrete structures," *NDT&E International*, **31**[4]:233–243, 1998.

Byler WH, Hays FR, "Fluorescence thermography," *NDT&E International*, **9**[3]:177–180, 1961.

Calais E, House WR, "IR imaging cuts industry losses," *Photonics Spectra*, **24**[6]:87–96, 1990.

Campanelli RW, Engblom JJ, "Effect of delaminations in graphite/PEEK composite plates on modal dynamic characteristics," *Composite Structures*, **31**[3]:195–202, 1995.

Cardone G, "Quantitative infrared thermography in thermo-fluid-dynamics," *Proc. QIRT-98, Eurotherm Seminar*, **60**:8–22, 1998.

Cardone G, Astarita T, Carlomagno GM, "IR heat transfer measurements on a rotating disk," *Proc. QIRT-94, Eurotherm Seminar*, **42**:146–151, 1994a.

Cardone G, de Luca L, Astarita T, Aymer de la Chevalerie D, Fonteneau A, "IR measurements of hypersonic viscous interaction," *Proc. QIRT-94, Eurotherm Seminar*, **42**:139–145, 1994b.

Cardone G, Astarita T, Carlomagno GM, "Heat transfer in a 180 deg turn channel," *Proc. QIRT-96, Eurotherm Seminar*, **50**:123–128, 1996.

Cardone G, Astarita T, Carlomagno GM, "Wall heat transfer in static and rotating 180° turn channels by quantitative infrared thermography," *Revue Générale de Thermique*, **37**[8]:644–652, 1998a.

Cardone G, Meola C, Carlomagno GM, "Convective heat transfer to a jet in crossflow," *Proc. QIRT-98, Eurotherm Seminar*, **60**:278–283, 1998b.

Carlomagno GM, Berardi PG, "Unsteady thermotopography in non-destructive testing," in Warren C, ed., *Proc. 3rd Biennial Infrared Information Exchange, IRIE '76*, pp. 33–40, 1976.

Carlslaw HS, Jaeger JC, *Conduction of Heat in Solids*, 2nd ed., Oxford University Press, New York, 1959.

Carlson B, "Pushing the medical and industrial digital x-ray image envelope—at last," *Advanced Imaging*, Nov., pp. 24–29, 1997.

Carter L, "Verification of plugged nozzles in reactor-building spray systems," in Allen LR, ed., *Thermosense XV, SPIE Proc.*, **1933**:36–42, 1993.

Cashdollar KL, Hertzberg M, "Infrared temperature measurements of gas and dust explosion," in Schooley JF, ed., *Temperature: Its Measurement and Control in Science and Industry*, Vol. 5, part 1, pp. 453–462, American Institute of Physics, New York, 1982.

Castro EH, Pfefferman JD, Gonzalez HJ, "Thermal studies from images obtained with a thermal scanning system," in Wurzbach RN, Burleigh DD, eds., *Thermosense XIX, SPIE Proc.*, **3056**:41–49, 1997.

Castro EH, Pfefferman JD, Gonzalez HJ, "Thermal analysis of images obtained with an airborne IR scanner in Argentina," in Snell JR, Wurzbach RN, eds., *Thermosense XX, SPIE Proc.*, **3361**:169–176, 1998.

Castro EH, Leone EO, Costanzo M, Luca R, "Thermographic study of an electrical distribution station," in Lemieux DH, Snell JR, eds., *Thermosense XXI, SPIE Proc.*, **3700**:273–282, 1999.

Cernuschi F, Lamperti M, Marchesi R, Russo A, "Characterization of spatial light distribution of flash lamp systems," *Proc. QIRT-96, Eurotherm Seminar*, **50**:15–19, 1996a.

Cernuschi F, Russo A, Piana GM, Mutti P, Viviani L, "Emissivity measurements at room temperature on polymeric and inorganic samples," *Proc. QIRT-96, Eurotherm Seminar*, **50**:41–45, 1996b.

Cernuschi F, Ludwig N, Teruzzi P, "Statistics-based procedure for defect sizing and experimental evaluation of convection phenomena by using video pulsed thermography," *Proc. QIRT-98, Eurotherm Seminar*, **60**:30–35, 1998a.

Cernuschi F, Ludwig N, Teruzzi P, Bottazzi G, "A critical analysis and possible modifications of two analytical models for defects sizing using video pulse thermography," *Proc. QIRT-98, Eurotherm Seminar*, **60**:205–211, 1998b.

Cesini C, Paroncini M, Ricci R, "Thermal distribution in circular slabs: a thermographic method," *Proc. QIRT-92, Eurotherm Seminar*, **27**:272–277, 1992.

Cesini G, Moro V, Ricci R, "A quantitative thermographic investigation of cooling of power electronic sources by forced convection cold plates," *Proc. QIRT-96, Eurotherm Seminar*, **50**:165–170, 1996.

Chambers JB, "Thermographic evaluation of bond lines and materials consistency of composites," *Thermosense VII, SPIE Proc.*, **520**, 1984.

Chapman CB II, "Some automotive applications of infrared thermography for quality improvements," in Wurzbach RN, Burleigh DD, eds., *Thermosense XIX, SPIE Proc.*, **3056**:117–123, 1997.

Chapnick NN, Fagin LL, "Infrared thermography for diagnostic evaluation of electronic modules," *Materials Evaluation*, **25**[7]:164–168, 172, 1967.

Charbonnier F, Lepoutre F, Roger JP, Lemoine A, Robert P, "The 'mirage' sensor in a industrial environment: optical and thermal losses determinations," in Thompson DO, Chimenti DE, eds., *Review of Progress in Quantitative Nondestructive Evaluation*, **8A**:1105–1110, 1989.

Charbonnier MJ, Lapostolle S, Simeonides G, "Comparative application of two infrared scanners to heat transfer measurements in a Mach 6 wind tunnel," *Proc. QIRT-94, Eurotherm Seminar*, **42**:152–157, 1994.

Charest A, "L'avenir est aux composites," *Plan* (magazine of the Professional Engineers Association of Québec, Canada), Mar., pp. 25–28, 1988.

Chatard JP, Angebault P, Tribolet P, "Sofradir infrared detectors," *Proc. QIRT-92, Eurotherm Seminar*, **27**:46–50, 1992.

Chen CH, ed., *Signal Processing Handbook*, Marcel Dekker, New York, 818 pp., 1988.

Chen J, Barrow R, "Novel applications of thermal imaging in the steel industry," in Semanovich SA, ed., *Thermosense XVII, SPIE Proc.*, **2473**:289–294, 1995.

Chen JS, Medioni G, "Detection, localization, and estimation of edges," *IEEE Transactions on Pattern Analysis and Machine Intelligence*, **11**[2]:191–198, 1989.

Chen L, Yang BT, Hu X-T, "Design principle for simultaneous emissivity and temperature measurements," *Optical Engineering*, **29**[12]:1445–1448, 1990.

Chiarantini L, Coppo P, "Infrared monitoring system for urban solid waste landfills: experimental activities for bio gas outflow modelling," *Proc. QIRT-94, Eurotherm Seminar*, **42**:305–309, 1994.

Chow CK, Kaneko T, "Automatic boundary detection of the left ventricle from cineorgiograms," *Computers and Biomedical Research*, **5**:388–410, 1972.

Chrysochoos A, Dupré JC, "An infrared set-up for continuum thermomechanics," *Proc. QIRT-92, Eurotherm Seminar*, **27**:129–134, 1992.

Chrzanowski K, "Remote temperature measurement of highly reflecting objects in outdoor conditions," *Proc. QIRT-92, Eurotherm Seminar*, **27**:32–38, 1992.

Chrzanowski K, Jankiewicz Z, "Accuracy analysis of measuring thermal imaging systems," *Proc. QIRT-94, Eurotherm Seminar*, **42**:53–59, 1994a.

Chrzanowski K, Jankiewicz Z, "Influence of changes of object-system distance on accuracy of remote temperature measurement," *Proc. QIRT-94, Eurotherm Seminar*, **42**:60–65, 1994b.

Church EL, Zavada JM, "Effects of surface microroughness in ellipsometry," *Journal of the Optical Society of America*, **66**:1136, 1976.

Cielo P, "Analysis of pulsed thermal inspection," in Moore DW, Matzkanin GA, eds., *Proc. 14th Symposium on Nondestructive Evaluation*, NTIAC, San Antonio, TX, Apr. 19–21, 1983.

Cielo P, "Pulsed photothermal evaluation of layered materials," *Journal of Applied Physics*, **56**[1]:230–234, 1984.

Cielo P, "NDE of industrial materials," *CIM Bulletin*, **82**[928]:81–89, 1989.

Cielo P, Maldague X, Rousset G, Jen CK, "Thermoelastic inspection of layered materials: dynamic analysis," *Materials Evaluation*, **43**[9]:1111–1116, 1985a.

Cielo P, Maldague X, Lewak R, Lamontagne M, "Recent progress on thermal NDE at IMRI," *Proc. 5th Infrared Information Exchange*, New Orleans, LA, Oct. 29–31, pp. 45–54, 1985b.

Cielo P, Lewak R, Balageas DL, "Thermal sensing for industrial quality control," in Kaplan H, ed., *Thermosense VIII, SPIE Proc.*, **581**:47–54, 1985c.

Cielo P, Jen CK, Maldague X, "The converging-surface-acoustic-wave technique: analysis and applications to nondestructive evaluation," *Canadian Journal of Physics*, **64**[9]:1324–1329, 1986a.

Cielo P, Lewak R, Maldague X, "Thermal methods of NDE," *Canadian Society of Nondestructive Testing Journal*, **7**[2]:30–49, 1986b.

Cielo P, Maldague X, Johar S, Lauzon B, "Some laser-based techniques for the characterization of sintered ceramics," *Materials Evaluation*, **44**[6]:770–774, 1986c.

Cielo P, Maldague X, Déom AA, Lewak R, "Thermographic nondestructive evaluation of industrial materials and structures," *Materials Evaluation*, **45**[6]:452–460, 1987a.

Cielo P, Maldague X, Krapez JC, Lewak R, "Optics-based techniques for the characterization of composites and ceramics," in Bussière JF, Monchalin JP, Ruud CO,

Green RE, eds., *Nondestructive Characterization of Materials*, Vol. II, pp. 733–744, Plenum Press, New York, 1987b.

Cielo P, Krapez JC, Lamontagne M, "Lumber moisture evaluation by a reflective cavity photothermal technique," *Revue de Physique Appliquée*, 23:1565–1576, 1988a.

Cielo P, Krapez J-C, Maldague X, "Enhanced thermographic imaging for subsurface flaw detection by full-field heating of the inspected surface," in Hess P, Pelzl J, eds., *Photoacoustic and Photothermal Phenomena*, Vol. 58 in Springer Series in Optical Sciences, pp. 404–407, Springer-Verlag, New York, 1988b.

Cielo P, *Optical Techniques for Industrial Inspection*, Academic Press, San Diego, CA, 606 pp., 1988.

Cielo P, Cole K, Remmers U, Fairlie M, Zouikin S, "Infrared monitoring of relubricant on rolled aluminum sheets," *Materials Evaluation*, 48[7]:898–903, 908, 1990.

Cielo P, Maldague X, Krapez JC, "Device for subsurface flaw detection in reflective materials by thermal-transfer imaging," U.S. patent 4,996,426, filed Feb. 26, 1991.

Cielo P, Krapez J, Lamontagne M, Thomson JC, Lamb MC, "Conical-cavity fiber optic sensor for temperature measurement in a steel furnace," in Eklund JK, ed., *Thermosense XIV*, SPIE Proc., 1682:142–154, 1992.

Clary JB, "Design tools for real-time signal processing systems," in Letellier JP, ed., *SPIE Proc. Real Time Signal Processing X*, 827:2–7, 1987.

Cline HE, Anthony TR, "Heat treating and melting material with a scanning laser or electron beam," *Journal of Applied Physics*, 48[9]:3895–3900, 1977.

Coates PB, "Multiwavelength pyrometry," *Metrologia*, 17:103–109, 1981.

Cockburn W, "Nondestructive testing of the human breast," in Lemieux DH, Snell JR, eds., *Thermosense XXI*, SPIE Proc., 3700:312–323, 1999.

Coester JY, "Technologies infrarouges," *L'Onde Électrique*, 68[2]:40–44, 1988.

Cohen E, "The design and application of the traversing infra-red inspection system: TIRIS," *Nondestructive Testing*, Apr., pp. 74–80, 1973.

Colantonio A, "Metal cladding envelope problems, retrofit solutions, and quality control investigations," in Eklund JK, ed., *Thermosense XIV*, SPIE Proc., 1682:64–73, 1992.

Colantonio A, "Air-leakage effects on stone cladding panels," in Semanovich SA, ed., *Thermosense XVII*, SPIE Proc., 2473:27–35, 1995.

Colantonio A, "Thermal performance patterns on solid masonry exterior walls of historic buildings," in Wurzbach RN, Burleigh DD, eds., *Thermosense XIX*, SPIE Proc., 3056:87–95, 1997.

Colantonio A, "Identification of convective heat loss on exterior cavity wall assemblies," in Lemieux DH, Snell JR, eds., *Thermosense XXI*, SPIE Proc., 3700:514–520, 1999a.

Colantonio A, "Infrared thermographic investigation procedures for four types of generic exterior wall assemblies," in Lemieux DH, Snell JR, eds., *Thermosense XXI*, SPIE Proc., 3700:458–462, 1999b.

Colbert F, "Problem-modeling diagnostic engines used with statistical analysis of infrared PdM inspections provide solutions for the management of facility-wide infrared PdM inspection programs," in Dinwiddie RB, Lemieux DH, eds., *Thermosense XXII*, SPIE Proc., 4020:71–79, 2000.

Collett NJ, Hobbs CP, Kenway-Jackson D, "Adapting transient thermography to industrial problems," *Insight: Non-Destructive Testing and Condition Monitoring*, **38**[3]:163–166, 1996.

Colombo G, Galfetti L, Salerno A, "The measurement of emissivity of charring ablative materials," *Proc. QIRT-94, Eurotherm Seminar*, **42**:108–112, 1994.

Colwell RN, ed., *Manual of Remote Sensing*, Vols. 1 and 2, 2nd ed., American Society for Photogrammetry and Remote Sensing, Falls Church, VA, 1983.

Connolly M, Copley D, "NDT solution: thermographic inspection of composite materials," *Materials Evaluation*, **48**[12]:1461–1463, 1990.

Coster M, Chermant JL, *Précis d'analyse d'images*, Presses du CNRS, Paris, 560 pp., 1989.

Couturier JP, Maldague X, "Pulsed phase thermography of aluminum specimens," in Wurzbach RN, Burleigh DD, eds., *Thermosense XIX, SPIE Proc.*, **3056**:170–175, 1997.

Couturier JP, Maldague X, "Pulsed phase thermography (PPT) applied to the inspection of wood panels," in Maldague X, ed., *Proc. 3rd International Workshop on Advances in Signal Processing for Nondestructive Evaluation of Materials, TONES* (ASNT), **3**:285–288, 1998.

Crabol P, "Intérêt des polymères pyroélectriques," *Contrôle Industriel Qualité*, **143**[Jan.], 1987.

Craig DM, Chapman CE, "NDI of impact damaged composite panels," *British Journal of Non-Destructive Testing*, **33**[2]:64–68, 1991.

Cramer KE, Winfree WP, "Thermal characterization of defects in aircraft structures via spatially controlled heat application," in Burleigh DD, Spicer JWM, eds., *Thermosense XVIII, SPIE Proc.*, **2766**:202–209, 1996.

Cramer KE, Winfree WP, "Thermographic detection and quantitative characterization of corrosion by application of thermal line source," in Snell JR, Wurzbach RN, eds., *Thermosense XX, SPIE Proc.*, **3361**:291–300, 1998.

Cramer KE, Winfree WP, "Application of the thermal line scanner to quantify material loss due to corrosion," in Dinwiddie RB, Lemieux DH, eds., *Thermosense XXII, SPIE Proc.*, **4020**:210–219, 2000.

Cramer KE, Winfree WP, Howell PA, Syed HI, Renouard KA, "Thermographic imaging of cracks in thin metal sheets," in Eklund JK, ed., *Thermosense XIV, SPIE Proc.*, **1682**:162–170, 1992.

Cramer KE, Howell PA, Winfree WP, "Quantitative thermal depth imaging of subsurface damage in insulating materials," in Allen LR, ed., *Thermosense XV, SPIE Proc.*, **1933**:188–196, 1993.

Cramer KE, Howell PA, Syed HI, "Quantitative thermal imaging of aircraft structures," in Semanovich SA, ed., *Thermosense XVII, SPIE Proc.*, **2473**:226–232, 1995.

Cronhoim M, "Thermography's impact on economic performance: minimizing the cost of downtime and maintenance," in Dinwiddie RB, Lemieux DH, eds., *Thermosense XXII, SPIE Proc.*, **4020**:99–106, 2000.

Crowther DJ, Favro LD, Kuo PK, Thomas RL, "Analytic calculations and numerical simulations of box-car thermal wave images of planar subsurface scatterers," in Thompson DO, Chimenti DE, eds., *Review of Progress in Quantitative Nondestructive Evaluation*, **11A**:417–424, 1998.

Csendes A, Szekely V, Rencz M, "Thermal mapping with liquid crystal method," *Microelectronic Engineering*, 31[1–4]:281–290, 1996.

Cuccurullo G, Di Maio L, "Velocity and thermal field thermography for thermoplastic polymers extrusion," *Proc. QIRT-96, Eurotherm Seminar*, 50:287–292, 1996.

Czerny M, *Z. Physik*, 53[1], 1929.

Czigany T, Ostgathe M, Karger-Kocsis J, "Damage development in GF/PET composite sheets with different fabric architecture produced of a commingled yarn," *Journal of Reinforced Plastics and Composites*, 17[3]:250–267, 1998.

d'Ambrosio G, Massa R, Migliore MD, Cavaccini G, Ciliberto A, Sabatino C, "Microwave excitation for thermographic NDE: an experimental study and some theoretical evaluations," *Materials Evaluation*, 53[4]:502–508, 1995.

Dance WE, Middlebrook JB, "Neutron radiographic nondestructive inspection for bonded composite structures," *Nondestructive Evaluation and Flaw Criticality for Composite Materials*, 696:57–71, 1978.

Danesi S, Salerno A, Wu D, Busse G, "Cooling down thermography: principle and results for NDE," in Snell JR, Wurzbach RN, eds., *Thermosense XX, SPIE Proc.*, 3361:266–274, 1998.

Daniels A, "Nondestructive pulsed infrared quantitative evaluation of metals," in Burleigh DD, Spicer JWM, eds., *Thermosense XVIII, SPIE Proc.*, 2766:185–201, 1996.

Daniels A, "Optimization of pulsed temporal stimulation for thermal nondestructive testing," in Snell JR, Wurzbach RN, eds., *Thermosense XX, SPIE Proc.*, 3361:254–265, 1998.

Danjoux R, van Schel E, Potier F, Beaudoin JL, Égee M, "Caractérisation x-y-z de matériaux minces par thermographie et sonde photothermique à balayage," *Contrôle Industriel Qualité*, 26[147]:54–58, 1987.

Dartois R, Égee M, Marx J, van Schel E, "Caractérisation des matériaux bi-couches semi-transparents par radiométrie photothermique," *Revue Générale de Thermique fr.*, 301:22–32, 1987.

Daryabeigi K, Wright RE Jr, Puram CK, Alderfer DW, "Directional emittance corrections for thermal infrared imaging," in Eklund JK, ed., *Thermosense XIV, SPIE Proc.*, 1682:325–335, 1992.

Datong W, Busse G, "Lock-in thermography for nondestructive evaluation of materials," *Revue Générale de Thermique*, 37[8]:693–703, 1998.

David DG, Marin JY, Tretout HR, "Automatic defects recognition in composite aerospace structures from experimental and theoretical analysis as part of an intelligent infrared thermographic inspection system," in Eklund JK, ed., *Thermosense XIV, SPIE Proc.*, 1682:182–193, 1992.

David DG, Trétout HR, Mann JY, Perez A, Delpet A, Desmarres JM, "Nondestructive testing (NDT) of satellite structures during their assembly by infrared stimulated thermography," in Snell JR, Jr., ed., *Thermosense XVI, SPIE Proc.*, 2245:164–172, 1994.

Dawson B, "Focusing on image enhancement," *ESD: Electronic System Design Magazine*, Mar., pp. 83–86, 1988 [in French].

Degiovanni A, "Une nouvelle technique d'identification de la diffusivité thermique pour la méthode «flash»," *Revue de Physique Appliquee*, 21:229–237, 1986.

Degiovanni A, "Conduction dans un "mur" multicouche avec sources: extension de la notion de quadripole," *International Journal of Heat and Mass Transfer*, **31**[3]:553–557, 1988.

Degiovanni A, Bendada A, Batsale JC, Maillet D, "Analytical simulation of a multi-dimensional temperature field produced by planar defects of any shape; application to non-destructive testing," *Proc. QIRT-94, Eurotherm Seminar*, **42**, 253–259, 1994.

de Halleux B, Dubois PM, Maldague X, Michaux D, "Auscultation des ouvrages d'art pour établir le tracé des armatures de précontrainte par voie thermographique," in Ballivy G, Rhazi J, eds., *5eme Colloque du CRIB*, pp. 29–43, 1998.

Delclaud Y, "Surface temperature measurements of thin films in the range −100°C to 100°C using infrared thermography," *Proc. QIRT-92, Eurotherm Seminar*, **27**:353–358, 1992.

Del Grande NK, Clark GA, "Buried remote detection technology for law enforcement," *SPIE Surveillance Technology*, **1479**:335, 1991.

Del Grande NK, Durbin PF, "Dual-band infrared imaging to detect corrosion damage within airframes and concrete structures," in Snell JR, Jr., ed., *Thermosense XVI, SPIE Proc.*, **2245**:202–209, 1994.

Del Grande NK, Durbin PF, "Precise thermal NDE quantifying structural damage," in Thompson DO, Chimenti DE, eds., *Review of Progress in Quantitative Non-destructive Evaluation*, **15**:525–531, 1996.

Délouard P, Marin JY, Avenas-Payan I, Tretout H, "Infrared thermography development for composite material evaluation," in Boogaard J, Van Dijk GM, eds., *Proc. 12th World Conference on Non-destructive Testing*, Amsterdam, Apr. 23–28, pp. 567–572, 1989.

Delpech PM, Balageas DL, "Mesure par thermographie infrarouge stimulée de résis-tances thermiques d'interface dans des structures bonnes conductrices de la chaleur," *Report of the Société Française des Thermiciens*, Jan. 9, 1991.

Delpech Ph, Bosher D, Déom A, Balageas D, "Utilisation de la transformation de Laplace pour la détermination des grandeurs thermiques," *Colloque de la Société Française des Thermiciens*, ONERA TP-42, 1990.

Delpech Ph M, Boscher DM, Lepoutre F, Déom AA, Balageas DL, "Time-resolved pulsed stimulated infrared thermography applied to carbon–carbon nondestructive evaluation," *Proc. QIRT-92, Eurotherm Seminar*, **27**:201–206, 1992.

Delpech PM, Krapez JC, Balageas DL, "Thermal defectometry using the temperature decay rate method," *Proc. QIRT-94, Eurotherm Seminar*, **42**:220–225, 1994.

Delpech PM, Boscher DM, Lepoutre F, Déom AA, Balageas DL, "Bonds NDE using stimulated infrared thermography," in Thompson DO, Chimenti DE, eds., *Review of Progress in Quantitative Nondestructive Evaluation*, **11A**:465–470, 1998.

de Luca L, Cardone C, "Experimental analysis of Goertler vortices in hypersonic wedge flow," in Eklund JK, ed., *Thermosense XIV, SPIE Proc.*, **1682**:271–281, 1992.

de Luca L, Cardone G, Carlomagno GM, "Fluid dynamics applications of IR imaging system," *Proc. QIRT-92, Eurotherm Seminar*, **27**:98–104, 1992.

De Mey G, Wiecek B, "Application of thermography for microelectronic design and modelling," *Proc. QIRT-98, Eurotherm Seminar*, **60**:23–29, 1998.

Deriche R, "Fast algorithms for low-level vision," *IEEE Transactions on Pattern Analysis and Machine Intelligence*, **12**[1]:78–87, 1990.

DeWitt DP, "Inferring temperature from optical radiation measurements," *Optical Engineering*, **25**[4]:569–601, 1986.

DeWitt DP, Kunz H, in *Temperature*, p. 54, American Institute of Physics, New York, 1972.

DeWitt DP, Nutter GD, *Theory and Practice of Radiation Thermometry*, Wiley, New York, 1138 pp., 1988.

Di Benedetto M, Huston CW, Sharp MW, Jones B, "Regional hypothermia in response to minor injury," *American Journal of Physical Medicine and Rehabilitation*, **75**[4]:270–277, 1996.

Dickstein PA, Spelt JK, Sinclair AN, Bushlin Y, "Investigation of nondestructive monitoring of the environmental degradation of structural adhesive joints," *Materials Evaluation*, **49**[12]:1498–1499, 1991.

Dillenz A, Busse G, Wu D, "Ultrasound lockin thermography: feasibility and limitations," in Lammasnieemi J, Wiggenhauser H, Busse G, Batchelor BC, Pölzleitner W, Dobmann G, eds., *SPIE Proc. Diagnostic Imaging, Technologies and Industrial Applications, Euro-Opto 1999*, **3827**:10–15, 1999.

Dillenz A, Zweschper T, Busse G, "Phase angle thermography with ultrasound burst excitation," in *QIRT 2000* [in press].

Ding K, "Test of jet turbine blades by thermography," *Optical Engineering*, **24**[6]:1055–1059, 1985.

Dinwiddie RB, Blau PJ, "Time-resolved tribo-thermography," in Lemieux DH, Snell JR, eds., *Thermosense XXI, SPIE Proc.*, **3700**[64]:358–368, 1999.

Dinwiddie RB, Lee K, "IR-camera methods for automotive brake system studies," in Snell JR, Wurzbach RN, eds., *Thermosense XX, SPIE Proc.*, **3361**:66–74, 1998.

Dixon J, "Radiation thermometry," *Journal of Physics E: Scientific Instruments*, **21**:425–436, 1988.

Dixon RD, Lassahn GD, DiGiallonardo A, "Infrared thermography of subsurface defects," *Materials Evaluation*, **30**[4]:73–77, 1972.

Dixon WJ, Massey FJ, *Introduction to Statistical Analysis*, McGraw-Hill, New York, 678 pp., 1983.

Doering ER, Basart JP, "Trend removal in x-ray images," in Thompson DO, Chimenti DE, eds., *Review of Progress in Quantitative Nondestructive Evaluation*, **7A**:785–794, 1988.

Domanski R, Wisniewski T, Rebow M, "Experimental study of natural convection in the melting of PCM in horizontal cylindrical annuli," *Proc. QIRT-96, Eurotherm Seminar*, **50**:177–182, 1996.

Dorey G, "Effects of defects on advanced composite performance," *Metals and Materials*, **4**[5]:286–289, 1988.

Dorlot JM, Baïlon JP, Masounave J, "Des Matériaux," Éd. École Polytechnique, Montréal, Québec, Canada, 467 pp., 1991.

Dresser D, "IR imaging for printed circuits boards (PCB) testing," *Advanced Imaging*, **5**[4]:46–48, 1990.

Dubois F, "Real-time image enhancement by simple video systems," in Ebbeni J, ed., *SPIE Proc. In-situ Industrial Applications of Optics*, **672**:1–6, 1986.

Dubrovskii AV, Kober VI, Mnatsakanyan EA, et al., "System for automatic experimental investigations of active thermovision inspection of materials and components," *Soviet Journal of Nondestructive Testing*, **25**[8]:611–614, 1990.

Duc E, "Discussion of different numerical models applied to air infiltration measurement in external walls," *Proc. QIRT-94, Eurotherm Seminar*, **42**:86–89, 1994.

Duc E, Jaworski J, "Air in- and ex-filtration through the joints of external walls," *Proc. QIRT-92, Eurotherm Seminar*, **27**:315–318, 1992.

Ducar RJ, "Pulsed thermographic inspection and application in commercial aircraft repair," in Lemieux DH, Snell JR, eds., *Thermosense XXI, SPIE Proc.*, **3700**:77–83, 1999.

Duda RO, Nitzan D, Barrett P, "Use of range and reflectance data to find planar surface regions," *IEEE Transactions on Pattern Analysis and Machine Intelligence*, **1**[3]:259–271, 1979.

Dufour M, Maldague X, "Prevention of spatter and molten particles emission on protective windows in welding applications," *Welding Journal*, **66**[6]:43–46, 1987.

Dufour M, Maldague X, Cielo P, "Environmental-noise analysis in active-vision systems for adaptive welding," in Cielo P, ed., *SPIE Proc. Optical Techniques for Industrial Inspection*, **665**:321–332, 1986.

Duke JC Jr, "Transportation infrastructure: an introduction," *Materials Evaluation*, **52**[4]:494–495, 1994.

Dukle NM, Hollingsworth DK, "Liquid crystal images of the transition from jet impingement convection to nucleate boiling. II: Nonmonotonic distribution of the convection coefficient," *Experimental Thermal and Fluid Science*, **12**[3]:288–297, 1996a.

Dukle NM, Hollingsworth DK, "Liquid crystal images of the transition from jet impingement convection to nucleate boiling. I: Monotonic distribution of the convection coefficient," *Experimental Thermal and Fluid Science*, **12**[3]:274–287, 1996b.

Dulski R, Sikorski Z, Niedziela T, "Modelling of infrared imaging for 3-D objects," *Proc. QIRT-98, Eurotherm Seminar*, **60**:200–204, 1998.

Dumoulin J, Reulet P, Grenier P, Plazanet M, Millan P, "Steady and unsteady wall heat transfer mapping by active infrared thermography at the mean aerodynamic reattachment point behind a backward-facing step," *Proc. QIRT-94, Eurotherm Seminar*, **42**:158–165, 1994.

Dumoulin J, Marchand M, Reulet P, Millan P, "Heat transfer identification induced by multihole cooling in combustion chambers," *Proc. QIRT-96, Eurotherm Seminar*, **50**:135–140, 1996.

Durrani TS, Rauf A, Boyle K, Lotti F, Baronti S, "Thermal imaging techniques for the nondestructive inspection of composite materials in real time," *Proc. ICASSP-87*, pp. 598–601, 1987.

Eads LG, Spring RW, "ASNT certification of thermographers at DuPont Company," in Burleigh DD, Spicer JWM, eds., *Thermosense XVIII, SPIE Proc.*, **2766**:132–135, 1996.

Edwin EH, Arnesen T, Hugosson GI, "Evaluation of thermal cracker operation by use of an infrared camera," in Snell JR, Wurzbach RN, eds., *Thermosense XX, SPIE Proc.*, **3361**:125–136, 1998.

Electronic Design, "Infrared camera spots malfunctions," **9**:12, 1961.

Elliot CT, "New detector for thermal imaging systems," *Electronics Letters*, **17**:312–313, 1981.

Elliott H, Derin H, Cristi R, "Applications of the Gibbs distribution to image segmentation," in Wegman EJ, DePriest DJ, eds., *Statistical Image Processing and Graphics*, pp. 3–24, Marcel Dekker, New York, 1986.

El-Ratal WH, Bee DC, "Infrared technology in automotive components research and development," in Snell JR, Wurzbach RN, eds., *Thermosense XX, SPIE Proc.*, **3361**:84–93, 1998.

Emeric PR, Winfree WP, "Thermal characterization of multilayer structures from transient thermal response," in Thompson DO, Chimenti DE, eds., *Review of Progress in Quantitative Nondestructive Evaluation*, **14**:475–482, 1995.

Ermanno G, Vavilov V, "Corrosion evaluation by thermal image processing and 3D modelling," *Revue Générale de Thermique*, **37**[8]:669–679, 1998.

Ervin J, Murawski C, MacArthur C, Chyu M, Bizzak D, "Temperature measurement of a curved surface using thermographic phosphors," *Experimental Thermal and Fluid Science*, **11**[4]:387–394, 1995.

Eshera MA, Don HS, Matsumoto K, Fu KS, "A syntactic approach for SAR image analysis," in Wegman EJ, DePriest DJ, eds., *Statistical Image Processing and Graphics*, pp. 71–92, Marcel Dekker, New York, 1986.

Esposti W, Meroni I, "Nondestructive testing: use of IR and acoustic methods in buildings pathology," in Semanovich SA, ed., *Thermosense XVII, SPIE Proc.*, **2473**:374–383, 1995.

Evans JP, Wurzbach RN, "Study of cost benefits of identification of non-problems with infrared thermography," in Snell JR, Jr., ed., *Thermosense XVI, SPIE Proc.*, **2245**:14–21, 1994.

Evans JP, Jarc TJ, Wurzbach RN, "Increasing predictive program effectiveness by integrating thermography and lubricant analysis," in Semanovich SA, ed., *Thermosense XVII, SPIE Proc.*, **2473**:82–90, 1995.

Falk V, Walther T, Diegeler A, Rauch T, Kitzinger H, Mohr FW, "Thermal-coronary-angiography (TCA) for interoperative evaluation of graft patency in coronary artery bypass surgery," *Proc. QIRT-96, Eurotherm Seminar*, **50**:348–353, 1996.

Farrow RL, Nagelberg AS, "Raman spectroscopy of surface oxides at elevated temperatures," *Applied Physics Letters*, **36**:945, 1980.

Favro LD, Han X, "Thermal wave material characterization and thermal wave imaging," in Birnbaum G, Auld BA, eds., *Sensing for Materials Characterization, Processing, and Manufacturing, ASNT TONES*, **1**:399–415, 1998.

Favro LD, Thomas RL, "Thermal hardware evolution using the NDI validation center," *Materials Evaluation*, **53**[7]:840–843, 1995.

Favro LD, Shepard SM, Kuo PK, Thomas RL, "Mechanisms for the generation and scattering of sound and thermal waves in thermoacoustic microscopes," in Heas P, Pelzl J, eds., *Photoacoustics and Photothermal Ph.*, **59**:371–376, 1988.

Favro LD, Ahned T, Crowther D, Jin HJ, Kuo PK, Thomas RL, Wang X, "Infrared thermal-wave studies of coatings and composites," in Baird GS, ed., *Thermosense XIII, SPIE Proc.*, **1467**:290–294, 1991.

Favro LD, Crowther DJ, Kuo PK, Thomas RL, "Inversion of pulsed thermal-wave images for defect sizing and shape recovery," in Eklund JK, ed., *Thermosense XIV, SPIE Proc.*, **1682**:178–181, 1992.

Favro LD, Crowther DJ, Kuo PK, Thomas RL, "Inversion of pulse-echo thermal wave images," in Allen LR, ed., *Thermosense XV, SPIE Proc.*, **1933**:138–141, 1993a.

Favro LD, Kuo PK, Thomas RL, Shepard SM, "Thermal wave imaging for aging aircraft inspection," *Materials Evaluation*, **51**[12]:1386–1389, 1993b.

Favro LD, Kuo PK, Thomas RL, "Thermal-wave imaging of composites and polymers," in Snell JR, Jr., ed., *Thermosense XVI, SPIE Proc.*, **2245**:90–94, 1994.

Favro LD, Ahmed T, Han X, Wang L, Wang X, Wang Y, Kuo PK, Thomas RL, "Thermal wave imaging of aircraft structures," in Thompson DO, Chimenti DE, eds., *Review of Progress in Quantitative Nondestructive Evaluation*, **14**:461–466, 1995a.

Favro LD, Han X, Kuo PK, Thomas RL, "Imaging the early time behavior of reflected thermal wave pulses," in Semanovich SA, ed., *Thermosense XVII, SPIE Proc.*, **2473**:162–166, 1995b.

Favro LD, Han X, Wang Y, Kuo PK, Thomas RL, "Pulse-echo thermal wave imaging," in Thompson DO, Chimenti DE, eds., *Review of Progress in Quantitative Nondestructive Evaluation*, **14**:425–429, 1995c.

Favro LD, Han X, Kuo PK, Thomas RL, "Measuring defect depths by thermal-wave imaging," in Burleigh DD, Spicer JWM, eds., *Thermosense XVIII, SPIE Proc.*, **2766**:236–239, 1996.

Favro LD, Ouyang Z, Wang L, Wang X, Zhang F, Thomas RL, "Infrared video lock-in imaging at high frequencies," in Wurzbach RN, Burleigh DD, eds., *Thermosense XIX, SPIE Proc.*, **3056**:184–188, 1997.

Favro LD, Jin HJ, Wang YX, Ahmed T, Wang X, Kuo PK, Thomas RL, "IR thermal wave tomographic studies of structural composites," in Thompson DO, Chimenti DE, eds., *Review of Progress in Quantitative Nondestructive Evaluation*, **11A**:447–452, 1998a.

Favro LD, Ouyang Z, Wang L, Zhang F, Zhang L, Thomas RL, "Fast infrared measurements of the thermal diffusivities of anisotropic materials," in Snell JR, Wurzbach RN, eds., *Thermosense XX, SPIE Proc.*, **3361**:248–253, 1998b.

Favro LD, Han X, Ouyang A, Sun G, Sui H, Thomas RL, "IR imaging of cracks excited by an ultrasonic pulse," in Dinwiddie RB, Lemieux DH, eds., *Thermosense XXII, SPIE Proc.*, **4020**:182–185, 2000.

Feest EA, "Exploitation of the metal matrix composites concept," *Metals and Materials*, **4**[5]:273–278, 1988.

Feigenbaum V, "Total quality: an international imperative," *Photonics Spectra*, **24**[8]:82–83, 1990.

Feit E, "Infrared thermography saves Shelburne Middle School time and money," *Materials Evaluation*, **45**[4]:400–401, 1987.

Feng C, Ahmad A, Li Y, Sarepaka RV, "Scanning telescope for the advanced polarized infrared imaging sensor (APIRIS)," in Semanovich SA, ed., *Thermosense XVII, SPIE Proc.*, **2473**:346–357, 1995.

Fennell HC, "Thermographic inspection and quality assurance of energy conservation procedures for electric buses," in Snell JR, Wurzbach RN, eds., *Thermosense XX, SPIE Proc.*, **3361**:200–211, 1998.

Figari A, "Photothermal method with arbitrary phase shift," *Proc. QIRT-92, Eurotherm Seminar*, **27**:377–381, 1992.

Finneson S, "Electrical fault location for surface-mounted feeders in metallic conduit," in Semanovich SA, ed., *Thermosense XVII, SPIE Proc.*, **2473**:99–107, 1995.

Fiorini AR, Fumero R, Marchesi R, "Cardio-surgical thermography," in Tesher AG, ed., *SPIE Proc. Applications of Digital Image Processing IV*, **359**:249–256, 1982.

Fisher WG, Meyer KE, Wachter EA, Perl DR, Kulowitch PJ, "Laser induced fluorescence imaging of thermal damage in polymer matrix composites," *Materials Evaluation*, **55**[6]:726, 1997.

Fitch SH, Morris JW, "Precision nondestructive testing of thermoelectric materials," *Nondestructive Testing*, **21**[5]:306–310, 1963.

Florin C, "Thermal testing methods as new tool in NDT," *Thermosense IX, SPIE Proc.*, **780**:76–83, 1987.

Florkowski M, Korendo Z, "IR trend analysis for HV/MV equipment diagnostics," *Proc. QIRT-98, Eurotherm Seminar*, **60**:245–249, 1998.

Fogarty DN, "Imaging of sealant injection into a main steam valve," in Snell JR, Wurzbach RN, eds., *Thermosense XX, SPIE Proc.*, **3361**:100–102, 1998.

Foley GM, Morse MS, Cezairliyan A, "Two-color microsecond pyrometer for 2000 to 6000 K," in Schooley JF, ed., *Temperature, Its Measurement and Control in Science and Industry*, pp. 447–452, American Institute of Physics, New York, 1982.

Forssander MI, "Thermographic inspection and heat flow simulation of midspan joints," in Eklund JK, ed., *Thermosense XIV, SPIE Proc.*, **1682**:43–53, 1992.

Fort C, Roux JM, Guidon M, "Les aspects thermiques des matériaux: élaboration, mise en forme, propriétés d'usage," Société Française des Thermiciens, meeting of May 3–5, 1988.

Foucher B, "Infrared machine vision: a new contender," in Lemieux DH, Snell JR, eds., *Thermosense XXI, SPIE Proc.*, **3700**:210–213, 1999.

Fourier J, "Théorie du mouvement de la chaleur dans les corps solides—1ère Partie," *Mémoires de l'Academie des Sciences*, **4**:185–555, 1824; **5**:153–246, 1826.

Fraedrich DS, "Methods in calibration and error analysis for infrared imaging radiometers," *Optical Engineering* **30**[11]:1764–1770, 1991.

Francou L, Herve P, "An FT-IR based instrument for measuring infrared diffuse reflectance," *Proc. QIRT-98, Eurotherm Seminar*, **60**:228–232, 1998.

Fredal D, Sega RM, Norgard JD, Bussey PE, "Hardware and software advancement for infrared detection of microwave fields," in Weathersby MR, ed., *SPIE Proc. Infrared Image Processing and Enhancement*, **781**:160–167, 1987.

Freeman H, "Survey of image processing applications in industry," in Cantoni V, Levialdi S, Musso G, eds., *Image Analysis and Processing*, pp. 1–9, Plenum Press, New York, 1986.

Frock BG, "Marr–Hildreth enhancement of NDE images," in Thompson DO, Chimenti DE, eds., *Review of Progress in Quantitative Nondestructive Evaluation*, **8A**:701–708, 1989.

Frumuselu D, Radu C, "IR thermography applied to ground-level reinforced concrete constructions belonging to electricity networks," *Insight: Non-Destructive Testing and Condition Monitoring*, **40**[7]:501–504, 1998.

Fu KS, Mui JK, "A survey on image segmentation," *Pattern Recognition*, **13**:3–16, 1981.

Fuchs EA, Mahin KW, Ortega AR, et al., "Thermal diagnosis for monitoring welding parameters in real time," in Baird GS, ed., *Thermosense XIII, SPIE Proc.*, **1467**:136–149, 1991.

Fuchs EA, Bertram LA, Bentley AE, Marburger SJ, "Evaluation of feedback parameters for weld process control," in Eklund JK, ed., *Thermosense XIV, SPIE Proc.*, **1682**:155–161, 1992.

Fulop GF, "U.S. market for infrared thermography equipment," in Semanovich SA, ed., *Thermosense XVII, SPIE Proc.*, **2473**:119–125, 1995.

Gadaj SP, Nowacki WK, Pieczyska EA, "Investigation of temperature distribution during plastic deformation of stainless steel," *Proc. QIRT-96, Eurotherm Seminar*, **50**:298–303, 1996.

Gadaj SP, Nowacki WK, Pieczyska EA, "Temperature evolution during simple shear test of steel," *Proc. QIRT-98, Eurotherm Seminar*, **60**:117–122, 1998.

Galmiche F, Maldague X, "Pulsed phase and thermal tomography: a comparison," *1998 Pan American Nondestructive Testing Conference*, Toronto, Ontario, Canada, 1998.

Galmiche F, Maldague X, "Active infrared thermography for land mine detection," in Lammasnieemi J, Wiggenhauser H, Busse G, Batchelor BC, Pölzleitner W, Dobmann G, eds., *SPIE: Diagnostic Imaging, Technologies and Industrial Applications, Euro-Opto 1999*, **3827**:146–154, 1999.

Galmiche F, Vallerand S, Maldague X, "Pulsed phase thermography with the wavelet transform," in Thompson DO, Chimenti DE, eds., *Review of Progress in Quantitative Nondestructive Evaluation*, Montréal, Québec, Canada, July 1999, [in press, 2000a].

Galmiche F, Vallerand S, Maldague X, "Wavelet transform applied to pulsed phased thermography," in Abbozzo LR, Carlomagno GM, Corsi C, eds., *Proc. 5th Workshop on Advances in Infrared Technology and Applications*, Atti della Fondazione G. Ronchi, Firenze, Italy [in press, 2000b].

Garner DL, Underwood HB, Porter WF, "Use of modern infrared thermography for wildlife population surveys," *Environmental Management*, **19**[2]:233, 1995.

Garner JM, "Infrared research services versus $$$: the bottom line," in Eklund JK, ed., *Thermosense XIV, SPIE Proc.*, **1682**:23–29, 1992.

Gartenberg E, "Retrospective on aerodynamic research with infrared imaging," *Proc. QIRT-92, Eurotherm Seminar*, **27**:63–85, 1992.

Gartenberg E, Roberts JAS, "Influence of temperature gradients on the measurement accuracy of IR imaging systems," in Semanovich SA, ed., *Thermosense XII, SPIE Proc.*, **1313**:218–221, 1990.

Garvie AM, Sorell GC, "Video-based analog tomography," *Review of Scientific Instruments*, **61**[1]:138–145, 1990.

Gauffre G, "New infrared detector and image quality," *SPIE. Proc. Image Detection and Quality*, **702**:39–46, 1986.

Gaussorgues G, "La thermographie dans les procédés industriels," *Proc. 3rd European Conference on Nondestructive Testing*, Firenza, Italy, Oct. 15–18, pp. 377–388, 1984a.

Gaussorgues G, *La Thermographie infrarouge*, Éditions Lavoisier, Paris, 482 pp., 1984b.

Gaussorgues G, *La Thermographie infrarouge*, 3rd ed., Tec & Doc Lavoisier, Paris, 540 pp., 1989.

Gayer A, Saya A, Shiloh A, "Automatic recognition of welding defects in real-time radiography," *Nondestructive Testing International*, **23**[3], 1990.

Gayo E, de Frutos J, "Interference filters as an enhancement tool for infrared thermography in humidity studies of building elements," *Infrared Physics and Technology*, **38**[4]:251–258, 1997.

Gayo E, Palomo A, Macias A, "Infrared thermography as a tool for studying the movement of water through some building materials. II: Evaporation process," *European Journal of Non-Destructive Testing*, **3**[2]:55–58, 1993.

Gayo E, De Frutos E, Palomo A, "Selective infrared thermography: application to detection of humidity in buildings," *Proc. QIRT-94, Eurotherm Seminar*, **42**:310–314, 1994.

Gayo E, Frutos J, Palomo A, Massa S, "Mathematical model simulating the evaporation processes in building materials: experimental checking through infrared thermography," *Building and Environment*, **31**[5]:469–475, 1996.

Geerkens J, Schmitz B, Goch G, "Photothermal investigations on advanced ceramics," *Proc. QIRT-96, Eurotherm Seminar*, **50**:102–107, 1996.

Geiger G, "NDE plays a critical role in advanced ceramics," *American Ceramic Society Bulletin*, **74**[11]:55–60, 1995.

Gerardi R, "Use of infrared thermography in aiding and monitoring weatherization work: lessons learned," in Snell JR, Jr., ed., *Thermosense XVI, SPIE Proc.*, **2245**:83–89, 1994.

Gericke OR, Vogel PEJ, "Infrared bond defect detection system," *Materials Evaluation*, **22**[2]:65–68, 1964.

Gershenson M, "Imaging through the diffusion equation," in Burleigh DD, Spicer JWM, eds., *Thermosense XVIII, SPIE Proc.*, **2766**:258–263, 1996.

Ghia V, Colibaba-Evulet A, "Combustion graphology used to improve emulsions of water-in-heavy fuel oil," *Proc. QIRT-96, Eurotherm Seminar*, **50**, 277–282, 1996.

Ghiardi GL, "Occupant thermal comfort evaluation," in Lemieux DH, Snell JR, eds., *Thermosense XXI, SPIE Proc.*, **3700**:324–331, 1999.

Gibkes J, Dietzel D, Pelzl J, Bein BK, "Characterization of NiTi shape memory alloys by means of photothermal radiometry," in *QIRT 2000*, Reims, France [in press].

Giesecke JL, "Substation component identification for infrared thermographers," in Wurzbach RN, Burleigh DD, eds., *Thermosense XIX, SPIE Proc.*, **3056**:153–163, 1997.

Gilmore JF, "Artificial intelligence in image processing," *SPIE Proc. Digital Image Processing*, **528**:192–201, 1985.

Girard G, Algazi VR, "Traitement numérique d'images infrarouges," *Traitement du Signal*, **2**[1]:29–43, 1985.

Gitzhofer F, Martin C, Fauchais P, "Contrôle par thermographie infrarouge de l'apparition de fissures dans un matériau céramique projeté par plasma et soumis à un cyclage thermique," *Revue Générale de Thermique Fr.*, **301**[Jan.]:63–69, 1987.

Goel NS, Gang F, "Simple method for pulse-echo thermal wave imaging of arbitrary-shaped subsurface scatterers in heterogeneous materials," *International Communications in Heat and Mass Transfer*, **23**[1]:45–54, 1996.

Goita K, Royer A, "Surface temperature and emissivity separability over land surface from combined TIR and SWIR AVHRR data," *IEEE Transactions on Geoscience and Remote Sensing*, **35**[3]:718–733, 1997.

Gonzalez RC, Wintz P, *Digital Image Processing*, Addison-Wesley, Reading, MA, 431 pp., 1977 (2nd ed., 1987).

Gonzalez RC, Woods RE, *Digital Image Processing*, Addison-Wesley, Reading, MA, 503 pp., 1992.

Goodman MA, "How ultrasound and infrared work together," in Dinwiddie RB, Lemieux DH, eds., *Thermosense XXII, SPIE Proc.*, **4020**:94–98, 2000.

Goodman SH, "Plastic processing," in Parker SP, ed., *Concise Encyclopedia of Science and Technology*, pp. 1341–1342, McGraw-Hill, New York, 1984.

Gopalsami N, Rose DN, "Thermal-wave characterization of diffusion coatings," *Materials Evaluation*, **48**[7]:890–893, 897, 1990.

Gorrill WS, "Industrial high-speed infrared pyrometer to measure the temperature of a soldered seam on a tin can," *Electronics*, **22**:112, 1949.

Goss AJ, "The pyroelectric vidicon: a review," *SPIE Proc. Passive Infrared Systems and Technology*, **807**:25–31, 1987.

Gostelow CR, Crocker RL, Saffari N, Bond LJ, "Improved ultrasonic methods for gas turbine NDE," in Thompson DO, Chimenti DE, eds., *Review of Progress in Quantitative Nondestructive Evaluation*, **5A**:145–148, 1986.

Govardhan SM, Chin BA, "Adaptive penetration control using infrared measured temperature gradients," in Allen LR, ed., *Thermosense XV, SPIE Proc.*, **1933**:93–100, 1993.

Graf RJ, Forister LM, Weil GJ, "Locating railroad track bed subsurface defects utilizing nondestructive remote sensing technologies," in Snell JR, Jr., ed., *Thermosense XVI, SPIE Proc.*, **2245**:188–195, 1994.

Gränicher H, in Parker SP, ed., *Concise Encyclopedia of Science and Technology*, McGraw-Hill, New York, 2065 pp., 1984.

Grecki M, Pacholik J, Wiecek B, Napieralski A, "Application of computer-based thermography to thermal measurements of integrated circuits and power devices," *Microelectronics Journal*, **28**[3]:337–347, 1997.

Green DR, "Emissivity independent infrared thermal testing method," *Materials Evaluation*, **23**[2]:79–85, 1965.

Green DR, "Thermal surface impedance method for nondestructive testing," *Materials Evaluation*, **25**[10]:231–236, 1967.

Green DR, "Principles and application of emittance-independent infrared nondestructive testing," *Applied Optics*, **7**[9]:1779, 1968.

Green DR, "High speed thermal image transducer for practical NDT applications," *Materials Evaluation*, **28**[5]:97–102, 110, 1970.

Green DR, "Thermal and infrared nondestructive testing of composites and ceramics," *Materials Evaluation*, **29**[11]:241–248, 1971.

Green DR, "Experimental electro-thermal method for nondestructively testing welds in stainless steel pipes," *Materials Evaluation*, **37**[11]:54–60, 1979.

Green DR, Hassberger JA, "Infrared electro-thermal examination of stainless steel," *Materials Evaluation*, **35**[3]:39–43, 50, 1977.

Green DR, Collingham RE, Fischer LH, "Infrared NDT of electrical fuel pin simulators used in LMFBR thermal hydraulic tests," *Materials Evaluation*, **31**[12]:247–254, 1973.

Green JE, "Computer methods for erythrocyte analysis," *IEEE Conference Record of the Symposium on Feature Extraction and Selection in Pattern Recognition*, Argonne, IL, Oct., **5–7**:100–109, 1970.

Green TE, Snell JR Jr, "Thermographic inspection of hydraulic systems," in Burleigh DD, Spicer JWM, eds., *Thermosense XVIII, SPIE Proc.*, **2766**:25–31, 1996.

Gregory VT, "Adaptative histogram equalization and its applications," in Tesher AG, ed., *Applications of Digital Image Processing IV, SPIE Proc.*, **359**:204–209, 1982.

Griffin DD, "Infrared microradiometry: precision and accuracy considerations applicable to microcircuit temperature measurements," *Materials Evaluation*, **26**[10]:215–220, 1968.

Griffith BT, Arasteh D, "Buildings research using infrared imaging radiometers with laboratory thermal chambers," in Lemieux DH, Snell JR, eds., *Thermosense XXI, SPIE Proc.*, **3700**:502–513, 1999.

Grimnes KH, "Thermography and indoor climate," in Dinwiddie RB, Lemieux DH, eds., *Thermosense XXII, SPIE Proc.*, **4020**:335–340, 2000a.

Grimnes KH, "Thermography in Norway: history and state of the art," in Dinwiddie RB, Lemieux DH, eds., *Thermosense XXII, SPIE Proc.*, **4020**:341–344, 2000b.

Grinzato EG, Mazzodli A, "Infrared detection of moist areas in monumental buildings based on the thermal inertia analysis," in Baird GS, ed., *Thermosense XIII, SPIE Proc.*, **1467**:75–82, 1991.

Grinzato EG, Vavilov V, "Corrosion evaluation by thermal image processing and 3D modelling," *Revue Générale de Thermique*, **37**:669–679, 1998.

Grinzato EG, Bressan C, Bison PG, Mazzoldi A, Baggio P, Bonacina C, "Evaluation of moisture content in porous material by dynamic energy balance," in Eklund JK, ed., *Thermosense XIV, SPIE Proc.*, **1682**:213–221, 1992.

Grinzato EG, Bison P, Marinetti S, Vavilov V, "Nondestructive evaluation of delaminations in fresco plaster using transient infrared thermography," *Research in Nondestructive Evaluation*, **5**[4]:257–274, 1994a.

Grinzato EG, Marinetti S, Bison PG, Manduchi G, "Application of neural networks to thermographic data reduction," *La thermographie infrarouge quantitative*, Société Française des Thermiciens, Dec. 7, 1994b.

Grinzato E, Marinetti S, Bison PG, Manduchi G, "Application of neural networks to thermographic data reduction for non-destructive evaluation," *Revue Générale de Thermique*, **34**[397]:17–27, 1995a.

Grinzato EG, Vavilov V, Bison PG, Marinetti S, Bressan C, "Methodology of processing experimental data in transient thermal nondestructive testing (NDT)," in Semanovich SA, ed., *Thermosense XVII, SPIE Proc.*, **2473**:167–178, 1995b.

Grinzato EG, Bison PG, Bressan C, Mazzoldi A, "NDE of frescoes by infrared thermography and lateral heating," *Proc. QIRT-98, Eurotherm Seminar*, **60**:64–70, 1998.

Grinzato EG, Peron F, Strada M, "Moisture monitoring of historical buildings by long-period temperature measurements," in Lemieux DH, Snell JR, eds., *Thermosense XXI, SPIE Proc.*, **3700**:471–482, 1999.

Grinzato EG, Bressan C, Peron F, Romagnoni P, Stevan AG, "Indoor climatic conditions of ancient buildings by numerical simulation and thermographic measurements," in Dinwiddie RB, Lemieux DH, eds., *Thermosense XXII, SPIE Proc.*, **4020**:314–323, 2000.

Gros XE, "Characterisation of low energy impact damages in composites," *Journal of Reinforced Plastics and Composites*, **15**[3]:267–282, 1996.

Gros XE, Strachan P, Lowden DW, "Theory and implementation of NDT data fusion," *Research in Nondestructive Evaluation*, **6**[4]:227–236, 1995.

Gross W, Hierl Th, Scheuerpflug H, Schirl U, Schulz MJ, "Quality control of heat pipelines and sleeve joints by infrared measurements," in Lemieux DH, Snell JR, eds., *Thermosense XXI, SPIE Proc.*, **3700**:63–69, 1999a.

Gross W, Scheuerpflug H, Zettner J, Hierl Th, Schulz MI, Karg F, "Defect localization in CuInSe$_2$ solar modules by thermal infrared microscopy," in Lemieux DH, Snell JR, eds., *Thermosense XXI, SPIE Proc.*, **3700**:70–76, 1999b.

Grover P, "NDT solution: infrared inspection of boilers and process heaters," *Materials Evaluation*, **49**[10]:1272–1274, 1991.

Grover PE, "Applying ANSI/IEEE/NEMA temperature standards to infrared inspections," in Eklund JK, ed., *Thermosense XIV, SPIE Proc.*, **1682**:101–107, 1992.

Grover P, Kelch CK, "Tech tips: infrared thermography identifies damaged capacitors," *Materials Evaluation*, **54**[5]:587, 1996; erratum, *Materials Evaluation*, **54**[11]:1284, 1996.

Grund MV, "Determination of thermocouple location in reentry vehicles by penetrating radiation," *Nondestructive Testing*, **18**[4]:258–262, 1960.

Gruss C, Balageas D, "Theoretical and experimental applications of the flying spot camera," *Proc. QIRT-92, Eurotherm Seminar*, **27**:19–24, 1992.

Guazzone LJP, Danjoux R, "Nondestructive investigation by thermal waves of new aeroengine components," in Semanovich SA, ed., *Thermosense XVII, SPIE Proc.*, **2473**:244–251, 1995.

Gubala M, Teague JR, Bums JM, Di Marco JS, "Field portable HgCdTe MWIR staring array imaging system," *Proc. SPIE*, **1050**:105–111, 1989.

Guglielmino E, La Rosa G, Oliveri SM, "Thermal infrared analysis of plastics under monoaxial loads," *Experimental Techniques*, **20**[1]:9–13, 1996.

Guilhem D, Garcin H, Seigneur A, Schlosser J, Vandelle P, "Infrared thermography on Tore-Supra, the French experimental Tokamak on nuclear controlled fusion," *Proc. QIRT-92, Eurotherm Seminar*, **27**:25–31, 1992.

Guitonny JB, Ozoki Z, Mansanares AM, Le Liboux M, Fournier D, Boccara AC, "Contrast enhancement in crack detection by stimulated differential infrared radiometry: modelisation and experiments," *Optics Communications*, **104**[1–3]:61–64, 1993.

Gulyaev YV, Markov AG, Koreneva LG, Zakharov PV, "Dynamical infrared thermography in humans," *IEEE Engineering in Medicine and Biology*, **14**[6]:766–771, 1995.

Haack A, Schreyer J, Jackel G, "State-of-the-art of non-destructive testing methods for determining the state of a tunnel lining," *Tunnelling and Underground Space Technology*, **10**[4]:413–431, 1995.

Haddon JF, "Generalised threshold selection for edge detection," *Pattern Recognition*, 21[3]:195–203, 1988.

Haferkamp H, Burmester I, "Fast and automatic thermographic material identification for the recycling process," in Snell JR, Wurzbach RN, eds., *Thermosense XX, SPIE Proc.*, **3361**:14–25, 1998.

Hagan MT, Demuth HB, Beale M, *Neural Networks Design*, PWS Publishing Company, Boston, 1996.

Hagemaier DJ, "Thermal insulation moisture detection," *Materials Evaluation*, 28[3]:55–60, 1970.

Halabe UB, Bidigalu GM, GangaRao HVS, Ross RJ, "Nondestructive evaluation of green wood using stress wave and transverse vibration techniques," *Materials Evaluation*, 55[9]:1013–1018, 1997.

Hall EL, *Computer Image Processing and Recognition*, Academic Press, San Diego, CA, 584 pp., 1979.

Hall JA, "Arrays and charged-coupled devices," *Applied Optics and Optical Engineering*, pp. 352–355, Academic Press, San Diego, CA, 1980.

Hameury J, "Determination of uncertainties for emissivity measurements in the temperature range [200°C–900°C]," *Proc. QIRT-94, Eurotherm Seminar*, **42**:113–117, 1994.

Hamit F, "Military infrared takes to the road: the Cadillac's oncoming 'night vision' option," *Advanced Imaging*, Oct., pp. 34–37, 1998.

Hammaker RC, Colsher RJ, "Advanced leak location–research evaluation demonstration (ALL-RED) project," in Snell JR, Jr., ed., *Thermosense XVI, SPIE Proc.*, **2245**:22–26, 1994.

Hammaker RC, Colsher RI, Miles JJ, Madding RP, "Evaluation of internal boiler components and gases using a high-temperature infrared (IR) lens," in Burleigh DD, Spicer JWM, eds., *Thermosense XVIII, SPIE Proc.*, **2766**:74–82, 1996.

Hamrelius T, "Accurate temperature measurement in thermography: an overview of relevant features, parameters and definitions," in Baird GS, ed., *Thermosense XIII, SPIE Proc.*, **1467**:448–457, 1991.

Hamrelius T, "Accurate temperature measurement in thermography," *Proc. QIRT-92, Eurotherm Seminar*, **27**:39–45, 1992.

Hamzah AR, Delpech P, Saintey MB, Almond DP, "Experimental investigation of defect sizing by transient thermography," *Insight: Non-Destructive Testing and Condition Monitoring*, **38**[3], 1996.

Han X, Favro LD, Kuo PK, Thomas RL, "Early-time pulse-echo thermal wave imaging," in Thompson DO, Chimenti DE, eds., *Review of Progress in Quantitative Nondestructive Evaluation*, **15A**:519–524, 1996.

Han X, Favro LD, Ahmed T, Ouyang Z, Wang L, Wang X, Zhang F, Kuo PK, Thomas RL, "Quantitative thermal wave imaging of corrosion on aircraft," in Thompson DO, Chimenti DE, eds., *Review of Progress in Quantitative Nondestructive Evaluation*, **16A**:353–356, 1998a.

Han X, Favro LD, Ahmed T, Wang X, Thomas RL, "Delamination depth determinations in composites using thermal wave imaging," in Thompson DO, Chimenti DE, eds., *Review of Progress in Quantitative Nondestructive Evaluation*, **18A**:593–596, 1998b.

Han X, Favro LD, Thomas RL, Chadwick MM, Caliskan AG, Griffith N, "Fast infrared imaging of static and dynamic crush tests of composite tubes," in Lemieux DH, Snell JR, eds., *Thermosense XXI, SPIE Proc.*, **3700**:48–53, 1999.

Hanks J, "A critical part of electronic inspection," *Photonics Spectra*, Jan., pp. 116–120, 1998.

Hansen RD, Olds TS, Richards DA, Richards CR, Leelarthaepin B, "Infrared thermometry in the diagnosis and treatment of heat exhaustion," *International Journal of Sports Medicine*, **17**[1]:66–70, 1996.

Happoldt PG, Ellingson WA, Gardiner TP, Krueger JA, "Defect detection in multilayered, plasma-sprayed zirconia by time-resolved infrared radiometry: a comparison between analytical and experimental methods," in Snell JR, Jr., ed., *Thermosense XVI, SPIE Proc.*, **2245**:210–219, 1994.

Haralambopoulos DA, Paparsenos GF, "Assessing the thermal insulation of old buildings: the need for in situ spot measurements of thermal resistance and planar infrared thermography," *Energy Conversion and Management*, **39**[1–2]:65–79, 1998.

Hardin RW, "Biometric recognition: photonics ushers in a new age of security," *Photonics Spectra*, Nov., pp. 88–100, 1997.

Harding JR, "Thermal imaging in the investigation of deep venous thrombosis," *Proc. QIRT-98, Eurotherm Seminar*, **60**:26–28, 1998.

Harding JR, Barnes KM, "Is DVT excluded by normal thermal imaging? An outcome study of 700 cases," *Proc. QIRT-98, Eurotherm Seminar*, **60**:32–35, 1998.

Harding JR, Wright AM, "Non-invasive imaging in the investigation of deep vein thrombosis in pregnancy," *Proc. QIRT-98, Eurotherm Seminar*, **60**:29–31, 1998.

Harding JR, Wertheim DF, Williams RJ, Melhuish JM, Banerjee D, Harding KG, "Infrared imaging in diabetic foot ulceration," *Proc. QIRT-98, Eurotherm Seminar*, **60**:36–39, 1998.

Harding KG, *Optical Considerations for Machine Vision*, Industrial Technology Institute, Ann Arbor, MI, 1988.

Hardisty H, Shirvani H, "Thermal imaging in electronics and rotating machinery," *British Journal of Non-Destructive Testing*, **36**[2]:73–78, 1994.

Harper BM, Norman TD, Exley DR, "Hydrogen fire detection using thermal imaging and its application to space launch vehicles," in Semanovich SA, ed., *Thermosense XII, SPIE Proc.*, **1313**:309–320, 1990.

Harris WJ, Woods DC, "Thermal stress studies using optical holographic interferometry," *Materials Evaluation*, **32**[3]:50–56, 1974.

Hartikainen J, Lehtiniemi R, Rantala J, Varis J, Luukkala M, "Fast infrared line-scanning method and its application," in Thompson DO, Chimenti DE, eds., *Review of Progress in Quantitative Nondestructive Evaluation*, **13A**:401–408, 1994.

Hartsock DL, Dinwiddie RB, Fash JW, Dalka T, Smith GH, Yi YB, Hecht R, "Development of a high-speed system for temperature mapping of a rotating target," in Dinwiddie RB, Lemieux DH, eds., *Thermosense XXII, SPIE Proc.*, **4020**:2–9, 2000.

Hayden HC, "Data smoothing routine," *Computers in Physics*, **1**[1]:74–75, 1987.

Heath DM, Winfree WP, "Thermal diffusivity measurement in carbon–carbon composites," in Thompson DO, Chimenti DE, eds., *Review of Progress in Quantitative Nondestructive Evaluation*, **8B**:1613–1619, 1989.

Heinrich H, "Thermography in Germany: state of the art," in Dinwiddie RB, Lemieux DH, eds., *Thermosense XXII, SPIE Proc.*, **4020**:310–313, 2000.

Heinrich H, Dahlem KH, "Quantitative infrared thermography in fire tests," *Proc. QIRT-98, Eurotherm Seminar*, **60**:223–227, 1998.

Heinrich H, Dahlem KH, "Thermography of glazings in fire tests," in Lemieux DH, Snell JR, eds., *Thermosense XXI, SPIE Proc.*, **3700**:494–501, 1999.

Henry RC, Hansman RJ, Balageas DL, "Study of heat transfer enhancement on surface protuberances using infrared technique," *Proc. QIRT-94, Eurotherm Seminar*, **42**:166–171, 1994.

Henz JA, Himasekhar K, "Design sensitivities of mold-cooling CAE software: an experimental verification," *Advances in Polymer Technology*, **15**[1]:1–16, 1996.

Herby G, "Traitements d'images infrarouge," *L'Onde Électrique*, **68**[2]:53–57, 1988.

Herman RA, "Aspects of using infrared for electronic circuit diagnosis," *Materials Evaluation*, **25**[9]:201–205, 1967.

Hermanson KS, Sandor BI, "Corrosion fatigue modeling via differential infrared thermography," *Experimental Techniques*, **22**[3]:19–21, 1998.

Hershel W, "Experiments on the solar, and on the terrestrial rays that occasion heat; with a comparative view of the laws to which light and heat, or rather the rays which occasion them, are subject, in order to determine whether they are the same or different," *Philosophical Transactions of the Royal Society of London*, **90**:293–437, 1800a.

Hershel W, "Investigation of the powers of the prismatic colours to heat and illuminate objects; with remarks, that prove the different refrangibility of radiant heat; to which is added, an inquiry into the method of viewing the sun advantageously, with telescopes of large apertures and high magnifying powers," *Philosophical Transactions of the Royal Society of London*, **90**:255, 1800b.

Hershey W, Kim JM, "Impulse response applications to nondestructive testing," in McGonnagle WJ, ed., *International Advances in Nondestructive Testing*, **15**:289–311, 1990.

Herve P, Morel A, "Thermography improvements using ultraviolet pyrometry," *Proc. QIRT-96, Eurotherm Seminar*, **50**:26–31, 1996.

Hetsroni G, Rozenblit R, Yarin LP, "Hot-foil infrared technique for studying the temperature field of a wall," *Measurement Science and Technology*, **7**[10]:1418–1427, 1996.

Hobbs C, "Quantitative measurement of thermal parameters over large areas using pulse video thermography," in Baird GS, ed., *Thermosense XIII, SPIE Proc.*, **1467**:264–277, 1991.

Hobbs C, Temple A, "Inspection of aerospace structures using transient thermography," *British Journal of Non-Destructive Testing*, **35**[4]:183–189, 1993.

Hockings C, "Back to basics: infrared equipment terminology," *Materials Evaluation*, **55**[9]:955–957, 1997.

Hogan H, "Image is everything," *Photonics Spectra*, Dec., pp. 82–88, 1998.

Holman JP, *Heat Transfer*, McGraw-Hill, New York, 1981.

Holmsten D, "Thermographic sensing for on-line industrial control," in Cielo P, ed., *SPIE Proc. Optical Techniques for Industrial Inspection*, **665**:75–87, 1986.

Holmsten D, Houis R, "High-resolution thermal scanning for hot-strip mills," in Semanovich SA, ed., *Thermosense XII, SPIE Proc.*, **1313**:322–331, 1990.

Holst GC, Pickard JW, "Analysis of observer minimum resolvable temperature responses," in Huber AJ, Triplett MJ, Wolverton JR, eds., *SPIE Proc. Imaging Infrared: Scene Simulation, Modeling, and Real Time Image Tracking*, **1110**:252–258, 1989.

Hook SJ, Abbott EA, Grove C, Kahle AB, Palluconi F, "Use of multispectral thermal infrared data in geological studies," in Rencz AN, ed., *Remote Sensing for the Earth Sciences*, Vol. 3, pp. 59–110, Wiley (in cooperation with the American Society for Photogrammetry and Remote Sensing), New York, 1998.

Horn BK, *Robot Vision*, pp. 243–269, MIT Press, Cambridge, MA, 1982.

Horn BKP, *Robot Vision*, McGraw-Hill, New York, 509 pp., 1986.

Houlbert AS, "Détection de défauts subsurfaciques dans les matériaux composites par thermographie infrarouge," *Report LEMTA* (Université de Nancy), **875**, 1991.

Howell PA, Winfree WP, "Numerical solutions for heat flow in adhesive lap joints," in Thompson DO, Chimenti DE, eds., *Review of Progress in Quantitative Nondestructive Evaluation*, **11A**:457–464, 1998.

Hsieh CK, Kassab AJ, "A general method for the solution of inverse heat conduction problems with partially unknown system geometries," *International Journal of Heat and Mass Transfer*, **29**[1]:47–58, 1986.

Huang Y, Jun X, Shih C, "Application of infrared technique to research on tensile test," *Materials Evaluation*, **38**[12]:76–79, 1980.

Huang Y, Li SY, Lin XR, Shih C, "Using the method of infrared sensing for monitoring fatigue process of metals," *Materials Evaluation*, **42**[8]:1020–1024, 1984.

Hudson B, "Modern techniques in non-destructive testing," *Metals and Materials*, **1**[2]:88–90, 1985.

Hudson RD, *Infrared System Engineering*, Wiley-Interscience, New York, 642 pp., 1969.

Hughett P, "Image processing software for real time quantitative infrared thermography," in Madding RP, ed., *Thermosense IX, SPIE Proc.*, **780**:176–183, 1987.

Humphries-Black H, "Infrared testing in art conservation," *Materials Evaluation*, **45**[4]:426–428, 430, 1987.

Hunter GB, Allemand CD, Eagar TW, "Multiwavelength pyrometry: an improved method," *Optical Engineering*, **24**[6]:1081–1085, 1985.

Hunter GB, Allemand CD, Eagar TW, "Prototype device for multiwavelength pyrometry," *Optical Engineering*, **25**[11]:1222–1231, 1986.

Hurley TL, "Infrared techniques for electric utilities," in Maldague XPV, ed., *Infrared Methodology and Technology*, pp. 265–317, Gordon and Breach, New York, 1994.

Hüttner R, Schollmeyer E, "Photothermal infrared radiometry applied to textile materials: general characterization and determination of moisture content," *Proc. QIRT-94, Eurotherm Seminar*, **42**:226–231, 1994.

Hüttner R, Schollmeyer E, "The on-line detection of moisture and moist coatings by means of thermal waves," *Proc. QIRT-96, Eurotherm Seminar*, **50**:215–219, 1996.

Ilyinski A, "Inverse conduction problems and quantitative infrared thermography," *Proc. QIRT-96, Eurotherm Seminar*, **50**:117–122, 1996.

Imbert B, Le Maoult Y, Quinard J, "Some problems in analysing premixed flames by infrared thermography," *Proc. QIRT-92, Eurotherm Seminar*, **27**:326–332, 1992.

Imhof RE, Birch DJS, Moksin MM, Webb J, Willson PH, Strivens TA, "Thermal wave NDE," *British Journal of Non-Destructive Testing*, **33**[4]:172–176, 1991.

Inagaki T, Okamoto Y, "Diagnosis of the leakage point on a structure surface using infrared thermography in near ambient conditions," *NDT&E International*, **30**[3]:135–142, 1997.

Inagaki T, Okamoto Y, Fan Z, Kurokawa K, "Temperature measurement and accuracy of bicolored radiometer applying pseudo graybody approximation," in Snell JR, Jr., ed., *Thermosense XVI, SPIE Proc.*, **2245**:274–285, 1994a.

Inagaki T, Suzuki K, Okamoto Y, Sato M, "Uncertainty analysis of surface temperature measurement using infrared radiometer," in Snell JR, Jr., ed., *Thermosense XVI, SPIE Proc.*, **2245**:262–273, 1994b.

Inagaki T, Ishii T, Iwamoto T, "On the NDT and E for the diagnosis of defects using infrared thermography," *NDT&E International*, **32**:247–257, 1999.

Incropera FP, DeWitt DP, *Fundamentals of Heat and Mass Transfer*, Wiley, New York, 919 pp., 1990.

IRC-160 User Manual, Cincinnati Electronics, Cincinnati, OH, 1993.

Israel S, "Why machine vision applications need frame grabbers with on-board memory," *Advanced Imaging*, Apr., pp. 8–11, 1998.

Jacobsen S, "Dual antenna concept for simultaneous thermography and hyperthermic heating," *Electronics Letters*, **34**[1]:94–95, 1998.

Jacoby MH, Lingenfelter DE, "Monitoring the performance of industrial computed tomography inspection systems," *Materials Evaluation*, **47**[Oct.]:1196–1199, 1989.

Jain R, Martin WN, Aggarwal JK, "Segmentation through the detection of changes due to motion," *Computer Vision, Graphics, and Image Processing*, **11**:13–34, 1979.

Jakowatz CV, Smiel AJ, Eichel PH, "Pyroelectric line scanner for remote IR imaging of vehicles," in Spiro IJ, ed., *Infrared Technology XIII, SPIE Proc.*, **819**:36–41, 1987.

Jambunathan K, Hartle SL, Ashforth-Frost S, Fontama VN, "Evaluating convective heat transfer coefficients using neural networks," *International Journal of Heat and Mass Transfer*, **39**[11]:2329–2332, 1996.

James PH, Welch CS, Winfree WP, "A numerical grid generation scheme for thermal stimulation in laminated structures," in Thompson DO, Chimenti DE, eds., *Review of Progress in Quantitative Nondestructive Evaluation*, **8A**:801–809, 1989.

Jarem JM, Pierluissi JH, Ng WW, "A transmittance model for atmospheric methane," in Spiro IJ, Mollicone RA, eds., *Infrared Technology X, SPIE Proc.*, **510**:94–100, 1984.

Jecic S, Goja S, "Recent development in thermoelastic stress analysis," *Proc. QIRT-98, Eurotherm Seminar*, **60**:108–111, 1998.

Jegou C, Brenier Y, "Thermal analysis at very high temperature," *Proc. QIRT-94, Eurotherm Seminar*, **42**:118–123, 1994.

Jen CK, Cielo P, Maldague X, "Non-contact ultrasonic characterization of piezoelectric ceramics," *Journal of the American Ceramic Society*, **68**[6]:C-146–C-148, 1985.

Jenkins FA, *Fundamentals of Optics*, 4th ed., McGraw-Hill, New York, 1976.

Joachim RJ, Pitasi MJ, "Infrared and thermal evaluation of a phased array antenna," *Materials Evaluation*, 29[9]:193–198, 204, 1971.

Johnson DE, Johnson JR, Hilburn JL, *Electric Circuit Analysis*, Prentice Hall, Upper Saddle River, NJ, 779 pp., 1992.

Johnson DE, Johnson RJ, Hilburn JL, Scott PD, *Electric Circuit Analysis*, 3rd ed., Prentice Hall, Upper Saddle River, NJ, 1997.

Johnson RB, Feng C, Fehribach JD, "On the validity and techniques of temperature and emissivity measurements," in Lucier RD, ed., *Thermosense X, SPIE Proc.*, **934**:202–206, 1988.

Johnson S, Neff R, "Improving thermographic analysis using image histories," in Wurzbach RN, Burleigh DD, eds., *Thermosense XIX, SPIE Proc.*, **3056**:164–166, 1997.

Jokinen P, Kauppinen TT, Varis T, Korpiola K, "Using IR thermography as a quality control tool for thermal spraying in the aircraft industry," in Dinwiddie RB, Lemieux DH, eds., *Thermosense XXII, SPIE Proc.*, **4020**:232–239, 2000.

Jones MW, Chandler JM, Bernstein S, "Inspection of low-density polyethylene using near-infrared imaging," in Semanovich SA, ed., *Thermosense XVII, SPIE Proc.*, **2473**:273–280, 1995.

Jones RW, White LM, "Infrared evaluation of multilayer boards," *Materials Evaluation*, 27[2]:37–41, 1969.

Jones TS, "Field condition limitations for thermography of marine composites," in Burleigh DD, Spicer JWM, eds., *Thermosense XVIII, SPIE Proc.*, **2766**:164–173, 1996.

Jones TS, Berger H, "Thermographic detection of impact damage in graphite–epoxy composites," *Materials Evaluation*, 50[12]:1446–1453, 1992.

Jones TS, Lindgren EA, "Thermographic inspection of marine composite structures," in Snell JR, Jr., ed., *Thermosense XVI, SPIE Proc.*, **2245**:173–175, 1994.

Jones TS, Berger H, Weaver E, "Large-area thermographic inspection of GRP composite marine vessel hulls," in Allen LR, ed., *Thermosense XV, SPIE Proc.*, **1933**:197–206, 1993.

Jones TS, Lindgren EA, Pergantis CG, "Infrared inspection of advanced army composites," in Semanovich SA, ed., *Thermosense XVII, SPIE Proc.*, **2473**:214–218, 1995.

Jónsson RH, Bjornsson S, "Increased image resolution of aerial thermography based on signal processing: distortion correction in line scanners," in Semanovich SA, ed., *Thermosense XVII, SPIE Proc.*, **2473**:360–373, 1995.

Jost SR, Meikleham VF, Myers TH, "InSb: a key material for IR detector applications," *Materials Research Society Symposium Proc.*, **90**:429–435, 1987.

Jouglar J, Mergui M, Vuillermoz PL, "Dynamical strain measurement by IRT," *Proc. QIRT-96, Eurotherm Seminar*, **50**:85–90, 1996.

Jung A, Zuber J, Sacha P, Lukasiewicz J, "Thermovision monitoring of the tuberculin reaction with children," *Proc. QIRT-96, Eurotherm Seminar*, **50**:339–342, 1996.

Kaasinen HI, "Assessing the quality of the waterproofing of bridges using thermography," *British Journal of Non-Destructive Testing*, 35[6]:301–304, 1993a.

Kaasinen HI, "Infrared thermography for assessing the quality of waterproofing of bridges under construction," in Allen LR, ed., *Thermosense XV, SPIE Proc.*, **1933**:55–60, 1993b.

Kaczmarek M, Nowakowski A, Siebert J, Rogowski J, "Intraoperative thermal coronary angiography: correlation between internal mammary artery (IMA) free flow and thermographic measurement during coronary grafting," *Proc. QIRT-98, Eurotherm Seminar*, **60**:250–257, 1998.

Kaiser JH, "Millimeter wave heating for thermographic inspection of carbon fiber–reinforced composites," *Materials Evaluation*, **52**[5]:597–599, 1994.

Kalus G, Bein BK, Pelzl J, Bosse H, Linnenbrügger A, "Characterizations of tribological protective films and friction wear by IR radiometry of thermal waves," *Proc. QIRT-96, Eurotherm Seminar*, **50**:203–208, 1996.

Kamoi A, Okamoto Y, Makishi O, "Minimum radiation temperature difference and detection limit of flaws by means of an infrared radiometer," in Lemieux DH, Snell JR, eds., *Thermosense XXI, SPIE Proc.*, **3700**:104–111, 1999.

Kamoi A, Okamoto Y, Eto M, "Experimental study on the effect of environmental fluctuations affecting thermal images of infrared radiometer," in Dinwiddie RB, Lemieux DH, eds., *Thermosense XXII, SPIE Proc.*, **4020**:374–381, 2000.

Kaplan H, "Practical applications of infrared thermal sensing and imaging equipment," *SPIE Proc.*, **TT13**, 137 pp., 1993.

Kaplan H, Friedman RF, "Two new portable infrared instruments for plant inspection," *Materials Evaluation*, **39**[2]:175–179, 1981.

Kaplan H, "Process control using IR sensors and scanners," *Photonics Spectra*, Dec., pp. 92–95, 1997a.

Kaplan H, "Gold rules the world of infrared," *Photonics Spectra*, Nov., pp. 54–56, 1997b.

Kaplan H, "Falling prices spawn IR training crisis," *Photonics Spectra*, Jan., pp. 73–74, 1998a.

Kaplan H, "Process control embraces thermography," *Photonics Spectra*, Feb., pp. 61–62, 1998b.

Kaplan H, "Benefits from aerospace programs," *Photonics Spectra*, June, pp. 108–113, 1998c.

Kaplan H, "Laser alignment helps appearance, safety, efficiency," *Photonics Spectra*, July, pp. 96–100, 1998d.

Kaplan H, "Optimizing combustion efficiency," *Photonics Spectra*, Apr., pp. 57–58, 1998e.

Kaplan H, "Smoke-penetrating vision systems save lives," *Photonics Spectra*, **32**[7]:63–64, 1998f.

Kaplan H, Zayicek PA, "Application of differential infrared thermography in power generation facilities," in Semanovich SA, ed., *Thermosense XVII, SPIE Proc.*, **2473**:67–74, 1995.

Kapur JN, Sahoo PK, Wong AKC, "A new method for gray-level picture thresholding using the entropy of the histogram," *Computer Vision, Graphics, and Image Processing*, **29**:273–285, 1985.

Karger-Kocsis J, "Assessment of the damage zone in a glass mat–reinforced thermoplastic polypropylene (GMT-PP) by infrared thermography," *Journal of Materials Science Letters*, **13**[19]:1422–1425, 1994.

Karger-Kocsis J, Fejes-Kozma Z, "Failure mode and damage zone development in a GMT-PP by acoustic emission and thermography," *Journal of Reinforced Plastics and Composites*, **13**[9]:768–792, 1994.

Karger-Kocsis J, Harmia T, Czigany T, "Comparison of the fracture and failure behavior of polypropylene composites reinforced by long glass fibers and by glass mats," *Composites Science and Technology*, 54[3]:287–298, 1995.

Karger-Kocsis J, Czigany T, Mayer J, "Fracture behaviour and damage growth in knitted carbon fibre fabric reinforced polyethylmethacrylate," *Plastics, Rubber and Composites Processing and Applications*, 25[3]:109–114, 1996.

Karjanmaa JM, "Thermal imaging and paper-finishing machines," in Allen LR, ed., *Thermosense XV, SPIE Proc.*, 1933:110–118, 1993.

Karpen W, Wu D, Steegmüller R, Busse G, "Depth profiling of orientation in laminates with local lockin thermography," *Proc. QIRT-94, Eurotherm Seminar*, 42:281–286, 1994.

Karpen W, Wu D, Busse G, "A theoretical model for the measurement of fiber orientation with thermal waves," *Research in Nondestructive Evaluation*, 11[4]:179–198, 1999.

Kassab AJ, Hsieh CK, "Application of infrared scanners and inverse heat conduction methods to infrared computerized axial tomography," *Review of Scientific Instruments*, 58[1]:89–95, 1987.

Kastek M, Madura H, Polakowski H, "Fast scanning pyrometer for temperature measurements of car wheels," in Sulej S, ed., *Proc. QIRT-98, Eurotherm Seminar*, 60:333–337, 1998a.

Kastek M, Madura H, Polakowski H, Sokolowski W, "IR camera and pyrometer used for woodworking control," *Proc. QIRT-98, Eurotherm Seminar*, 60:195–199, 1998b.

Kastelic JP, Cook RB, Coulter GH, Saacke RG, "Ejaculation increases scrotal surface temperature in bulls with intact epididymides," *Theriogenology*, 46[5]:889–892, 1996.

Kauppinen TT, Häkkilä A, Hekkanen M, "Renovation concepts for private houses: the use of thermography as a supporting method," in Eklund JK, ed., *Thermosense XIV, SPIE Proc.*, 1682:54–63, 1992.

Kauppinen TT, Hyartt J, Sasi L, "Thermal performance of prefabricated multistory houses in Tallinn, Estonia, based on IR survey," in Wurzbach RN, Burleigh DD, eds., *Thermosense XIX, SPIE Proc.*, 3056:59–70, 1997.

Kauppinen TT, Alamaki P, Lilja J, Ruotsalainen K, "Thermography in the condition monitoring of a refractory lining," in Lemieux DH, Snell JR, eds., *Thermosense XXI, SPIE Proc.*, 3700:214–226, 1999.

Kauppinen TT, Maierhofer C, Wiggenhauser H, Arndt D, "Use of cooling down thermography in locating below-surface defects of building facades," in Dinwiddie RB, Lemieux DH, eds., *Thermosense XXII, SPIE Proc.*, 4020:345–359, 2000.

Kehoe L, Kelly PV, Crean GM, "Laser irradiated transient thermography inspection of iron–zinc alloy coatings on steel substrates," *Proc. QIRT-96, Eurotherm Seminar*, 50:209–214, 1996.

Kelch CK, Grover PE, "Using thermography to detect misalignment in coupled equipment," in Burleigh DD, Spicer JWM, eds., *Thermosense XVIII, SPIE Proc.*, 2766:91–100, 1996.

Kennedy HV, "Modeling second-generation thermal imaging systems," *Optical Engineering*, 30[11]:1771–1778, 1991.

Kenning DBR, Youyou Y, "Pool boiling heat transfer on a thin plate: features revealed by liquid crystal thermography," *International Journal of Heat and Mass Transfer*, 39[15]:3117–3137, 1996.

Kiiskinen HT, Pakarinen PI, "Infrared thermography for examination of paper structure," in Snell JR, Wurzbach RN, eds., *Thermosense XX, SPIE Proc.*, **3361**:228–233, 1998.

Kiiskinen HT, Pakarinen PI, Luontama M, Laitinen A, "Using infrared thermography as a tool to analyze curling and cockling of paper," in Eklund JK, ed., *Thermosense XIV, SPIE Proc.*, **1682**:134–141, 1992.

Kiliski S, "Magnetic disk storage for video images," *Advanced Imaging*, **6**[5]:32–34, 1991.

Kimata M, Denda M, Yutani N, Iwade S, Tsubouchi N, "High density Schottky-barrier infrared image sensor," *Thermosense X, SPIE Proc.*, **930**, 1988.

Kingslake R, *Applied Optics and Optical Engineering*, Vol. II, Academic Press, San Diego, CA, pp. 212–213, 1965.

Kirkwood JJ, "Behavioral observations in thermal imaging of the big brown bat, *Eptesicus fuscu*," in Baird GS, ed., *Thermosense XIII, SPIE Proc.*, **1467**:369–371, 1991.

Kleinfeld JM, "Finite element analysis as a tool for thermography," in Lemieux DH, Snell JR, eds., *Thermosense XXI, SPIE Proc.*, **3700**:6–13, 1999a.

Kleinfeld JM, "Examination of a carton sealing line using a thermographic scanner," in Lemieux DH, Snell JR, eds., *Thermosense XXI, SPIE Proc.*, **3700**:237–244, 1999b.

Knehans A, Ledford J, "Impact of aerial infrared roof moisture scans on the U.S. Army's ROOFER program," in Allen LR, ed., *Thermosense XV, SPIE Proc.*, **1933**:67–75, 1993.

Kobayashi A, Ueda S, "Development of television camera for gas pipeline inspection," *5th Pan Pacific Conference on Nondestructive Testing*, Vancouver, British Columbia, Canada, Apr. pp. 391–402, 1987.

Kocanda A, Czyzewski P, Kita K, Pregowski P, "IR thermography application in studying phenomena in warm extrusion tooling," *Proc. QIRT-98, Eurotherm Seminar*, **60**:238–244, 1998.

Kocsis JK, Kozma ZF, "Failure mode and damage zone development in a GMT-PP by acoustic emission and thermography," *NDT&E International*, **30**[2]:110, 1997.

Koehler H, "Infrared detectors continue to diversify," *Laser Focus World*, **27**[3]:A31–A34, 1991.

Koelzer J, Oesterschulze E, Deboy G, "Thermal imaging and measurement techniques for electronic materials and devices," *Microelectronic Engineering*, **31**[4]:251–270, 1996.

Kofanov YuN, Uvaisov SU, Arlov BL, Pjatnitskaja GA, Segen AV, "Thermovision control, interferometry and computer modeling for increasing reliability and quality of electronic products," *Proc. QIRT-96, Eurotherm Seminar*, **50**:394–400, 1996.

Kogelnik H, in Parker SP, ed., *Concise Encyclopedia of Science and Technology*, McGraw-Hill, New York, 2065 pp., 1984.

Kohler A, Hoffmann R, Platz A, Bino M, "Diagnostic value of duplex ultrasound and liquid crystal contact thermography in preclinical detection of deep vein thrombosis after proximal femur fractures," *Archives of Orthopaedic and Trauma Surgery*, **117**[1–2]:39–42, 1998.

Koppel T, Lahdeniemi M, Ekholm A, "Application of IR thermography for unsteady fluid-flow research," in Snell JR, Wurzbach RN, eds., *Thermosense XX, SPIE Proc.*, **3361**:40–48, 1998.

Koppel T, Ainola L, Ekholm A, Lähdeniemi M, "Heat transfer investigation in pipe by IR thermography," in Dinwiddie RB, Lemieux DH, eds., *Thermosense XXII, SPIE Proc.*, **4020**:267–275, 2000.

Kortüm G, *Reflectance Spectroscopy*, Springer-Verlag, Berlin, 1969.

Kos A, De Mey G, Boone E, "Experimental verification of the temperature distribution on ceramic substrates," *Journal of Physics D: Applied Physics*, **27**[10]:2163–2166, 1994.

Koskelainen L, "Predictive maintenance of district heating networks by infrared measurement," in Eklund JK, ed., *Thermosense XIV, SPIE Proc.*, **1682**:89–97, 1992.

Koskelainen L, "Database system for managing thermography files of district heating networks," in Allen LR, ed., *Thermosense XV, SPIE Proc.*, **1933**:2–8, 1993.

Kosonocky WF, "Review of Schottky-barrier imager technology," in Dereniak EL, Sampson RE, eds., *SPIE Proc. Infrared Detectors and Focal Plane Arrays*, **1308**:2–26, 1990.

Kotelly G, "Vision '98 reflects German upswing in image processing," *Advanced Imaging*, Nov., pp. 17–18, 1998.

Kourous HE, Shabestari BN, Luster SD, Sacha IP, "Online industrial thermography of die casting tooling using dual-wavelength IR imaging," in Snell JR, Wurzbach RN, eds., *Thermosense XX, SPIE Proc.*, **3361**:218–227, 1998.

Kovacevic R, Mohan R, Beardsley HE, "Monitoring of thermal energy distribution in abrasive waterjet cutting using infrared thermography," *Transactions of the American Society of Mechanical Engineers*, **118**[4]:555–563, 1996.

Kraft GD, Toy WN, *Microprogrammed Control and Reliable Design of Small Computers*, Prentice Hall, Upper Saddle River, NJ, 1981.

Kraft GD, Wing NT, *Microprogrammed Control and Reliable Design of Small Computers*, Prentice Hall, Upper Saddle River, NJ, 428 pp., 1981.

Krapez JC, "Contribution à la caractérisation des défauts de type délaminage ou cavité par thermographie stimulée," Ph.D. thesis, École Centrale des Arts et Manufacture de Paris, 533 pp., Mar. 19, 1991.

Krapez JC, "Thermal ellipsometry: a tool applied for in-depth resolved characterization of fibre orientation in composites," in Thompson DO, Chimenti DE, eds., *Review of Progress in Quantitative Nondestructive Evaluation*, **14**:533–540, 1996.

Krapez JC, "Compared performances of four algorithms used for modulation thermography," *Proc. QIRT-98, Eurotherm Seminar*, **60**:148–153, 1998.

Krapez JC, Balageas D, "Early detection of thermal contrast in pulsed stimulated infrared thermography," *Proc. QIRT-94, Eurotherm Seminar*, **42**:260–266, 1994.

Krapez JC, Cielo P, "Thermographic nondestructive evaluation: data inversion procedures. I: 1D analysis," *Research in Nondestructive Evaluation*, **3**[2]:81–100, 1991.

Krapez JC, Cielo P, Maldague X, Utracki LA, "Optothermal analysis of polymer composites," *Polymer Composites*, **8**[6]:396–407, 1987.

Krapez JC, Cielo P, Lamontagne M, "Reflective-cavity infrared temperature sensors: an analysis of sperical, conical and double wedge geometries," in Lettington AH, ed., *SPIE Proc. Infrared Technologies and Applications*, **1320**:186–201, 1990.

Krapez JC, Maldague X, Cielo P, "Thermographic nondestructive evaluation: data inversion procedures. II: 2-D analysis and experimental results," *Research in Nondestructive Evaluation*, **3**:101–124, 1991.

Krapez JC, Boscher DM, Delpech PhM, Déom AA, Gardette G, Balageas DL, "Time-resolved pulsed stimulated infrared thermography applied to carbon–epoxy non-destructive evaluation," *Proc. QIRT-92, Eurotherm Seminar*, **27**:195–200, 1992.

Krapez JC, Lepoutre F, Balageas D, "Early detection of thermal contrast in pulsed stimulated thermography," *Journal de Physique IV, Colloque C7*, **4**:47–50, 1994.

Krapez JC, Gardette G, Balageas D, "Thermal ellipsometry in steady-state or by lock-in thermography: application for anisotropic materials characterization," *Proc. QIRT-96, Eurotherm Seminar*, **50**:257–262, 1996.

Krapez JC, Legrandjacques L, Lepoutre F, Balageas DL, "Optimization of the photo-thermal camera for crack detection," *Proc. QIRT-98, Eurotherm Seminar*, **60**:305–310, 1998.

Kraska T, Zwolenik S, Wiecek B, "Examples of thermografic examination in pediatric orthopaedics," *Proc. QIRT-98, Eurotherm Seminar*, **60**:61–65, 1998.

Kraus K, "Blooming of second and third generation image intensifier tubes," M.Sc. thesis, Université Laval, 67 pp., 1997.

Krishnakumar N, Sitharama S, Hoylder R, Lybanon M, "Feature labelling in infrared oceanographic images," *Image and Vision Computing*, **8**[2]:142–147, 1990.

Krishnan G, Walters D, "Segmenting, intersecting and incomplete boundaries," in Trivedi MM, ed., *Applications of Artificial Intelligence VI, SPIE Proc.*, **937**:550–556, 1988.

Kröse BJA, Van Der Smargt PP, *An Introduction to Neural Networks*, 7th ed., Chaps. 3–5, University of Amsterdam, Faculty of Mathematics and Computer Science, 1995.

Krummar UKP, ed., *Parallel Architectures and Algorithms for Image Understanding*, Academic Press, San Diego, CA, 565 pp., 1991.

Kruse PW, "Thermal images move from military to market place," *Photonics Spectra*, **29**[3]:103–108, 1995.

Kruszka L, Nowacki WK, Oliferuk W, "Application of infrared thermography for determining the temperature distribution in Taylor's impact test," *Proc. QIRT-92, Eurotherm Seminar*, **27**:150–153, 1992.

Kulkarni MR, Brady RP, "Preliminary characterization of thermal diffusivity for carbon–carbon composites through pulsed video thermal imaging," *Journal of Testing and Evaluation*, **24**[5]:275–278, 1996.

Kulowitch PJ, Perez IM, Granata DM, "Flash infrared thermography for nondestructive testing (NDT) I/E of naval aircraft," in Semanovich SA, ed., *Thermosense XVII, SPIE Proc.*, **2473**:252–262, 1995.

Kuo PK, Feng ZJ, Ahmed T, Favro LD, Thomas RL, Hartikainen J, "Parallel thermal wave imaging using a vector lock-in video technique," *Optical Sciences*, **58**:415–419, 1987.

Kuo PK, Ahmed T, Favro LD, Jin H-J, Thomas RL, "Synchronous thermal wave IR video imaging for nondestructive evaluation," *Journal of Nondestructive Evaluation*, **8**[2]:97–106, 1989.

Kurilenko G, "Predicting crack resistance by infrared thermography," *Proc. QIRT-96, Eurotherm Seminar*, **50**:91–95, 1996.

Kurilenko G, Pshenichny A, "The investigation of metals' damage through thermal field kinetics," *Proc. QIRT-92, Eurotherm Seminar*, **27**:145–149, 1992.

Kutzscher EW, Zimmerman KH, Botkin JL, "Thermal and infrared methods for nondestructive testing of adhesive-bonded structures," *Materials Evaluation*, **26**[7]:143–148, 1968.

Kvernes I, Espeland M, Norholm O, "Plasma spraying of alloys and ceramics," *Scandinavian Journal of Metallurgy*, **17**:8–16, 1988.

LaBorde TC, Ferguson JH, Dubinsky M, Kirsch PJ, "Reimbursement for unproven therapies: the case of thermography," *Journal of the American Medical Association*, **270**[21]:2558–2559, 1993.

Lähdeniemi M, Ekholm A, Santamäki O, "Thermal imaging and frequency analysis," *Proc. QIRT-96, Eurotherm Seminar*, **50**:283–286, 1996a.

Lähdeniemi M, Ekholm A, Santamäki O, "IR frequency analysis in paper industry," in Burleigh DD, Spicer JWM, eds., *Thermosense XVIII, SPIE Proc.*, **2766**:2–4, 1996b.

Laine A, "Detection of failures in plastic composites using thermography," in Eklund JK, ed., *Thermosense XIV, SPIE Proc.*, **1682**:207–212, 1992.

Lajournade JB, Schanne G, "Le contrôle d'aspect de surface des billettes à haute température," *Proc. 3rd European Conference on Nondestructive Testing*, pp. 199–211, 1984.

Lan TTN, Haupt K, Seidel U, Walther HG, "Reconstruction of thermal defects from photothermal images," *Proc. QIRT-94, Eurotherm Seminar*, **42**:267–272, 1994.

Lan TTN, Son DT, Walther HG, "Photothermal characterization of surface hardened steel," *Proc. QIRT-96, Eurotherm Seminar*, **50**:220–226, 1996.

Lancaster WC, Thomson SC, Speakman JR, "Wing temperature in flying bats measured by infrared thermography," *Journal of Thermal Biology*, **22**[2]:109–116, 1997.

Lang D, "Initiation et propagation des endommagements dans les composites stratifiés carbone-époxy," *Matériaux et Techniques*, Apr.–May pp. 17–22, 1988.

Lanius MA, "Infrared applications for steam turbine condenser systems," in Dinwiddie RB, Lemieux DH, eds., *Thermosense XXII, SPIE Proc.*, **4020**:107–114, 2000.

Lao KQ, "Temperature measurement using video cameras," in Spiro IJ, ed., *Infrared Technology XIII, SPIE Proc.*, **819**:311–317, 1987.

Largouët Y, "Utilisation des réseaux de neurones pour le contrôle non-destructif par thermographie de phase pulsée," LVSN Report, Université Laval, 77 pp., 1997.

Largouët Y, Darabi A, Maldague X, "Depth evaluation in pulsed phase thermography with neural network," in Thompson DO, Chimenti DE, eds., *Review of Progress in Quantitative Nondestructive Evaluation*, **18A**:611–617, 1998a.

Largouët Y, Vallerand S, Maldague X, "Pulsed phase thermography of aluminum laminates: neural network investigation," *Proc. QIRT-98, Eurotherm Seminar*, **60**:167–171, 1998b.

Larish J, "A/D conversion revolution for CMOS sensors?" *Advanced Imaging*, Sept. p. 71, 1998.

Laszczynska J, Kaczanowski R, Wojcik K, Sacha P, Lukasiewicz J, Kowalski W, "Human body skin surface distribution as measured by infrared thermography in altitude hypoxia conditions," *Proc. QIRT-96, Eurotherm Seminar*, **50**:335–338, 1996.

Lau C, ed., *Neural Networks*, IEEE Press, New York, 327 pp., 1991.

Lau SK, Almond DP, Patel M, Corbett J, Quigley MBC, "Analysis of transient thermal inspection," in Lettington AH, ed., *SPIE Proc. Infrared Technology and Applications*, **1320**:178–185, 1990.

Laurendeau D, "Acquisition automatique et traitement mathématique de formes anatomiques," M.Sc. thesis, Université Laval, 1982.

Laurendeau D, Poussart D, "A segmentation algorithm for extracting 3D edges from range data," *Compint-85 (IEEE)*, pp. 765–767, 1985.

Laurin TC, "Editorial comment: no alternative to quality," *Photonics Spectra*, **24**[8]:14, 1990.

Lawnson RN, "Implications of surface temperature in the diagnosis of breast cancer," *Canadian Medical Association Journal*, **75**:309, 1956.

Lecky JE, "Using MMX technology to speed up machine vision algorithms, part 1," *Advanced Imaging*, Apr., pp. 15–19, 1998.

Lee DH, "Thermal analysis of integrated-circuit chips using thermographic imaging techniques," *IEEE Transactions on Instrumentation and Measurement*, **43**[6]:824–829, 1994.

Lee DJ, Krile TF, Mitra S, "Digital registration technique for sequential fundus images," in Tesher AD, ed., *Applications of Digital Image Processing X, SPIE Proc.*, **829**:293–300, 1987.

Lee JS, "Digital image enhancement and noise filtering by use of local statistics," *IEEE Transactions on Pattern Analysis and Machine Intelligence*, **2**[2]:165–168, 1980.

Lee TS, Shie JS, Li HT, "Application studies of a simulated low-density room-temperature IRFPA," in Snell JR, Wurzbach RN, eds., *Thermosense XX, SPIE Proc.*, **3361**:26–34, 1998.

Lefebvre M, Gil S, Brunet D, Natonek E, Baur C, Gugerli P, Pun T, "Computer vision and agricultural robotics for disease control: the potato operation," *Computers and Electronics in Agriculture*, **9**[1]:85–102, 1993.

Lehtiniemi RK, "Bibliography of the application of infrared thermography to electronics," in Lemieux DH, Snell JR, eds., *Thermosense XXI, SPIE Proc.*, **3700**:202–209, 1999.

Lehtiniemi RK, Rantala J, "Infrared thermography in electronics applications," in Lemieux DH, Snell JR, eds., *Thermosense XXI, SPIE Proc.*, **3700**:112–120, 1999.

Lehtiniemi R, Rantala J, Hartikainen J, "A photothermal line-scanning system for NDT of plasma-sprayed coatings of nuclear power plant components," *Research in Nondestructive Evaluation*, **6**[2]:99, 1995.

Lehtiniemi R, Fager CM, Rantala J, "True temperature measurement of electronics through infrared transparent materials," *Proc. QIRT-98, Eurotherm Seminar*, **60**:178–183, 1998a.

Lehtiniemi R, Hartikainen J, Rantala J, Varis J, Luukkala M, "Fast photothermal inspection of plasma-sprayed coatings of primary circulation seal rings of a nuclear reactor," in Thompson DO, Chimenti DE, eds., *Review of Progress in Quantitative Nondestructive Evaluation*, **11A**:441–446, 1998b.

LeMieux DH, "Use of infrared thermal imaging instrumentation in vehicle fire propagation studies," in Lemieux DH, Snell JR, eds., *Thermosense XXI, SPIE Proc.*, **3700**:332–342, 1999.

Le Niliot C, Gallet P, "Infrared thermography applied to the resolution of inverse heat conduction problems: recovery of heat line sources and boundary conditions," *Revue Générale de Thermique*, **37**[8]:629–643, 1998.

Leon-Garcia A, *Probability and Random Processes for Electrical Engineering*, 2nd ed., Secs. 3.3 and 6.2, Addison-Wesley, Toronto, Ontario, Canada, 596 pp., 1994.

Lepoutre F, Roger J-P, "Mesures thermiques par effet mirage," *Revue Générale de Thermique Fr.*, **301**:15–21, 1987.

Leslie JR, Wait JR, "Detection of overheated transmission line joints by means of a bolometer," *Transactions of the American Institute of Electrical Engineers*, **68**:64, 1949.

Lesniak JR, Bazile DI, "Forced-diffusion thermography technique and projector design," in Burleigh DD, Spicer JWM, eds., *Thermosense XVIII, SPIE Proc.*, **2766**:210–217, 1996.

Lesniak JR, Boyce BR, "Differential thermography applied to structural integrity assessment," in Semanovich SA, ed., *Thermosense XVII, SPIE Proc.*, **2473**:179–189, 1995.

Lesniak JR, Bazile DJ, Zickel MJ, "NDT solution: coating tolerant thermography for the detection of cracks in structures," *Materials Evaluation*, **55**[9]:961–965, 1997a.

Lesniak JR, Bazile DJ, Zickel MJ, "Coating-tolerant thermography for the detection of cracks in structures," in Wurzbach RN, Burleigh DD, eds., *Thermosense XIX, SPIE Proc.*, **3056**:235–241, 1997b.

Lesniak JR, Bazile DJ, Zickel MJ, "Theory and application of coating tolerant thermography," in Snell JR, Wurzbach RN, eds., *Thermosense XX, SPIE Proc.*, **3361**:325–330, 1998.

Leszczynski KW, Shalev S, "A robust algorithm for contrast enhancement by local histogram modification," *Image and Vision Computing*, **7**[3]:205–209, 1989.

Levesque P, Balageas DL, "Single-sided interferometric EMIR method for NDE of structures," *Proc. QIRT-98, Eurotherm Seminar*, **60**:36–42, 1998.

Levesque P, Déom A, Balageas D, "Nondestructive evaluation of absorbing materials using microwave stimulated infrared thermography," *Proc. QIRT-92, Eurotherm Seminar*, **27**:302–307, 1992a.

Levesque P, Lisiecki B, Kubin L, Caron P, Déom A, Balageas D, "Infrared thermography of plastic instabillties in a single crystal superalloy," *Proc. QIRT-92, Eurotherm Seminar*, **27**:135–140, 1992b.

Levine BF, Choi KK, Bethea CG, Malik J, "New 10 mm infrared detector using intersubband in resonant tunneling GaAlAs superlattices," *Applied Physics Letters*, **50**:16, 1987.

Levinstein H, "Infrared detectors," in Kingslake R, ed., *Applied Optics and Optical Engineering*, Vol. II, Chap. 8, pp. 311–347, Academic Press, San Diego, CA, 1965.

Levit AD, "Programming and graphics support for infrared thermal plotters," *Materials Evaluation*, **26**[9]:180–186, 1968.

Lewak R, "Infrared techniques in the nuclear power industry," in Maldague XPV, ed., *Infrared Methodology and Technology*, International Advances in Nondestructive Testing Monograph Series, Chap. 9, pp. 319–266, Gordon and Breach, New York, 1994.

Li Q, Liasi E, Simon DL, Du R, Bujas-Dimitrijevic J, Chen A, "Heating of industrial sewing machine needles: FEA model and verification using IR radiometry," in Lemieux DH, Snell JR, eds., *Thermosense XXI, SPIE Proc.*, **3700**:347–357, 1999.

Li S, Lu F, "New XT relation of blackbody radiation and its experimental investigation," in Snell JR, Jr., ed., *Thermosense XVI, SPIE Proc.*, **2245**:296–303, 1994.

Liddicoat TJ, Mansi MV, "An infrared radiometer using uncooled pyroelectric detectors for scientific and general use," in Hall PR, Seeley J, eds., *SPIE Proc. Infrared Systems: Design and Testing*, **916**:63–68, 1988.

Liebman S, "A 'no DSP' image processing board design choice: the FPGA alternative," *Advanced Imaging*, Sept., pp. 10–12, 1997.

Liebman S, "Infrared imaging's hot market: predictive maintenance," *Advanced Imaging*, Mar., p. 57, 1998.

Lin HM, Willson AN, "Median filters with adaptive length," *IEEE Transactions on Circuits and Systems*, **35**[6]:675–690, 1988.

Lindermeir E, Tank V, Haschberger P, "Contactless measurement of the spectral emissivity and temperature of surfaces with a Fourier transform infrared spectrometer," in Eklund JK, ed., *Thermosense XIV, SPIE Proc.*, **1682**:354–364, 1992.

Lineberry M, "Image segmentation by edge tracing," in Tesher AG, ed., *Applications of Digital Image Processing IV, SPIE Proc.*, **359**:361–367, 1982.

Linkous A, McKnight B, "Using thermography to detect and measure wall thinning," in Allen LR, ed., *Thermosense XV, SPIE Proc.*, **1933**:26–30, 1993.

Linnander B, "NDT solution: using infrared technology to speed product development," *Materials Evaluation*, **52**[5]:557–558, 1994.

Litvinenkoa SV, Kilchitskayaa SS, Skryshevskya VA, Strikhaa VI, Laugier A, "Application of dynamical optical reflection thermography (DORT) for detecting of dark current inhomogeneity in semiconductor devices," *Applied Surface Science*, **137**[1–4]:45–49, 1999.

Liu D, Tang Y, Yu D, Ke C, Huang D, "Distributed fiber optic temperature sensor based on Raman scattering of data fiber," in Dinwiddie RB, Lemieux DH, eds., *Thermosense XXII, SPIE Proc.*, **4020**:293–301, 2000.

Ljungberg SÅ, "Infrared techniques in buildings and structures: operation and maintenance," in Maldague XPV, ed., *Infrared Methodology and Technology*, International Advances in Nondestructive Testing Monograph Series, Chap. 6, pp. 211–252, Gordon and Breach, New York, 1994a.

Ljungberg SÅ, "Thermographic inspection of a tennis court building with modern steel construction: constructional mistakes and bad craftsmanship," in Snell JR, Jr., ed., *Thermosense XVI, SPIE Proc.*, **2245**:62–70, 1994b.

Ljungberg SÅ, "Infrared survey of 50 buildings constructed during 100 years: thermal performances and damage conditions," in Semanovich SA, ed., *Thermosense XVII, SPIE Proc.*, **2473**:36–52, 1995.

Ljungberg SÅ, "Infrared thermography as a diagnostic tool to indicate sick-house syndrome: a case study," in Burleigh DD, Spicer JWM, eds., *Thermosense XVIII, SPIE Proc.*, **2766**:32–49, 1996.

Ljungberg SÅ, "Information potential using IR technology for condition monitoring of reheating furnaces within steel and iron industry," in Wurzbach RN, Burleigh DD, eds., *Thermosense XIX, SPIE Proc.*, **3056**:133–145, 1997.

Ljungberg SÅ, Kulp TJ, McRae TC, "State of the art and future plans for IR imaging of gaseous fugitive emission," in Wurzbach RN, Burleigh DD, eds., *Thermosense XIX, SPIE Proc.*, **3056**:2–19, 1997.

Lott LA, Malik RK, "Ultrasonic inspectability improvements in austenitic stainless steel welds after thermal-mechanical processing," *Materials Evaluation*, **41**[6]:738–742, 1983.

Loubet D, "Mesures et suivi de l'endommagement des matériaux composites," Report 872-930-110 of the Société Aérospatiale, Paris, Oct. 1987.

Lozano-Garciá DF, Fernández, Johannsen CJ, "Assessment of regional biomass–soil relationship using vegetation index," *IEEE Transactions on Geoscience and Remote Sensing*, **29**[2]:331–339, 1991.

Lozinski A, Wang F, Uusimäki A, Leppävuori S, "Thick-film pyroelectric linear array," *Sensors and Actuators A: Physical*, **68**[1–3]:290–293, 1998.

Lu YJ, Hsu YH, Maldague X, "Vehicle classification using infrared image analysis," *Journal of Transportation Engineering*, **118**[Mar.–Apr.]:223–240, 1992.

Lu Y, Kuo PK, Favro DL, Thomas RL, "A dipole thermal wave source and mirage detection," in Thompson DO, Chimenti DE, eds., *Review of Progress in Quantitative Nondestructive Evaluation*, **16A**:379–382, 1998.

Lucier RD, "So now what? Things to do if your IR program stops producing results," in Baird GS, ed., *Thermosense XII, SPIE Proc.*, **1467**:59–62, 1991a.

Lucier RD, "Back to basics: recent work in infrared thermography: history, results, and feedback," *Materials Evaluation*, **49**[7]:856–859, 1991b.

Lucier RD, "Trends in infrared thermography," *Materials Evaluation*, **49**[9]:1162, 1991c.

Lucier RD, "Predictive maintenance for the '90s: an overview," in Eklund JK, ed., *Thermosense XIV, SPIE Proc.*, **1682**:35–42, 1992.

Lucier RD, Hammaker RG, Singh A, "Essential goals and elements for EPRI's infrared technical evaluation (IRITE) project," in Allen LR, ed., *Thermosense XV, SPIE Proc.*, **1933**:31–35, 1993.

Ludwig N, Rosina E, "Moisture detection through thermographic measurements of transpiration," in Wurzbach RN, Burleigh DD, eds., *Thermosense XIX, SPIE Proc.*, **3056**:78–86, 1997.

Lulay KF, Safai M, "Optimization of thermographic NDT using finite element analysis," in Snell JR, Jr., ed., *Thermosense XVI, SPIE Proc.*, **2245**:106–110, 1994.

Luong MP, "Infrared thermography of fatigue in metals," in Eklund JK, ed., *Thermosense XIV, SPIE Proc.*, **1682**:222–233, 1992.

Luong MP, "Infrared scanning of failure processes in wood," in Semanovich SA, ed., *Thermosense XVII, SPIE Proc.*, **2473**:298–311, 1995.

Luong MP, "Infrared thermographic scanning of fatigue in metals," *International Journal of Fatigue*, **19**[3]:266, 1997.

Luong MP, "Fatigue limit evaluation of metals using an infrared thermographic technique," *Mechanics of Materials*, **28**[1–4]:155–163, 1998.

Luong MP, Parganin D, "Infrared scanning of damage in leather," in Wurzbach RN, Burleigh DD, eds., *Thermosense XIX, SPIE Proc.*, **3056**:189–200, 1997.

Lüthi T, Zogmal O, "Nondestructive materials characterization by thermal methods," *Success of Materials by Combination, SAMPE '96*, pp. 28–30, Society for the Advancement of Materials and Process Engineering, Anaheim, California, 1996.

Lybaert P, Feldheim V, Lebrun I, "Thermography measurement of the local heat transfer distribution for flow around a surface-mounted obstacle," *Proc. QIRT-94, Eurotherm Seminar*, **42**:172–179, 1994.

Lyon BR Jr, Orlove CL, Peters DL, "Relationship between current load and temperature for quasi-steady state and transient conditions," in Dinwiddie RB, Lemieux DH, eds., *Thermosense XXII, SPIE Proc.*, **4020**:62–70, 2000.

Mack RT, "ASNT *Thermal/Infrared Handbook:* a new resource for testing service providers," in Allen LR, ed., *Thermosense XV, SPIE Proc.*, **1933**:226–237, 1993.

Maclachlan-Spicer JW, Kerns WD, Aamodt LC, Murphy JC, "Time-resolved infrared radiometry (TRIR) of multilayer organic coatings using surface and subsurface heating," in Baird GS, ed., *Thermosense XIV, SPIE Proc.*, **1467**:311–321, 1991.

Maclachlan-Spicer JW, Kerns WD, Aamodt LC, Murphy JC, "Source patterning in time-resolved infrared radiometry of composite structures," in Eklund JK, ed., *Thermosense XIV, SPIE Proc.*, **1682**:248–259, 1992.

Maclachlan-Spicer JW, Kerns WD, Aamodt LC, Murphy JC, "Time-resolved infrared radiometry (TRIR) for characterization of impact damage in composite materials," in Thompson DO, Chimenti DE, eds., *Review of Progress in Quantitative Nondestructive Evaluation*, **11A**:433–440, 1998.

MacNamara NA, Hammett AE, "Development of a comprehensive IR inspection program at a large commercial nuclear utility," in Eklund JK, ed., *Thermosense XIV, SPIE Proc.*, **1682**:30–34, 1992.

MacNamara NA, Zayicek PA, "Evaluation of IR technology applied to cooling tower performance," in Lemieux DH, Snell JR, eds., *Thermosense XXI, SPIE Proc.*, **3700**:252–267, 1999.

Madding RP, *Thermographic Instruments and Systems*, manual from the Department of Engineering and Applied Science, University of Wisconsin–Extension, 132 pp., 1979.

Madding RP, "High-voltage switchyard thermography case study," in Snell JR, Wurzbach RN, eds., *Thermosense XX, SPIE Proc.*, **3361**:94–99, 1998.

Madding RP, "Emissivity measurement and temperature correction accuracy considerations," in Lemieux DH, Snell JR, eds., *Thermosense XXI, SPIE Proc.*, **3700**:393–401, 1999a.

Madding RP, "Thermographer-friendly equipment design for predictive maintenance: baseline thermograms, thermal modeling, and emissivity," in Lemieux DH, Snell JR, eds., *Thermosense XXI, SPIE Proc.*, **3700**:2–5, 1999b.

Madding RP, Lyon BR Jr, "Wind effects on electrical hot spots: some experimental IR data," in Dinwiddie RB, Lemieux DH, eds., *Thermosense XXII, SPIE Proc.*, **4020**:80–84, 2000.

Madding RP, MacNamara NA, "Steam-leak cost estimation using thermographically acquired pipe temperature data," in Wurzbach RN, Burleigh DD, eds., *Thermosense XIX, SPIE Proc.*, **3056**:146–152, 1997.

Madrid A, "On the estimation of hardware failure rates using IR thermography," in Semanovich SA, ed., *Thermosense XII, SPIE Proc.*, **1313**:30–46, 1990.

Madrid A, "Quantitative evaluation of aging in bearings and electric brushes using infrared thermography," *Proc. QIRT-92, Eurotherm Seminar*, **27**:160–165, 1992.

Madura H, Polakowski H, Wiecek B, "Spectral emissivity evaluation for material used in microelectronics," *Proc. QIRT-96, Eurotherm Seminar*, **50**:52–57, 1996.

Mahoney BJ, Sandor SI, "Quantifying matrix cracking in composites by a thermoelastic method," *Proc. QIRT-92, Eurotherm Seminar*, **27**:171–178, 1992.

Maillet D, Didierjean S, Houlbert AS, Degiovanni A, "Nondestructive transient thermal evaluation of delaminations inside a laminate: a thermal processing technique of thermal images," *Proc. QIRT-92, Eurotherm Seminar*, **27**:212–217, 1992.

Maillet D, Houlbert AS, Didierjean S, Lamine AS, Degiovanni A, "Non-destructive thermal evaluation of delaminations in a laminate. II: The experimental Laplace transforms method," *Composites Science and Technology*, **47**[2]:155–172, 1993.

Mäkipää E, Tanttu JT, Virtanen H, "IR-based method for copper electrolysis short circuit detection," in Wurzbach RN, Burleigh DD, eds., *Thermosense XIX, SPIE Proc.*, **3056**:100–109, 1997.

Maldague X, *Nondestructive Evaluation of Materials by Infrared Thermography*, Springer-Verlag, London, 1993.

Maldague X, ed., "Advances in signal processing for nondestructive evaluation of materials," *Proc. 2nd International Research Workshop*, NATO-OTAN Advanced Sciences Institute Series, Ser. E: Applied Sciences, **262**, 1994a.

Maldague X, ed., *Infrared Methodology and Technology*, Gordon and Breach, New York, 1994b.

Maldague X, "3D sensing: overview with a thermal non-destructive testing perspective," *Proc. QIRT-94, Eurotherm Seminar*, **42**:195–213, 1994c.

Maldague X, "Infrared thermography: a useful tool to the Canadian industry," *Canadian Society of Nondestructive Testing Journal*, **15**[July–Aug.]:22–35, 1994d.

Maldague X, "Instrumentation for the infrared," in Maldague XPV, ed., *Infrared Methodology and Technology*, International Advances in Nondestructive Testing Monograph Series, Chap. 3, pp. 53–101, Gordon and Breach, New York, 1994e.

Maldague X, "Detekcja 3D: przeglad perspektywicznych, nieniszczacych metod dadan," in Pregowski P, ed., *TTP'96*, Warsaw, pp. 33–54, Nov. 26–29, 1996.

Maldague X, ed., Advances in Signal Processing for Nondestructive Evaluation of Materials, *Proc. 3rd International Research Workshop, Quebec City, TONES (ASNT)*, **3**, 1998a.

Maldague X, "NDT by infrared thermography: principles with applications," in Birnbaum G, Auld BA, eds., *Sensing for Material Characterization, Processing, and Manufacturing, TONES (ASNT)*, **1**:385–398, 1998b.

Maldague X, "Pipe inspection by infrared thermography: NDT solution," *Materials Evaluation*, **57**[9]:899–902, 1999.

Maldague X, "Applications of infrared thermography in nondestructive evaluation," in Rastogi P, ed., *Trends in Optical Nondestructive Testing*, pp. 591–609, Elsevier, New York, 2000.

Maldague X, Couturier JP, "Review of pulsed phase thermography," in Abbozzo LR, Carlomagno GM, Corsi C, eds., *Proc. 4th Workshop on Advances in Infrared Technology*, Atti della Fondazione G. Ronchi, Firenze, Italy, **53**:271–286, 1997.

Maldague X, Dufour M, "A dual-imager and its applications for active vision robot welding, surface inspection and two-color pyrometry," *Optical Engineering*, **28**[8]:872–880, 1989.

Maldague X, Fortin L, "Aspects of thermal modelling for a thermographic inspection station," *16th Annual Review of International Advances in Nondestructive Testing*, pp. 161–192, Gordon and Breach, New York, 1991.

Maldague X, Marinetti S, "Pulse phase infrared thermography," *Journal of Applied Physics*, **79**[Mar.]:2694–2698, 1996.

Maldague X, Pelletier JF, "Two-dimensional correction of thermograms for depth and orientation in pulse thermography inspection," in Zhang S, ed., 9th International Conference on Photoacoustic and Photothermal Phenomena, *Progress in Natural Science*, **6**[suppl.]:166–168, 1996.

Maldague X, Poussart D, Laurendeau D, April R, "Tridimensional form acquisition apparatus," in Cielo P, ed., *SPIE Proc. Optical Techniques for Industrial Inspection*, **665**:200–208, 1986a.

Maldague X, Cielo P, Jen CK, "NDT applications of laser-generated focused acoustic waves," *Materials Evaluation*, **44**[9]:1120–1124, 1986b.

Maldague X, Cielo P, Cole K, Vaudreuil G, "Detection of rolled-in surface scale in steel sheets by optical and thermal inspection techniques," in Penny CM, Caulfield HJ, eds., *Optical Techniques for Industrial Measurement and Control*, ICALEO-86 Conference, Springer-Verlag, Berlin, 1987a.

Maldague X, Cielo P, Ashley PJ, Farahbakhsh B, "Thermographic NDT of aluminium laminates," *Canadian Society of Nondestructive Testing Journal*, **8**[4]:44–50, 1987b.

Maldague X, Krapez JC, Cielo P, Poussart D, "Processing of thermal images for the detection and enhancement of subsurface flaws in composite materials," in Chen CH, ed., *Signal Processing and Pattern Recognition in Nondestructive Evaluation of Materials*, Vol. F44 of NATO ASI Series, pp. 257–285, Springer-Verlag, Berlin, 1988.

Maldague X, Krapez JC, Cielo P, Poussart D, "Inspection of materials and structures by infrared thermography: signal processing techniques for defect enhancement and characterization," *Canadian Society of Nondestructive Testing Journal*, **10**[1]:28–36, 1989a.

Maldague X, Krapez JC, Cielo P, Poussart D, "Infrared thermographic inspection by internal temperature perturbation techniques," in Boogaard J, Van Dijk, GM, eds., *Proc. 12th World Conference on Non-Destructive Testing*, Amsterdam, Apr. 23–28, pp. 561–566, 1989b.

Maldague X, Cielo P, Poussart D, Craig D, Bourret R, "Transient thermographic NDE of turbine blades," in Semanovich SA, ed., *Thermosense XII, SPIE Proc.*, **1313**:161–171, 1990a.

Maldague X, Krapez J-C, Poussart D, "Thermographic nondestructive evaluation (NDE): an algorithm for automatic defect extraction in infrared images," *IEEE Transactions on Systems, Man, and Cybernetics*, **20**[3]:722–725, 1990b.

Maldague X, Cielo P, Poussart D, "Thermographic nondestructive evaluation (NDE) of turbine blades: methods and image processing," *Industrial Metrology Journal*, **1**[2]:139–153, 1990c.

Maldague X, Krapez JC, Cielo P, "Subsurface flaw detection in reflective materials by thermal-transfer imaging," *Optical Engineering*, **30**[1]:117–125, 1991a.

Maldague X, Krapez J-C, Cielo P, "Temperature recovery and contrast computations in NDE thermographic imaging systems," *Journal of Nondestructive Evaluation*, **10**[1]:19–30, 1991b.

Maldague X, Fortin L, Picard J, "Applications of tridimensional heat calibration to a thermographic NDE station," in Baird GS, ed., *Thermosense XIII, SPIE Proc.*, **1467**:239–251, 1991c.

Maldague X, Nouah A, Fortin L, Robitaille F, Picard J, "Non-planar inspection with infrared thermography," in Hallai C, Kulcsar P, eds., *Non-Destructive Testing-92*, **2**:725–729, 1992a.

Maldague X, Vavilov V, Côté J, Poussart D, "Thermal tomography for NDT of industrial materials," *Canadian Society of Nondestructive Testing Journal*, **13**[May/June]: 22–32, 1992b.

Maldague X, Barker E, Nouah A, Boisvert E, Dufort B, Fortin L, Laurendeau D, "On methods for shape correction and reconstruction in themographic NDE," in Abbozzo LR, Carlomagno GM, Corsi C, eds., *Advanced Infrared Technology and Applications Workshop*, Atti della Fondazione G. Ronchi, Firenze, Italy, **49**[1]:91–98, 1993.

Maldague X, Barker E, Nouah A, Boisvert E, Dufort B, Fortin L, "On methods for shape correction and reconstruction in thermographic NDT," in Maldague X, ed., *Advances in Signal Processing for Nondestructive Evaluation of Materials*, NATO-OTAN Advanced Sciences Institute Series, Ser. E: Applied Sciences, **262**:209–224, 1994a.

Maldague X, Pitre L, Moussa M, Laurendeau D, "A robotic application of active thermographic NDT," *Proc. 6th European Conference on Nondestructive Testing*, 1[Oct.]:663–667, 1994b.

Maldague X, Shiryaev VV, Boisvert É, Vavilov V, "Transient thermal nondestructive testing (NDT) of defects in aluminum panels," in Semanovich SA, ed., *Thermosense XVII, SPIE Proc.*, **2473**:233–243, 1995.

Maldague X, Couturier JP, Marinetti S, Salerno A, Wu D, "Advances in pulse phase thermography," *Proc. QIRT-96, Eurotherm Seminar*, **50**:377–382, 1996a.

Maldague X, Couturier JP, Wu D, Salerno A, "Advances in pulse phase thermography," in Balageas D, Busse G, Carlomagno C, eds., *Proc. QIRT-96, Eurotherm Seminar*, **50**:377–382, 1996b.

Maldague X, Marinetti S, Busse G, Couturier JP, "Possible applications of pulse phase thermography," in Zhang S, ed., 9th International Conference on Photoacoustic and Photothermal Phenomena, *Progress in Natural Science*, **6**[suppl.]:80–82, 1996c.

Maldague X, Marinetti S, Couturier JP, Busse G, "Applications of pulse phase thermography," in Thompson DO, Chimenti DE, eds., *Review of Progress in Quantitative Nondestructive Evaluation*, **16A**:339–344, 1996d.

Maldague X, Couturier JP, Salerno A, Wu D, "Phase analysis in pulsed thermography" *Canadian Society of Nondestructive Testing Journal*, **19**[July–Aug.]:5–10, 1997.

Maldague X, Largouët Y, Couturier JP, "Depth study in pulsed phase thermography using neural networks: modelling, noise, experiments," *Revue Générale de Thermique*, **37**[Sept.]:704–708, 1998.

Mallick PK, *Fiber-Reinforced Composites*, Marcel Dekker, New York, 1986.

Mann JM, Schmerr LW, Moulder JC, "Neural network inversion of uniform-field eddy current data," *Materials Evaluation*, **49**[Jan.]:34–39, 1991.

Mansour TM, "Nondestructive thickness measurement of phosphate coatings by infrared absorption," *Materials Evaluation*, **41**[3]:302–308, 1983.

Mansoor AK, Allemand C, Eagar TW, "Noncontact temperature measurement. I: Interpolation based techniques," *Review of Scientific Instruments*, **62**[2]:392–402, 1991a.

Mansoor AK, Allemand C, Eagar TW, "Noncontact temperature measurement. II: Least squares based techniques," *Review of Scientific Instruments*, **62**[2]:403–409, 1991b.

Mao Z, Strickland RN, "Image sequence processing for target estimation in forward-looking infrared imagery," *Optical Engineering*, **27**[7]:541–549, 1988.

Marche PP, "New generation infrared camera," in Lettington AH, ed., *SPIE Proc. Infrared Technology and Applications*, **1320**:159–163, 1990.

Marengo S, Comotti P, "Studies of catalysts and catalytic reactions by infrared thermography," *Proc. QIRT-94, Eurotherm Seminar*, **42**:9–12, 1994.

Marincic A, Tmusic R, "Some problems in application of integrated radiation thermopile sensor," *Proc. QIRT-98, Eurotherm Seminar*, **60**:299–304, 1998.

Marinetti S, Maldague X, Prystay M, "Technical paper: calibration procedure for focal plane array cameras and noise equivalent material loss for quantitative thermographic NDT," *Materials Evaluation*, **55**[3]:407–412, 1997.

Marinetti S, Muscio A, Bison PG, Grinzato EG, "Modeling of thermal nondestructive evaluation techniques for composite materials," in Dinwiddie RB, Lemieux DH, eds., *Thermosense XXII, SPIE Proc.*, **4020**:164–173, 2000.

Marr D, Hildreth E, "Theory of edge detection," *Proceedings of the Royal Society of London, Series B*, **207**:187–217, 1980.

Martinelli NS, Seoane R, "Automotive night vision system," in Lemieux DH, Snell JR, eds., *Thermosense XXI, SPIE Proc.*, **3700**:343–346, 1999.

Martiny M, Schiele R, Gritsch M, Schulz A, Wittig S, "In situ calibration for quantitative infrared thermography," *Proc. QIRT-96, Eurotherm Seminar*, **50**:3–8, 1996.

Masters RC, Simon DL, "Weld electrode cooling study," in Lemieux DH, Snell JR, eds., *Thermosense XXI, SPIE Proc.*, **3700**:190–193, 1999.

Matini A, Norgard JD, Sega RM, Ayres B, Harrison MG, Komar R, Pohle H, Prather WF, Smith M, "Thermal images of magnetic fields near conductive surfaces," in Allen LR, ed., *Thermosense XV, SPIE Proc.*, **1933**:286–297, 1993.

Matsueda H, "Photonic emission from human body controlled by will," in Dinwiddie RB, Lemieux DH, eds., *Thermosense XXII, SPIE Proc.*, **4020**:252–256, 2000.

May TH, Stamm GL, Blodgett JA, "Application of digital processing to calibrated infrared imagery," *SPIE Proc. Infrared Systems*, **256**:55–61, 1980.

Mayer R, Henkes RAWM, van Ingen JL, "Wall-shear stress measurement with quantitative IR thermography," *Proc. QIRT-96, Eurotherm Seminar*, **50**:153–158, 1996.

McCarthy DC, "Airborne multispectral radiometer images fire and ice," *Photonics Spectra*, **32**[10]:26–29, 1998.

McCleary D, Harvey R, Semanovich SA, "Improvements in new residential construction using infrared thermography and blower door," in Burleigh DD, Spicer JWM, eds., *Thermosense XVIII, SPIE Proc.*, **2766**:50–54, 1996.

McClendon RM, "Thermography in the electronic communications industry; feature," *Materials Evaluation*, **56**[9]:1072–1074, 1998.

McComb J, Niebla HE, "Infrared thermography and overloaded neutral conductors," in Lemieux DH, Snell JR, eds., *Thermosense XXI, SPIE Proc.*, **3700**:268–272, 1999.

McCullough LD, Green DR, "Electrothermal nondestructive testing of metal structures," *Materials Evaluation*, **30**[4]:87–91, 1972.

McCullough RW, "Laser technology in nondestructive testing: an introduction," *Materials Evaluation*, **53**[12]:1331, 1995.

McGonnagle W, Park F, "Nondestructive testing," *International Science and Technology*, July, p. 14, 1964.

McLaughin PV, *Naval Air Systems Command*, report Air Task A310310G/ 0518/ 4F41460000, Washington, DC, 1985.

McLaughin PV, "Defect detection and quantification in laminated composites by EATF (passive) thermography," in Thompson DO, Chimenti DE, eds., *Review of Progress in Quantitative Nondestructive Evaluation*, **7A**:125–1132, 1988.

McLaughlin PV, McAssey EV, Koert DN, Deitrich RC, "NDT of composites by thermography," *Proc. DARPA/AFWAL of Progress in Quantitative Nondestructive Evaluation*, pp. 60–68, 1981.

McMillan A, "Expanding the use of infrared thermography through streamlined processes," in Wurzbach RN, Burleigh DD, eds., *Thermosense XIX, SPIE Proc.*, **3056**:167–169, 1997.

McMullan PC, "Case study of commercial building envelope air leakage detection using infrared imaging," in Eklund JK, ed., *Thermosense XIV, SPIE Proc.*, **1682**:74–83, 1992.

McMullan PC, "Masonry building envelope analysis," in Allen LR, ed., *Thermosense XV, SPIE Proc.*, **1933**:43–54, 1993.

McNamara DK, Ahearn JS, "Adhesive bonding of steel for structural applications," *International Materials Review*, **32**[6]:292–306, 1987.

Medri G, Ricci R, "Thermomechanical evaluation of polypropylene fracture resistance," *Proc. QIRT-92, Eurotherm Seminar*, **27**:166–170, 1992.

Mehrotra Y, "Aircraft subsystems inspection: objective and easy … then, why skimp?" in Dinwiddie RB, Lemieux DH, eds., *Thermosense XXII, SPIE Proc.*, **4020**:220–231, 2000.

Meinders ER, van der Meer ThH, Hanjalic K, "Application of infrared image restoration to improve the accuracy of surface temperature measurements," *Proc. QIRT-96, Eurotherm Seminar*, **50**:32–40, 1996.

Mendonsa RA, "IR imager helps preserve the Alamo," *Photonics Spectra*, July, p. 22, 1998.

Mengers P, "Micro-computer based digital image processing for real-time radiography," in Thompson DO, Chimenti DE, eds., *Review of Progress in Quantitative Nondestructive Evaluation*, **5A**:825–833, 1986.

Meola C, Carlomagno GM, Riegel E, Salvato F, "Infrared thermography on testing an anti-icing device," *Proc. QIRT-94, Eurotherm Seminar*, **42**:186–194, 1994a.

Meola C, De Luca L, Carlomagno GM, "Thermal measurements in a single axisymmetric jet impinging normal to a flat plate," *Proc. QIRT-94, Eurotherm Seminar*, **42**:180–185, 1994b.

Meola C, Cardone G, Carlomagno GM, Marino L, "Heat transfer in separated and reattached flow regions over a circular cylinder," *Proc. QIRT-96, Eurotherm Seminar*, **50**:171–176, 1996.

Mérienne E, Hakem K, Egée M, "Photothermal radiometry apparatus using pseudo-random excitation for nondestructive evaluation of layered materials," *Materials Evaluation*, **52**[2]:312, 1994.

Meroni I, Esposti V, "Energy assessment of building envelopes through NDT methods," in Wurzbach RN, Burleigh DD, eds., *Thermosense XIX, SPIE Proc.*, **3056**:50–58, 1997.

Meyer-Arendt JR, *Introduction to Classical and Modern Optics*, Prentice Hall, Upper Saddle River, NJ, 558 pp., 1972.

Mikolajczyk Z, Wiecek B, Michalak M, "Thermovision method in stress analysis of textile materials," *Proc. QIRT-98, Eurotherm Seminar*, **60**:140–147, 1998.

Milazzo M, Ludwig N, Poldi G, "Moisture detection in walls trough measurement of temperature," *Proc. QIRT-98, Eurotherm Seminar*, **60**:91–96, 1998.

Miles JJ, Hammaker RC, Madding RP, Sunderland JE, "Analysis of thermal radiation in coal-fired furnaces," in Wurzbach RN, Burleigh DD, eds., *Thermosense XIX, SPIE Proc.*, **3056**:20–32, 1997.

Miles JJ, Hammaker RG, Madding RP, Sunderland JE, "Radiometric imaging of internal boiler components inside a gas-fired commercial boiler," in Snell JR, Wurzbach RN, eds., *Thermosense XX, SPIE Proc.*, **3361**:103–117, 1998.

Milne JM, Carter P, "A study into the generation of temperature gradients in materials for detecting and measuring the depth of sub-surface cracks," report Harwell UK AERE R 12382, N87 22991, Oct. 1986.

Milne JM, Carter P, "A transient thermal method of measuring the depths of sub-surface flaws in metals," *British Journal of Non-Destructive Testing*, **30**[5]:333–336, 1988.

Milovanovic D, Marincic A, Barbaric Z, Petrovic G, "Statistical analysis of computer-generated thermal images based on overall modeling of line-scanning process," *Proc. QIRT-94, Eurotherm Seminar*, **42**:13–18, 1994.

Milovanovic D, Wiecek B, Marincic A, Barbaric Z, Petrovic G, "Statistical analysis techniques for aerial infrared images in wavelets transform domain," *Proc. QIRT-96, Eurotherm Seminar*, **50**:368–376, 1996.

Milovanovic D, Wiecek B, Marincic A, Petrovic G, Barbaric Z, "A comparative study of advanced frequency-domain coding techniques in compression of infrared line-scan images," *Proc. QIRT-98, Eurotherm Seminar*, **60**:342–348, 1998.

Minor LG, Sklonsky J, "The detection and segmentation of blobs in infrared images," *IEEE Transactions on Systems, Man, and Cybernetics*, **11**[3]:194–201, 1981.

Misiti M, Misiti Y, Oppenheim G, Poggi JM, *Matlab Wavelet Toolbox User's Guide*, MatWorks, Natick, MA, 1997.

Mitra S, Nutter BS, Krile TF, Brown RH, "Automated method for fundus image registration and analysis," *Applied Optics*, **27**[6]:1107–1112, 1988.

Monchalin JP "Optical detection of ultrasound," *IEEE Transactions on Ultrasonics, Ferroelectrics and Frequency Control*, **33**, Sept. 1986.

Monti R, *Flow Visualization and Digital image Processing*, Lecture Series 1986-09, von Karman Institute for Fluid Dynamics, Rhode Saint Genèse, Belgium, 1986.

Monti R, Mannara G, "NDT of honeycomb structures by computerized thermographic systems," *Acta Astronautica*, **12**[6]:405–414, 1985.

Monti R, Mannara G, "The computerized thermography for NDT in aerospace applications," *Proc. 4th European Conference on Nondestructive Testing*, Sept. 13–17, pp. 1266–1279, Sept. 1987.

Moore PO, McIntire P, eds., *Nondestructive Testing Overview*, 2nd ed., American Society for Nondestructive Testing, Columbus, Ohio, 1996.

Moreno JB, "Infrared inspection of bearings on slow-moving equipment," in Lemieux DH, Snell JR, eds., *Thermosense XXI, SPIE Proc.*, **3700**:297–299, 1999.

Morey JL, "Microlens arrays sharpen the details," *Photonics Spectra*, Dec., pp. 110–114, 1997.

Morgan WT, "Thermographic inspections of air distribution systems, Infrared Engineering Services," in Eklund JK, ed., *Thermosense XIV, SPIE Proc.*, **1682**:84–88, 1992.

Morgan WT, "Integration of infrared thermography into various maintenance methodologies," in Allen LR, ed., *Thermosense XV, SPIE Proc.*, **1933**:9–13, 1993.

Morisseau P, Huet J, Pauton M, "Dimensional and geometry analysis of turbine blades by the use of computerized tomography (CT)," in Farley JM, Nichols RW, eds., *Proc. 4th European Conference on Nondestructive Testing*, **2**:1257–1265, 1987.

Moropoulou A, Koui M, Avdelidis NP, Kakaras K, "Inspection of airport runways and asphalt pavements using long-wave infrared thermography," in Dinwiddie RB, Lemieux DH, eds., *Thermosense XXII, SPIE Proc.*, **4020**:302–309, 2000.

Moussa M, Maldague X, "Lateral thermal modeling in transient infrared thermography," in Abbozzo LR, Carlomagno GM, Corsi C, eds., *Proc. 3rd International Workshop on Advanced Infrared Technology and Applications*, Atti della Fondazione G. Ronchi, Firenze, Italy, **51**:115–125.

Muralidhar C, Arya NK, "Evaluation of defects in axisymmetric composite structures by thermography," *NDT&E International*, **26**[4]:189–193, 1993.

Murphy CG, Wilburn JB Jr, "Transfer function evaluation of thermal imaging systems at the U.S. Army Electronic Proving Ground," *Materials Evaluation*, **34**[8]:185–188, 1976.

Murphy JC, Wetsel GC Jr, "Photothermal methods of optical characterization of materials," *Materials Evaluation*, **44**[10]:1224–1230, 1986.

Murphy JC, Aamodt LC, Maclachlan Spicer JW, "Principles of photothermal detection in solids," in Mandelis A, ed., *Principles and Perspectives of Photothermal and Photoacoustic Phenomena*, pp. 41–94, Elsevier, New York, 1992.

Murphy JC, Spicer JWM, Osiander R, Kerns WD, Aamodt LC, "Quantitative nondestructive evaluation of coatings by thermal wave imaging," in Thompson DO, Chimenti DE, eds., *Review of Progress in Quantitative Nondestructive Evaluation*, **13A**:417–425, 1994.

Muscio A, Corticelli MA, Tartarini P, "Theoretical, numerical, and experimental investigation of a one-side measurement technique for thermal diffusivity," in Dinwiddie RB, Lemieux DH, eds., *Thermosense XXII, SPIE Proc.*, **4020**:143–151, 2000.

Na'ama G, Rispler S, Sideman S, Shofty R, Beyar R, "Thermographic imaging in the beating heart: a method for coronary flow estimation based on a heat transfer model," *Medical Engineering and Physics*, **20**[6]:443–451, 1998.

Nacitas M, Levesque P, Balageas DL, "Lock-in infrared thermography applied to the characterization of electromagnetic fields," in Balageas D, Busse G, Carlomagon GM, eds., *Proc. QIRT-94, Eurotherm Seminar*, **42**:287–292, 1994.

Nagarajan S, Chin BA, "Infrared image analysis for on-line monitoring of arc misalignment in gas tungsten arc welding processes," *Materials Evaluation*, **48**[12]:1469–1472, 1990.

Nagarajan S, Chin BA, "Infrared techniques for real-time weld quality control," in Maldague X, ed., Chap. 10, *Infrared Methodology and Technology*, Gordon and Breach, New York, 1994.

Nagarajan S, Chen WH, Groom KN, Chin BA, "Infrared sensors for seam tracking and penetration depth control," in Grover CP, ed., *Optical Testing and Metrology II, SPIE Proc.*, **954**:568–573, 1988.

Nagarajan S, Chen WH, Groom KN, Chin BA, "Infrared sensing for adaptive arc welding," *Welding Journal*, **68**[11]:462s–466s, 1989.

Nalwa VS, *A Guided Tour of Computer Vision*, Addison-Wesley, Reading, MA, 361 p., 1993.

Nana L, Farré J, Giovannini A, "Infrared temperature measurement of vaporizing droplets," *Proc. QIRT-92, Eurotherm Seminar*, **27**:105–110, 1992.

Nana L, Farré I, Giovannini A, Sabatier P, Naudin N, "Infrared system for methanol droplet temperature measurement," in Allen LR, ed., *Thermosense XV, SPIE Proc.*, **1933**:240–251, 1993.

Nandhakumar N, Aggarwal JK, "Integrated analysis of thermal and visual images for scenes interpretation," *IEEE Transactions on Pattern Analysis and Machine Intelligence*, **10**[4]:469–481, 1988.

Nandhakumar N, Karthik S, Aggarwal JK, "Unified modeling of non-homogeneous 3D objects for thermal and visual image synthesis," *Pattern Recognition*, **27**[10]:1303–1316, 1994.

Narendra PM, "A separable median filter for image noise smoothing," *IEEE Transactions on Pattern Analysis and Machine Intelligence*, 3[1]:20–29, 1981.

Naudin N, Farré J, Lavergne C, "Infrared system for methanol droplets temperature measurement on a monodisperse jet," in Semanovich SA, ed., *Thermosense XVII, SPIE Proc.*, **2473**:338–345, 1995.

Nazif AM, Levine MD, "Low level image segmentation: an expert system," *IEEE Transactions on Pattern Analysis and Machine Intelligence*, **6**[5]:555–577, 1984.

Nelson MD, Johnson JF, Lomheim TS, "General noise processes in hybrid infrared focal plane arrays," *Optical Engineering*, **30**[11]:1682–1700, 1991.

Netzelmann U, Walle G, "High-speed pulsed thermography of thin metallic coatings," *Proc. QIRT-98, Eurotherm Seminar*, **60**:81–85, 1998.

Netzelmann U, Walle G, Dobmann G, "Real-time 3D-representation of time-resolved infrared thermographic data," *Proc. QIRT-92, Eurotherm Seminar*, **27**:239–242, 1992.

Neuer G, "Emissivity measurements on graphite and composite materials in the visible and infrared spectral range," *Proc. QIRT-92, Eurotherm Seminar*, **27**:359–366, 1992.

Newitt J, "Back to basics: application-specific thermal imaging," *Materials Evaluation*, **45**[5]:500, 502, 504, 1987.

Newman WM, Sproull RF, *Principles of Interactive Computer Graphics*, McGraw-Hill, New York, 541 pp., 1979.

Newnham P, Abrate S, "Finite element analysis of heat transfer in anisotropic solids: application to manufacturing problems," *Journal of Reinforced Plastics and Composites*, **12**[8]:854–864, 1993.

Newport R, "Infrared electrical inspection myths," in Wurzbach RN, Burleigh DD, eds., *Thermosense XIX, SPIE Proc.*, **3056**:124–132, 1997.

Ng D, Williams WD, "Full-spectrum multiwavelength pyrometry for nongray surfaces," in Eklund JK, ed., *Thermosense XIV, SPIE Proc.*, **1682**:260–270, 1992.

Nicholas JR, Young RK, "Recommendations for strengthening the infrared technology component of any condition monitoring program," in Lemieux DH, Snell JR, eds., *Thermosense XXI, SPIE Proc.*, **3700**:227–236, 1999.

Nichols JT, "Temperature measuring," U.S. patent 2,008,793, July 1935 (description of a radiometer for use in steel-rolling mills to indicate the uniformity with which the strip is heated).

Nickolls J, "The design of the MasPar MP-1, a cost effective massively parallel computer," *Proc. IEEE Compcon*, Feb. 1990.

Nicodemus FE, "Radiometry," in Kingslake R, ed., *Applied Optics and Optical Engineering*, Vol. IV, Chap. 8, pp. 263–307, Academic Press, San Diego, CA, 1967.

Niro LA, "Yesteryears: Herschel's discovery of infrared solar rays," *Materials Evaluation*, **45**[4]:434–435, 1987.

Norgard JD, Metzger D, Sega R, Harrison M, Komar R, Pohle H, Schmelzel A, Smith M, Stupic J, Seifert M, Cleary J, "Infrared measurements of electromagnetic fields," *Proc. QIRT-92, Eurotherm Seminar*, **27**:308–314, 1992a.

Norgard JD, Sega RM, Seifert M, Cleary JC, Harrison MC, "Infrared detection of free-field and cavity perturbations of electromagnetic probe measurements," in Eklund JK, ed., *Thermosense XIV, SPIE Proc.*, **1682**:282–295, 1992b.

Norgard JD, Sadler J, Baca E, Prather W, Sega R, Seifert R, "High power microwave antenna design using infrared imaging techniques," *Proc. QIRT-94, Eurotherm Seminar*, **42**:19–23, 1994a.

Norgard JD, Sadler J, Sega RM, Baca EA, Prather W, "Infrared measurements of waveguide modes and radiation patterns of beveled-cut circular waveguide microwave aperture antennas," in Snell JR, Jr., ed., *Thermosense XVI, SPIE Proc.*, **2245**:286–295, 1994b.

Norgard JD, Seifert M, Sega RM, Pesta A, "Empirical calibration of infrared images of electromagnetic fields," in Semanovich SA, ed., *Thermosense XVII, SPIE Proc.*, **2473**:332–337, 1995.

Norton PR, "Infrared image sensors," *Optical Engineering*, **30**[11]:1649–1663, 1991.

Nouah A, Maldague X, "Exploitation de la vision dans le spectre visible et infrarouge pour l'évaluation non-destructive des matériaux," Bogdadi G., ed., *Proc. Industrial Automation Conference*, Montréal, Quebec, Canada, June 1–3, 1992.

Nouah A, Maldague X, Robitaille F, "Shape correction in transient thermography inspection of non-planar components," *Proc. QIRT-92, Eurotherm Seminar*, **27**:224–228, 1992.

Nutter BS, Mitra S, Krile TF, "Image registration for a PC-based system," in Tesher AG, ed., *SPIE Proc. Applications of Digital Image Processing XI*, **829**:214–221, 1987.

O'Brien T, "Money woes slow photonics fight on crime," *Photonics Spectra*, Nov., pp. 78–79, 1997.

Oermann RJ, *Radiometry using thermal images. II: Technical details*, technical report ERL-0423-TR, Electronics Reseach Laboratory, Defence Science and Technology Organisation, Department of Defence, Salisbury, South Australia, 15 pp., 1987.

Offermann S, Bissieux C, Beaudoin JL, "Thermoelastic stress analysis with standard thermographic equipment by means of statistical noise rejection," *Research in Nondestructive Evaluation*, 7[4]:239–252, 1996a.

Offermann S, Bissieux C, Beaudoin JL, Frick H, "Quantitative stress analysis by means of standard infrared thermographic equipment," *Proc. QIRT-96, Eurotherm Seminar*, 50:79–84, 1996b.

Offermann S, Bicanic D, Krapez JC, Balageas D, Gerkema E, Chirtoc M, Egee M, Keijzer K, Jalink H, "Infrared transient thermography for non-contact, nondestructive inspection of whole and dissected apples and of cherry tomatoes at different maturity stages," *Instrumentation Science and Technology*, 26[2–3]:145–155, 1998a.

Offermann S, Bissieux C, Beaudoin JL, "Optical and thermal restoration applied to thermoelastic stress analysis by IR thermography," *Proc. QIRT-98, Eurotherm Seminar*, 60:123–128, 1998b.

Offermann S, Bissieux C, Beaudoin JL, "Statistical treatment applied to infrared thermoelastic analysis of applied and residual mechanical stresses," *Revue Générale de Thermique*, 37[8]:718–724, 1998c.

Ogasawara N, Shiratori M, "Application of infrared thermography to fracture mechanics," in Wurzbach RN, Burleigh DD, eds., *Thermosense XIX, SPIE Proc.*, 3056:201–215, 1997.

Ogawa K, Shoji T, Abe I, Hashimoto H, "In situ NDT of degradation of thermal barrier coatings using impedance spectroscopy," *Materials Evaluation*, 58[3]:476–481, 2000.

O'Gorman L, "A note on histogram equalization for optimal intensity range utilization," *Computer Vision, Graphics, and Image Processing*, 41:229–232, 1988.

Öhman C, "Practical methods for improving thermal measurements," in Grot RA, Wood JT, eds., *Thermosense IV, SPIE Proc.*, 313:204–212, 1981.

Okamoto Y, Kaminaga F, Inagaki T, Numao T, Fukuzawa K, Ichikawa H, "Remote sensing study of detecting of flaws in structural material," in Allen LR, ed., *Thermosense XV, SPIE Proc.*, 1933:215–225, 1993.

Okamoto Y, Inagaki T, Suzuki T, Kurokawa T, "Use of radiometer to reform and repair an old living house to passive solar one," in Snell JR, Jr., ed., *Thermosense XVI, SPIE Proc.*, 2245:40–51, 1994a.

Okamoto Y, Inagaki Y, Tsuyuzaki N, Chen W, "Thermal measurement techniques of IC package boards by means of infrared radiometer," in Snell JR, Jr., ed., *Thermosense XVI, SPIE Proc.*, 2245:231–240, 1994b.

Okamoto Y, Liu C, Fan Z, Inagaki T, "IR detection limit of underground structure by thermal image technique," *Proc. QIRT-94, Eurotherm Seminar*, 42:315–320, 1994c.

Okamoto Y, Fan Z, Liu C, Inagaki T, "Thermal image study of detecting near-underground structures by means of infrared radiometer," in Semanovich SA, ed., *Thermosense XVII, SPIE Proc.*, 2473:281–288, 1995.

Okamoto Y, Inagaki T, Liu C, Miyata K, "Detection of flashing temperature spots of dry friction interface by means of infrared radiometer," *Proc. QIRT-96, Eurotherm Seminar*, 50:326–334, 1996.

Okamoto Y, Liu C, Agu H, Inagaki T, "Remote-sensing visible and infrared method of locating underground objects using thermal image technique," in Wurzbach RN, Burleigh DD, eds., *Thermosense XIX, SPIE Proc.*, 3056:33–40, 1997.

Okamoto Y, Kamoi A, Ishii T, "Thermal analysis on internal and surface flaws by means of an infrared radiometer," *Proc. QIRT-98, Eurotherm Seminar*, **60**:71–76, 1998.

Oliferuk W, "Investigation of metal deformation using thermography," *Proc. QIRT-98, Eurotherm Seminar*, **60**:134–139, 1998.

Oliver DW, Brown J, Cueman K, Czechowski J, Eberhard J, "XIM: X-ray inspection module for automatic high speed inspection of turbine blades and automated flaw detection and classification," in Thompson DO, Chimenti DE, eds., *Review of Progress in Quantitative Nondestructive Evaluation*, **5A**:817–823, 1986.

Orlove GL, "Nondestructive evaluation of steam traps," in Lemieux DH, Snell JR, eds., *Thermosense XXI, SPIE Proc.*, **3700**:283–288, 1999.

Ortolano DJ, "Resistance thermometry," *Standardization News*, **16**[5]:52–55, 1988.

Osiander R, Spicer JWM, Murphy JC, "Thermal imaging of subsurface microwave absorbers in dielectric materials," in Snell JR, Jr., ed., *Thermosense XVI, SPIE Proc.*, **2245**:111–119, 1994.

Osiander R, Spicer JWM, Murphy JC, "Microwave-source time-resolved infrared radiometry for monitoring of curing and deposition processes," in Semanovich SA, ed., *Thermosense XVII, SPIE Proc.*, **2473**:194–204, 1995a.

Osiander R, Spicer JM, Murphy JC, "Thermal nondestructive evaluation using microwave sources," *Materials Evaluation*, **53**[8]:942–948, 1995b.

Osiander R, Spicer JWM, Murphy JC, "Analysis methods for full-field time-resolved infrared radiometry," in Burleigh DD, Spicer JWM, eds., *Thermosense XVIII, SPIE Proc.*, **2766**:218–227, 1996.

Osiander R, Spicer JWM, Amos M, "Thermal inspection of SiC/SiC ceramic matrix composites," in Snell JR, Wurzbach RN, eds., *Thermosense XX, SPIE Proc.*, **3361**:339–349, 1998.

Ottesen DK, Nagelberg AS, *Thin Solid Film*, **73**:347, 1980.

Ouyang Z, Favro LD, Thomas RL, "Progress in the rapid, contactless measurement of thermal diffusivity," in Thompson DO, Chimenti DE, eds., *Review of Progress in Quantitative Nondestructive Evaluation*, **18A**:627–630, 1998a.

Ouyang Z, Wang L, Wang X, Zhang F, Favro LD, Kuo PK, Thomas RL, "Lock-in thermal wave imaging," in Thompson DO, Chimenti DE, eds., *Review of Progress in Quantitative Nondestructive Evaluation*, **16A**:383–388, 1998b.

Padé O, Mandelis A, "Computational thermal-wave slice tomography with back-propagation (BP) and transmission (T) reconstructions," in Thompson DO, Chimenti DE, eds., *Review of Progress in Quantitative Nondestructive Evaluation*, **13A**:409–416, 1994.

Pajani D, "Thermographie IR: quelle longueur d'onde choisir I," *Mesures*, May 25, pp. 71–76, 1987a.

Pajani D, "Thermographie IR: quelle longueur d'onde choisir II," *Mesures*, June 9, pp. 77–80, 1987b.

Pajani D, "La thermographie infrarouge dans l'industrie du verre," *Contrôle Industriel Qualité*, **152-bis**[Oct.]:55–62, 1987c.

Pajani D, "La thermographie IR sur les bains d'électrolyse," *Mesures*, **11**[634]:45–47, 1991.

Pao JF, Zhang XR, Gan CM, Gu ZC, "Determination of thermal diffusivity of orthogonal crystal by mirage detection," in Thompson DO, Chimenti DE, eds., *Review of Progress in Quantitative Nondestructive Evaluation*, **18A**:597–604, 1998.

Parker RC, Marshall PR, "The measurement of the temperature of sliding surfaces with particular references to railway brake blocks and shoes," *Proc. Institute of Mechanical Engineers*, **158**:209, 1948.

Parker SP, ed., *Concise Encyclopedia of Science and Technology*, McGraw-Hill, New York, 2065 pp., 1984.

Parker WJ, Jenkins RJ, Butler CP, Abbott GL, "Flash method of determining thermal diffusivity, heat capacity, and thermal conductivity," *Journal of Applied Physics*, **32**[9]:1679–1684, 1961.

Pas G, "Broadening IR applications through using spectral filters," *Proc. QIRT-98, Eurotherm Seminar*, **60**:172–177, 1998.

Patel PM, Lau SK, Almond DP, "A review of image analysis techniques applied in transient thermographic nondestructive testing," *Journal of Nondestructive Evaluation*, **6**:343–364, 1991.

Patorski JA, Bauer GS, Dementjev S, "Two-dimensional and dynamic (2DD) method of visualization of the flow characteristics in a convection boundary layer using infrared thermography," in Dinwiddie RB, Lemieux DH, eds., *Thermosense XXII, SPIE Proc.*, **4020**:240–251, 2000.

Pau LF, "Integrated testing and algorithms for visual inspection of integrated circuits," *IEEE Transactions on Pattern Analysis and Machine Intelligence*, **5**[6]:602–608, 1983.

Peacock GR, "Review of noncontact process temperature measurements in steel manufacturing," in Lemieux DH, Snell JR, eds., *Thermosense XXI, SPIE Proc.*, **3700**:171–189, 1999.

Peacock GR, "Thermal imaging of liquid steel and slag in a pouring stream," in Dinwiddie RB, Lemieux DH, eds., *Thermosense XXII, SPIE Proc.*, **4020**:50–61, 2000.

Peacock RG, "Radiation thermometry," *Standardization News*, **16**[5]:31–35, 1988.

Pearce J, Schuize M, Ryu Z, "Improving the accuracy of inferred temperatures in small spot size experiments," *Proc. QIRT-92, Eurotherm Seminar*, **27**:13–18, 1992.

Pelletier JF, Maldague X, "Infrared thermography: nonplanar inspection," in Burleigh DD, Spicer JWM, eds., *Thermosense XVIII, SPIE Proc.*, **2766**:264–275, 1996.

Pelletier JF, Maldague X, "Shape from heating: a two-dimensional approach for shape extraction in infrared images," *Optical Engineering*, **36**[Feb.]:371–375, 1997.

Pelletier JF, Grinzato E, Dessi R, Maldague X, "Shape and uneven heating correction for NDT on cylinders by thermal methods," *Proc. QIRT-96, Eurotherm Seminar*, **50**:263–268, 1996.

Pelligrini PW, "A comparison of iridium silicide and platinum silicide photodiodes," in Buser RG, Warren FB, eds., *SPIE Proc. Infrared Sensors and Sensor Fusion*, **782**:93–98, 1987.

Peng W, Almond D, "NDE of CFRP composites by transient thermography," *Proc. QIRT-98, Eurotherm Seminar*, **60**:86–90, 1998.

Pennington KS, Moorhead RJ, II, eds., "Image processing algorithms and techniques," *SPIE Proc.*, **1244**:444, 1990.

Peralta SB, Ellis SC, Christofides C, Mandelis A, Sang H, Farahbakhsh B, "Photo-pyroelectric measurement of the thermal diffusivity of recrystallized high purity aluminum," *Research in Nondestructive Evaluation*, 3[2]:69–80, 1991.

Perch-Nielsen T, Sorensen JC, "Guidelines to thermographic inspection of electrical installations," in Snell JR, Jr., ed., *Thermosense XVI, SPIE Proc.*, **2245**:2–13, 1994.

Pereira P, Carmona C, Pineyro I, Sabini G, Torres M, Macedo N, "Thermographic assessment of psoriasic plaques," *European Journal of Dermatology*, **5**[7]:578–580, 1995.

Perez IM, Santos RP, Kulowitch PJ, Ryan M, "Calorimetric modeling of thermographic data," in Snell JR, Wurzbach RN, eds., *Thermosense XX, SPIE Proc.*, **3361**:75–83, 1998.

Perez IM, Davis WR, Kulowitch PJ, Dersch MF, "Thermographic modeling of water entrapment," in Lemieux DH, Snell JR, eds., *Thermosense XXI, SPIE Proc.*, **3700**:32–39, 1999.

Pergantis CC, "Thermographic field-inspection study on Ml main battle tank track systems," in Eklund JK, ed., *Thermosense XIV, SPIE Proc.*, **1682**:239–247, 1992.

Perl A, "Reasons for poor acceptance of thermography in medical community," *Proc. 9th Conference of the IEEE Engineering in Medicine and Biology Society*, Boston, Nov. 13–16, 1987.

Photonics Spectra, **31**[11]:185, 1997.

Photonics Spectra, **32**:66, 1998.

Photonics Spectra, **34**:114, 2000a.

Photonics Spectra, **34**:115, 2000b.

Pica S, Scarpetta G, "Direct detection of temperature maps on electronic devices surface by using an infrared radiometric microscope," *Proc. QIRT-94, Eurotherm Seminar*, **42**:24–29, 1994.

Pica S, Scarpetta G, "Thermal instability observation in power transistors by radiometric detection of temperature maps," in Semanovich SA, ed., *Thermosense XVII, SPIE Proc.*, **2473**:91–98, 1995.

Pieczyska EA, Gadaj SP, Nowacki WK, "Thermoelastic and thermoplastic effects during loading and unloading of an austenitic steel," *Proc. QIRT-98, Eurotherm Seminar*, **60**:112–116, 1998.

Pietola M, Varrio JP, "Usefulness of high-resolution thermography in fault diagnosis of fluid power components and systems," in Burleigh DD, Spicer JWM, eds., *Thermosense XVIII, SPIE Proc.*, **2766**[19]:101–109, 1996.

Pina JF, Oppenheim JW, "Infrared analysis of telemetry amplifiers and discriminators," *Materials Evaluation*, **28**[4]:88–96, 1970.

Pitre L, Maldague X, Laurendeau D, "Range finder based guidance of Puma robot for the infrared inspection of non-planar material," in Bhargava V, ed., *Proc. Canadian Conference on Electrical and Computer Engineering*, 1993.

Plotnikov YA, Winfree WP, "Advanced image processing for defect visualization in infrared thermography," in Snell JR, Wurzbach RN, eds., *Thermosense XX, SPIE Proc.*, **3361**:331–338, 1998.

Plotnikov YuA, Winfree WP, "Visualization of subsurface defects in composites using a focal plane array infrared camera," in Lemieux DH, Snell JR, eds., *Thermosense XXI, SPIE Proc.*, **3700**:26–31, 1999.

Poloszyk S, Rozanski L, "Influence of the radiation diffraction in image converter of the thermograph upon its metrological parameters," *Proc. QIRT-96, Eurotherm Seminar,* **50**:20–25, 1996.

Poncet E, Béreiziat D, Grangeot G, Batsale JC, "Thermal diffusivity estimation with averaged infrared thermography," *Proc. QIRT-98, Eurotherm Seminar,* **60**:101–107, 1998.

Poropat GV, "Effect of system point spread function, apparent size, and detector instantaneous field of view on the infrared image contrast of small objects," *Optical Engineering,* **32**[10]:2598–2607, 1993.

Potet P, Bathias C, Degrigny B, "Quantitative characterization of impact damage in composite materials: a comparison of computerized vibrothermography and x-ray tomography," *Materials Evaluation,* **46**[8]:1050–1054, 1988.

Potet P, Garçon L, Balageas D, Boscher D, Déom A, "Appareillage industriel de contrôles non destructifs par thermographie infrarouge stimulée," *Journée d'Étude Société Française des Thermiciens,* Jan. 18, report ONÉRA t.p. 1989-16, 12 pp., 1989.

Poussart D, Laurendeau D, "3D sensing for industrial computer vision," in Sanz JLZ, ed., *Advances in Machine Vision: Applications and Architectures,* **3**:122–159, 1988.

Prabhu DR, Winfree WP, "Neural network based processing of thermal NDE data for corrosion detection," in Thompson DO, Chimenti DE, eds., *Review of Progress in Quantitative Nondestructive Evaluation,* **12**, 1993.

Prabhu DR, Howell PA, Syed HI, Winfree WP, "Application of artificial neural networks to thermal detection of disbonds," in Thompson DO, Chimenti DE, eds., *Review of Progress in Quantitative Nondestructive Evaluation,* **11B**:1331–1338, 1992.

Prati J, "Detecting hidden exfoliation corrosion in aircraft wing skins using thermography," in Dinwiddie RB, Lemieux DH, eds., *Thermosense XXII, SPIE Proc.,* **4020**:200–209, 2000.

Pratt WK, *Digital Image Processing,* Wiley, New York, 698 pp., 1991.

Predmesky RL, Zaluzec MJ, "Infrared in automotive applications (invited paper)," in Wurzbach RN, Burleigh DD, eds., *Thermosense XIX, SPIE Proc.,* **3056**:110–116, 1997.

Pregowski P, "Atmospheric effects in infrared thermography," *Proc. QIRT-92, Eurotherm Seminar,* **27**:338–342, 1992.

Pregowski P, "Rescaling of thermographic camera readouts based on the results of contact measurements during unsteady process," *Proc. QIRT-94, Eurotherm Seminar,* **42**:66–71, 1994.

Pregowski P, Swiderski W, "Experimental determination of the transmission of the atmosphere based on thermographic measurements," *Proc. QIRT-96, Eurotherm Seminar,* **50**:363–367, 1996.

Pregowski P, Swiderski W, Walczak RT, "Surface and volume effects in thermal signatures of buried mines: experiment and modelling," *Proc. QIRT-98, Eurotherm Seminar,* **60**:233–237, 1998.

Pregowski P, Swiderski W, Walczak WT, Usowicz B, "Role of time and space variability of moisture and density of sand for thermal detection of buried objects: modeling and experiments," in Lemieux DH, Snell JR, eds., *Thermosense XXI, SPIE Proc.,* **3700**:444–457, 1999.

Pron H, Henry JF, Offermann S, Bissieux C, Beaudoin JL, "Analysis of stress influence on thermal diffusivity by photothermal infrared thermography," *Proc. QIRT-98, Eurotherm Seminar*, **60**:129–133, 1998.

Provost D, "Infrared thermography at EDF: a common technique for high-voltage lines but new in monitoring and diagnosis of PWR plant components," in Burleigh DD, Spicer JWM, eds., *Thermosense XVIII, SPIE Proc.*, **2766**:83–90, 1996.

Prystay M, Loong CA, Nguyen K, "Optimization of cooling channel design and spray patterns in aluminum die casting using infrared thermography," in Burleigh DD, Spicer JWM, eds., *Thermosense XVIII, SPIE Proc.*, **2766**:15–24, 1996a.

Prystay M, Wang H, Garcia-Rejon A, "Application of thermographic temperature measurements in injection molding and blow molding of plastics," in Burleigh DD, Spicer JWM, eds., *Thermosense XVIII, SPIE Proc.*, **2766**:5–14, 1996b.

Pugh R, Huff R, "Transfer of infrared thermography predictive maintenance technologies to Soviet-designed nuclear power plants: experience at Chernobyl," in Lemieux DH, Snell JR, eds., *Thermosense XXI, SPIE Proc.*, **3700**:300–311, 1999.

Qaddoumi N, Carriveau G, Ganchev S, Zoughi R, "Microwave imaging of thick composite panels with defects," *Materials Evaluation*, **53**[8]:926–929, 1995.

Qaddoumi N, Shroyer A, Zoughi R, "Microwave detection of rust under paint and composite laminates," *Research in Nondestructive Evaluation*, **9**[4]:201–212, 1997.

Qin YW, Bao N, "Thermographic nondestructuve testing (NDT) technique for delaminated defects in composite structures," in Semanovich SA, ed., *Thermosense XVII, SPIE Proc.*, **2473**:219–225, 1995.

Qin YW, Bao NK, "Infrared thermography and its application in the NDT of sandwich structures," *Optics and Lasers in Engineering*, **25**[2–3]:205–211, 1996.

Qin Y, Ji H, Chen J, Li H, "Thermography applied to acupuncture and qi-gong," in Wurzbach RN, Burleigh DD, eds., *Thermosense XIX, SPIE Proc.*, **3056**:270–275, 1997.

Quinn MT, Hribar JR, Ruiz RL, Hawkins GF, "Thermographic detection of buried disbonds," in Thompson DO, Chimenti DE, eds., *Review of Progress in Quantitative Nondestructive Evaluation*, **7A**:1117–1123, 1988.

Raczkowski JW, Wegrewicz A, Pala W, "The appliance of the thermovision computerised system for the evaluation of the corrective exercises efficiency in the adolescent idiopathic scoliosis treatment," *Proc. QIRT-98, Eurotherm Seminar*, **60**:15–17, 1998.

Raghavan S, Wikle HC III, Chin BA, "Adaptive control of arc welding using infrared sensing," in Allen LR, ed., *Thermosense XV, SPIE Proc.*, **1933**:76–82, 1993.

Raghu S, Staub FW, "Obtaining the surface temperature distribution in a shock wave–boundary layer interaction region using a liquid crystal technique," *Experimental Thermal and Fluid Science*, **9**[3]:283–288, 1994.

Ramelot D, Ludovicy JM, Stolz C, Fischbach JP, "Capteurs industriels pour applications basses températures," *Revue Générale de Thermique Fr*, **322**[Oct.]:517–524, 1988.

Rantala J, Hartikainen J, "Numerical estimation of the spatial resolution of thermal NDT techniques based on flash heating," *Research in Nondestructive Evaluation*, **3**[3]:125–140, 1991.

Rantala J, Wu D, Busse G, "Amplitude-modulated lock-in vibrothermography for NDE of polymers and composites," *Research in Nondestructive Evaluation*, **7**[4]:215–228, 1996a.

Rantala J, Wu D, Salerno A, Busse G, "Lock-in thermography with mechanical loss angle heating at ultrasonic frequencies," *Proc. QIRT-96, Eurotherm Seminar*, **50**:389–393, 1996b.

Rantala J, Wu D, Busse G, "NDT of polymer materials using lock-in thermography with water-coupled ultrasonic excitation," *NDT&E International*, **31**[1]:43–49, 1998.

Rapaport S, "The secret art of frame grabbing," *Photonics Spectra*, Apr., pp. 137–140, 1988.

Rastogi P, Inaudi D, *Trends in Optical Nondestructive Testing and Inspection*, Elsevier, Amsterdam, 633 pp., 2000.

Ravi J, Madding RP, "Nondestructive method of high-intensity discharge lamp temperature measurement," in Dinwiddie RB, Lemieux DH, eds., *Thermosense XXII, SPIE Proc.*, **4020**:281–288, 2000.

Ravich LE, "Evaluation of electronic images," *Laser Focus/Electro-Optics*, June, pp. 145–155, 1988.

Ravindra NM, Tong FM, Amin S, Shah J, Kosonocky WF, McCaffrey NJ, Manikopoulos CN, Singh B, Soydan R, White LK, Zanzucchi P, Hoffman D, Markham JR, Liu S, Kinsella K, Lareau RT, Casas LM, Monahan T, Eckart DW, "Development of emissivity models and induced transmission filters for multiwavelength imaging pyrometry," in Snell JR, Jr. ed., *Thermosense XVI, SPIE Proc.*, **2245**:304–319, 1994.

Reichenbach SE, Park SK, Narayanswamy R, "Characterizing digital image acquisition devices," *Optical Engineering*, **30**[2]:171–177, 1991.

Reigl M, Gapp M, Schmitz B, Stein J, Goch G, Seidel U, Walther HG, "Investigations of subsurface structures and buried inhomogeneities by photothermal inspection," *Proc. QIRT-94, Eurotherm Seminar*, **42**:232–237, 1994.

Rencz AN, *Remote Sensing for the Earth Sciences*, Vols 1 and 3, Wiley (in cooperation with the American Society for Photogrammetry and Remote Sensing), New York, 1998.

Renius O, "Laser illumination for infrared nondestructive testing," *Materials Evaluation*, **31**[5]:80–84, May 1973.

Reulet P, Marchand M, Millan P, "Experimental characterisation of the convective heat transfer in a vortex–wall interaction," *Proc. QIRT-98, Eurotherm Seminar*, **60**:212–222, 1998.

Reungoat D, Tournerie B, "Temperature measurement by infrared thermography in lubricated contact radiometric analysis," *Proc. QIRT-94, Eurotherm Seminar*, **42**:30–36, 1994.

Reynolds PM, "A review of multicolour pyrometry for temperatures below 1500°C," *British Journal of Applied Physics*, **15**:579–589, 1964.

Reynolds WN, "Quality control of composite materials by thermography," *Metals and Materials*, **1**[2]:100–102, 1985.

Reynolds WN, "Thermographic methods applied to industrial materials," *Canadian Journal of Physics*, **64**:1150–1154, 1986.

Reynolds WN, "Inspection of laminates and adhesive bonds by pulse-video thermography," *Nondestructive Testing International*, **21**[4]:229–232, 1988.

Reynolds WN, Wells GM, "Video-compatible thermography," *British Journal of Non-Destructive Testing*, Jan., pp. 40–44, 1984.

Rhim HC, Büyüköztürk O, Blejer DJ, "Remote radar imaging of concrete slabs with and without a rebar," *Materials Evaluation*, **53**[2]:295, 1995.

Ribeiro VS, Erfer LN, "Noncontact IR thermometers in modern maintenance," in Lemieux DH, Snell JR, eds., *Thermosense XXI, SPIE Proc.*, **3700**:289–296, 1999.

Richardson CH, Schafer RW, "Application of mathematical morphology to Flir images," in Hsing TR, ed., *Visual Communications and Image Processing II, SPIE Proc.*, **845**:249–252, 1987.

Rigney MP, Franke EA, "Machine vision applications of low-cost thermal infrared camera," in Dinwiddie RB, Lemieux DH, eds., *Thermosense XXII, SPIE Proc.*, **4020**:15–26, 2000.

Ring EFJ, Plassmann P, "BTHERM: a computer system for quantitative infrared medical imaging," *Proc. QIRT-98, Eurotherm Seminar*, **60**:52–56, 1998.

Rioux M, Bechthold G, Taylor D, Duggan M, "Design of a large depth of view three-dimensional camera for robot vision," *Optical Engineering*, **26**[12]:1245–1250, 1987.

Ritter R, Schmitz B, "Photothermal inspections of adhesion strengths and detection of delaminations," *Proc. QIRT-96, Eurotherm Seminar*, **50**:197–202, 1996.

Roberts GT, East RA, "Liquid crystal thermography for heat transfer measurement in hypersonic flows: a review," *Journal of Spacecraft and Rockets*, **33**[6]:761–768, 1996.

Roberts CC Jr, "Thermography: an often misused term (proceedings only)," in Eklund JK, ed., *Thermosense XIV, SPIE Proc.*, **1682**:108–114, 1992.

Roberts CC Jr, "Infrared thermal analysis of ski and snowboard binding systems," in Burleigh DD, Spicer JWM, eds., *Thermosense XVIII, SPIE Proc.*, **2766**:110–120, 1996a.

Roberts CC Jr, "Remote temperature sensing as a means of maintaining ski lift towers," in Burleigh DD, Spicer JWM, eds., *Thermosense XVIII, SPIE Proc.*, **2766**:121–131, 1996b.

Robertson JS, *Engineering Mathematics with Maple*, McGraw-Hill, New York, 1996.

Robinson K, "IR cameras save the ship," *Photonics Spectra*, Jan., pp. 20–24, 1998.

Robinson K, Wheeler MD, "Sports professionals muster lasers, cameras, fiber optics and photodiodes to improve performance," *Photonics Spectra*, Dec., pp. 76–85, 1997.

Rodriguez AA, Mitchell OR, "Image segmentation by background extraction refinements," in Casasent DP, ed., *SPIE Proc. Intelligent Robots and Computer Vision VIII: Algorithms and Techniques*, **1192**:122–134, 1989.

Roellig TL, Werner MW, Becklin EE, "Thermal emission from Saturn's ring at 380 microns," *Icarus*, **73**[3]:574–583, 1988.

Rogovsky AJ, "Ultrasonic and thermographic methods for NDE of composite tubular parts," *Materials Evaluation*, **43**[5]:547–555, 1985.

Roney JE, "Steel surface temperature measurement in industrial furnaces by compensation for reflected radiation errors," in Schooley JF, ed., *Temperature, Its Measurement and Control in Science and Industry*, pp. 485–489, American Institute of Physics, New York, 1982.

Rosencwaig A, "Thoretical aspects of photoacoustic spectroscopy," *Journal of Applied Physics*, **49**[5]:2905–2910, 1978.

Rosenfeld A, *Picture Processing by Computer*, Academic Press, San Diego, CA, 1969.

Rosina E, Ludwig N, Rosi L, "Optimal environmental conditions to detect moisture in ancient buildings: case studies in northern Italy," in Snell JR, Wurzbach RN, eds., *Thermosense XX, SPIE Proc.*, **3361**[29]:188–199, 1998.

Rossi R, Creus M, Lluesma ET, "Distribution and temperatures in odontology acupuncture," in Dinwiddie RB, Lemieux DH, eds., *Thermosense XXII, SPIE Proc.*, **4020**:276–280, 2000.

Rossignol N, *Thermographie élémentaire*, LVSN report, Université Laval, June 2000.

Roth DJ, Bodis JR, Bishop C, "Thermographic imaging for high-temperature composite materials: a defect detection study," *Research in Nondestructive Evaluation*, **9**[3]:147–170, 1997.

Rounds EM, Sutty G, "Segmentation based on second order statistics," *Optical Engineering*, **19**[6]:936–940, 1980.

Rousset G, Lévesque D, Bertrand L, Maldague X, Cielo P, "Pulsed photothermoelastic quantitative evaluation of flaws in laminates," *Canadian Journal of Physics*, **64**[9]:1293–1296, 1986.

Rozlosnik AE, "Infrared thermography and ultrasound for both testing and analyzing valves," in Snell JR, Wurzbach RN, eds., *Thermosense XX, SPIE Proc.*, **3361**:137–155, 1998.

Rozlosnik AE, "Bringing up-to-date applications of infrared thermography to surveillance, safety, and rescue," in Dinwiddie RB, Lemieux DH, eds., *Thermosense XXII, SPIE Proc.*, **4020**:387–406, 2000.

Rozlosnik AE, Lardone VM, "Infrared thermography for process control and predictive maintenance purposes in a steel wire drawing machine," in Lemieux DH, Snell JR, eds., *Thermosense XXI, SPIE Proc.*, **3700**:141–163, 1999.

Ruddock W, "Infrared thermography and global positioning system integration in location-sensitive situations," in Burleigh DD, Spicer JWM, eds., *Thermosense XVIII, SPIE Proc.*, **2766**:65–73, 1996.

Russell SS, Henneke EG II, "Vibrothermographic inspection of a glass-fiber epoxy machine part," *Materials Evaluation*, **49**[7]:870–874, 1991.

Ryu ZM, "Measurement of point spread function of thermal imager," in Baird GS, ed., *Thermosense XIII, SPIE Proc.*, **1467**:469–474, 1991.

Sadat AB, "Machining of graphite epoxy composite material," *SAMPE Quarterly*, **19**[2]:1–4, 1988.

Sadek HM, "How to prepare a written practice for infrared/thermal nondestructive examination method," in Snell JR, Jr., ed., *Thermosense XVI, SPIE Proc.*, **2245**:34–39, 1994.

Safabakhsh R, "Processing infrared images for high speed power line inspection," in McIntosh GB, ed., *Thermosense XI, SPIE Proc.*, **1094**:75–82, 1989.

Safai M, "Thermography evaluation of metal bonding materials," in Eklund JK, ed., *Thermosense XIV, SPIE Proc.*, **1682**:234–238, 1992.

Safai M, King TR, Greegor RB, "Thermal response measurements of aluminum and graphite epoxy due to irradiation by 256-MeV protons," in Burleigh DD, Spicer JWM, eds., *Thermosense XVIII, SPIE Proc.*, **2766**:307–315, 1996.

Saghri A, Hou HS, Tescher AG, "Personal computer based image processing with halftoning," *Optical Engineering*, **25**[3]:499–504, 1986.

Sahoo PK, Soltani S, Wong AKC, Chen Y, "A survey of thresholding techniques," *Computer Vision, Graphics, and Image Processing*, 41:233–260, 1988.

Sahr CA, "Thermal imaging of railroad cars used for molten iron transport," in Lemieux DH, Snell JR, eds., *Thermosense XXI, SPIE Proc.*, **3700**:194–201, 1999.

Saintey MB, Almond DP, "Defect sizing by transient thermography. II: A numerical treatment," *Journal of Physics D: Applied Physics*, 28[12]:2539–2546, 1995.

Saintey MB, Almond DP, "Mathematical modelling of transient thermography and defect sizing," in Thompson DO, Chimenti DE, eds., *Review of Progress in Quantitative Nondestructive Evaluation*, **15A**:503–509, 1996.

Saintey MB, Almond DP, "An artificial neural network interpreter for transient thermography image data," *NDT&E International*, 30[5]:291–295, 1997.

Sakagami T, "Abstracts of the 2nd Symposium on Thermographic NDT&E Techniques," in Lemieux DH, Snell JR, eds., *Thermosense XXI, SPIE Proc.*, **3700**:121–131, 1999.

Sakagami T, Kubo S, "Development of a new crack identification method based on singular current field using differential thermography," in Lemieux DH, Snell JR, eds., *Thermosense XXI, SPIE Proc.*, **3700**:369–376, 1999.

Sakagami T, Ogura K, "Thermographic NDT based on transient temperature field under Joule effect heating," in Snell JR, Jr., ed., *Thermosense XVI, SPIE Proc.*, **2245**:120–130, 1994.

Sakagami T, Ogura K, "Overview of recent Japanese activities in thermographic NDT," in Wurzbach RN, Burleigh DD, eds., *Thermosense XIX, SPIE Proc.*, **3056**:216–219, 1997.

Sakagami T, Ogura K, Shoda M, "Thermal sensing and imaging of the dry-sliding contact surface using IR thermomicroscope," in Semanovich SA, ed., *Thermosense XVII, SPIE Proc.*, **2473**:263–272, 1995.

Sakagami T, Ogura K, Kubo S, "Damage inspection of CFRP using resistive-heating thermographic NDT," in Burleigh DD, Spicer JWM, eds., *Thermosense XVIII, SPIE Proc.*, **2766**:178–184, 1996.

Sakagami T, Ogura K, Kubo S, Lesniak JR, Boyce BR, Sandor BI, "Visualization of contact stress distribution using infrared stress measurement system," in Wurzbach RN, Burleigh DD, eds., *Thermosense XIX, SPIE Proc.*, **3056**:250–259, 1997.

Sakagami T, Madhavan V, Harish G, Krishnamurthy K, Ju Y, Farris TN, Chandrasekar S, "Full-field IR measurement of subsurface grinding temperatures," in Snell JR, Wurzbach RN, eds., *Thermosense XX, SPIE Proc.*, **3361**:234–247, 1998.

Sakagami T, Kubo S, Komiyama T, Suzuki H, "Proposal for a new thermographic nondestructive testing technique using microwave heating," in Lemieux DH, Snell JR, eds., *Thermosense XXI, SPIE Proc.*, **3700**:99–103, 1999.

Sakagami T, Kubo S, Teshima Y, "Fatigue crack identification using near-tip singular temperature field measured by lock-in thermography," in Dinwiddie RB, Lemieux DH, eds., *Thermosense XXII, SPIE Proc.*, **4020**:174–181, 2000.

Salerno A, Wu D, Busse G, Rantala J, "Thermographic inspection with ultrasonic excitation" in Thompson DO, Chimenti DE, eds., *Review of Progress in Quantitative Nondestructive Evaluation*, **16A**:345–352, 1996.

Salerno A, Dillenz A, Wu D, Rantala J, Busse G, "Progress in ultrasound lockin thermography," *Proc. QIRT-98, Eurotherm Seminar*, **60**:154–160, 1998a.

Salerno A, Wu D, Busse G, Malter U, "Glass fiber airplane inspected with infrared lockin thermography," in Thompson DO, Chimenti DE, eds., *Review of Progress in Quantitative Nondestructive Evaluation*, **16A**:357–364, 1998b.

Salerno A, Wu D, Busse G, Rantala J, "Thermographic inspection with ultrasonic excitation," in Thompson DO, Chimenti DE, eds., *Review of Progress in Quantitative Nondestructive Evaluation*, **16A**:345–352, 1998c.

Salnick A, Mandelis A, Othonos A, Christofides C, "Noncontact lifetime reconstruction in continuously inhomogeneous semiconductors: generalized theory and experimental photothermal results for ion-implanted Si," in Thompson DO, Chimenti DE, eds., *Review of Progress in Quantitative Nondestructive Evaluation*, **16A**:371–378, 1998.

Sanmartin ML, "Matériau sandwich nouveau," *Matériaux et Techniques*, June, pp. 3–7, 1988.

Santey MB, Almond DP, "An artificial neural network interpreter for transient thermography image data," *NDT&E International*, **30**[5]:291–295, 1997.

Saptzin VM, Kober VI, Vavilov VP, "New method of digital modulative adaptive auto-calibration of infrared imaging devices," *Proc. QIRT-92, Eurotherm Seminar*, **27**:3–7, 1992.

Saridis GN, Brandin DM, "An automatic surface inspection system for flat rollest steel," *Automatica*, **15**:505–520, 1979.

Sarle WS, "Neural Network FAQ, part 1 of 7: Introduction," *ftp://ftp.sas.com/pub/ neural/FAQ.html*, June 2000.

Sasaki R, Uematsu Y, Yamada M, Saeki H, "Application of infrared thermography and a knowledge-based system to the evaluation of the pedestrian-level wind environment around buildings," *Journal of Wind Engineering and Industrial Aerodynamics*, **67– 68**[Apr.]:873–883, 1997.

Sass DT, White JD, Gonda TG, Jones JC, "Tracking rotating components for complete construction of simulated images," in Snell JR, Wurzbach RN, eds., *Thermosense XX, SPIE Proc.*, **3361**:212–217, 1998.

Sato M, Kato K, Watarai T, Miyata K, Inagaki T, Okamoto Y, "Detection of high-temperature spots on reciprocating interface under dry friction by means of an infrared radiometer," in Burleigh DD, Spicer JWM, eds., *Thermosense XVIII, SPIE Proc.*, **2766**:295–306, 1996.

Sawicki P, Stein R, Wiecek B, "Directional emissivity correction by photogrammetric 3D object reconstruction," *Proc. QIRT-98, Eurotherm Seminar*, **60**:327–332, 1998.

Sayers CM, "Detectability of defects by thermal non-destructive testing," *British Journal of Non-Destructive Testing*, Jan., pp. 28–33, 1984.

Schachter BL, Davis LS, Rosenfeld A, "Some experiments in image segmentation by clustering of local feature values," *Pattern Recognition*, **11**:19–28, 1979.

Schaper M, Haferkamp H, Niemeyer M, Pelz C, Viets R, "Thermal investigation of compound cast steel tools," in Lemieux DH, Snell JR, eds., *Thermosense XXI, SPIE Proc.*, **3700**:164–170, 1999.

Schaufler ER, "A look at the inside of the American Society for Nondestructive Testing (ASNT) level III certification program," in Semanovich SA, ed., *Thermosense XVII, SPIE Proc.*, **2473**:126–130, 1995.

Schivley JD, "Solar-stimulated flow and water incursion detection in fiberglass structures," in Snell JR, Jr., ed., *Thermosense XVI, SPIE Proc.*, **2245**:183–187, 1994.

Schlicht O, Wölfel HP, "Determination of time dependent crack contact behaviour by thermoelastic stress analysis," *Proc. QIRT-96, Eurotherm Seminar*, **50**:96–101, 1996.

Schmalz H, "Infrared visualized air turbulence," in Semanovich SA, ed., *Thermosense XII, SPIE Proc.*, **1313**:278–281, 1990.

Schmalz HH, "Infrared air flow studies of chemical hood performance," in Eklund JK, ed., *Thermosense XIV, SPIE Proc.*, **1682**:336–340, 1992.

Schmidt J, "Thermomechanical coupling as a criterion of the yield point of spheroidal graphite cast iron," *Proc. QIRT-94, Eurotherm Seminar*, **42**:90–95, 1994.

Schmitz B, Reick M, Goch G, Steiner R, "Photothermal detection of surface defects and thermal changes in near-surface layers," *Proc. QIRT-94, Eurotherm Seminar*, **42**:293–297, 1994.

Schneider H, Gruner H, Kohlhage H, Deppe GJ, Ehlert KP, "Optical system for on-line detection of longitudinal surface cracks in hot continuously cast rounds," *Proc. 3rd European Conference on Nondestructive Testing*, Firenza, Italy, Oct. 15–18, pp. 188–198, 1984.

Schneider K, Mauser W, "Processing and accuracy of Landsat Thermatic Mapper data for lake surface temperature measurement," *International Journal of Remote Sensing*, **17**[11]:2027–2041, 1996.

Schott JR, Biegel JD, "Comparison of modelled and empirical atmospheric propagation data," in Spiro IJ, Mollicone RA, eds., *Infrared Technology IX, SPIE Proc.*, **430**:35–50, 1987.

Schuler F, Rampp F, Martin J, Wolfrum J, "TACCOS: a thermography-assisted combustion control system for waste incinerators," *Combustion and Flame*, **99**[2]:431–439, 1994.

Schumacher DH, "Measuring microbond integrity with an infrared microradiometer," *Materials Evaluation*, **26**[12]:257–260, 1968.

Scott IG, "Bridging the gap," in McGonnagle WJ, ed., *International Advances in Nondestructive Testing*, **15**:1–8, 1990.

Scribner DA, Sarkady KA, Caufield JT, Kruer MR, Katz G, Gridley CI, "Nonuniformity correction for staring IR focal plane arrays using scene-based techniques," in Dereniak EL, Sampson RE, eds., *SPIE Proc. Infrared Detectors and Focal Plane Arrays*, **1308**:224–233, 1990.

Segal E, Thomas G, Rose J, "Hope for solving the adhesive bond nightmare?" *Proc. 12th Symposium on Nondestructive Evaluation*, San Antonio, TX, pp. 269–281, 1979.

Seidel U, Walther HG, "Estimation of depth, width, and magnitude of material inhomogeneities from photothermal measurements," in Burleigh DD, Spicer JWM, eds., *Thermosense XVIII, SPIE Proc.*, **2766**:240–248, 1996.

Seidel U, Haupt K, Walther HG, Burt J, Bein BK, "Analysis of the detectability of buried inhomogeneities by means of photothermal microscopy," *Journal of Applied Physics*, **75**[9]:4396–4401, 1994.

Seiler MR, Haselwood JL, Stockum LA, "Infrared imaging of microwave sources," in Eklund JK, ed., *Thermosense XIV, SPIE Proc.*, **1682**:296–307, 1992.

Selman JJ, Miller JT, "Evaluation of a prototype thermal wave imaging system for nondestructive evaluation of composite and aluminum aerospace structures," in Allen LR, ed., *Thermosense XV, SPIE Proc.*, **1933**:178–187, 1993.

Shann T, Oakley JP, "Novel approach to boundary finding," *Image and Vision Computing*, **8**[1]:32–36, 1990.

Sheng Y, "The transforms and applications handbook," Chap. 10 in *Wavelet Transform*, CRC Press, Boca Raton, FL, 1996.

Shepard SM, Ahmed T, "Characterization of active thermographic system performance," in Lemieux DH, Snell JR, eds., *Thermosense XXI, SPIE Proc.*, **3700**:388–392, 1999.

Shepard SM, Sass DT, "Analysis of high frequency thermal phenomena with a commercial imaging radiometer" in Semanovich SA, ed., *Thermosense XII, SPIE Proc.*, **1313**:298–301, 1990a.

Shepard SM, Sass DT, "Thermal imaging of repetitive events at above-frame-rate frequencies," *Optical Engineering*, **29**[2]:105–109, 1990b.

Shepard SM, Sass DT, Imirowicz T, "Enhanced temporal resolution with a scanning imaging radiometer," *Optical Engineering*, **30**[11]:1716–1719, 1991.

Shepard SM, Imirowicz TP, Sass DT, "Comparison of optomechanical and focal plane array methods for enhanced temporal resolution," in Eklund JK, ed., *Thermosense XIV, SPIE Proc.*, **1682**:308–314, 1992a.

Shepard SM, Sass DT, Imirowicz TP, "Methods for achieving enhanced temporal resolution in IR image acquisition," *Proc. QIRT-92, Eurotherm Seminar*, **27**:8–12, 1992b.

Shepard SM, Ahmed T, Favro LD, Thomas RL, Kuo PK, "Comparison of scanning and focal-plane-array cameras for IR thermal wave imaging," in Allen LR, ed., *Thermosense XV, SPIE Proc.*, **1933**:142–147, 1993.

Shepard SM, Favro LD, Thomas RL, "Thermal wave nondestructive testing (NDT) of ceramic coatings," in Semanovich SA, ed., *Thermosense XVII, SPIE Proc.*, **2473**:190–193, 1995.

Shepard SM, Ahmed T, Caudhry BC, "Resolution criteria for subsurface features in active thermography," in Thompson DO, Chimenti DE, eds., *Review of Progress in Quantitative Nondestructive Evaluation*, **18A**:605–610, 1998a.

Shepard SM, Chaudhry BB, Predmesky RL, Zaluzec MJ, "Pulsed thermographic inspection of spot welds," in Snell JR, Wurzbach RN, eds., *Thermosense XX, SPIE Proc.*, **3361**:320–324, 1998b.

Shepard SM, Lhota JR, Rubadeux BA, Ahmed T, "Onward and inward: extending the limits of thermographic NDE," in Dinwiddie RB, Lemieux DH, eds., *Thermosense XXII, SPIE Proc.*, **4020**:194–199, 2000.

Shepherd FD, Yang AC, "Silicon Schottky retinas for infrared imaging," *IEDM Technical Digest*, Dec., pp. 310–313, 1973.

Shepherd FD, Moorey JM, "Design considerations for IR staring-mode cameras," in Wight R, ed., *SPIE Proc. Electro-optical Imaging System Integration*, **762**:35–50, 1987.

Shorrocks NM, Porter SG, Whatmore WB, Parsons AD, Gooding JN, Pedder DJ, "Uncooled infrared thermal detector arrays," in Lettington AH, ed., *SPIE Proc. Infrared Technology and Applications*, **1320**:88–94, 1990.

Shushan A, Meninberg Y, Levy I, Kopeika NS, "Infrared image sensors," *Optical Engineering*, **30**[11]:1709–1715, 1991.

Siebes G, Johnson KR, McAffee D, "Experience with an imaging infrared radiometer in a simulated space environment," in Burleigh DD, Spicer JWM, eds., *Thermosense XVIII, SPIE Proc.*, **2766**:316–322, 1996.

Siegel R, Howell JR, *Thermal Radiation Heat transfer*, McGraw-Hill, New York, 814 pp., 1972.

Silk MG, "Weld inspection methods," *Metals and Materials*, **5**[4]:192–196, 1989.

Silverman BG, Daley WR, Rubin JD, "The use of infrared ear thermometers in pediatric and family practice offices," *Public Health Reports*, **113**[3]:268–272, 1998.

Simeonides G, Vermeulen JP, Zemsch S, "Application of quantitative infrared thermography in the VKI Mach 6 hypersonic wind tunnel," *Proc. QIRT-92, Eurotherm Seminar*, **27**:111–116, 1992.

Simons AJ, McClean IP, Stevens R, "Phosphors for remote thermograph sensing in lower temperature ranges," *Electronics Letters*, **32**[3]:253–2541, 1996.

Simpson WA, Deeds WE, Cheng CC, Dodd CV, "An infrared microscope system for the detection of internal flaws in solids," *Materials Evaluation*, **28**[9]:205–211, 1970.

Sinclair KF, Zagorites HA, Zimmermann KA, "Scintillography using thermal neutrons," *Materials Evaluation*, **33**[3]:56–60, 1975.

Sippach HG, "Infrared imaging with vidicons and charge injection devices," *Materials Evaluation*, **33**[8]:209–212, 1975.

Smart AE, "Trend in optical sensors for hostile environments," *SPIE Proc. International Symposium on Optical Engineering and Industrial Sensing*, Dearborn MI, June 26–30, 1988.

Smith M, "Non-destructive testing," *Building Research and Information*, **23**[1]:11–13, 1995.

Smith RL, "Recent developments in nondestructive testing," *Metals and Materials*, **3**[4]:187–191, 1987.

Smither RK, "Thermal imaging of synchrotron beams on sillcon crystals," *Proc. QIRT-92, Eurotherm Seminar*, **27**:296–301, 1992.

Sneeringer JW, Hacke KP, Roehrs RJ, "Practical problems related to the thermal infrared nondestructive testing of a bonded structure," *Materials Evaluation*, **29**[4]:88–92, 1971.

Snell R Jr, "Problems commonly encountered in quantitative thermo graphic electrical inspections," *Proc. QIRT-94, Eurotherm Seminar*, **42**:37–43, 1994.

Snell JR, Jr, "Problems inherent to quantitative thermographic electrical inspections," in Semanovich SA, ed., *Thermosense XVII, SPIE Proc.*, **2473**:75–81, 1995.

Snell JR, Jr, "Developing written inspection procedures for thermal/infrared thermography," in Burleigh DD, Spicer JWM, eds., *Thermosense XVIII, SPIE Proc.*, **2766**:136–144, 1996.

Snell JR, Jr, Renowden J, "Improving the results of thermographic inspections of electrical transmission and distribution lines," in Dinwiddie RB, Lemieux DH, eds., *Thermosense XXII, SPIE Proc.*, **4020**:115–127, 2000.

Snell JR, Jr, Spring RW, "Developing operational protocol for thermographic inspection programs," in Eklund JK, ed., *Thermosense XIV, SPIE Proc.*, **1682**:12–22, 1992.

Snyder WC, Schott JR, "Combined aerial and ground technique for assessing structural heat loss," in Snell JR, Jr., ed., *Thermosense XVI, SPIE Proc.*, **2245**:71–82, 1994.

Soucy M, Laurendeau D, "Generating non-redundant surface representation of 3D objects using multiple range views," in *IEEE Proc. ICPR*, Atlantic City, NJ, June 16–21, pp. 198–200, 1990.

Sparrow EM, Chess RD, *Radiation Heat Transfer*, Sec. 2-2, McGraw-Hill, New York, 1978.

Spicer JB, Champion JL, Osiander R, Spicer JWM, "Time resolved shearographic and thermographic NDE methods for graphite epoxy/honeycomb composite," *Materials Evaluation*, **54**[10]:1210–1213, 1996.

Spicer JWM, Kerns WD, Aamodt LC, Osiander R, Murphy JC, "Time-resolved infrared radiometry (TRIR) using a focal-plane array for characterization of hidden corrosion," in Allen LR, ed., *Thermosense XV, SPIE Proc.*, **1933**:148–159, 1993.

Spicer JWM, Osiander R, Chang Y, Hildebrand R, "Time-resolved microwave thermoreflectometry for infrastructure inspection," in Thompson DO, Chimenti DE, eds., *Review of Progress in Quantitative Nondestructive Evaluation*, **15A**:541–547, 1996.

Spicer JWM, Wilson DW, Osiander R, Thomas J, Oni BO, "Evaluation of high-thermal-conductivity graphite fibers for thermal management in electronics applications," in Lemieux DH, Snell JR, eds., *Thermosense XXI, SPIE Proc.*, **3700**:40–47, 1999.

Spiegel MR, *Mathematical Handbook of Formulas and Tables*, Schaum Series, McGraw-Hill, New York, 1974.

Spiro JJ, ed., *Selected Papers on Radiometry*, SPIE Milestone Series, Vol. 14, 692 pp., 1990.

Spiro IJ, Schlessinger M, *Infrared technology fundamentals*, in Thompson BJ, ed., Vol. 22 of Optical Engineering Series, Marcel Dekker, New York, 1989.

Spring RW, "American Society for Nondestructive Testing (ASNT) certification of thermographers," in Semanovich SA, ed., *Thermosense XVII, SPIE Proc.*, **2473**:131–134, 1995.

Spring RW, Snell JR, JR, "ASNT central certification program," in Wurzbach RN, Burleigh DD, eds., *Thermosense XIX, SPIE Proc.*, **3056**:96–99, 1997.

Staaf Ö, Ribbing CG, Andersson SK, "Broadband emission factors: temperature variation for nongray samples," in Burleigh DD, Spicer JWM, eds., *Thermosense XVIII, SPIE Proc.*, **2766**:334–345, 1996.

Stansfield SA, "ANGY: a rule-based expert system for automatic segmentation of coronary vessels from digital subtracted angiograms," *IEEE Transactions on Pattern Analysis and Machine Intelligence*, **8**[2]:188–199, 1986.

Steckenrider JS, Ellingson WA, Rothermel SA, "Full-field characterization of thermal diffusivity in continuous-fiber ceramic composite materials and components," in Semanovich SA, ed., *Thermosense XVII, SPIE Proc.*, **2473**:205–213, 1995.

Stein MC, Heller WG, "Fractal methods for flaw detection in NDE imagery," in Thompson DO, Chimenti DE, eds., *Review of Progress in Quantitative Nondestructive Evaluation*, **8A**:689–700, 1989.

Stiefeld S, Yoshimura RH, "A capability and limitation study of thermography of carbon–carbon cones," *Materials Evaluation*, **29**[12]:281–287, 1971.

Stillwell PFTC, "Thermal imaging," *Journal of Physics E: Scientific Instruments*, **14**:1113–1118, 1981.

Stirckland RN, Gerber MR, "Estimation of ship profiles from a time sequence of forward-looking infrared images," *Optical Engineering*, **25**[8]:995–1000, 1986

Stockton GR, Allen LR, "Using infrared thermography to determine the presence and correct placement of grouted cells in single-width concrete masonry unit (CMU) walls," in Lemieux DH, Snell JR, eds., *Thermosense XXI, SPIE Proc.*, **3700**:483–493, 1999.

Storozhenko VA, "Quantitative infrared thermography application for thermal defectometry," *Proc. QIRT-92, Eurotherm Seminar*, **27**:283–290, 1992.

Stovicek D, "Better maintenance through thermography," *Power Transmission Design*, June, pp. 48–52, 1987.

Strmiska RG, "Five important power line component anomalies," in Dinwiddie RB, Lemieux DH, eds., *Thermosense XXII, SPIE Proc.*, **4020**:85–90, 2000.

Sugimoto T, Kaya K, Okumura H, Wakoh M, "Development of eddy current inspection system for detection of surface defects on hot steel slabs," *Automation in Mining and Metal Processing*, 401–405, 1986.

Sundberg J, "Visualizing airflow using IR-techniques," *Proc. QIRT-92, Eurotherm Seminar*, **27**:333–337, 1992.

Sundberg J, "Use of thermography to register air temperatures in cross sections of rooms and to visualize the airflow from air-supply diffusers," in Allen LR, ed., *Thermosense XV, SPIE Proc.*, **1933**:61–66, 1993.

Svaic S, "Infrared thermography and the numerical heat transfer analysis," *Proc. QIRT-92, Eurotherm Seminar*, **27**:121–128, 1992.

Svaic S, Sundov I, "IR thermography and heat treatment of metals," *Proc. QIRT-94, Eurotherm Seminar*, **42**:96–102, 1994.

Svaic S, Sundov I, "Mathematical models for simulation of cooling processes using infrared surface-temperature measurement," in Semanovich SA, ed., *Thermosense XVII, SPIE Proc.*, **2473**:152–161, 1995.

Svedamar B, "Sewage leak detection: a novel method to detect underground sewer defects," *Proc. 5th Infrared Information Exchange*, New-Orleans, LA, Oct., **1**:79–83, 1985.

Syed HI, Cramer KE, "Corrosion detection in aircraft skin," in Allen LR, ed., *Thermosense XV, SPIE Proc.*, **1933**:160–165, 1993.

Syed HI, Winfree WP, Cramer KE, "Processing infrared images of aircraft lapjoints," in Eklund JK, ed., *Thermosense XIV, SPIE Proc.*, **1682**:171–177, 1992.

Szeloch RF, Gotszalk TP, Rangelow I, Borkowicz Z, "Black silicon as secondary standard of the black body," *Proc. QIRT-98, Eurotherm Seminar*, **60**:338–341, 1998.

Takala J, Lahdeniemi M, Tanttu JT, "Infrared monitoring of plant damage and herbivore invasions," in Semanovich SA, ed., *Thermosense XVII, SPIE Proc.*, **2473**:358–359, 1995.

Tam KC, "Multispectral limited-angle image reconstruction," *IEEE Transactions on Nuclear Science*, **30**[1]:697–701, 1983.

Taniguchi M, Takagi T, "Holographic pattern measuring system and its application to deformation analysis of printed circuit board due to heat of mounted parts," *IEEE Transactions on Instrumentation and Measurement*, **43**[2]:326–331, 1994.

Taylor JO, Dupont HM, "Inspection of metallic thermal protection systems for the X-33 launch vehicle using pulsed infrared thermography," in Snell JR, Wurzbach RN, eds., *Thermosense XX, SPIE Proc.*, **3361**:301–310, 1998.

Teich AC, "Back to basics: thermography for power generation and distribution," *Materials Evaluation,* **50**[6]:658, 660, 662, 1992.

Tenek LH, Henneke EG, "Flaw dynamics and vibro-thermographic thermoelastic NDE of advanced composite materials" in Baird GS, ed., *Thermosense XIII, SPIE Proc.,* **1467**:252–263, 1991.

Tenek LH, Henneke EG II, Gunzburger MD, "Vibration of delaminated composite plates and some applications to non-destructive testing," *Composite Structures,* **23**[3]:253–262, 1993.

Tervo M, Kiukaanniemi E, Kauppinen T, "Applications of aerial thermography in peat production in Finland," in Allen LR, ed., *Thermosense XV, SPIE Proc.,* **1933**:101–109, 1993.

Theilen DA, Christofersen RJ, Dods BG, Emahiser DC, Robles BH, "Infrared thermographic inspection of superplastically formed/diffusion-bonded titanium structures," in Allen LR, ed., *Thermosense XV, SPIE Proc.,* **1933**:174–177, 1993.

Theuwissen AJP, "Dynamic range: buyers need comparable specifications," *Photonics Spectra,* Nov., pp. 161–163, 1997.

Thévenaz L, "Distributed deformation sensors," in Rastogi P, ed., *Trends in Optical Nondestructive Testing,* pp. 447–458, Elsevier, Amsterdam, 2000.

Thomas LC, *Fundamentals of Heat Transfer,* Prentice Hall, Upper Saddle River, NJ, 702 pp., 1980.

Thomas ME, Wayland PS, Terry DH, "Imaging pyrometry of oxides," in Snell JR, Wurzbach RN, eds., *Thermosense XX, SPIE Proc.,* **3361**:2–13, 1998.

Thomas RE, Kennedy EM, Herring BE, "Computer-aided algorithm for evaluating thermograms of the hand," *Computers and Industrial Engineering,* **26**[3]:501–509, 1994.

Thomas RL, Favro LD, "From photoacoustic microscopy to thermal-wave imaging," *MRS Bulletin,* **21**[10]:47–52, 1996.

Thomas RL, Favro LD, Crowther DJ, Kuo PK, "Inversion of thermal wave infrared images," *Proc. QIRT-92, Eurotherm Seminar,* **27**:278–282, 1992.

Thompson KG, Crisman EM, "Thermographic inspection of solid-fuel rocket booster field joint components," *Materials Evaluation,* **48**[9]:1096–1099, 1990.

Thompson LW, "A thermal method for measuring wall thickness and detection of incomplete core removal in investment castings," *Materials Evaluation,* **29**[5]:105–111, 1971.

Thornhill KL, Bitting HC, Lee RB III, Paden JD, Pandey K, Priestley KJ, Thomas S, Wilson S, "Spectral characterizations of the clouds and the Earth's radiant energy system (CERES) thermistor bolometers using Fourier transform spectrometer (FTS) techniques," in Snell JR, Wurzbach RN, eds., *Thermosense XX, SPIE Proc.,* **3361**:55–65, 1998.

Tissot JL, Rothan F, Vedel C, Vilain M, Yon JJ, "Uncooled IRFPA developments review," in Balageas D, Busse G, Carlomagno C, eds., *Proc. QIRT-98, Eurotherm Seminar,* **60**:292–298, 1998.

Tobiasson W, Greatorex A, "Use of an infrared scanner and a nuclear meter to find wet insulation in a ballasted roof," in Snell JR, Jr., ed., *Thermosense XVI, SPIE Proc.,* **2245**:52–61, 1994.

Tom VT, Wolfe GJ, "Adaptive histogram equalization and its applications," in Tesher AG, ed., *Applications of Digital Image Processing IV, SPIE Proc.*, **359**:204–209, 1982.

Tonner PD, Tosello G, "Computed tomography scanning for location and sizing of cavities in valve castings," *Materials Evaluation*, **44**[2]:203–208, 1986.

Torgunakov VG, Sukhanov MS, Vavilov VP, Yamanaev NM, "The rotating cement kiln 3D computer model oriented toward solving thermal nondestructive testing problems," *Proc. QIRT-98, Eurotherm Seminar*, **60**:97–100, 1998.

Tossell DA, "Numerical analysis of heat input effects in thermography," *Journal of Nondestructive Evaluation*, **6**[2]:101–107, 1987.

Tossell DA, "The analysis of heat input effects in passive thermographic NDE," in Thompson DO, Chimenti DE, eds., *Review of Progress in Quantitative Nondestructive Evaluation*, **8A**:1763–1770, 1989.

Tou JT, Gonzalez RC, *Pattern Recognition Principles*, Addison-Wesley, Reading, MA, 431 p., 1974.

Touloukian YS, DeWitt DP, "Thermal radiative properties," *Thermophysical Properties of Matter*, Vol. 7, IFI/Plenum, New York, 1540 pp., 1970.

Tower JR, "Staring PtSi IR cameras: more diversity, more applications," *Photonics Spectra*, **25**[2]:103–106, 1991a.

Tower JR, "PtSi thermal imaging systems: a new level of performance and maturity," *Photonics Spectra*, **25**[4]:91–97, 1991b.

Traycoff RB, "Computerized infrared thermography: clinical applications and diagnostic value," *Proc. 9th Conference of the IEEE Engineering in Medicine and Biology Society*, Boston, Nov. 13–16, 1987.

Traycoff RB, "Medical applications of infrared thermography," in Maldague XPV, ed., *Infrared Methodology and Technology*, International Advances in Nondestructive Testing Monograph Series, Chap. 14, pp. 469–482, Gordon and Breach, New York, 1994.

Tremblay M, Poussart D, "MAR: an integrated system for focal plane edge-tracking with parallel analog filtering and built-in primitives for image acquisition and analysis," *Proc. 10th International Conference on Pattern Recognition*, Atlantic City, NJ, **2**:292–298, 1990.

Tretout H, "Applications industrielle de la thermographie infrarouge au contrôle non destructif de pièces en matériaux composites," *Revue Générale de Thermique Fr.*, **301**[Jan.]:47–53, 1987.

Tretout H, Marin JY, "Transient thermal technique for infrared nondestructive testing of composite materials," in Baker LR, Masson A, eds., *SPIE Proc. IR Technology and Applications*, **590**:277–284, 1985.

Tretout H, David D, Marin JY, de Mol R, "Thermally stimulated infrared thermography for composite ceramic and metallic materials inspection," post-deadline paper (unpublished) presented at the Thermosense XIII SPIE Conference, Apr. 3–5, 1991.

Tretout H, David D, Marin JY, Dessendre M, Couet M, Avenas-Payan I, "An evaluation of artificial neural networks applied to infrared thermography inspection of composite aerospace structures," in Thompson DO, Chimenti DE, eds., *Review of Progress in Quantitative Nondestructive Evaluation*, **14**:827–834, 1995.

Tretout H, David D, Marin JY, Dessendre M, Couet M, Avenas-Payan I, "An evaluation of artificial neural networks applied to infrared thermography inspection of composite aerospace structures," *NDT&E International*, **30**[5]:330, 1997.

Trezek GJ, Balk S, "Provocative techniques in thermal NDT imaging," *Materials Evaluation*, **34**[8]:172–176, 1976.

Trivedi MM, *Selected Papers on Digital Image Processing*, SPIE Milestone Series, Vol. 17, 720 pp., 1990.

Troitsky O, "Pulsed thermal nondestructive testing of layered materials," in Allen LR, ed., *Thermosense XV, SPIE Proc.*, **1933**:309–313, 1993.

Tsuor Y, Weeks T, Trubiano R, Pelligrini PW, Yew TR, "IR Si Schottky barrier infrared diodes with 10 mm cutoff wavelength," *IEEE Electronic Devices Letters*, **9**[2], 1988.

Turck J, "Les bases de la détection infrarouge," *L'onde Électrique*, **68**[2]:36–39, 1988.

Türler D, Orlando E, "Predicting the geometry and location of defects in adhesive and spot-welded lap joints using steady-state thermographic techniques," in Lemieux DH, Snell JR, eds., *Thermosense XXI, SPIE Proc.*, **3700**:54–62, 1999.

Ulaby F, *Fundamentals of Applied Electromagnetics*, Prentice Hall, Upper Saddle River, NJ, 426 pp., 1999.

Usuki K, Kanekura T, Aradono K, Kanzaki T, "Effects of nicotine on peripheral cutaneous blood flow and skin temperature," *Journal of Dermatological Science*, **16**[3]: 73–181, 1998.

Vacelet H, "Contrôles non destructif des assemblages collés," *Proc. 3rd European Conference on Nondestructive Testing*, Firenza, Italy, Oct. 15–18, pp. 172–191, 1984.

Vaidya UK, Raju PK, "Monitoring processing stages of carbon–carbon composites using vibration based nondestructive evaluation," *Materials Evaluation*, **52**[6]:682–688, 1994.

Vallerand S, "Détection et caractérisation de défauts par thermographie de phase pulsée," M.Sc. thesis, Université Laval, 2000.

Vallerand S, Maldague X, "Defect characterization in pulsed thermography: a statistical method compared with Kohonen and perceptron neural networks," *NDT&E International*, **33**[5]:307–315, 2000.

Vallerand S, Darabi A, Maldague X, "Defect detection in pulsed thermography: a comparison of Kohonen and perceptron neural networks," in Wurzbach RN, Burleigh DD, eds., *Thermosense XXI, SPIE Proc.*, **3700**:20–25, 1999.

Vanier JG, Gay DE, "Thermographic inspection of mechanical system," in Courville GE, ed., *Thermosense V, SPIE Proc.*, **371**:60–66, 1982.

Vanzetti R, "Fundamentals of infrared radiation," *Materials Evaluation*, **23**[1]:48–54, 1965.

Varis J, Rantala J, "Thermal nondestructive evaluation of copper products using an infrared line scanning technique," *Proc. QIRT-98, Eurotherm Seminar*, **60**:77–80, 1998.

Varis J, Hartikainen J, Lehtiniemi R, Luukkala M, "A simple transportable imaging system for fast thermal nondestructive testing," *Proc. QIRT-92, Eurotherm Seminar*, **27**:235–238, 1992.

Varis J, Rantala J, Lehtiniemi R, Hartikainen J, Luukkala M, "Thermal measurement equipment for detecting delaminations in carbon fiber tubes," in Thompson DO,

Chimenti DE, eds., *Review of Progress in Quantitative Nondestructive Evaluation*, **13A**:677–683, 1994.

Varis J, Lehtiniemi R, Hartikainen J, Rantala J, "Transportable infrared line scanner based equipment for thermal nondestructive testing," *Research in Nondestructive Evaluation*, **6**[2]:85–98, 1995a.

Varis J, Oksanen M, Rantala J, Luukkala M, "Observations on image formation in the line scanning thermal imaging method," in Thompson DO, Chimenti DE, eds., *Review of Progress in Quantitative Nondestructive Evaluation*, **14**:447–452, 1995b.

Varis J, Rantala J, Hartikainen J, "A numerical study on the effects of line heating in layered anisotropic carbon fiber composites," *Research in Nondestructive Evaluation*, **6**[2]:69–84, 1995c.

Varis J, Nurminen A, Tuominen J, Autere A, Rantala J, "Determination of glaze thickness on ceramic substrate using an infrared camera," *Proc. QIRT-96, Eurotherm Seminar*, **50**:293–297, 1996.

Varis J, Lehtiniemi R, Rantala J, "Detection of vertical surface cracks in unidirectional carbon fibre composites with an infrared line scanning technique," *Proc. QIRT-98, Eurotherm Seminar*, **60**:59–63, 1998.

Varis J, Lehtiniemi RK, Vuohelainen R, "Thermal inspection of solder quality of electronic components," in Lemieux DH, Snell JR, eds., *Thermosense XXI, SPIE Proc.*, **3700**:93–98, 1999.

Varrio J, Kesanto S, Heikkinen T, "Use of an infrared sensing technique in the regular monitoring of incinerators and a kiln," in Semanovich SA, ed., *Thermosense XVII, SPIE Proc.*, **2473**:108–118, 1995.

Vavilov VP, "Infrared non-destructive testing of bonded structures: aspects of theory and practice," *British Journal of Non-Destructive Testing*, July, pp. 175–183, 1980.

Vavilov VP, "Effects of the size of the scanning spot and the frequency characteristics of the photoreceiver on the sensitivity of active thermal testing," *Soviet Journal of Nondestructive Testing*, **22**[1]:269–271; translated from *Defektoskopiya*, **4**[Apr.]:4–57, 1984.

Vavilov VP, "Dynamic thermal tomography: perspective field of thermal NDT," in Semanovich SA, ed., *Thermosense XI, SPIE Proc.*, **1313**:178–182, 1990.

Vavilov VP, "Infrared techniques for material analysis and nondestructive testing," in Maldague XPV, ed., *Infrared Methodology and Technology*, International Advances in Nondestructive Testing Monograph Series, Chap. 5, pp. 131–210, Gordon and Breach, New York, 1992a.

Vavilov VP, "Thermal nondestructive testing: short history and state-of-art," *Proc. QIRT-92, Eurotherm Seminar*, **27**:179–194, 1992b.

Vavilov VP, "Transient thermal NDT: conception in formulae," *Proc. QIRT-92, Eurotherm Seminar*, **27**:229–234, 1992c.

Vavilov VP, "Subjective remarks on the terminology used in thermal/infrared nondestructive testing," in Burleigh DD, Spicer JWM, eds., *Thermosense XVIII, SPIE Proc.*, **2766**:276–283, 1996.

Vavilov VP, "Infrared thermographic surveying of building debris: Tomsk High Military School of Communication Engineering catastrophe case study," in Snell JR, Wurzbach RN, eds., *Thermosense XX, SPIE Proc.*, **3361**:164–168, 1998.

Vavilov VP, "Accuracy of thermal NDE numerical simulation and reference signal evolutions," in Lemieux DH, Snell JR, eds., *Thermosense XXI, SPIE Proc.*, **3700**:14–19, 1999.

Vavilov VP, "Three-dimensional analysis of transient thermal NDT problems by data simulation and processing," in Dinwiddie RB, Lemieux DH, eds., *Thermosense XXII, SPIE Proc.*, **4020**:152–163, 2000.

Vavilov VP, Maldague X, "Dynamic thermal tomography: new promise in the IR thermography of solids," in Eklund JK, ed., *Thermosense XIV, SPIE Proc.*, **1682**:194–206, 1992.

Vavilov VP, Maldague X, "Optimization of heating protocol in thermal NDT: back to the basics," *Research in Nondestructive Evaluation*, **6**:1–18, 1994.

Vavilov VP, Marinetti S, "Pulsed phase thermography and Fourier analysis thermal tomography," *Russian Journal of Nondestructive Testing*, **35**[2]:134–145, 1999 [translated from *Defektoskopiya*].

Vavilov VP, Taylor R, "Theoretical and practical aspects of the thermal nondestructive testing of bonded structures," in Sharpe RS, ed., *Research Techniques in Nondestructive Testing*, **5**:238–279, 1982.

Vavilov VP, Ahmed T, Jin HJ, Favro RL, Thomas LD, "Experimental thermal tomography of solids by using the pulsed one-side heating," *Soviet Journal of Nondestructive Testing*, **12**:60–66, 1990.

Vavilov VP, Degiovanni A, Didierjean S, Maillet D, Sengulye AA, Houlbert AS, "Thermal testing and tomography of carbon epoxy plastics," *Soviet Journal of Nondestructive Testing*, **16**, 1991.

Vavilov VP, Maldague X, Picard J, Thomas RL, Favro LD, "Dynamic thermal tomography: new NDE technique to reconstruct inner solids structure by using multiple IR image processing," in Thompson DO, Chimenti DE, eds., *Review of Progress in Quantitative Nondestructive Evaluation*, **11A**:425–432, 1992a.

Vavilov VP, Bison PG, Bressan C, Grinzato E, Marinetti S, "Some new ideas in dynamic thermal tomography," *Proc. QIRT-92, Eurotherm Seminar*, **27**:259–265, 1992b.

Vavilov VP, Maldague X, Côté J, "Thermogram processing by infrared thermography and tomography," *Defktoskopiya*, Feb., pp. 56–64, 1992c. [in Russian].

Vavilov VP, Poletika MF, Pushnyh VA, Shipulin AV, Reino VV, "Forecasting of tools' wear resistance by measurement of their thermal diffusivity," in Eklund JK, ed., *Thermosense XIV, SPIE Proc.*, **1682**:341–346, 1992d.

Vavilov VP, Maldague X, Dufort B, Fobitaille F, Picard J, "Thermal non-destructive testing of carbon–epoxy composites: detailed analysis and data processing", *NDT&E International*, **26**[Apr.]:85–95, 1993a.

Vavilov VP, Maldague X, Dufort B, Ivanov AI, "Adaptive thermal tomography algorithm," in Allen LR, ed., *Thermosense XV, SPIE Proc.*, **1933**:166–173, 1993b.

Vavilov VP, Kourtenkov D, Grinzato E, Bison P, Marinetti S, Bressan C, "Inversion of experimental data and thermal tomography using "Termo.Heat" and "Termidge" software," *Proc. QIRT-94, Eurotherm Seminar*, **42**:273–280, 1994a.

Vavilov VP, Pushnykh VA, Tsipilev V, Shipulin AV, Grinzato EC, "New results for prediction of cutting tools wearing resistance using one- and two-side transient thermal NDT method," in Snell JR, Jr., ed., *Thermosense XVI, SPIE Proc.*, **2245**:196–201, 1994b.

Vavilov VP, Grinzato E, Bison PG, Marinetti S, "Peculiarities of thermal inspection of materials with short observation time," *Proc. QIRT-94, Eurotherm Seminar*, **42**:44–52, 1994c.

Vavilov VP, Grinzato E, Bison PG, Marinetti S, Bales MJ, "Surface transient temperature inversion for hidden corrosion characterisation: theory and applications," *International Journal of Heat and Mass Transfer*, **39**[2]:355–371, 1996a.

Vavilov VP, Marinetti S, Grinzato E, Bison PG, Anoshkin I, Kauppinen T, "Transient thermographic detection of buried defects: attempting to develop the prototype basic inspection procedure," *Proc. QIRT-96, Eurotherm Seminar*, **50**:239–244, 1996b.

Vavilov VP, Bison PG, Grinzato EG, "Statistical evaluation of thermographic NDT performance applied to CFRP," in Burleigh DD, Spicer JWM, eds., *Thermosense XVIII, SPIE Proc.*, **2766**:174–177, 1996c.

Vavilov VP, Grinzato E, Bison PG, Marinetti S, Bressan C, "Thermal characterization and tomography of carbon fiber reinforced platics using individual identification technique," *Materials Evaluation*, **54**[5]:604–610, 1996d.

Vavilov VP, Kauppinen T, Grinzato E, "Thermal characterization of defects in building envelopes using long square pulse and slow thermal wave techniques," *Research in Nondestructive Evaluation*, **9**[4]:181–200, 1997a.

Vavilov VP, Anoshkin IA, Kourtenkov DC, Trofimov CD, Kauppinen TT, "Quantitative evaluation of building thermal performance by IR thermography inspection data," in Wurzbach RN, Burleigh DD, eds., *Thermosense XIX, SPIE Proc.*, **3056**:71–77, 1997b.

Vavilov VP, Almond DP, Busse G, Grinzato E, Krapez JC, Maldague X, Marinetti S, Peng W, Shirayev V, Wu D, "Infrared thermographic detection and characterisation of impact damage in carbon fibre composites: results of the round robin test," *Proc. QIRT-98, Eurotherm Seminar*, **60**:43–52, 1998a.

Vavilov VP, Marinetti S, Grinzato E, Bison PG, "Thermal tomography characterization and pulse-phase thermography of impact damage in CFRP, or why end users are still reluctant about practical use of transient IR thermography," in Snell JR, Wurzbach RN, eds., *Thermosense XX, SPIE Proc.*, **3361**:275–281, 1998b.

Vavilov VP, Shiryaev VV, Grinzato E, "Detection of hidden corrosion in metals by using transient infrared thermography," *Insight: Non-Destructive Testing and Condition Monitoring*, **40**[6]:408–410, 1998c.

Verdugo RJ, Ochoa JL, "Use and misuse of conventional electrodiagnosis, quantitative sensory testing, thermography, and nerve blocks in the evaluation of painful neuropathic syndromes," *Muscle and Nerve*, **16**[10]:1056–1062, 1993.

Vermeulen JP, Baudoin B, "Study of free convection by infrared thermography over a constant heat-flux heated plate," *Proc. QIRT-92, Eurotherm Seminar*, **27**:86–91, 1992.

Viets R, Breuer M, Haferkamp H, Krussel T, Niemeyer M, "Solidification process and infrared image characteristics of permanent mold castings," in Lemieux DH, Snell JR, eds., *Thermosense XXI, SPIE Proc.*, **3700**:132–140, 1999.

Vikstršm M, "Thermography of foam-core sandwich structures," *Materials Evaluation*, **47**[7]:802, 804, 1989.

Villa-Aleman E, Garrett AJ, Kurzeja RJ, Pendergast MM, "Aerial thermography studies of power plant heated lakes," in Dinwiddie RB, Lemieux DH, eds., *Thermosense XXII, SPIE Proc.*, **4020**:367–373, 2000.

Vogel J, Auersperg J, Dost M, Faust W, Michel B, "Experimental and numerical investigations of thermomechanical field coupling effects during crack evolution," *Proc. QIRT-96, Eurotherm Seminar*, **50**:108–116, 1996.

Volinia M, "Integration of qualitative and quantitative infrared surveys to study the plaster conditions of Valentino Castle," in Dinwiddie RB, Lemieux DH, eds., *Thermosense XXII, SPIE Proc.*, **4020**:324–334, 2000.

von Wolfersdorf J, Hoecker R, Sattelmayer T, "Hybrid transient step-heating heat transfer measurement technique using heater foils and liquid-crystal thermography," *Journal of Heat Transfer, Transactions ASME*, **115**[2]:319–324, 1993.

Waggener W, "Digital recording for electro-optical imagery," *SPIE Proc. Electro-Optical Imaging Systems Integration*, **762**:69–85, 1987.

Waggoner J, "Mapping buried pipe systems with thermal IR," *Photonics Spectra*, **25**[2–3], 1991.

Walle G, "Nondestructive characterization of the tendency to chilling in cast iron using pulsed video thermography," *Proc. QIRT-96, Eurotherm Seminar*, **50**:320–325, 1996.

Walle G, Burgschweiger G, Netzelmann U, "Numerical modelling of the defect response in pulsed video thermography on samples with finite optical penetration," *Proc. QIRT-94, Eurotherm Seminar*, **42**:238–246, 1994.

Walle G, Meyendorf N, Vetterlein T, Netzelmann U, Becker D, Bruns H, "Spatially resolved measurement of the spectral emissivity of high-temperature components by multi-channel thermography," *Proc. QIRT-98, Eurotherm Seminar*, **60**:311–315, 1998.

Wallin B, "Temperature measurement on and inside lamps," in Snell JR, Jr., ed., *Thermosense XVI, SPIE Proc.*, **2245**:241–251, 1994.

Wallin B, "Very fast thermal measurements by means of fast line scanning," in Lemieux DH, Snell JR, eds., *Thermosense XXI, SPIE Proc.*, **3700**:425–435, 1999.

Walther HG, Karpen W, "Monitoring of paint adhesion on polymers using photothermal detection," *Proc. QIRT-92, Eurotherm Seminar*, **27**:388–392, 1992.

Walther HG, Seidel U, "Some remarks on definition, resolution and contrast in photo thermal imaging," *Proc. QIRT-92, Eurotherm Seminar*, **27**:251–258, 1992.

Wandelt M, Roetzel W, "Lock-in thermography as a measurement technique in heat transfer," *Proc. QIRT-96, Eurotherm Seminar*, **50**:189–196, 1996.

Wang H, Payzant FA, "Infrared imaging of temperature distribution in a high-temperature x-ray diffraction furnace," in Lemieux DH, Snell JR, eds., *Thermosense XXI, SPIE Proc.*, **3700**:377–387, 1999.

Wang H, Dinwiddie RB, Jiang L, Liaw PK, Brooks CR, Klarstrom DL, "Application of high-speed IR imaging during mechanical fatigue tests," in Dinwiddie RB, Lemieux DH, eds., *Thermosense XXII, SPIE Proc.*, **4020**:186–193, 2000.

Wang Y, Telenkov S, Favro LD, Kuo PK, Thomas RL, "Thermal imaging for the analysis of energy balance during crack propagation," in Thompson DO, Chimenti DE, eds., *Review of Progress in Quantitative Nondestructive Evaluation*, **16A**:365–370, 1998.

Wang YQ, Kuo PK, Favro LD, Thomas RL, "A novel "flying-spot" infrared camera for imaging very fast thermal-wave phenomena," in Murphy JC, Maclachlan-Spicer JW, Aamodt LC, Royce BSH, eds., *Photoacoustic and Photothermal Phenomena II*, **62**:24–26, 1989.

Wang YQ, Chen P, Kuo PK, Favro LD, Thomas RL, "Flying laser spot thermal wave IR imaging," in Thompson DO, Chimenti DE, eds., *Review of Progress in Quantitative Nondestructive Evaluation*, **11A**:453–456, 1998.

Watwe AA, Hollingsworth DK, "Liquid crystal images of surface temperature during incipient pool boiling," *Experimental Thermal and Fluid Science*, **9**[1]:22–33, 1994.

Weil G, "Computer-aided infrared analysis of bridge deck delaminations," *Proc. 5th Infrared Information Exchange*, **1**:85–93, 1985.

Weil GJ, "Nondestructive testing of the concrete roof shell at the Seattle Kingdome," in Snell JR, Wurzbach RN, eds., *Thermosense XX, SPIE Proc.*, **3361**:177–187, 1998.

Weil GJ, Graf RJ, "Infrared thermographic detection of buried grave sites," in Eklund JK, ed., *Thermosense XIV, SPIE Proc.*, **1682**:347–353, 1992.

Weil GJ, Rowe TJ, "Nondestructive testing and repair of the concrete roof shell at the Seattle Kingdome," *NDT&E International*, **31**[6]:389–400, 1998.

Weil GJ, Graf RJ, Forister LM, "Nondestructive remote sensing of hazardous waste sites," in Allen LR, ed., *Thermosense XV, SPIE Proc.*, **1933**:252–261, 1993.

Weiss SA, "High-speed cameras improve vehicle performance and safety," *Photonics Spectra*, July, pp. 102–109, 1998.

West LM, Segerström T, "Commercial applications in aerial thermography: power line inspection, research, and environmental studies," in Dinwiddie RB, Lemieux DH, eds., *Thermosense XXII, SPIE Proc.*, **4020**:382–386, 2000.

Wheeler MD, "Raytheon amber closes facility, merges with Hughes," *Photonics Spectra*, Mar., p. 66, 1998.

White G, Torrington G, "Back to basics: crack detection and measurement using laser pulse heating and thermal microscopy," *Materials Evaluation*, **53**[12]:1332–1334, 1995.

White GB, Safai M, Torrington GK, "Calibration issues affecting the operation of infrared microscopes over large temperature ranges," in Burleigh DD, Spicer JWM, eds., *Thermosense XVIII, SPIE Proc.*, **2766**:357–365, 1996.

White S, Burleigh DD, "Thermographic techniques for arc-jet testing," in Allen LR, ed., *Thermosense XV, SPIE Proc.*, **1933**:262–273, 1993.

Wiecek B, "Quantitative approach into the heat transfer by convection in microelectronics," *Proc. QIRT-96, Eurotherm Seminar*, **50**:183–188, 1996.

Wiecek B, "Modelling of conjugate heat transfer in microelectronics with variable fluid and substrate parameters," *Proc. QIRT-98, Eurotherm Seminar*, **60**:284–291, 1998.

Wiecek B, Grecki M, "Advanced image processing in thermography," *Proc. QIRT-94, Eurotherm Seminar*, **42**:72–80, 1994.

Wiecek B, Madura H, "Radiative and convective heat transfer in microelectronics," *Proc. QIRT-96, Eurotherm Seminar*, **50**:58–66, 1996.

Wiecek B, Pacholik J, "Technical methods of emissivity correction in thermography," *Proc. QIRT-94, Eurotherm Seminar*, **42**:124–132, 1994.

Wiecek B, Zwolenik S, "Multichannel thermography systems for real-time and transient thermal process application," *Proc. QIRT-98, Eurotherm Seminar*, **60**:322–326, 1998.

Wiecek B, Grecki M, Pacholik J, "Computer-based thermographic system," *Proc. QIRT-92, Eurotherm Seminar*, **27**:57–62, 1992a.

Wiecek B, Grecki M, Pacholik J, "Thermal measurements of power semiconductor devices using thermographic system," *Proc. QIRT-92, Eurotherm Seminar*, **27**:291–295, 1992b.

Wiecek B, De Baetselier E, De Mey G, "Active thermography application for solder thickness measurement in surface mounted device technology," *Microelectronics Journal*, **29**:223–228, 1998a.

Wiecek B, De Baetselier E, De Mey G, "Active thermography application for solder thickness measurement in surface mounted device technology," *Microelectronics Journal*, **29**[45]:223–228, 1998b.

Wiecek B, Michalski A, Napiòrkowski K, "IR microscope measurement of the plasma spreading in thyristors," *Proc. QIRT-98, Eurotherm Seminar*, **60**:191–194, 1998c.

Wiecek B, Zwolenik S, Jung A, Zuber J, Kalicki B, "Thermal and radiological image processing in pneumonia monitoring," *Proc. MIRT-98, Eurotherm Seminar*, **60**:57–60, 1998d.

Wiederhold PR, "Infrared pyrometer for temperature monitoring of train wheels and jet engine rotors," *Materials Evaluation*, **32**[11]:239–243, 248, 1974.

Wilburn DK, "Survey of infrared inspection and measurement techniques," *Materials Research and Standards*, **1**[Oct.]:528, 1961a.

Wilburn DK, "Radiographic penetrameters: an infrared standard," *Materials Evaluation*, **22**[10]:471–472, 478, 1961b.

Wild W, Büscher KA, Wiggenhauser H, "Amplitude-sensitive modulation thermography to measure moisture in building materials," in Snell JR, Wurzbach RN, eds., *Thermosense XX, SPIE Proc.*, **3361**:156–163, 1998.

Wildi T, *Understanding Units*, Book Society, Agincourt (Canada), 110 pp., 1973.

Will JE, Norgard J, Stubenrauch C, Seifert M, "Infrared imaging techniques for the measurement of complex near-field antenna patterns," *Proc. QIRT-96, Eurotherm Seminar*, **50**:67–72, 1996a.

Will JE, Norgard JD, Stubenrauch C, MacReynolds K, Seifert M, Sega RM, "Phase measurements of electromagnetic fields using infrared imaging techniques and microwave holography," in Burleigh DD, Spicer JWM, eds., *Thermosense XVIII, SPIE Proc.*, **2766**:323–333, 1996b.

Will JE, Norgard JD, Seifert MF, "Phaseless measurements of antenna near fields from infrared images using holographic phase retrieval," in Wurzbach RN, Burleigh DD, eds., *Thermosense XIX, SPIE Proc.*, **3056**:260–269, 1997.

Williams JH, Mansouri S, Lee S, "One-dimensional analysis of thermal nondestructive detection of delamination and inclusion flaws," *Br J NDT* (May): 113–118, 1980.

Williams JH Jr, Felenchak BR, Nagem RJ, "Quantitative geometric characterization of two-dimensional flaws via liquid crystals thermography," *Materials Evaluation*, **41**[2]:190–201, 218, 1983.

Williams TL, "The MFT of thermal imaging cameras: its relevance and measurement," *Proc. QIRT-92, Eurotherm Seminar*, **27**:51–56, 1992.

Williams TL, Davidson NT, "Recent advances in testing of thermal imager," in Huber AJ, Triplett MJ, Wolverton JR, eds., *SPIE Proc. Imaging, Imaging Infrared: Scene Simulation, Modeling, and Real Image Tracking*, **1110**:220–231, 1989.

Williamson R, *Introduction to Differential Equations and Dynamic Systems*, McGraw-Hill, New York, 1997.

Wilson J, "Thermal analysis of the bottle forming process," in Baird GS, ed., *Thermosense XIII, SPIE Proc.*, **1467**:219–228, 1991.

Winfree WP, Cramer KE, "Computational pulse shaping for thermographic inspections," in Burleigh DD, Spicer JWM, eds., *Thermosense XVIII, SPIE Proc.,* **2766**:228–235, 1996.

Winfree WP, Cramer KE, "Reduction of thermal data using neural networks," in Dinwiddie RB, Lemieux DH, eds., *Thermosense XXII, SPIE Proc.,* **4020**:128–136, 2000.

Winfree WP, Heath DM, "Thermal diffusivity imaging of aerospace materials and structures," in Snell JR, Wurzbach RN, eds., *Thermosense XX, SPIE Proc.,* **3361**:282–290, 1998.

Winfree WP, Crews BS, Howell PA, "Comparison of heating protocols for detection of disbonds in lap joints," in Thompson DO, Chimenti DE, eds., *Review of Progress in Quantitative Nondestructive Evaluation,* **11A**:471–478, 1998.

Wirahadikusumah R, Abraham DM, Iseley T, Prasanth RK, "Assessment technologies for sewer system rehabilitation," *Automation in Construction,* **7**[4]:259–270, 1998.

Wong ETW, "A study of the effect of the linespread function on the performance of an infrared scanning system," M.S. thesis, University of Florida, 1982.

Wood T, "Description of a facility for evaluating infrared imaging systems for building applications," in Courville GE, ed., *Thermosense V, SPIE Proc.,* **371**:246–249, 1982.

Woodmansee WE, "Cholesteric liquid crystals and their application to thermal nondestructive testing," *Materials Evaluation,* **24**[10]:564–566, 571–572, 1966.

Woolaway JT, "New sensor technology for the 3 to 5 μm imaging band," *Photonics Spectra,* **25**[2]:113–119, 1991.

Wright, RE Jr, Puram CK, Daryabeigi K, "Desirable features of an infrared imaging system for aerodynamic research," in Eklund JK, ed., *Thermosense XIV, SPIE Proc.,* **1682**:315–324, 1992.

Wu D, "Lockin thermography for defect characterization in veneered wood," *Proc. QIRT-94, Eurotherm Seminar,* **42**:298–304, 1994.

Wu D, Busse G, "Lock-in thermography for nondestructive evaluation of material," *Revue Générale de Thermique,* **37**:693–703, 1998.

Wu D, Karpen W, Busse G, "Lockin thermography for multiplex photothermal nondestructive evaluation," *Proc. QIRT-92, Eurotherm Seminar,* **27**:371–376, 1992.

Wu D, Steegmüller R, Karpen W, Busse G, "Characterization of CFRP with lockin thermography," in Thompson DO, Chimenti DE, eds., *Review of Progress in Quantitative Nondestructive Evaluation,* **14**:439–446, 1995.

Wu D, Hamann H, Salerno A, Busse G, "Lock-in thermography for imaging of modulated flow in blood vessels," *Proc. QIRT-96, Eurotherm Seminar,* **50**:343–347, 1996a.

Wu D, Rantala J, Karpen W, Zenzinger G, Schönbach B, Rippel W, Steegmüller R, Diener L, Busse G, "Applications of lockin-thermography methods," in Thompson DO, Chimenti DE, eds., *Review of Progress in Quantitative Nondestructive Evaluation,* **15**:511–518, 1996b.

Wu D, Salerno A, Malter U, Aoki R, Kochendörfer R, Kächele PK, Woithe K, Pfister K, Busse G, "Inspection of aircraft structural components using lock-in thermography," *Proc. QIRT-96, Eurotherm Seminar,* **50**:251–256, 1996c.

Wu D, Salerno A, Sembach J, Maldague X, Rantala J, Busse G, "Wood-based products inspection by lockin thermography," *Euholz-Symposium,* Braunschweig, Germany, Oct. 1–2, 1996d.

Wu D, Wu CY, Busse G, "Investigation of resolution in lock-in thermography: theory and experiment," *Proc. QIRT-96, Eurotherm Seminar*, **50**:269–276, 1996e.

Wu D, Zenzinger G, Karpen W, Busse G, "Nondestructive inspection of turbine blades with lock-in thermography," *Materials Science Forum*, Vols. 210–213, pp. 289–294, Transtec Publications, Switzerland, 1996f.

Wu D, Salerno A, Schonbach B, Hallin H, Busse C, "Phase-sensitive modulation thermography and its applications for NDE," in Wurzbach RN, Burleigh DD, eds., *Thermosense XIX, SPIE Proc.*, **3056**:176–183, 1997a.

Wu D, Salerno A, Sembach J, Maldague X, Rantala J, Busse C, "Lock-in thermographic inspection of wood particle boards," in Wurzbach RN, Burleigh DD, eds., *Thermosense XIX, SPIE Proc.*, **3056**:230–234, 1997b.

Wullnik J, van der Stel J, "The application of thermal imaging in metal industry," *Proc. QIRT-96, Eurotherm Seminar*, **50**:315–319, 1996.

Wurzbach RN, "Diagnostic monitoring by infrared imaging of avian embryos," in Snell JR, Wurzbach RN, eds., *Thermosense XX, SPIE Proc.*, **3361**:49–54, 1998a.

Wurzbach RN, "Infrared thermography of fan and compressor systems in a predictive maintenance program," in Snell JR, Wurzbach RN, eds., *Thermosense XX, SPIE Proc.*, **3361**:118–124, 1998b.

Wurzbach RN, Hammaker RG, "Role of comparative and qualitative thermography in predictive maintenance," in Eklund JK, ed., *Thermosense XIV, SPIE Proc.*, **1682**:3–11, 1992.

Wurzbach RN, Hart JE, "Increasing maintainability and operability of emergency diesel generators with thermographic inspections," in Allen LR, ed., *Thermosense XV, SPIE Proc.*, **1933**:14–25, 1993.

Wurzbach RN, Seith DA, "Infrared monitoring of power plant effluents and heat sinks to optimize plant efficiency," in Dinwiddie RB, Lemieux DH, eds., *Thermosense XXII, SPIE Proc.*, **4020**:360–366, 2000.

Wyss P, Lüthi Th, Primas R, Zogmal O, "Factors affecting the detectability of voids by infrared thermography," *Proc. QIRT-96, Eurotherm Seminar*, **50**:227–232, 1996.

Xiong Y, "Neuro-vision for 3D machine vision in intelligent manufacturing system," in Dinwiddie RB, Lemieux DH, eds., *Thermosense XXII, SPIE Proc.*, **4020**:38–49, 2000.

Xu MH, Cheng JC, Zhang SY, "Inversion for multi-parameter depth-profiles: thermal conductivity and thermal impedance," in Thompson DO, Chimenti DE, eds., *Review of Progress in Quantitative Nondestructive Evaluation*, **18A**:619–626, 1998.

Yamada M, Uematsu Y, Sasaki R, "Visual technique for the evaluation of the pedestrian-level wind environment around buildings by using infrared thermography," *Aerodynamics*, **65**[1–3]:261–271, 1996.

Yamauchi T, Okumura S, Noguchi M, "Application of thermography to the deforming process of paper materials," *Journal of Materials Science*, **28**[17]:4549–4552, 1993.

Yang L, Geng W, Jiang L, Zou L, Hong J, Lu Y, Cui Z, "Approximate solution to the inverse problem in thermal non-destructive testing by using Coons surface," *Insight: Non-Destructive Testing and Condition*, **39**[8]:563–565, 1997.

Yanisov VV, Yanisova LK, "The revealability of defects in optical synthesis of thermograms from x-ray photographs of different aspects," *Sov J NDT* **12**:1–9, Plenum, New York, 1984. Translated from Defektoskopiya.

Yeh CY, Ranu E, Zoughi R, "A novel microwave method for surface crack detection using higher order waveguide modes," *Materials Evaluation*, **52**[6]:676–681, 1994.

Yoon HW, Davies MA, Burns TJ, Kennedy MD, "Calibrated thermal microscopy of the tool-chip interface in machining," in Dinwiddie RB, Lemieux DH, eds., *Thermosense XXII, SPIE Proc.*, **4020**:27–37, 2000.

Young HT, Chou TL, "Investigation of edge effect from the chip-back temperature using IR thermographic techniques," *Journal of Materials Processing Technology*, **52**[2–4]:213–224, 1995.

Yutani N, Kimata M, Denda M, Iwade S, Tsubouchi N, "IrSi Schottky barrier infrared image sensors," *Proc. IEDM*, **6**[3], 1987.

Zahorszki F, Lyons AR, "Online slag detection in steelmaking," in Dinwiddie RB, Lemieux DH, eds., *Thermosense XXII, SPIE Proc.*, **4020**:10–14, 2000.

Zalameda JN, Farley GL, Smith BT, "Field deployable nondestructive impact damage assessment methodology for composite structures," *Journal of Composites Technology and Research*, **16**[2]:161–169, 1994.

Zalameda JN, Winfree WP, Heath DM, Thornhill JA, "Heat source characterization for quantitative thermal nondestructive evaluation," in Thompson DO, Chimenti DE, eds., *Review of Progress in Quantitative Nondestructive Evaluation*, **18A**:585–592, 1998.

Zanio K, "HgCdTe on Si for hybrid and monolithic FPAs," in Dereniak EL, Sampson RE, eds., *SPIE Proc. Infrared Detectors and Focal Plane Arrays*, **1308**:180–193, 1990.

Zayicek PA, Shepard SM, "IR technique for detection of wall thinning in service water piping," in Wurzbach RN, Burleigh DD, eds., *Thermosense XIX, SPIE Proc.*, **3056**:242–249, 1997.

Zee MR, "Measurement of point spread function of thermal imager," in Baird GS, ed., *Thermosense XIII, SPIE Proc.*, **1467**:469–474, 1991.

Zhang XR, Gan CM, Zhou JY, Wang FC, Ying MF, "Investigation of thermal diffusivity of composite material by mirage effect," in Thompson DO, Chimenti DE, eds., *Review of Progress in Quantitative Nondestructive Evaluation*, **14**:467–474, 1995.

Zielinski M, Milewski S, Gawlikowski J, Ksiazek A, "The infrared thermography in the imaging diagnosis of inflammatory state of the epicondilitis of the bone of arm to application in surgery," *Proc. QIRT-98, Eurotherm Seminar*, **60**:46–51, 1998.

Zinko H, Perers B, "TX model: a quantitative heat-loss analysis of district heating pipes by means of IR surface-temperature measurements," in Semanovich SA, ed., *Thermosense XVII, SPIE Proc.*, **2473**:15–26, 1995.

Zoughi R, "Microwave and millimeter wave nondestructive testing: a succinct introduction," *Research in Nondestructive Evaluation*, **7**[2–3]:71–74, 1995.

Zweschper Th, Wu D, Busse G, "Detection of loose rivets in aeroplane components using lockin thermography," *Proc. QIRT-98, Eurotherm Seminar*, **60**:161–166, 1998.

Zweschper Th, Wu D, Busse G, "Detection of tightness of mechanical joints using lockin thermography," in Lammasnieemi J, Wiggenhauser H, Busse G, Batchelor BC, Pölzleitner W, Dobmann G, eds., *SPIE Proc. Diagnostic Imaging, Technologies and Industrial Applications*, **3827**:16–21, Euro-Opto 1999, 1999.

Zwolenik S, Wiecek B, "Multichannel thermography systems for real-time and transient thermal process applications," in Balageas DL, Carlomagno G, Busse G, eds., *Proc. QIRT-98, Eurotherm Seminar*, **60**:322–326, 1998.

Computer Model

This computer modeling program is discussed in Chapter 3. See also *mod_therm.m* Matlab script in Appendix F.

```
/*****************************************************/
/*
                        SIMUL.C

Computation of the temperature distribution in a CFRP specimen

        radial conductivity ra_conduct:.05w/KgC
        axial   conductivity ax_conduct:.01w/cmC;
        specific heat: 1000 J/KgC
        mass density:  2000 Kg/m3
        Thermal losses geometric factor:
            Frad:5.67*10e-12w/cm2;
            Fcon:10e-3

        Program adapted in Borland "C" language by Mr. Ahmed NOUAH
        from Engineering Dept, Université Laval, Québec city,
        Québec Canada G1K 7P4.

        Special thanks are acknowledged to Dr. Paolo CIELO from
        Industrial Materials Research Institute, Boucherville,
        Canada for his ideas and program examples on heat transfer
        modelling.

        This computer program work on PC or Sun type of computer.
        Results are displayed on the computer screen. On a PC, the
        output can be redirected in a file using the following
        command:

                    simul > filename

                    where "simul" is modelling program name
                    and "filename" the name of the file to save
                    results

******************************************************/
```

```c
#include <stdio.h>
#include <math.h>
#include <alloc.h>
#define pi 3.1415
#define Sqr(a) ((a) * (a))
#define Sqr4(a) ((a)*(a)*(a)*(a))
main()
{
/* thermal parameters */

#define rad_nodes 20      /* number of radial nodes */
#define axe_nodes 30      /* number of axial nodes */
#define def_nod_pos 11    /* number of axial defect nodes */

int i,j,l,iter;
double ***temp,**rhd,**cp;
double **qr,**qz,*kint;
double *dslat,*dslong;
double dr,dz,dt,tim,timlc,timl,timc,source_power;
double temp_var,red_loss,conv_loss,ra_conduct,ax_conduct;

/*memory allocation */
kint=(double*)malloc(rad_nodes*sizeof(double));
dslat=(double*)malloc(rad_nodes*sizeof(double));
dslong=(double*)malloc(rad_nodes*sizeof(double));

rhd = (double**)malloc(rad_nodes*sizeof(double*));
cp  = (double**)malloc(rad_nodes*sizeof(double*));
qr  = (double**)malloc(rad_nodes*sizeof(double*));
qz  = (double**)malloc(rad_nodes*sizeof(double*));
temp=(double***)malloc(rad_nodes*sizeof(double**));

for(i=0;i<=rad_nodes;i++){
     temp[i]=(double**)malloc(axe_nodes*sizeof(double*));
     for(j=0;j<=axe_nodes;j++)
     temp[i][j]=(double*)malloc(2*sizeof(double));

     rhd[i] =(double*)malloc(axe_nodes*sizeof(double));
     cp[i] = (double*)malloc(axe_nodes*sizeof(double));
     qr[i] = (double*)malloc(axe_nodes*sizeof(double));
     qz[i] = (double*)malloc(axe_nodes*sizeof(double));
             }

for(i=0;i<rad_nodes;i++){
     kint[i]=0.01;
for(j=0;j<axe_nodes;j++){
     for(l=0;l<2;l++)
            temp[i][j][l]=0.00001;
     rhd[i][j]=2.0;
     cp[i][j]=1.0;
     qr[i][j]=0.00000001;
     qz[i][j]=0.00000001;
     }}

/* Defect parameters */
for(i=0;i<5;i++) kint[i]=0.00024;/* defect is air*/

/* Modelling geometry */
dr=0.2;
dz=0.02;
dt=2.5e-2;
```

```
for(i=0;i<rad_nodes;i++){
      dslat[i]=2.0*pi*(i+1)*dr*dz;
      dslong[i]=pi*(Sqr((i+1)*dr)-Sqr((i)*dr));
      }

/* beginning of time iteration loop
    "tim" is the current time
    "timc" indicates when a display of results is to be done
*/
tim=0.0;  /* starting time */
timc=0.4; /* first time to display result */
timlc=log10(timc);
for(iter=0;iter<=1200;iter++){/*1200 iterations*/
      tim+=dt;
      timl=log10(tim);

/* Thermal perturbation source  */
/* Application of thermal pulse for 2s on the surface (Z=1)    */
source_power=1.0; /*    power is in W/cm^2 */
if(tim>2.0) source_power=0.0;
for(i=0;i<rad_nodes;i++){
      temp_var=(source_power*dt)/(rhd[i][0]*cp[i][0]*dz);
      temp[i][0][0]+=temp_var;
      temp[i][0][1]=temp[i][0][0];
      }

/* heat propagation through the specimen */
for(i=0;i<rad_nodes-1;i++)
for(j=0;j<axe_nodes-1;j++){
      ra_conduct=0.05;/* thermal conductivity */
      ax_conduct=0.01;
      if(j==def_nod_pos) ax_conduct=kint[i];

      qr[i][j]=ra_conduct*dslat[i]*(temp[i][j][0]-temp[i+1][j]
      [0])/dr;
      qz[i][j]=ax_conduct*dslong[i]*(temp[i][j][0]-temp[i][j+1]
      [0])/dz;

      temp[i][j][1]-=(qr[i][j]+qz[i][j])*dt/
            (rhd[i][j]*cp[i][j]*dslong[i]*dz);

      temp[i+1][j][1]+=qr[i][j]*dt/
            (rhd[i+1][j]*cp[i+1][j]*dslong[i+1]*dz);

      temp[i][j+1][1]+=qz[i][j]*dt/
            (rhd[i][j+1]*cp[i][j+1]*dslong[i]*dz);
      }
for(i=0;i<axe_nodes;i++){
      temp[rad_nodes-1][i][0]=temp[rad_nodes-2][i][0];
      temp[rad_nodes-1][i][1]=temp[rad_nodes-2][i][1];
      }

/* Computation of surface losses */
for(i=0;i<rad_nodes;i++){
      red_loss=dslong[i]*
            (5.67e-12)*(Sqr4(temp[i][0][1]+273.0) - Sqr4(273.0));
      conv_loss=dslong[i]*(1.0e-3)*temp[i][0][1];
      temp[i][0][1]-=(red_loss+conv_loss)
      *dt/(rhd[i][0]*cp[i][0] *dslong[i]*dz);
      }
/* Transfer of the current temperature distribution
      into the matrix of the next iteration  */
```

```
for(i=0;i<rad_nodes;i++)
for(j=0;j<axe_nodes;j++){
     temp[i][j][0]=temp[i][j][1];
     }
/* printout of results on the PC screen
   and compute next time for printout,
   to get same values as for Figure 2.7 use "timlc=timlc+0.01"
   instead */
if(timl>=timlc){
     timlc=timlc+0.1;
     printf("\n\n Time is: %f sec,  Pass # %d\n",tim, iter);
     for(j=0;j<rad_nodes;j++){
     for(i=0;i<rad_nodes-10;i++){
          printf("%3.3f",temp[i][j][1]);
          }
          printf("\n");
          }

     /*printf("\n\n radial flux \n");
     for(j=0;j<rad_nodes;j++){
     for(i=0;i<rad_nodes;i++){
          printf("%g",qr[i][j]*10.0/dslat[i]);
          }
          printf("\n");
          }
     printf("\n\n axial flux\n");
     for(j=0;j<rad_nodes;j++){
     for(i=0;i<rad_nodes;i++){
          printf("%g",qz[i][j]*10.0/dslong[i]);
          }
          printf("\n");
          }*/
     }
}/*End of the iterations */

printf("\nconductivity of the %d subsurface elements\n",
def_nod_pos);
for(i=0;i<rad_nodes;i++) printf("%f",kint[i]);

/* deallocate memory*/
free(dslong);free(kint);free(cp);free(dslat);
free(rhd);free(qr);free(qz);free(temp);return(0);
}
```

Smoothing Routine

In this appendix we give the routines necessary to smooth data. The method is discussed in Section 5.5.5. The code is written in C language. See also *smooth.m* Matlab script in Appendix F. We give below an example of a program that can be used to smooth an image. It smooths rows and columns separately (notice that only functions for smoothing are given in the module SMOOTH.C below).

```
#define GAUSS MAXCOL /* for the sliding Gaussian */

main()
{

/* -------- VARIABLES OF THE PROGRAM ---------------------- */

/* include here the list of all variables used by the program */

printf("Program to smooth images");
input_parameters(); /* ask the user to input all the parameters
*/
 /* read the image to smooth from disk */
read_disk(matresul, filename);

 /* Processing on ROWS */
    init_gauss(y_gauss, &gauss_dim, Rlig);
    a_zero(lisse); /* set elements of vector "lisse" to zero */
    for (row = 0; row <= nb_rows; row++)
      {
      get_row(tempo, matresul, row); /* extract 1 row from image
      */
      lissage_gauss(lisse, tempo, dimlig, gauss_dim, y_gauss,
      nb_passe);
      save_lig(lisse, matresul, row); /* save smoothed row in the
      image */
      } /* for */
 /* Processing on COLUMNS */
{
a_zero(lisse); /* set elements of vector "lisse" to zero */
init_gauss(y_gauss, &gauss_dim, Rcol); /* another parameter for
the Gaussian */
```

```
      for (column = 0; column < MAXCOL; column++)
        {
        get_column(tempo, matresul, column, nb_lignes);
        lissage_gauss(lisse, tempo, dimcol, gauss_dim, y_gauss,
        cnb_passe);
        save_col(lisse, matresul, column);
        } /* for */
} /* if */
  /* Save the smoothed image */
save_disk(matresul, filename);
printf("\n ------- End of the smoothing program ----------- \n");
} /* end of the program */

/
**********************************************************************
*
*                         S M O O T H . C
*
*   Module for smoothing raw data based on a sliding Gaussian.
*   function "lissage_gauss()" which performs the smoothing on a
*   vector having a maximum of MAXCOL elements.
*   The main program should call "lissage_gauss()" routine for
*   smoothing one row/column of the image at a time.
*   All parameters are passed to the routine. Prior to the call,
*   the Gaussian is first initialized with function "init_gauss"
*   so that the summation of all the elements is one. A uniform
*   (all elements equal) vector will come out intact of the
*   smoothing process.
*
*   Xavier Maldague + Gilles Filion
*
**********************************************************************/

#include "c:\archives\disdef.h"
#define GAUSS MAXCOL   /* number of elements for the Gaussian */

/* --- functions available in this module --- */
void init_gauss(double [GAUSS], int *, double);
void lissage_gauss(double [MAXCOL], double [MAXCOL], int, int,
                   double [GAUSS], int);

/* ---------------------------------------------------------- */

void init_gauss(y_gauss, pgauss_dim, R)
double y_gauss [GAUSS];    /* vector containing Gaussian Bell */
int    *pgauss_dim;        /* nb elements of Gaussian         */
double R;                  /* parameter for Gaussian width    */
/*
** Function which computes a Gaussian and save it in a vector.
** Parameter C is adjusted to insure the summation of all
** elements is one despite the small and discrete number of
** points.
*/
{
/* initialization of vectors and parameters for the      **
   smoothing by sliding Gaussian                          */
double total, i_maxfl;
int i, j, i_max;
double C = 0.39894228; /* (2pi)^-0.5 */
double square;
```

```
/* initialization of the vector containing the Gaussian */
for (i = 0; i < GAUSS; i++) y_gauss [i] = 0.0;
i_maxfl = 7.0 *  R + 1.0; /* in real to be more accurate */
/* +0.5 round truncation due to the integer number of points */
i_max   = (int) i_maxfl + 0.5;

/* Determine "C" to insure summation of all elements is one */
total = 0.0;
for (i = 0; i < i_max; ++i)
    {
    square = (double) (i_maxfl/2 - i) / R;
    total  +=  exp(-.5 * square * square);
    }
C = R / total;
/* printf("The parameter C equals %lf \n", C); */

for (i = 0; i < i_max; ++i)
    {
    square = (double) (i_maxfl/2 - i) / R;
    y_gauss[i] = C * exp(-.5 * square * square) / R;
    }
*pgauss_dim = i_max;

total = 0.0;
for (i = 0; i < i_max; ++i)
    total += y_gauss[i];
/* printf("Summation all elements of Gaussian %lf\n", total); */

} /* init_gauss */

/* -------------------------------------------------------------- */

void lissage_gauss(lisse, brut, dimension, gauss_dim, y_gauss,
nb_passe)
double  lisse [MAXCOL], brut [MAXCOL];/* vector raw + smoothed */
int     dimension; /* nb of elements to consider in the vector */
int     gauss_dim; /* nb elements to consider in the Gaussian  */
double  y_gauss [GAUSS];    /* Gaussian */
int     nb_passe;
{
double H, sortie, rebus [MAXCOL];
int i, j, index, passe;

for (passe = 1; passe <= nb_passe; passe++)
{
for (i = 0; i < dimension; i++)
    {
    H = lisse[i];
    for (j = 0; j < gauss_dim; j++)
        {
        index = i - gauss_dim / 2+j;
        if (index < 0)
           H += (brut[0] - lisse[0]) * y_gauss[j];
        else
           if (index >= dimension)
              H += (brut[dimension-1] - lisse[dimension-1])
                   * y_gauss[j];
           else
              H += (brut[index] - lisse[index])*y_gauss[j];
```

```
        }/* inner for */
    rebus[i] = H;
    } /* outer for */
for (i = 0; i < dimension; ++i)
    lisse[i] = rebus[i];
} /* for-passe */

} /* lissage_gauss */

/* ------------------------------------------------------------ */
```

Parabola Computations

The problem we want to solve is to fit a parabola to a data set (x_i, y_i). The parabola is given by the following function:

$$y = a + bx + cx^2 \tag{C.1}$$

Bevington [1969, p. 137, eq. (8.6)] exposes the least squares method for fitting quadratic function to N data points. He demonstrates that

$$a = \frac{1}{\Delta} \begin{vmatrix} \sum y_i & \sum x_i & \sum x_i^2 \\ \sum x_i y_i & \sum x_i^2 & \sum x_i^3 \\ \sum x_i^2 y_i & \sum x_i^3 & \sum x_i^4 \end{vmatrix}$$

$$b = \frac{1}{\Delta} \begin{vmatrix} N & \sum y_i & \sum x_i^2 \\ \sum x_i & \sum x_i y_i & \sum x_i^3 \\ \sum x_i^2 & \sum x_i^2 y_i & \sum x_i^4 \end{vmatrix}$$

$$c = \frac{1}{\Delta} \begin{vmatrix} N & \sum x_i & \sum y_i \\ \sum x_i & \sum x_i^2 & \sum x_i y_i \\ \sum x_i^2 & \sum x_i^3 & \sum x_i^2 y_i \end{vmatrix} \tag{C.2}$$

$$\Delta = \begin{vmatrix} N & \sum x_i & \sum x_i^2 \\ \sum x_i & \sum x_i^2 & \sum x_i^3 \\ \sum x_i^2 & \sum x_i^3 & \sum x_i^4 \end{vmatrix}$$

This formula is valid only if uncertainties are the same through the data set; otherwise, individual standard deviations must be considered [Bevington, 1969, p. 137, eq. (8.5)].

Higher-Order Gradient Computations Based on the Roberts Gradient

In this appendix we derive a higher-order gradient approximation with respect to Roberts gradient [eq. (10.52)]. We first consider the unidimensional case of a discrete function $f(x)$ for which we want to compute the first derivative $f'(x)$. We use a Taylor series:

$$f(x - h) = f(x) - hf'(x) + \frac{h^2}{2}f''(x) - \frac{h^3}{6}f^{(3)}(x) + \frac{h^4}{24}f^{(4)}(x) - \cdots$$

$$f(x + h) = f(x) + hf'(x) + \frac{h^2}{2}f''(x) + \frac{h^3}{6}f^{(3)}(x) + \frac{h^4}{24}f^{(4)}(x) + \cdots$$

$$f(x + 2h) = f(x) + 2hf'(x) + \frac{4h^2}{2}f''(x) + \frac{8h^3}{6}f^{(3)}(x) + \frac{16h^4}{24}f^{(4)}(x) + \cdots$$

$$f(x - 2h) = f(x) - 2hf'(x) + \frac{4h^2}{2}f''(x) - \frac{8h^3}{6}f^{(3)}(x) + \frac{16h^4}{24}f^{(4)}(x) - \cdots$$

If we multiply each line by a, b, c, d, respectively, and compute the summation, we obtain (limiting ourselves to fourth-order derivatives)

$$
\begin{aligned}
af(x - h) &+ bf(x + h) + cf(x + 2h) + df(x - 2h) \\
&= +f(x)(a + b + c + d) + hf'[(b - a) + 2(c - d)] \\
&\quad + \frac{h^2}{2}f''[(b + a) + 4(c + d)] + \frac{h^3}{6}f^{(3)}[(b - a) + 8(c - d)] \\
&\quad + \frac{h^4}{24}f^{(4)}[(b + a) + 16(c + d)]
\end{aligned}
\tag{D.1}
$$

629

Since we want $f'(x)$, we set to zero the multiplicative terms of $f(x)$, $f''(x)f^{(3)}(x)$, and $f^{(4)}(x)$ in eq. (D.1). This leads to the following equations:

$$a + b + c + d = 0$$

$$a + b + 4c + 4d = 0$$

$$-a + b + 8c - 8d = 0$$

$$a + b + 16c + 16d = 0$$

which solves for $a = 8$, $b = -8$, $c = 1$, and $d = -1$. If we substitute these values in eq. (D.1), we find that

$$[8f(x - h) - 8f(x + h) + f(x + 2h) - f(x - 2h)]$$
$$= hf'(x)[-8 - 8 + 2 + 2] = -12hf'(x)$$

and thus:

$$f'(x) = \frac{1}{12}h[-f(x - 2h) + 8f(x - h) - 8f(x - h) + f(x + 2h)] \qquad \text{(D.2)}$$

If we take one pixel step ($h = 1$) and neglect the scale factor, we obtain

$$f'(x) = f(x - 2) - 8f(x - 1) + 8f(x + 1) - f(x + 2) \qquad \text{(D.3)}$$

To compute the gradient image $|G|$, we apply eq. (D.3) in a cross fashion on both image f rows and columns as shown in Figure 10.12b: For $\forall i$, $i = 2, 3, \ldots, \text{Maxcol-3}$ and for $\forall j, j = 2, 3, \ldots, \text{Maxrow-3}$, compute

$$|G| = \left\{ \begin{array}{l} [f(i, j - 2) - 8f(i, j - 1) + 8f(i, j + 1) - f(i, j + 2)]^2 \\ + [f(i - 2, j) - 8f(i - 1, j) + 8f(i + 1, j) - f(i + 2, j)]^2 \end{array} \right\}^{1/2} \qquad \text{(D.4)}$$

Properties of Metals and Nonmetals

TABLE E.1 Properties of Metals

Metal	Properties at 20°C				Thermal Conductivity k (W m⁻¹ °C⁻¹)									
	ρ (kg m⁻³)	C_p (kJ kg⁻¹ °C⁻¹)	k (W m⁻¹ °C⁻¹)	α (m² s⁻¹ ×10⁵)	−100°C −148°F	0°C 32°F	100°C 212°F	200°C 392°F	300°C 572°F	400°C 752°F	600°C 1112°F	800°C 1472°F	1000°C 1832°F	1200°C 2192°F
Aluminum														
Pure	2,707	0.896	204	8.418	215	202	206	215	228	249				
Al–Cu (Duralumin), 94–96% Al, 3–5% Cu, trace Mg	2,787	0.883	164	6.676	126	159	182	194						
Al–Si (Silumin, cooper-bearing), 86.5% Al, 1% Cu	2,659	0.867	137	5.933	119	137	144	152	161					
Al–Si (Alusil), 78–80% Al, 20–22% Si	2,627	0.854	161	7.172	144	157	168	175	178					
Al–Mg–Si, 97% Al, 1% Mg, 1% Si, 1% Mn	2,707	0.892	177	7.311		175	189	204						
Lead	11,373	0.13	35	2.343	36.9	35.1	33.4	31.5	29.8					
Iron														
Pure	7,897	0.452	73	2.034	87	73	67	62	55	48	40	36	35	36
Wrought iron, 0.5% C	7,849	0.46	59	1.626		59	57	52	48	45	36	33	33	33
Steel (C mas ≈ 1.5%)														
Carbon steel														
C ≈ 0.5%	7,833	0.465	54	1.474		55	52	48	45	42	35	31	29	31
1.0%	7,801	0.473	43	1.172		43	43	42	40	36	33	29	28	29
1.5%	7,753	0.486	36	0.970		36	36	36	35	33	31	28	28	29
Nickel steel														
Ni ≈ 0%	7,897	0.452	73	2.026										
20%	7,933	0.46	19	0.526										
40%	8,169	0.46	10	0.279										
80%	8,618	0.46	35	0.872										

	ρ	c	λ	a										
Invar 36% Ni	8,137	0.46	10.7	0.286										
Chrome steel														
Cr = 0%	7,897	0.452	73	2.026	87	73	67	62	55	48	40	36	35	36
1%	7,865	0.46	61	1.665		62	55	52	47	42	36	33	33	
5%	7,833	0.46	40	1.110		40	38	36	36	33	29	29	29	
20%	7,689	0.46	22	0.635		22	22	22	22	24	24	26	29	
Cr–Ni (chrome–nickel)														
15% Cr, 10% Ni	7,865	0.46	19	0.527										
18% Cr, 8% Ni (V2A)	7,817	0.46	16.3	0.444	16.3	16.3	17	17	19	19	22	27	31	
20% Cr, 10% Ni	7,833	0.46	15.1	0.415										
25% Cr, 20% Ni	7,865	0.46	12.8	0.361										
Tungsten steel														
W = 0%	7,897	0.452	73	2.026										
1%	7,913	0.448	66	1.858										
5%	8,073	0.435	54	1.525										
10%	8,314	0.419	48	1.391										
Copper														
Pure	8,954	0.3831	386	11.234	407	386	379	374	369	363	353			
Aluminum bronze, 95% Cu, 5% Al	8,666	0.410	83	2.330										
Bronze, 75% Cu, 25% Sn	8,666	0.343	26	0.859										
Red brass, 85% Cu, 9% Sn, 6% Zn	8,714	0.385	61	1.804		59	71							
Brass, 70% Cu, 30% Zn	8,522	0.385	111	3.412	88		128	144	147	147				
German silver, 62% Cu, 15% Ni, 22% Zn	8,618	0.394	24.9	0.733	19.2	26	31	40	45	48				
Constantan, 60% Cu, 40% Ni	8,922	0.410	22.7	0.612	21	22.2								

(Continued)

TABLE E.1 (Continued)

Metal	Properties at 20°C ρ (kg m⁻³)	C_p (kJ kg⁻¹ °C⁻¹)	k (W m⁻¹ °C⁻¹)	α (m² s⁻¹ ×10⁵)	Thermal Conductivity k (W m⁻¹ °C⁻¹) −100°C −148°F	0°C 32°F	100°C 212°F	200°C 392°F	300°C 572°F	400°C 752°F	600°C 1112°F	800°C 1472°F	1000°C 1832°F	1200°C 2192°F
Magnesium														
Pure	1,746	1.013	171	9.708	178	171	168	163	157					
Mg–Al (electrolytic), 6–8% Al, 1–2% Zn	1,810	1.00	66	3.605		52	62	74	83					
Molybdenum	10,220	0.251	123	4.790	138	125	118	114	111	109	106	102	99	92
Nickel														
Pure (99.9%)	8,906	0.4459	90	2.266	104	93	83	73	64	59				
Ni, Cr 90% Ni, 10% Cr	8,666	0.444	17	0.444		17.1	18.9	20.9	22.8	24.6				
80% Ni, 20% Cr	8,314	0.444	12.6	0.343		12.3	13.8	15.6	17.1	18.0	22.5			
Silver														
Purest	10,524	0.2340	419	17.004	419	417	415	412						
Pure (99.9%)	10,525	0.2340	407	16.563	419	410	415	374	362	360				
Tin, pure	7,304	0.2265	64	3.884	74	65.9	59	57						
Tugsten	19,350	0.1344	163	6.271		166	151	142	133	126	112	76		
Zinc, pure	7,144	0.3843	112.2	4.106	114	112	109	106	100	93				

Source: Adapted from J. P. Holman (1981, Table A-2, p. 535).
ρ = density; C_p = specific heat; α = thermal diffusivity.

TABLE E.2 Properties of Nonmetals

Substance	Temperature (°C)	k (W m^{-1} °C^{-1})	ρ (kg m^{-3})	C_p (kJ kg^{-1} °C^{-1})	α (m^2 s^{-1} × 10^7)
		Structural and Heat-Resistant Materials			
Asphalt	20–55	0.74–0.076			
Brick					
Building brick, common	20	0.69	1600	0.84	5.2
Face		1.32	2000		
Carborundum brick	600	18.5			
	1400	11.1			
Chrome brick	200	2.32	3000	0.84	9.2
	550	2.47			9.8
	900	1.99			7.9
Diatomaceous earth, molded and fired	200	0.24			
	870	0.31			
Fireclay brick	500	1.04	2000	0.96	5.4
Burned at 2426°F	800	1.07			
	1100	1.09			
Burned at 2642°F	500	1.28	2300	0.96	5.8
	800	1.37			
	1100	1.40			
Missouri	200	1.00	2600	0.96	4.0
	600	1.47			
	1400	1.77			

(Continued)

TABLE E.2 *(Continued)*

Substance	Temperature (°C)	k (W m^{-1} °C^{-1})	ρ (kg m^{-3})	C_p (kJ kg^{-1} °C^{-1})	α (m^2 s^{-1} × 10^7)
Magnesite	200	3.81		1.13	
	650	2.77			
	1200	1.90			
Cement, portland		0.29	1500		
Mortar	23	1.16			
Concrete, cinder	23	0.76			
Stone 1–2–4 mix	20	1.37	1900–2300	0.88	8.2–6.8
Glass, window	20	0.78 (avg)	2700	0.84	3.4
Corosilicate	30–75	1.09	2200		
Plaster, gypsum	20	0.48	1440	0.84	4.0
Metal lath	20	0.47			
Wood lath	20	0.28			
Stone					
Granite		1.73–3.98	2640	0.82	8–18
Limestone	100–300	1.26–1.33	2500	0.90	5.6–5.9
Marble		2.07–2.94	2500–2700	0.80	10–13.6
Sandstone	40	1.83	2160–2300	0.71	11.2–11.9
Wood (across the grain)					
Balsa, 8.8 lb/ft^3	30	0.055	140		
Cypress	30	0.097	460		
Fir	23	0.11	420	2.72	0.96
Maple or oak	30	0.166	540	2.4	1.28
Yellow pine	23	0.147	640	2.8	0.82
White pine	30	0.112	430		

Insulating materials

Material						
Asbestos						
Loosely packed	−45	0.149				
	0	0.154	470–570	0.816	3.3–4	
	100	0.161				
Asbestos-cement boards	20	0.74				
Sheets	51	0.166				
Felt						
40 laminations/in.	38	0.057				
	150	0.069				
	260	0.083				
20 laminations/in.	38	0.078				
	150	0.095				
	260	0.112				
Corrugated, 4 plies/in.	38	0.087				
	93	0.100				
	150	0.119				
Asbestos cement		2.08				
Balsam wool, 2.2 lb/ft³	32	0.04	35			
Cardboard						
Corrugated		0.064				
Celotex	32	0.048				
Corkboard, 10 lb/ft³	30	0.043	160			
Cork						
Regranulated	32	0.045	45–120	1.88	2–5.3	
Ground	32	0.043	150			
Diatomaceous earth (Sil-o-cel)	0	0.061	320			

(Continued)

TABLE E.2 *(Continued)*

Substance	Temperature (°C)	k (W m^{-1} °C^{-1})	ρ (kg m^{-3})	C_p (kJ kg^{-1} °C^{-1})	α (m^2 s^{-1} × 10^7)
Felt					
Hair	30	0.036	130–200		
Wool	30	0.052	330		
Fiber, insulating board	20	0.048	240		
Glass wool, 1.5 lb/ft^3	23	0.038	24	0.7	22.6
Insulex, dry	32	0.064			
		0.144			
Kapok	30	0.035			
Magnesia, 85%	38	0.067	270		
	93	0.071			
	150	0.074			
	204	0.080			
Rock wool					
10 lb/ft^3	32	0.040	160		
Loosely packed	150	0.067	64		
	260	0.087			
Sawdust	23	0.059			
Silica aerogel	32	0.024	140		
Wood shavings	23	0.059			

Source: Adapted from J. P. Holman (1981, Table A-2, p. 536).
k = thermal conductivity; ρ = density; C_p = specific heat; α = thermal diffusivity.

Matlab M-Scripts Available

Refer to the table indicated for a list of parameters and examples of use. The latest versions of these scripts* are available online.† No special toolboxes were called in developing these scripts (except for defect.m). Scripts are listed in the following order:

Name of Matlab M-Script	**profil.m**
(Table 5.1)	**color_to_bw.m**
read_uc.m	**Name of Matlab M-Script**
write_uc.m	**(Table 3.9)**
plot_img.m	**mod_therm.m**
planck.m	**Name of Matlab M-Script**
grid_sm.m	**(Table 6.1)**
low_bit.m	**defect.m**
read_fl.m	(optional toolbox use)
mat_fl_to_uc.m	**moment.m**
(uses internally	**logsig.m**
function	**Name of Matlab M-Script**
calc_echel.fm)	**(Table 10.7)**
black.m	**timezones2.m**
matresul.m	**Plotsh3.m**
(uses internally	(uses internally
function **black.m**)	function **sh3.m**)
poin_pro.m	**plotsh4.m**
median_filter.m	(uses internally
kernel.m	function **sh4.m**)
stretch.m	**plotsh5.m**
enh_pro.m	**Plot_sinc.m**
snr.m	uses internally function
smooth.m	**sinc_x.m**
ctrst.m	**f.m**

*Special thanks to Valérie Lavigne for her help with the Matlab scripts.
† http://irndt.gel.ulaval.ca/

Scripts listed in the following order:

READ_UC.M

```
%**********
% MATLAB M-File to read 'raw type' of files of unsigned char
% A header is account for.
% Adjustable parameters are:
%  Header: number of bytes before the image
%  MaxRow: number of pixels per row
%  Maxcol: number of pixels per row
%
% To call this function type in the Matlab window
% once in the directory of interest (file name is d_1234.uc):
%  m=read_uc('d_1234.uc');
%
% Date : 17.3.99
%
% Last modification: 19.05.2000
%
%**********

function out = read_uc(file)

fid = fopen(file,'r');
Header = 105; Maxcol = 105; Maxrow = 68;
fread(fid,Header,'uchar');
out1 = fread(fid,[Maxcol,Maxrow],'uchar');
fclose(fid);

% ***** Correct for Matlab display (transpose)
out=out1';

% ***** Display the image
colormap(gray);
imagesc(out);
colorbar;
```

WRITE_UC.M

```
%**********
% MATLAB M-File to write a 'raw type' of file of unsigned char
% No header is account for.
% Adjustable parameter is:
%  Headersize: number of bytes before the image
%  (Header is filled with blank spaces)
%
% To call this function type in the Matlab window
% once in the directory of interest (file name is d_1234.uc
% and mat is the image to write on disk):
%  count=write_uc('d_1234.uc', mat);
% Return value count indicates the number of bytes saved.
%
% Date : 10.4.99
%
% Last modification: 23.05.2000
```

```
%
%*********

function out = write_uc(file,mat)

% ***** Correct for saving (transpose)
mat1=mat';

% ***** Save on disk
fid = fopen(file,'w');
Headersize = 105;
Header=zeros(1,Headersize);
fwrite(fid,Header,'uchar')
out = fwrite(fid,mat1,'uchar')
fclose(fid);
```

PLOT_IMG.M

```
%**********
% MATLAB M-File to draw a 3-D plot of an image'mat'
%
% To call this function type in the Matlab window
% once the image 'mat' was made available:
%  plot_image(mat);
%
% Date : 10.4.99
%
% Last modification: 01.06.2000
%
%*********

function out = plot_img(mat)
[Maxcol, Maxrow]=size(mat);
x=[1:Maxrow]; y=[1:Maxcol];
out=mesh(x,y,mat);
rotate3d;                % enable rotation of the image
```

PLANCK.M

```
%*****************************
% Display Planck's law
% 22.5.97, last revision 14.6.2000

c1=1.191044e8;
c2=1.438769e4;

% lambda from 0.1 to 150 microns, interval of 0.01 microns
l=[0.1:0.01:150];

% spectral radiance computation
t=5800; % temperature in kelvin
l5800=(c1.*(l.^(-5))) ./ ((exp(c2./(l.*t))) - 1);
t=2000;
l2000=(c1.*(l.^(-5))) ./ ((exp(c2./(l.*t))) - 1);
t=1000;
l1000=(c1.*(l.^(-5))) ./ ((exp(c2./(l.*t))) - 1);
```

```
t=800;
l800=(c1.*(l.^(-5))) ./ ((exp(c2./(l.*t))) - 1);
t=300;
l300=(c1.*(l.^(-5))) ./ ((exp(c2./(l.*t))) - 1);
t=100;
l100=(c1.*(l.^(-5))) ./ ((exp(c2./(l.*t))) - 1);
t=50;
l50=(c1.*(l.^(-5))) ./ ((exp(c2./(l.*t))) - 1);

% display the results
loglog(l,l5800, l,l2000,l,l1000,l,l800,l,l300,l,l100,l,l50)
AXIS([0.1,150,0.0001,100000000])
title('Planck''s law')
xlabel('wavelength')
ylabel('spectral radiance')
text(1.5,10000000,'7000 K')
% text(1.5,11000000,'5800 K')
% text(1.4,3e+005,'2000 K')
% text(0.38,1.8e+005,'2000 K')
text(0.9,10e+003,'1000 K')
text(2.2,190,'800 K')
text(3,16,'300 K')
text(4.7,0.18,'150 K')
text(10.5,0.0366,'100 K')
text(25,0.002,'50 K')

% hold current graph
hold;

% draw the yellow rectangle of the visible spectrum
x=[0.4,0.4,0.8,0.8];
y=[1e-4,1e8,1e8,1e-4];
patch(x,y,'y')

% display the location of the visible spectrum
text(.11,230,'Locus of ')
text(.11,50,'visible')
text(.11,14,'spectrum')
x=[.15,.43];y=[5e+002,5.9859e+003];
line(x,y);

% display results over the rectangle
loglog(l,l5800, 'r',l,l2000,l,l1000,l,l800,l,l300,l,l100,l,l50)
text(0.48,2.5e+005,'2000 K')
text(0.26,2.6e6,'5800 K')

% add grid lines to the current axes
grid

% display line of maximums
t=20;      % 20 K
la=2898/t;
s= (c1*(la^(-5))) / ((exp(c2/(la*t))) - 1);
x(1)=la; y(1)=s;

t=10000;        % 10000 K
la=2898/t;
s= (c1*(la^(-5))) / ((exp(c2/(la*t))) - 1);
x(2)=la; y(2)=s;
line(x,y)

% display text about the line of max
```

```
text(12,3.4e5,'Locus of maxima')
x=[2.43,11.32];y=[2.1886e+004,3.1730e+005];
line(x,y);

% spectral radiance computation at 7000 K
t=7000;
17000=(c1.*(1.^(-5))) ./ ((exp(c2./(1.*t))) - 1);
loglog(1,17000,'y');

% spectral radiance computation at 150 K
t=150;
1150=(c1.*(1.^(-5))) ./ ((exp(c2./(1.*t))) - 1);
loglog(1,1150,'y');

% add the file name
text(15, 1.1e-005,'[Planck.m]')

% gets the x and y coordinates on the plot
% [X,Y]=ginput(1)
```

GRID_SM.M

```
%**********
% MATLAB M-File to reduce the grid size of an image
% To call this function type in the Matlab window
% once the image mat has been made available:
%  mat1=grid_sm(mat,block);
% mat1 is the image which was sampled by blocks of size
%  ''blocks''
%
% Date : 17.3.99
%
% Last modification: 02.06.2000
%
%*********

function out = grid_sm(mat,block)

[mx my]=size(mat);
out=zeros(mx,my);

for x=1:block:mx,
   for y=1:block:my,
      out(x:x+block-1,y:y+block-1)=mat(x,y);
   end
end

% ***** Display the image
colormap(gray);
imagesc(out);
colorbar;
```

LOW_BIT.M

```
%**********
% MATLAB M-File to reduce the bit per pixel of an image
% To call this function type in the Matlab window
```

```
% once the image mat has been made available:
%   mat1=low_bit (mat,ratio);
% mat1 is the result image whith reduced bit per pixel
%
% For example with a ratio of:   the number of bits per pixel
%              is reduced to:
%                          128     ->   1
%                           64          2
%                           32          3
%                           16          4
%                            8          5
%                            4          6
%                            2          7
%
% Date : 11.4.9
%
% Last modification: 30.05.2000
%
%*********

function out = low_bit(mat,ratio)

out=floor(mat./ratio);
```

READ_FL.M

```
%***********
% MATLAB M-File to read files of 'float type' of data (4 bytes
% per pixel): 'float32', floating-point on 32 bits.
% A header is account for, if any.
% Adjustable parameters are:
%   Header: number of bytes before the image
%   MaxRow: number of pixels per row
%   Maxcol: number of pixels per row
%
% To call this function type in the Matlab window
% once in the directory of interest (file name is d_1234.uc):
%   m=read_fl('d_1234.uc');
%
% Date : 13.5.99
%
% Last modification: 02.06.2000
%
%*********

function out = read_fl(file)


fid = fopen(file,'r');
Header = 0; Maxcol = 105; Maxrow = 68;
fread(fid,Header,'uchar');
out1 = fread(fid,[Maxcol,Maxrow],'float32');
fclose(fid);


% ***** Correct for Matlab display (transpose)
out=out1';
```

```
% ***** Display the image
colormap(gray);
imagesc(out);
colorbar;
```

MAT_FL_TO_UC.M

```
%**********
% MATLAB M-File to convert a matrix of 'float type' of data (4
  bytes
% per pixel): 'float32', floating-point on 32 bits to unsignedc.
%
% Adjustable parameters are:
% N_HAUT = 255;  maximum value in saved image
% N_BAS  =   0; minimum value in saved image
%
% Note: The [50,255] span allows for comfortable viewing on the
% screen while the [0,255] span is to dark.
%
% To call this function type in the Matlab window
% once in the directory of interest (file name is d_1234.uc):
%   mat_uc=mat_fl_to_uc(mat);
%
% Date : 13.5.99
%
% Last modofication: 01.06.2000
%
%*********

function out = mat_fl_to_uc(mat)

N_HAUT = 255;
N_BAS  =   0;

[mx my]=size(mat); % size of the image

% 1st determine min and max of image
maxv=max(mat(:));
minv=min(mat(:));

% 2nd compute slope and origin ordinate
m_niv = (N_HAUT - N_BAS) / (maxv - minv);
ord_niv = N_HAUT - (m_niv * maxv);

% 3rd conversion in level 0-255
nivfloat = m_niv.*mat + ord_niv;
out = round(nivfloat);
```

CALC_ECHEL.M

```
%**********
% MATLAB M-File that performs conversion from float to
% 0-255 values. Function used by 'mat_fl_to_uc.m'
% Inputs are:
```

```
%     t_cal          = float to convert
%     tpmin, tpmax   = min and max value span of the float
%                       to convert
%     m_niv          = slope of the regression line
%     ord_niv        = origin ordinate of the regression line
%
% Adjustable parameters are: none
%
% Date : 13.5.99
%
%*********

function out = calc_echel(t_cal, tpmin, tpmax, m_niv, ord_niv,
N_NEG, N_POS)

if (t_cal < tpmin)
        niveau = N_NEG; % cas des zeros
elseif (t_cal > tpmax)
        niveau = N_POS;

else
        nivfloat = (m_niv * t_cal) + ord_niv;
        %puis on arondi le niveau a une valeur entiere
        niveau = round(nivfloat);

end
 out = niveau;
```

BLACK.M

```
%**********
% MATLAB M-File to write a black color map
%
% Black is defined as '0'
% useful to plot 3D plots in 'black'
% To call this function type in the Matlab window
%       black(64);
% Return value count is the new all '0' color map matrix.
% To restore original color map:
%       colormap(gray);
%       colormap(jet);
%
% Date : 6.09.1999
%
% Last modification: 23.05.2000
%
%*********

function out = black(num)

out=zeros(num,3);                    % matrix num x 3 filled with zeros
colormap(out);
```

MATRESUL.M

```
%**********
% MATLAB M-File to extract data from
```

```
% a 'MATRESUL.DIS' file
% There are 2 adjustable parameters
% corresponding to the rows of active data within
% the matresul.dis file (either or not converted in uc).
% To call this function type in the Matlab window:
%                 mat2=matresul(mat, 52,67);
%
% l1 and l2 found with commands mat(1,52), etc.
% A 3D plot is provided with black color map.
%
% Date : 6.9.1999
%
% Last modification: 01.06.2000
%
%*********

function out = matresul(mat,l1,l2)

s=size(mat);
Maxrow=s(1);

out(1:Maxrow,1:l2-l1+1)=mat(1:Maxrow,1+l1:l2+1);

% ***** Display the image
%colormap(gray);
% color map is black
black(64);
% imagesc(out);
plot_img(out);
% colorbar;
```

POIN_PRO.M

```
%**********
% poin_pro.m
%
% MATLAB M-File for point processing routine
% with parameters a and b. Return the processed
% image, based on figure 5.18. Images are of
% unsigned characters type.
%
% Adjustable parameter is:
% Maxcolor: maximum value of a pixel
%       (255 if on 8 bits)
%
% Example of use:
%   mat1=poin_pro(mat,10,25);
%
% with b=Maxcolor, a becomes the threshold, eq.(5.15).
%
% Date : 16.05.2000
%
% Last modification: 02.06.2000
%
%**********

function mat1 = pp2(mat,a,b)
```

```
Maxcolor=255;

% ***** Size of matrix and initialization of mat1
[mrow, mcol]=size(mat);
mat1=zeros(mrow,mcol);

% ***** Values needed for point processing
m1=b/a;
m2=(Maxcolor-b)/(Maxcolor-a);
c=b-m2*a;

% ***** Point processing

sm=find(mat<=a);          % find all data smaller or equal to a
gr=find(mat>a);           % find all data greater than a

if b~=Maxcolor
    mat1(sm)=m1.*mat(sm);
else      % a becomes a threshold
    mat1(sm)=0;
end

% values given by a line through
% (a,b) and (Maxcolor, Maxcolor)
mat1(gr)=m2.*mat(gr)+c;

% ***** Display the image
colormap(gray);
imagesc(mat1);
colorbar;
```

MEDIAN_FILTER.M

```
%**********
% median_filter.m
%
% MATLAB M-File for noise suppresion through
% median filtering. Return the processed image.
% Images are of unsigned characters type.
% Parameter 'size' defines the size of the kernel
% of interest and must be a odd number. If size
% is even, the fonction uses size plus one.
%
% Example of use:
%   mat1=median_filter(mat,3);
%
% Date : 17.05.2000
%
% Last modification: 02.06.2000
%
%**********

function mat1 = median_filter(mat,size)

% ***** Verification of variable size
if mod(size,2)~=1,    % remainder after division by two
    disp('warning: size should be odd');   % display message
    size=size+1;
```

```
end

% ***** Size of matrix and initialization of mat1
[mrow, mcol]=size(mat);
mat1=zeros(mrow,mcol);


fsize=floor(size/2);    % round towards lowest nearest integer

% ***** Noise suppresion through median filtering
for i=1:mrow,
    for j=1:mcol,

% ***** Number of elements around (i,j)

% ***** Left side (with (i,j))
if j>fsize  % general case
    l=fsize+1;
else
    l=j;  % sides and corners case
end

% ***** Right side
if (mcol-j)>fsize  % general case
    r=fsize;
else
    r=mcol-j;   % sides and corners case
end

% ***** Up side (with (i,j))
if i>fsize  % general case
    u=fsize+1;
else
    u=i;  % sides and corners case
end

% ***** Down side
if (mrow-i)>fsize    % general case
    d=fsize;
else
    d=mrow-i;         % sides and corners case
end

% ***** Kernel of interest
ker=mat(i-u+1:i+d,j-l+1:j+r);

% ***** Attribution of new value to (i,j)
mat1(i,j)=median(ker(:));  % find the median in kernel

    end
end

%mat1=mat_fl_to_uc(mat1);

% ***** Display the image
colormap(gray);
imagesc(mat1);
colorbar;
```

KERNEL.M

```
%**********
% kernel.m
%
% MATLAB M-File for image enhancement through the
% use of a kernel of values, eq. (5.16). Return the
% enhanced image. Images are of unsigned characters type.
%
% Example of use:
%   mat1=kernel(mat,2,array);
%
% with array defined previously as for a 2x2
% Roberts gradient: array = [1 0; 0 -1];
%
% Date : 17.05.2000
%
% Last modification: 02.06.2000
%
%**********

function mat1 = kernel(mat,ascale,array)

% ***** Size of matrix and initialization of mat1
[mrow, mcol]=size(mat);
mat1=zeros(mrow,mcol);

% ***** Size of array
[arow, acol]=size(array);
array=ascale.*array;

% ***** Number of elements around (i,j)
frow=floor(arow/2);                    % round towards smallest
fcol=floor(acol/2);                    % nearest integer

crow=ceil(arow/2);                     % round towards largest
ccol=ceil(acol/2);                     % nearest integer

% ***** Image enhancement
for i=1:mrow,
   for j=1:mcol,

% ***** Left side (with (i,j))
if j>fcol    % general case
   l=ccol;
else
   l=j;  % sides and corners case
end

% ***** Right side
if (mcol-j)>fcol  % general case
   r=fcol;
else
   r=mcol-j;   % sides and corners case
end

% ***** Up side (with (i,j))
if i>frow    % general case
   u=crow;
```

```
else
   u=i;   % sides and corners case
end

% ***** Down side
if (mrow-i)>frow   % general case
   d=frow;
else
   d=mrow-i;    % sides and corners case
end

% ***** Kernel of interest
ker=array(1:u+d,1:l+r).*mat(i-u+1:i+d,j-l+1:j+r);

% ***** Attribution of new value to (i,j)
mat1(i,j)=sum(ker(:));       % sum all elements of ker

   end
end

%mat1=mat_fl_to_uc(mat1);

% ***** Display the image
colormap(gray);
imagesc(mat1);
colorbar;
```

STRETCH.M

```
%**********
% stretch.m
%
% MATLAB M-File for histogram stretching through
% linear remapping of gray level of image mat
% within available gray level range gmin to gmax,
% eq.(5.17). Images are of unsigned characters type.
%
% Example of use:
%   mat1=stretch(mat,0,255);
%
% Date : 18.05.2000
%
% Last modification: 02.06.2000
%
%**********

function mat1 = stretch(mat,gmin,gmax)

% ***** Size of matrix
[mrow, mcol]=size(mat);

% ***** Maximum and minimum values within mat
maxv=max(max(mat));   % find the largest component of mat
minv=min(min(mat));   % find the smallest component of mat

% ***** Values needed for histogram stretching
m=(gmax-gmin)/(maxv-minv);
```

```
b=gmax-(m*maxv);

% ***** Histogram stretching
mat1=m.*mat+b;

% ***** Display the image
colormap(gray);
imagesc(mat1);
colorbar;
```

ENH_PRO.M

```
%**********
% enh_pro.m
%
% MATLAB M-File for image enhancement with parameters k1
% and k2. Return the enhanced image, based on eq.(5.20).
% (eq.(5.19) can be obtained by choosing k2=1). Images
% are of unsigned characters type.
% Adjustable parameter is:
%  Size:        local mean is calculated over Size pixels wide
%     window centered in (i,j)
%
% Example of use:
%   mat1=enh_pro(mat,5,0.5);
%
% Date : 18.05.2000
%
% Last modification: 02.06.2000
%
%**********

function  mat1=enh_pro(mat,k1,k2)

% ***** Size of matrice and initialization
[mrow, mcol]=size(mat);
mat1=zeros(mrow,mcol);
Size=3;
fsize=floor(Size/2);     % round towards lowest nearest integer

% ***** Image enhancement
for i=1:mrow,
   for j=1:mcol,

% ***** Number of elements around (i,j)

% ***** Left side (with (i,j))
if j>fsize            % general case
   l=fsize+1;
else
   l=j;     % sides and corners case
end

% ***** Right side
if (mcol-j)>fsize              % general case
   r=fsize;
else
```

```
    r=mcol-j;    % sides and corners case
end

% ***** Up side (with (i,j))
if i>fsize    % general case
    u=fsize+1;
else
    u=i;    % sides and corners case
end

% ***** Down side
if (mrow-i)>fsize                % general case
    d=fsize;
else
    d=mrow-i;                    % sides and corners case
end

% ***** Computation of the local mean in kernel of interest
ker(1:u+d,1:l+r)=mat(i-u+1:i+d,j-l+1:j+r);
locm=mean(ker(:));    % average of ker

% ***** Attribution of new value to (i,j)
mat1(i,j)=(k2-k1)*locm+k1*mat(i,j);

    end
end

%mat1=mat_fl_to_uc(mat1);

% ***** Display the image
colormap(gray);
imagesc(mat1);
colorbar;
```

SNR.M

```
%**********
% snr.m
%
% MATLAB M-File for computation of the signal
% to noise ratio between two images mat1 and mat2
% recorded one after another, based on eq.(5.13).
% Images are of unsigned characters type.
%
% Example of use:
%   snr(mat1,mat2);
%
% Date : 17.05.2000
%
% Last modification: 01.06.2000
%
%**********

function snr(mat1,mat2)

% ***** Size of matrice
[mrow, mcol]=size(mat1);
```

```
% ***** Average power image
sumofApower2=sum(mat1(:).^2);
% element-by-element power 2 then sum all elements
avpi=sumofApower2/(mrow*mcol);

% ***** Average power noise
N=abs(mat1-mat2);        % noise, absolute value of mat1 minus mat2
stdN=std(N(:))        % standard deviation of N

avpn=0.5*(stdN^2);

% ***** SNR signal to noise ratio
SNR=10*log10(avpi/avpn)
```

SMOOTH.M

```
%**********
% smooth.m
%
% MATLAB M-File to smooth out noise from image
% mat with parameter B (for Gaussian width),
% based on eq.(5.31). Images are of unsigned
% characters type.
% Adjustable parameter is:
%  Nbpasse: number of iterations for smoothing
%
% Example of use:
%  mat1=smooth(mat,4);
%
% Date : 25.05.2000
%
% Last modification: 02.06.2000
%
%**********

function mat1=smooth(mat,B)

% ***** Size of matrix and initialization
[mrow, mcol]=size(mat);
mat0=zeros(mrow,mcol);
mat1=zeros(mrow,mcol);
Nbpasse=2;

% ***** Initialization of the Gaussian
imax=7*B+1;
i=[1:imax];            % array with values needed for i
expo=exp((-0.5)*(((i-imax/2)./B).^2));

% ***** Compute the Gaussian, eq.(5.29) and (5.30)
f=expo./sum(expo);

% ***** Processing on rows
smoothed=zeros(mrow,mcol);      % vector of mcol zeros,
initialixaton of smoothed

for it=1:Nbpasse    % for all iterations

   for omega=1:mcol    % for all elements of the row
      value=smoothed(:,omega);
```

```
    for i=1:imax    % for all elements of the gaussian
              index = omega - round(imax/2) + i;

    if index < 1
       value = value + (mat(:,1)-smoothed(:,1)).*f(i);
      elseif index > mcol
       value = value + (mat(:,mcol)-smoothed(:,mcol)).*f(i);
      else
       value = value + (mat(:,index)-smoothed(:,index)).*f(i);
      end % if

    end % for imax

  mat0(:,omega)=value;
  end % for omega

end % for Nbpasse

% ***** Processing on columns
smoothed=zeros(mrow,mcol);              % vector of mcol zeros,
initialixaton of smoothed

for it=1:Nbpasse        % for all iterations

  for omega=1:mrow        % for all elements of the row
    value=smoothed(omega,:);

    for i=1:imax            % for all elements of the gaussian
              index = omega - round(imax/2) + i;

    if index < 1
       value = value + (mat0(1,:)-smoothed(1,:)).*f(i);
      elseif index > mrow
       value = value + (mat0(mrow,:)-smoothed(mrow,:)).*f(i);
      else
       value = value + (mat0(index,:)-smoothed(index,:)).*f(i);
      end % if

    end % for imax

  mat1(omega,:)=value;
  end % for omega

end % for Nbpasse

%mat1=mat_fl_to_uc(mat1);

% ***** Display the image
colormap(gray);
imagesc(mat1);
colorbar;
```

CTRST.M

```
%**********
% ctrst.m
%
% MATLAB M-File for computation of the thermal contrast
```

```
% from a sequence of images (of unsigned characters type)
% located in the working directory. Return a floating
% point contrast image.
%
% Parameter TYPE allows to compute the various contrast
% discussed depending of provided value (Section 5.4):
% (1) absolute contrast, (2) running contrast,
% (3) normalized contrast, (4) standard contrast.
% Defect-free zone within the image is defined with
% rectangle having respectively upper left and bottom-right
% coordinates (x1,y1) and (x2,y2). The matrice mat is the
% termal image for which the contrast is being computed and
% mat0 is the image at time tm or t0. For TYPE 1 and 2,
% mat0 is not used and can be any matrice.
%
% Example of use:
%   mat_ctr=ctrst(4,2,2,50,10,mat,mat0);
%
% Date : 26.05.2000
%
% Last modification: 31.05.2000
%
%**********

function mat1 = ctrst(TYPE,x1,y1,x2,y2,mat,mat0)

% ***** Defect-free area
Ts=mean(mean(mat(x1:x2,y1:y2)));     % mean value of the defect-
free area

% ***** Contrast computation

switch TYPE

case 1,
% ***** Absolute contrast

    % Ca(t)=deltaT(t)=Tdef(t)-Ts(t)
    mat1 = mat - Ts;

    % ***** Display the image
    colormap(gray);
    imagesc(mat1);
    colorbar;

case 2,
% ***** Running contrast

    % Cr(t)=Ca(t)/Ts(t)=deltaT(t)/Ts(t)
    mat1 = (mat - Ts) / Ts;

    % ***** Display the image
    colormap(gray);
    imagesc(mat1);
    colorbar;

case 3,
% ***** Normalized contrast
```

```
% ***** Defect-free area of image at tm
Tsm=mean(mean(mat0(x1:x2,y1:y2)));   % mean value of the defect-
free area

% Cn(t)=Ti(t)/Ti(tm)-Ts(t)/Ts(tm), Ti=Tdef
mat1 = mat./mat0 - Ts/Tsm;

% ***** Display the image
colormap(gray);
imagesc(mat1);
colorbar;

case 4,
% ***** Standard contrast

    % ***** Defect-free area of image at tm
    Ts0=mean(mean(mat0(x1:x2,y1:y2))); % mean value of the defect-
    free area

    % Cs(t)=(Ti(t)-Ti(t0))/(Ts(t)-Ts(t0))
    if Ts-Ts0~=0,
       mat1=(mat-mat0) / (Ts-Ts0);
    else
       disp('error: division by zero')
    end

    % ***** Display the image
    colormap(gray);
    imagesc(mat1);
    colorbar;

% ***** wrong type number
otherwise,
   disp('Wrong type')

end
```

PROFIL.M

```
%***********
% MATLAB M-File to plot a profile from
% a matrix
% There is two adjustable parameters
% corresponding to the row to be plotted (nrow)
% and the value to be displayed for the line being
% plotted.
% after calling the function, 'mat' is displayed
% with superimposition of the row of interest.
% Then, to plot the profile, type the command:
%               plot(prof);
% To call this function type in the Matlab window:
%               prof=profil(mat,row_number, value);
%
% Matrix 'mat'stays unaffected.
%
% Date : 19.1.2000
%
```

```
% Last modification: 02.06.2000
%
%*********

function out = profil(mat,nrow,value)

out=mat(nrow,:);
mat(nrow,:)=value;

colormap(gray);
imagesc(mat);
colorbar;
% hold on
% plot(out);
```

COLOR_TO_BW.M

```
%**********
% color_to_bw.m
%
% Routine to convert a color image into a black and
% white one based on a color scale included in the
% image and located in the rectangle with upper left
% corner x1,y1 and lower right corner x2,y2. x is the
% number of the row and y is the number of the column.
% If x1, y1, x2 and y2 are not provided, the function
% will ask the user to click on the corners of the scale.
% Provided are the three splitted channels on 8 bits:
% red (matr), green (matg), blue (matb). Returned
% matrix mat is the black and white image coded on
% unsigned char (0 to 255).
%
% Example of use:
%   mat=color_to_bw(matr, matg, matb, x1, y1, x2, y2);
%
% Adjustable parameter Number is used to decide the number of
% elements taken on each side to average the scale. If
% you want the scale to be kept unchanged, set Number to 0.
%
% This routine is useful to convert color scanned
% printed thermograms whose black and white version is
% not available. Splitted channel images are easily
% obtained from the original image in software such as
% Paint Shop Pro among others.
%
% Date : 08.06.2000
% Last modification: 18.07.2000
%
%**********

function mat=color_to_bw(matr, matg, matb, varargin);

Number=2;

% ***** Get the coordinates of the scale
if isempty(varargin)
   colormap(gray);
   imagesc(matr);                  % display the image
```

```
    disp('Click on the upper left corner of the scale');
    [y1,x1]=ginput(1);                      % upper left corner

    disp('Click on the lower right corner of the scale');
    [y2,x2]=ginput(1);                      % lower right corner

    x1=round(x1);               % round towards nearest integer
    x2=round(x2);
    y1=round(y1);
    y2=round(y2);
else
    x1=varargin{1};
    y1=varargin{2};
    x2=varargin{3};
    y2=varargin{4};
end

% ***** Get the color scale
if y2-y1 > x2-x1                    % horizontal scale

scale(1,:)=sum(matr(x1:x2,y1:y2))./(x2-x1+1);
scale(2,:)=sum(matg(x1:x2,y1:y2))./(x2-x1+1);
scale(3,:)=sum(matb(x1:x2,y1:y2))./(x2-x1+1);

else    % vertical scale

scale(1,:)=sum(matr(x1:x2,y1:y2)')./(y2-y1+1);
scale(2,:)=sum(matg(x1:x2,y1:y2)')./(y2-y1+1);
scale(3,:)=sum(matb(x1:x2,y1:y2)')./(y2-y1+1);

end

[srow scol]=size(scale);
scale(4,:)=mat_fl_to_uc(1:scol);        & vector 0-255, same length
as scale

scale=scale';                % transpose to use the function sum

% ***** Improve the color scale
if Number > 0
    for col=1:scol

    first=col-Number;
    last=col+Number;

    if first < 1
            first=1;
    elseif last > scol
            last=scol;

    end

    total=last-first+1;
    newscale(col,1:3)=sum(scale(first:last,1:3))./(total);
    % sum from first to last and divide by total to find average

    end
    scale(:,1:3)=newscale;                   % save the result in scale
```

```
end

% ***** Convert the image
[mrow mcol]=size(matr);
mat=zeros(mrow,mcol);              % initialization

for i=1:mrow                       % for all pixels in the image
for j=1:mcol

% find the position of the pixel in the scale (nearest absolute
   value)
vecred=abs(scale(:,1)-matr(i,j));
vecgreen=abs(scale(:,2)-matg(i,j));
vecblue=abs(scale(:,3)-matb(i,j));

vector=vecred+vecgreen+vecblue;
minvalue=min(vector);
f=find(vector==minvalue);

mat(i,j)=scale(f(1),4);            % store the corresponding black
and white value in mat

end
end

clear matr;
clear matg;
clear matb;

% ***** Display the image
colormap(gray);
imagesc(mat);
colorbar;
```

DEFECT.M

```
%**********
% defect.m
%
% Implementation of the automatic defect detection
% procedure on unsigned character image mat with
% parameter MND. Max_def is the maximum number of
% defects that can be found in an image. If parameter
% cross is set to 1, returned image mat1 is equal to
% mat with only crosses indicated at defect center.
% Any other value of cross returns ma1 as a map of
% defect(s) detected in mat (0 value at background
% and 1 value at defect location). mat_out is the
% matrix containing:
%  first column:     row of defect center
%  second column:    column of defect center
%  third column:     value of defect center
%
% Example of use:
%  [mat1, mat_out]=defect(mat,50,10,1);
%
% Adjustable parameters are:
```

```
% Border_width:            width in pixels of the border
%  around the image that is not searched for defects
% Cross_size:       width on each side of the cross in pixels
% Alpha:        used to find the minimum value of a defect
%
% Note: if you have the Image Processing Toolbox,
%  you may add a parameter (any number will do) after
%  cross to enable the use of functions from this
%  toolbox. You may obtain better results with these
%  functions.
% Example of use with the Image Processing Toolbox:
%  [mat1, mat_out]=defect(mat,50,10,1,1);
%
% Date : 30.05.2000
% Last modification: 19.06.2000
%
%**********

function [mat1, varargout] = defect(mat,MND,Max_def,cross,
varargin)

Border_width=4;       % minimum value is 1 pixel to have enough
space for crosses
Cross_size=2;       % must be smaller or equal to Border_witdh, if
=0 only a point
Alpha=1.0;       % parameter for the minimum value of a defect
% the equation is: min= Alpha*(mean_value+Alpha*std_dev) where
  mean_value
% is the mean value of mat and std_dev is the standard deviation
  of mat
% if the value of Alpha is too high, no defect will be found


% ***** Size of the matrix
[mrow mcol]=size(mat);
total=mrow*mcol;

% ***** Sort pixels of matrix mat
[mat1d_ord,ind1d]=sort(mat(:));

mat1d_ord=flipud(mat1d_ord);       % reverse order from largest to
smallest
ind1d=flipud(ind1d);       % reverse order from largest to
smallest

[x y]= ind2sub([mrow mcol],ind1d);       % get x,y coodinates from
ind1d

% ***** Minimum value of a defect
mean_value=mean(mat(:));             % mean value of mat
std_dev=std(mat(:));             % standard deviation of mat
min_value=Alpha*(mean_value + Alpha*std_dev);
last_higher=max(find(mat1d_ord>min_value));

if isempty(last_higher)       % no defect in the image
    display('There is no defect');
    mat1=mat;
    varargout={[]};
else
```

```
% ***** Find defect centers
n_pix_ini(1)=1;               % initialization of first defect pixel
i=0;
j=1:total;
pixel=0;
total_limit=ones(total,1);

while (pixel<=last_higher) & (i<Max_def )
   i=i+1;
   xi(i)=x(n_pix_ini(i));
   yi(i)=y(n_pix_ini(i));
   distance=sqrt((x-xi(i)).^2+(y-yi(i)).^2);
   limit_dist=(distance > MND);
% limit_dist = 0 if distance<d and 1 if distance>MND
   total_limit=total_limit & limit_dist;
% total_limit = 0 if limit_dist = 0 or distance<d for defect
  found before

   indice=( j > n_pix_ini(i))';
% indice=0 for pixels higher than n_pix_ini and 1 for pixels
  lower

   pixel=min(find(total_limit & indice));

   if isempty(pixel)
      break;
    elseif (pixel<=last_higher) & (i<Max_def )
      n_pix_ini(i+1)=pixel;
   end

end

% eliminate border around the image
up=find(xi<=Border_width);
down=find(mrow-xi<=Border_width);
left=find(yi<=Border_width);
right=find(mcol-yi<=Border_width);

n_pix_ini(up)=0;
n_pix_ini(down)=0;
n_pix_ini(left)=0;
n_pix_ini(right)=0;
f=find(n_pix_ini > 0);
n_pix_ini=n_pix_ini(f);
xi=xi(f);
yi=yi(f);

[r nbr]=size(f);

mat_out=[xi' yi' mat1d_ord(n_pix_ini)];
varargout={mat_out};

% ***** Parameter cross

if cross==1
% ***** Add crosses to mat
mat1=mat;
```

```
s=Cross_size;                    % size of the cross

% ***** Draw the crosses on each defect center
for i=1:nbr

mat1(xi(i)-s:xi(i)+s,yi(i))=0;
mat1(xi(i),yi(i)-s:yi(i)+s)=0;

end


else
% ***** Threshold value of each defect

[val_histo,histo]=hist(mat(:),256);
for i=1:nbr

    % variation of the number of aglomerated pixels function
    % to the value of the thresholding

    agl_nbr=cumsum(fliplr(val_histo(histo < mat1d_ord
    (n_pix_ini(i))))));

    % computation of the angle formed by 3 points of the curb
    % distant of 20 pixels each
    ind=21:(size(agl_nbr,2)-20);

    agl_nbr_n=256*(agl_nbr/max(agl_nbr));
    y1=abs(agl_nbr_n(ind+20)-agl_nbr_n(ind));
    y2=abs(agl_nbr_n(ind-20)-agl_nbr_n(ind));
    teta=pi+atan(y2/20)-atan(y1/20);

    pt_inflexion=max(find(teta == min(teta))+20);
    thres(i)=histo(pt_inflexion);

end

% ***** Extract image with defects
if isempty(varargin)
value=min(thres);
mat1=(mat>value);
mat1=median_filter(mat1,3);

else
for i=1:nbr
    im_thres=(mat > thres(i));

    % Use of the Image Processing Toolbox to clear up images
    label=bwlabel(im_thres,4);
    label2=(label == label(int16(xi(i)),int16(yi(i))));
    im(:,:,i)=bwmorph(label2,'close');
end

mat1=(sum(im,3) ~= 0);


end        % if Toolbox
end        % if cross
end        % if no defect
```

```
% ***** Display the image
colormap(gray);
imagesc(mat1);
colorbar;
```

MOMENT.M

```
%**********
% moment.m
%
% Return the moment image in float from the image
% sequence present on disk in directory dir.
% Computation proceeds in the time period ta to
% tb (coded in the name of the file having the
% prefix pre and extension ext). incre is the
% time increment between images. option parameters
% are: 1- add all images, 2- average of all images,
% 3- add all images and convert to unsigned char.
%
% A header is account for.
% Adjustable parameters are:
%  Header: number of bytes before the image
%  MaxRow: number of pixels per row
%  Maxcol: number of pixels per row
%
% Example of use:
%  mat=moment('G:\images\','d_','uc',200,2000,10,2);
%
% Date : 31.05.2000
%
% Last modification: 01.06.2000
%
%**********

function mat = moment(dir,pre,ext,ta,tb,incre,option)

% ***** Initialization of mat
Header = 105; Maxcol = 105; Maxrow = 68;
mat=zeros(Maxrow,Maxcol);

numfiles=0;

% ***** Summation of images eq.(6.3) from time ta to tb
for t=ta:incre:tb

    t=num2str(t);    % convert number t to a character string

    % ***** Concatenate the name and location of the file
    % into a single string
    file=strcat(dir,pre,t,'.',ext);

    % ***** As in chapter 5 function read_uc.m
    fid = fopen(file,'r');

    if fid==-1,
        error=strcat('Cannot open:',file);
        disp(error);
```

```
    else
        fread(fid,Header,'uchar');
        out1 = fread(fid,[Maxcol,Maxrow],'uchar');
        fclose(fid);
        out=out1';   % correct for Matlab display (transpose)
        numfiles=numfiles+1;

    % ***** Summation of images
    mat=mat+out;

    end

end

if option==2
    mat=mat./numfiles;
elseif option==3
    mat=mat_fl_to_uc(mat);
end

% ***** Display the image
colormap(gray);
imagesc(mat);
colorbar;
```

LOGSIG.M

```
% Display Logsig function, in black color
% 1.2.2000

n=[-10:0.1:10];
f=1./(1+exp(-n));
plot(n,f,'-k');
```

TIMEZONES2.M

```
%**********
% timezones2.m
%
% Matlab M-file to compute a logarithmic time scale.
%
% NM is the maximum of images. q is the image
% acquisition + temporaray storage rate. N is the
% number of zones wanted. to is the starting of
% acquisition and tf is the end of the experiment.
% Parameter p is specifies for witch case the time
% zones are computed:
%  1: dijoint case
%   all other values: continuous case
%
% Example of use:
%   out=timezones2(284,0.27,10,1,100,1)
%
% Date : 06.06.2000
```

```
%
% Last modification: 06.06.2000
%
%**********

function out = timezones2(NM,q,N,to,tf,p)

if p==1       % disjoint case

   % ***** Zones of temporal scale
   mu=(tf/to)^(1/(N-1))     % eq.(10.33)
   ti=to*mu.^(0:N-1);       % eq.(10.33)

   % ***** Lambda computation, eq.(10.35) and (10.36)
   factor=(to/q)*((1-mu^N)/(1-mu)+N);

   i=1
   lambda=factor/(factor-NM)
   NMest=sum(ceil((ti./q).*(1-1/lambda)+1))

   while (NMest > NM) & (i< 50)

      lambda=factor/(factor-NMest);
      NMest=sum(floor((ti./q).*(1-1/lambda)+1));
      i=i+1;

   end

   i
   lambda
   NMest

   % ***** Low borders of time zones
   ti_=ti/lambda;  % zone i spans from ti_ to ti, eq.(10.34)

else  % continuous case

   mu=to/(to-q)     % eq.(10.46)

   tf=min(tf,to+NM*q)     % eq.(10.47)

   N=round(1+log(tf/to)/log(mu))        % eq.(10.48)

   % zone i spans from ti_ to ti,
   ti=to*mu.^(0:N-1);    % eq.(10.33)
   ti_=ti/mu;       % eq.(10.34) and (10.42)

end

zone=0:N-1;

out=[zone' ti_' ti'];
```

PLOTSH3.M

```
% plot sh values for figure 10.20(3)
% (temperature rise versus square root of time)
```

```
% in STEP - HEATING
% 11 - 04 -2000
% g is mismatch factor Gamma
% la = L/sqrt(thermal diffusivity)
% symbol is to overprint on the solid line
%         a symbol. Ex: ko is black (k) with
%         circles (o). Type 'help plot' in
%         Matlab command window for a complete
%         list.
%
% Example of use:
%                         Plotsh3(0.5,1,'ko');
% (maintain the plot)  hold
% (replot)             Plotsh3(0.5,1,'ko');
% (plot next curve)    Plotsh3(-0.5,1,'ks');
%
% This function uses 'sh3() function that performs
% the computations.

function out =Plotsh3(g,la,symbol)

t=0;
inc_t= 0.04; % plot for t 0 to 2=sqrt(4)
for n=1:100
   result(n)=sh3((t),g,la);
   %result(n)=sqrt(t); % semi-infinite case
   tt(n)=sqrt(t);
   t=t+inc_t;
end;
plot(tt,result,'k');
%hold;
% and now lets plot spare points

ttz(1)=tt(1);
resultz(1)= result(1);

ttz(2)=tt(25);
resultz(2)= result(25);

ttz(3)=tt(75);
resultz(3)= result(75);

plot(ttz,resultz,symbol);
```

SH3.M

```
% Computation Step-Heating for figure 3 (sh3.m)
% (temperature rise versus square root of time)
% in STEP - HEATING
% Function used by Plotsh3.m
% Can be called individually with the following parameters:
%  t is time
% g is mismatch factor Gamma
% la = L/sqrt(thermal diffusivity)
% Example of use:
%                    sh3(2,0.5, 4)
%      returns 1.4142
```

```
%
% 11 April 2000

function out = sh(t,g,la)
sqrpi=sqrt(pi);
sumsh=0;
t=sqrt(t);
max=10; % number of terms summed
for n=1:max
   nn=(la*n)/t;
   nn2= nn*nn;
   result = exp(-nn2)-(nn*sqrpi*erfc(nn));
   result = (2*((-g)^n))* result;
   sumsh=sumsh+result;
end;
sumsh=(sumsh+1)*(t);
out=sumsh;
```

PLOTSH4.M

```
% plot sh values for figure 10.21 (4)
% (normalized temperature rise versus square
% root of normalized time)
% in STEP - HEATING
%
% g is mismatch factor Gamma
% symbol is to overprint on the solid line
%         a symbol. Ex: ko is black (k) with
%         circles (o). Type 'help plot' in
%         Matlab command window for a complete
%         list.
%
% Example of use:
%                         Plotsh4(1,'ko');
% (maintain the plot)  hold
% (replot)             Plotsh4(1,'ko');
% (plot next curve)    Plotsh4(-1,'ks');
%
% This function uses 'sh4() function that performs
% the computations.

% 11 - 04 -2000

function out = plotsh4(g,symbol)

t=0;
inc_t= 0.025; % plot for t 0 to 2.5
for n=1:100
   result(n)=sh4(sqrt(t),g);
   %result(n) = 0; % semi-infinite case
   tt(n)=sqrt(t);
   t=t+inc_t;
end;
plot(tt,result,'k');
%hold;
% and now lets plot spare points
```

```
ttz(1)=tt(1);
resultz(1)= result(1);

ttz(2)=tt(25);
resultz(2)= result(25);

ttz(3)=tt(75);
resultz(3)= result(75);

plot(ttz,resultz,symbol);
```

SH4.M

```
% Computation Step-Heating for figure 4 sh4.m
% (normalized temperature rise versus square
% root of normalized time)
% in STEP - HEATING
% Function used by Plotsh4.m and Plotsh5.m
% Can be called individually with the following parameters:
%   t is time
% g is mismatch factor Gamma
% Example of use:
%                 sh4(2,-1)
%      returns 0.9202
%
% 11 April 2000

function out = sh(t,g)
sqrpi=sqrt(pi);
sumsh=0;
max=10; % number of terms summed
for n=1:max
    nn=n/t;
    nn2= nn*nn;
    result = exp(-nn2)-(nn*sqrpi*erfc(nn));
    result = (2*((-g)^n))* result;
    sumsh=sumsh+result;
end;

out=sumsh;
```

PLOTSH5.M

```
% plot sh values for figure 10.22 (5)
% (normalized temperature rise versus
% thermal mismatch factor)
% in STEP - HEATING
%
% t is the normalized time
% g is mismatch factor Gamma
% symbol is to overprint on the solid line
%         a symbol. Ex: ko is black (k) with
%         circles (o). Type 'help plot' in
%         Matlab command window for a complete
```

```
%          list.
%
% Example of use:
%                          Plotsh5(1,'ko');
% (maintain the plot)  hold
% (replot)             Plotsh5(1,'ko');
% (plot next curve)    Plotsh5(2,'ks');
%
% This function uses 'sh4() function that performs
% the computations.

% 11 - 04 -2000

function out = plotsh(t,symbol)

g=-1;
inc_g= 0.01; % plot for g: -1 to +1
for n=1:200
   result(n)=sh4(sqrt(t),g);
   % result(n)= 0;
   gg(n)=g;
   g=g+inc_g;
end;
plot(gg,result,'k');

% and now lets plot spare points

ggz(1)=gg(25);
resultz(1)= result(25);

ggz(2)=gg(75);
resultz(2)= result(75);

ggz(3)=gg(125);
resultz(3)= result(125);

ggz(4)=gg(150);
resultz(4)= result(150);

plot(ggz,resultz,symbol);
```

SINC_X.M

```
% Display sinc x
% 14-4-2000
%
% Example of use: Plot_sinc(1,1,20,'-k');
% parameters: A:       pulse amplitude
%             DT:      pulse duration
%             length:  length of the plot from -length to + length
%             symbol:  to plot the curve with, for instance -k

%                      plot a continuous line in black

% plot_sinc(1,10,0.5,'k');
% plot_sinc(1,1,10,'k');
% plot_sinc(1,0.005,100,'k')
```

```
function out = Plot_sinc(A,DT,length,symbol)

g = -(length);
inc_g = length/1000;
for n=1:2*(length/(inc_g))
   x=pi*DT*g;
   result(n)= A*DT*sinc_x(x);
   %   result(n)= g;
   gg(n)=g;
   g=g+inc_g;
end;
plot(gg,result,symbol);
grid;
```

SINC_X.M

```
% Compute sinc x = sin x/x [sinc_x.m]
% 14-4-2000
%
function out = sinc_x(x)

out = (sin(x))/x;
```

F_.M

```
% Planck law for chapter 2 problems and
% exercises.
% return the spectral radiance
% t, 1 are input invariables
% t - in kelvin
% 1 - lambda in microns
% and s is the output variable

function [s] = f(t,1)

c1=1.191044e8;
c2=1.438769e4;
s=(c1*(1^(-5)))/((exp(c2/(1*t))) - 1)
```

Index

Lightning Source UK Ltd.
Milton Keynes UK
UKOW031613160413

209326UK00001B/5/A

90 0923041 0